Introductory Algebra
A TEXT/WORKBOOK

THIRD EDITION

Charles D. Miller
American River College

Margaret L. Lial
American River College

SCOTT, FORESMAN AND COMPANY
Glenview, Illinois
London, England

To the Student

If you need further help with algebra, you may want to get a copy of the *Student's Solutions Manual* that goes with this book. It contains solutions to the odd-numbered exercises that are not already solved at the back of the textbook, plus solutions to all chapter review exercises, cumulative review exercises, and chapter tests. Your college bookstore either has this book or can order it for you.

Cover: Yaacov Agam. *Florida.* 1977. Oil on aluminum.
99 × 127 cm. Private collection, Miami Beach.

Copyright © 1987, 1983, 1979 Scott, Foresman and Company.
All Rights Reserved.
Printed in the United States of America.

ISBN: 0-673-18461-7

Preface

Introductory Algebra: A Text/Workbook, Third Edition, is designed for a first course in algebra for college students. The only prerequisite is some knowledge of arithmetic, and the first two sections review the key ideas of fractions and decimals that students will need in algebra. The book is suitable for a traditional lecture class or individualized instruction.

KEY FEATURES

Objectives Each section begins with a list of skills that students should learn in that section. The objectives are keyed to the appropriate discussions in the text by numbered symbols such as ▶.

Margin problems After a new idea or technique has been explained and illustrated, students are instructed to work a problem in the margin of the page. Answers to the problems are given on the same page for immediate reinforcement. Any difficulties can be identified quickly by the use of these problems.

Examples More than 650 worked-out examples clearly illustrate concepts and techniques. Daily experience in our mathematics learning center has guided us in developing explanations that are clear, consistent, and accessible to students. Colored shading is used to identify pertinent steps within examples, and explanatory side comments are printed in color. For clarity, the end of each example is indicated with the symbol ◀.

Word problems A problem-solving approach is presented to introduce students to word problems. A list of steps for solving word problems is first given in Chapter 3, yet even Chapter 1 includes some simple word problems. In this way, students see word problems early and gradually improve their problem-solving skills. Throughout the text students are given practice in translating words into the symbols of algebra.

Workbook format The format of the exercise sets and the perforated pages allow the instructor to use the materials as homework assignments or quizzes to be handed in. Students avoid excessive writing, and instructors can grade assignments quickly.

Pedagogical use of second color Colored boxes set off key definitions, formulas, procedures, and cautionary comments, helping students review readily. Colored side comments within examples explain the structure of the problem.

Flexibility This textbook is suitable for lecture classes or self-paced classes. The text is carefully laid out to match standard courses, and sections are concise enough to be covered in one class period. The extensive package of supplementary materials allows each instructor to design a unique curriculum.

EXERCISES

Graded exercises The range of difficulty in the exercise sets affords ample practice with drill exercises. Students are eased gradually through problems of increasing difficulty to those that will challenge outstanding students. More than 5100 exercises keyed to examples are included. A few optional calculator exercises have been integrated throughout the book and are indicated by a colored problem number.

Chapter review exercises Extensive review sets at the end of each chapter, more than 875 problems in all, provide further opportunity for mastery of the material before students take an examination. These exercises are keyed to appropriate sections in the text.

Chapter tests Sample tests, of a length comparable to that of actual classroom tests, help students prepare for examinations.

Cumulative review exercises Sets of exercises that review previously studied topics are included after Chapters 2, 4, 6, and 8. A sample final examination after Chapter 10 provides a thorough review of the entire book.

Supplementary exercises A few sets of supplementary exercises, designed to review difficult or confusing topics, have been included. For example, a set in Chapter 6 requires students to distinguish between performing operations on rational expressions and solving equations involving rational expressions.

Answers and solutions Answers to all odd-numbered section exercises as well as answers to every review and test problem are given at the back of the book. A separate section presents solutions to alternate odd-numbered section exercises, providing a source of extra worked-out examples for students.

NEW CONTENT HIGHLIGHTS

Based on the comments of users of the previous edition, the third edition has been revised to include the following elements.

- ▶ A revised treatment of negative exponents to make the topic more accessible to students (Section 4.1).
- ▶ Expanded presentation of solving for specified variables to give students more help with this difficult topic (Sections 3.5 and 6.6).
- ▶ Examples of arithmetic fractions next to algebraic fractions to smooth the transition (throughout Chapter 6).

- ▶ Two separate sections on slope and the equations of a line to provide slower-paced coverage of these key ideas (Sections 7.4 and 7.5).
- ▶ Applications using the Pythagorean formula in the sections on applications of quadratic equations (Sections 5.6) and finding roots (Section 9.1).
- ▶ A few challenging problems added to the review exercises that require students to extend the ideas presented in the text.

SUPPLEMENTS

Introductory Algebra has an extensive supplemental package that includes testing materials, solutions, software, and other electronic media.

Instructor's Guide The Instructor's Guide includes suggestions for using the text in a mathematics laboratory; short-answer and multiple-choice versions of a diagnostic placement test; answers to even-numbered exercises; two instructional units with exercises covering compound inequalities and an introduction to functions, both of which may be duplicated for classroom use; six forms of chapter tests for each chapter, including two multiple-choice forms; two forms of a final examination; and a lengthy set of additional exercises, providing 10 to 20 exercises keyed to each textbook objective, which can be used as an additional source of questions for tests, quizzes, or student review of difficult topics.

Computer-Assisted Testing System (CATS) A testing system that provides more than 7500 questions organized by objectives for both Apple and IBM computers is available to users of the book. The system features an editing capability that allows instructors to add their own problems.

Computer-Assisted Instruction A set of computer-assisted tutorials, based on a mastery learning approach, has been developed for this text. Available for Apple and IBM computers, the package of ten diskettes not only gives additional worked-out examples and practice exercises, but also requires students to demonstrate mastery of the skills associated with a particular concept or objective.

Videotapes A set of ten professional-quality videotapes featuring computer-generated graphics is available at low cost to users of the book. The tapes discuss the ideas within each chapter that cause students the most difficulty.

Audiotapes A set of audiotapes covering the material in each section in the textbook is available at no charge to users of the book. Students needing help with a particular topic, or those who have missed class, will find these tapes helpful.

Student's Solutions Manual This book contains solutions to half of the odd-numbered section exercises (those not included at the back of the textbook) as well as solutions to all chapter review exercises, chapter tests, and cumulative review exercises.

Instructor's Solutions Manual Available at no charge to instructors, this book includes solutions to all the margin problems in the textbook and solutions to the even-numbered section exercises. The two solutions manuals plus the solutions given at the back of the textbook provide detailed, worked-out solutions to all the exercises and margin problems in the book.

ACKNOWLEDGMENTS

We would like to thank the many users of the second edition of this book who were kind enough to share suggestions with us. This revision has benefited from their comments. In particular, we would like to thank the people who reviewed all or part of the manuscript and made many helpful suggestions: Nancy D. Avery, Durham Technical Institute; Janice Beach, State University of New York, College at Plattsburgh; Francine Bortzel, Seton Hall University; Bruce John Carroll, Evergreen Valley College; Leonard Fellman, College of Alameda; James Fryxell, College of Lake County; Herb Garrett, Long Beach City College; Janice McFatter, Gulf Coast Community College; Gloria J. Mills, Tarrant County Junior College; Jack W. Rotman, Lansing Community College; Kathleen Schaffer, State University of New York, College at Plattsburgh; Mary Selander, Virginia Western Community College; and Thomas L. Van Wingen, Grand Rapids Junior College. We also would like to thank the following people who helped us check the answers: Ellen Credille; James Hodge, College of Lake County; William Schooley, William Rainey Harper College; Marjorie Seachrist; and Alexa Stiegemeier, Elgin Community College.

At Scott, Foresman we received help beyond the call of duty from Bill Poole, Linda Youngman, and Janet Tilden.

Charles D. Miller
Margaret L. Lial

Contents

Diagnostic Pretest   xi

1 Number Systems   1

1.1 Fractions   1
1.2 Decimals and Percents   9
1.3 Symbols   19
1.4 Exponents and Order of Operations   23
1.5 Variables and Equations   29
    Chapter 1 Review Exercises   37
    Chapter 1 Test   41

2 Operations with Real Numbers   43

2.1 Real Numbers and the Number Line   43
2.2 Addition of Real Numbers   51
2.3 Subtraction of Real Numbers   59
2.4 Multiplication of Real Numbers   65
2.5 Division of Real Numbers   71
2.6 Properties of Addition and Multiplication   79
2.7 Simplifying Expressions   87
    Chapter 2 Review Exercises   93
    Chapter 2 Test   97

Cumulative Review Exercises   99

3 Solving Equations and Inequalities   103

3.1 The Addition Property of Equality   103
3.2 The Multiplication Property of Equality   109
3.3 Solving Linear Equations   115
3.4 From Word Problems to Equations   123

3.5 Formulas 131
3.6 Ratios and Proportions 139
3.7 The Addition Property of Inequality 147
3.8 The Multiplication Property of Inequality 153
Chapter 3 Review Exercises 159
Chapter 3 Test 167

4 Exponents and Polynomials 169

4.1 Exponents 169
4.2 Power Rules for Exponents 177
4.3 An Application of Exponents: Scientific Notation 183
4.4 Polynomials 187
4.5 Multiplication of Polynomials 195
4.6 Products of Binomials 201
4.7 Dividing a Polynomial by a Monomial 207
4.8 The Quotient of Two Polynomials 211
Chapter 4 Review Exercises 217
Chapter 4 Test 221

Cumulative Review Exercises 223

5 Factoring 227

5.1 Factors; The Greatest Common Factor 227
5.2 Factoring Trinomials 235
5.3 More on Factoring Trinomials 241
5.4 Special Factorizations 249
Supplementary Factoring Exercises 255
5.5 Solving Quadratic Equations by Factoring 257
5.6 Applications of Quadratic Equations 263
Chapter 5 Review Exercises 271
Chapter 5 Test 277

6 Rational Expressions 279

6.1 The Fundamental Property of Rational Expressions 279
6.2 Multiplication and Division of Rational Expressions 285

6.3 Least Common Denominators 291

6.4 Addition and Subtraction of Rational Expressions 295

6.5 Complex Fractions 301

6.6 Equations Involving Rational Expressions 307

Supplementary Exercises on Rational Expressions 315

6.7 Applications of Rational Expressions 317

Chapter 6 Review Exercises 325

Chapter 6 Test 329

Cumulative Review Exercises 331

7 Graphing Linear Equations 335

7.1 Linear Equations in Two Variables 335

7.2 Graphing Ordered Pairs 341

7.3 Graphing Linear Equations 347

7.4 The Slope of a Line 357

7.5 Equations of a Line 365

7.6 Graphing Linear Inequalities in Two Variables 373

Chapter 7 Review Exercises 381

Chapter 7 Test 385

8 Linear Systems 389

8.1 Solving Systems of Linear Equations by Graphing 389

8.2 Solving Systems of Linear Equations by Addition 397

8.3 Two Special Cases 405

8.4 Solving Systems of Linear Equations by Substitution 409

8.5 Applications of Linear Systems 417

8.6 Solving Systems of Linear Inequalities 425

Chapter 8 Review Exercises 429

Chapter 8 Test 433

Cumulative Review Exercises 437

9 Roots and Radicals 441

- 9.1 Finding Roots 441
- 9.2 Products and Quotients of Radicals 447
- 9.3 Addition and Subtraction of Radicals 455
- 9.4 Rationalizing the Denominator 459
- 9.5 Simplifying Radical Expressions 465
- 9.6 Equations with Radicals 473
- Chapter 9 Review Exercises 481
- Chapter 9 Test 485

10 Quadratic Equations 487

- 10.1 Solving Quadratic Equations by the Square Root Method 487
- 10.2 Solving Quadratic Equations by Completing the Square 491
- 10.3 Solving Quadratic Equations by the Quadratic Formula 497
- Supplementary Exercises on Quadratic Equations 505
- 10.4 Graphing Quadratic Equations in Two Variables 507
- Chapter 10 Review Exercises 513
- Chapter 10 Test 517

Final Examination 519

Appendices 523

- Appendix A Sets 523
- Appendix B Complex Solutions of Equations 529

Symbols 535

Table 1 Selected Powers of Numbers 535

Table 2 Powers and Roots 536

Answers to Selected Exercises 537

Solutions to Selected Exercises 561

Index 608

name _____ date _____ hour _____

DIAGNOSTIC PRETEST

Ask your instructor whether or not you should work this pretest. It is designed to tell what material in the course may already be familiar to you. The actual course begins on page 1.

1. Write $\dfrac{96}{144}$ in lowest terms. 1. __2/3__

2. Add: $\dfrac{3}{4} + \dfrac{11}{12} + \dfrac{7}{8}$. 2. __7/8__

3. Divide: $\dfrac{15}{16} \div \dfrac{25}{24}$ 3. __9/10__

Simplify each of the following.

4. $2(5 + 6) + 7 \cdot 3 - 4^2$ 4. __27__

5. $\dfrac{4(5 + 3) + 3}{2(3) - 1}$ 5. __35/5__ __=7__

6. $|-5|$ 6. _____

7. $-|-14|$ 7. _____

Add.

8. $-2 + (-9)$ 8. __−11__

9. $-8 + 15 + 13 + (-4 + 9)$ 9. __25__

Subtract.

10. $-7 - (-15)$ 10. __8__

11. $8 - [-7 - (-4)]$ 11. __19__

Solve each equation.

12. $4r + 5r - 3 + 8 - 3r = 5r + 12 + 8$ 12. _____

13. $4(k - 3) - k = k - 6$ 13. _____

Diagnostic Pretest

14. _____

14. The perimeter of a rectangle is 68 meters. The length of the rectangle is 7 meters more than the width. Find the width of the rectangle.

15. _____

15. Solve: $3z + 2 - 5 > -z + 7 + 2z$.

16. _____

16. Solve: $-\dfrac{2}{3}y \leq 6$.

17. _____625_____

17. Evaluate 5^4. Name the base and the exponent.

Simplify and write each expression in Exercises 18 and 19 without negative exponents.

18. _____

18. $(2x^7)(-4x^8)$

19. _____

19. $\left(\dfrac{3^2 y^{-2}}{4^{-1} y^3}\right)^{-3}$

20. _____

20. Subtract $6x^3 - 4x^2 + 2$ from $11x^3 + 2x^2 - 8$.

21. _____

21. Multiply $3p - 5$ and $2p + 6$.

22. _____

22. Divide $2x + 3$ into $8x^3 - 4x^2 - 14x + 15$.

Write each of the following in prime factored form.

23. _____

23. 50

24. _____

24. 320

Factor each trinomial in Exercises 25–27.

25. _____

25. $m^2 + 9m + 14$

26. _____

26. $12a^2 - ab - 20b^2$

27. _____

27. $121z^2 - 44z + 4$

28. _____

28. Solve the quadratic equation $x^2 - 5x = -6$.

29. _____

29. The product of two consecutive odd integers is 1 less than five times their sum. Find the integers.

Diagnostic Pretest

30. Find the product: $\dfrac{x^2 + 3x}{x^2 - 3x - 4} \cdot \dfrac{x^2 - 5x + 4}{x^2 + 2x - 3}$.

31. Add: $\dfrac{x}{x^2 - 1} + \dfrac{x}{x + 1}$.

32. Solve: $\dfrac{2}{r^2 - r} = \dfrac{1}{r^2 - 1}$.

Graph each of the following.

33. $4x - 5y = 20$

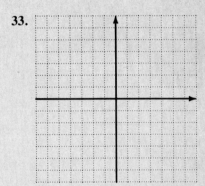

34. $x = 3$

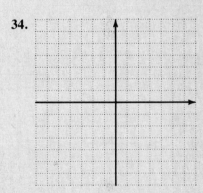

35. $2x - 5y \geq 10$

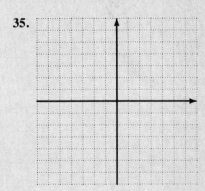

36. Find the slope of the line through the points $(8, -3)$ and $(-4, 7)$.

Solve each of the following systems of equations.

37. $2x + 3y = -15$
    $5x + 2y = 1$

Diagnostic Pretest

38. $3x - y = 4$
    $-9x + 3y = -12$

Find each of the following roots.

39. $-\sqrt{1024}$

40. $\sqrt[3]{216}$

Simplify each of the following.

41. $\sqrt{72}$

42. $\sqrt[3]{40}$

43. $2\sqrt{12} + 3\sqrt{75}$

44. $\sqrt{18} \cdot \sqrt{75}$

Rationalize each denominator.

45. $\dfrac{14}{\sqrt{7}}$

46. $\dfrac{-2}{3 - \sqrt{2}}$

Solve each equation in Exercises 47–49.

47. $3\sqrt{y} = \sqrt{y+8}$

48. $z^2 = 2z + 1$

49. $\dfrac{1}{10}t^2 = \dfrac{2}{5} - \dfrac{1}{2}t$

50. Graph $y = (x - 2)^2$.

Number Systems   1

1.1 FRACTIONS

As preparation for the study of algebra, this section begins with a brief review of arithmetic.

▶ In everyday life, the numbers seen most often are the **whole numbers,**

$$0, 1, 2, 3, 4, 5, \ldots$$

and the **fractions,** such as

$$\frac{1}{3}, \frac{2}{3}, \text{ and } \frac{11}{12}.$$

In a fraction, the number on top is called the **numerator** and the number on the bottom is the **denominator.**

A fraction is a **proper fraction** if the numerator is smaller than the denominator; otherwise it is an **improper fraction.**

EXAMPLE 1 The numbers $\frac{3}{4}$, $\frac{7}{8}$, $\frac{9}{10}$, and $\frac{125}{126}$ are proper fractions, and $\frac{5}{4}$, $\frac{17}{15}$, and $\frac{28}{3}$ are improper fractions. ◀*

An improper fraction can be written as a **mixed number,** the sum of a whole number and a proper fraction. For instance, the improper fraction $\frac{4}{3}$ can be written as $1\frac{1}{3}$, since $1\frac{1}{3}$ equals $\frac{3}{3} + \frac{1}{3} = \frac{4}{3}$. In algebra, the improper fraction form is usually preferred.

Work Problem 1 at the side.

▶ In the statement $2 \times 9 = 18$, the numbers 2 and 9 are called **factors** of 18. Other factors of 18 include 1, 3, 6, and 18. The result of the multiplication, 18, is called the **product.**

The number 18 is **factored** by writing it as the product of two or more numbers. For example, 18 can be factored in several ways, as $6 \cdot 3$, or $18 \cdot 1$, or $9 \cdot 2$, or $3 \cdot 3 \cdot 2$. In algebra, raised dots are used instead of the $\times$ symbol to indicate multiplication.

A whole number (except 1) is **prime** if it has only itself and 1 as factors. (By agreement, the number 1 is not a prime number.) The first dozen primes are listed here.

$$2, 3, 5, 7, 11, 13, 17, 19, 23, 29, 31, 37.$$

In algebra, it is often useful to find all the **prime factors** of a number—those factors that are prime numbers. For example, the only prime factors of 18 are 2 and 3.

---

*This symbol, ◀, indicates the end of an example.

Objectives

▶ Understand the terminology of fractions.

▶ Learn the definition of *factor.*

▶ Write fractions in lowest terms.

▶ Multiply and divide fractions.

▶ Write a fraction as an equivalent fraction with a given denominator.

▶ Add and subtract fractions.

1. Which of these fractions are proper fractions?

$$\frac{1}{2}, \frac{5}{3}, \frac{7}{4}, \frac{10}{9}, \frac{8}{15}, \frac{13}{13}$$

ANSWER

1. $\frac{1}{2}$ and $\frac{8}{15}$ are proper since the numerator is smaller than the denominator in each; the others are improper.

1.1 Fractions

2. Write each number as the product of prime factors.

   (a) 90

   (b) 48

3. Write each fraction in lowest terms.

   (a) $\dfrac{8}{14}$

   (b) $\dfrac{35}{42}$

   (c) $\dfrac{9}{18}$

   (d) $\dfrac{12}{20}$

EXAMPLE 2 Write each number as the product of prime factors.

(a) 35

Write 35 as the product of the prime factors 5 and 7, or as
$$35 = 5 \cdot 7.$$

(b) 24

Divide by the smallest prime, 2, to get
$$24 = 2 \cdot 12.$$

Now divide 12 by 2 to find factors of 12.
$$24 = 2 \cdot 2 \cdot 6$$

Since 6 can be written as $2 \cdot 3$,
$$24 = 2 \cdot 2 \cdot 2 \cdot 3,$$

where all factors are prime. ◀

Work Problem 2 at the side.

▶ Prime factors are used to write fractions in **lowest terms**. A fraction is in lowest terms when the numerator and the denominator have no factors in common (other than 1). Use the following steps to write a fraction in lowest terms.

Writing a Fraction in Lowest Terms

Step 1 Write the numerator and the denominator as the product of prime factors.

Step 2 Divide the numerator and the denominator by the **greatest common factor,** the product of all factors common to both.

EXAMPLE 3 Write each fraction in lowest terms.

(a) $\dfrac{10}{15} = \dfrac{2 \cdot 5}{3 \cdot 5} = \dfrac{2}{3}$

Since 5 is the only common factor of 10 and 15, dividing both numerator and denominator by 5 gives the fraction in lowest terms.

(b) $\dfrac{15}{45} = \dfrac{3 \cdot 5}{3 \cdot 3 \cdot 5} = \dfrac{1}{3}$

The factored form shows that 3 and 5 are the common factors of both 15 and 45. Dividing both 15 and 45 by $3 \cdot 5 = 15$ gives $\dfrac{15}{45}$ in lowest terms as $\dfrac{1}{3}$. ◀

Work Problem 3 at the side.

▶ Multiplication of fractions is defined next.

Multiplying Fractions

To multiply two fractions, multiply the numerators and multiply the denominators.

ANSWERS
2. (a) $2 \cdot 3 \cdot 3 \cdot 5$ (b) $2 \cdot 2 \cdot 2 \cdot 2 \cdot 3$
3. (a) $\dfrac{4}{7}$ (b) $\dfrac{5}{6}$ (c) $\dfrac{1}{2}$ (d) $\dfrac{3}{5}$

EXAMPLE 4 Find the product of $\frac{3}{8}$ and $\frac{4}{9}$, and write it in lowest terms.

$$\frac{3}{8} \cdot \frac{4}{9} = \frac{3 \cdot 4}{8 \cdot 9} \quad \text{Multiply numerators}$$
$$\text{Multiply denominators}$$
$$= \frac{3 \cdot 4}{2 \cdot 4 \cdot 3 \cdot 3} \quad \text{Factor}$$
$$= \frac{1}{6} \quad \text{Divide by the greatest common factor} \blacktriangleleft$$

Work Problem 4 at the side.

Two fractions are **reciprocals** of each other if their product is 1. For example, $\frac{3}{4}$ and $\frac{4}{3}$ are reciprocals since

$$\frac{3}{4} \cdot \frac{4}{3} = 1.$$

The numbers $\frac{7}{11}$ and $\frac{11}{7}$ are reciprocals also. Reciprocals are used in dividing fractions.

Dividing Fractions

To divide two fractions, multiply the first fraction and the reciprocal of the second.

The reason this method works will be explained in a later chapter. The answer to a division problem is called the **quotient.** For example, the quotient of 20 and 10 is 2, since $20 \div 10 = 2$.

EXAMPLE 5 Find the following quotients, and write them in lowest terms.

(a) $\frac{3}{4} \div \frac{8}{5} = \frac{3}{4} \cdot \frac{5}{8} = \frac{3 \cdot 5}{4 \cdot 8} = \frac{15}{32}$

↑ Multiply by reciprocal of second fraction

(b) $\frac{3}{4} \div \frac{5}{8} = \frac{3}{4} \cdot \frac{8}{5} = \frac{3 \cdot 8}{4 \cdot 5} = \frac{24}{20} = \frac{6}{5}$

(c) $\frac{2}{30} \div \frac{6}{15} = \frac{2}{30} \cdot \frac{15}{6} = \frac{2 \cdot 15}{30 \cdot 6} = \frac{30}{180} = \frac{1}{6}$

(d) $\frac{5}{8} \div 10 = \frac{5}{8} \div \frac{10}{1} = \frac{5}{8} \cdot \frac{1}{10} = \frac{1}{16}$ ◀

↑ Write 10 as $\frac{10}{1}$

Work Problem 5 at the side.

▶ All the fractions in an addition or subtraction problem must be written with the same denominator. If some of the fractions have different denominators, rewrite all the fractions with a new common denominator. For example, to rewrite $\frac{3}{4}$ as a fraction with a denominator

4. Find each product, and write it in lowest terms.

(a) $\frac{5}{8} \cdot \frac{2}{10}$

(b) $\frac{3}{4} \cdot \frac{2}{3}$

(c) $\frac{1}{10} \cdot \frac{12}{5}$

(d) $\frac{7}{9} \cdot \frac{12}{14}$

5. Find each quotient, and write it in lowest terms.

(a) $\frac{9}{10} \div \frac{3}{5}$

(b) $\frac{3}{4} \div \frac{9}{16}$

(c) $\frac{1}{2} \div \frac{1}{8}$

(d) $\frac{2}{3} \div 6$

ANSWERS

4. (a) $\frac{1}{8}$ (b) $\frac{1}{2}$ (c) $\frac{6}{25}$ (d) $\frac{2}{3}$
5. (a) $\frac{3}{2}$ (b) $\frac{4}{3}$ (c) 4 (d) $\frac{1}{9}$

6. Write as fractions having the given denominators.

(a) $\dfrac{9}{10} = \dfrac{}{40}$

(b) $\dfrac{4}{5} = \dfrac{}{60}$

(c) $\dfrac{1}{7} = \dfrac{}{49}$

(d) $\dfrac{3}{2} = \dfrac{}{8}$

7. Add. Write sums in lowest terms.

(a) $\dfrac{1}{3} + \dfrac{1}{3}$

(b) $\dfrac{4}{7} + \dfrac{1}{7}$

(c) $\dfrac{1}{9} + \dfrac{5}{9}$

(d) $\dfrac{1}{4} + \dfrac{1}{4}$

ANSWERS

6. (a) $\dfrac{36}{40}$ (b) $\dfrac{48}{60}$ (c) $\dfrac{7}{49}$ (d) $\dfrac{12}{8}$

7. (a) $\dfrac{2}{3}$ (b) $\dfrac{5}{7}$ (c) $\dfrac{2}{3}$ (d) $\dfrac{1}{2}$

---

of 32, find the number that can be multiplied by 4 to give 32. Since $4 \cdot 8 = 32$, use the number 8. Keep the value of the original fraction, $\dfrac{3}{4}$, the same by multiplying $\dfrac{3}{4}$ by the fraction $\dfrac{8}{8}$, which equals 1.

$$\dfrac{3}{4} = \dfrac{3}{4} \cdot \dfrac{8}{8} = \dfrac{3 \cdot 8}{4 \cdot 8} = \dfrac{24}{32}$$

↑ Multiplying by 1 does not change the value

EXAMPLE 6 Write each of the following as fractions having the given denominators.

(a) $\dfrac{5}{8} = \dfrac{}{72}$

Since 8 must be multiplied by 9 to get 72, multiply $\dfrac{5}{8}$ by $\dfrac{9}{9}$.

$$\dfrac{5}{8} = \dfrac{5}{8} \cdot \dfrac{9}{9} = \dfrac{5 \cdot 9}{8 \cdot 9} = \dfrac{45}{72}$$

(b) $\dfrac{2}{3} = \dfrac{}{18}$

Since $3 \times 6 = 18$, multiply by $\dfrac{6}{6}$.

$$\dfrac{2}{3} = \dfrac{2}{3} \cdot \dfrac{6}{6} = \dfrac{2 \cdot 6}{3 \cdot 6} = \dfrac{12}{18}$$ ◂

Work Problem 6 at the side.

▶ The result of adding two numbers is called the *sum* of the numbers. For example, since $2 + 3 = 5$, the sum of 2 and 3 is 5. The sum of two fractions is found as follows.

Adding Fractions

The **sum** of two fractions with the same denominator is found by adding their numerators.

EXAMPLE 7 Add. Write sums in lowest terms.

(a) $\dfrac{3}{7} + \dfrac{2}{7} = \dfrac{3+2}{7} = \dfrac{5}{7}$ Denominator does not change

(b) $\dfrac{2}{10} + \dfrac{3}{10} = \dfrac{2+3}{10} = \dfrac{5}{10} = \dfrac{1}{2}$ Lowest terms ◂

Work Problem 7 at the side.

When the two fractions to be added do not have the same denominator, first rewrite them with a common denominator and then use the rule above.

EXAMPLE 8 Add.

(a) $\dfrac{1}{2} + \dfrac{1}{3}$

Number Systems

These fractions cannot be added until both have the same denominator. The smallest number that both 2 and 3 can be divided into is 6, so the **least common denominator** is $2 \cdot 3 = 6$. Write both $\frac{1}{2}$ and $\frac{1}{3}$ as fractions with a denominator of 6.

$$\frac{1}{2} = \frac{1}{2} \cdot \frac{3}{3} = \frac{1 \cdot 3}{2 \cdot 3} = \frac{\mathbf{3}}{\mathbf{6}} \quad \text{and} \quad \frac{1}{3} = \frac{1}{3} \cdot \frac{2}{2} = \frac{1 \cdot 2}{3 \cdot 2} = \frac{\mathbf{2}}{\mathbf{6}}$$

Now add.

$$\frac{1}{2} + \frac{1}{3} = \frac{3}{6} + \frac{2}{6} = \frac{3+2}{6} = \frac{5}{6}$$

(b) $\dfrac{3}{10} + \dfrac{5}{12}$

Find the **least common denominator** by first writing each denominator as the product of prime factors.

$$10 = 2 \cdot 5 \quad \text{and} \quad 12 = 2 \cdot 2 \cdot 3$$

For the factors of the least common denominator, take each prime the *greatest* number of times it appears in any of the factored forms. Since there are two 2s in the factored form of 12, the least common denominator has two factors of 2. The primes 3 and 5 each appear only once, so the least common denominator is

$$2 \cdot 2 \cdot 3 \cdot 5 = 60.$$

Now write each fraction with a denominator of 60.

$$\frac{3}{10} = \frac{3 \cdot 6}{\mathbf{10 \cdot 6}} = \frac{18}{60} \quad \text{and} \quad \frac{5}{12} = \frac{5 \cdot 5}{\mathbf{12 \cdot 5}} = \frac{25}{60}$$

Finally,

$$\frac{3}{10} + \frac{5}{12} = \frac{18}{60} + \frac{25}{60} = \frac{18+25}{60} = \frac{43}{60}.$$

(c) $3\dfrac{1}{2} + 2\dfrac{3}{4}$

Change both mixed numbers to improper fractions, as follows.

$$3\frac{1}{2} = 3 \mathbf{+} \frac{1}{2} = \frac{3}{1} + \frac{1}{2} = \frac{6}{2} + \frac{1}{2} = \frac{6+1}{2} = \frac{7}{2}$$

Also,

$$2\frac{3}{4} = 2 \mathbf{+} \frac{3}{4} = \frac{8}{4} + \frac{3}{4} = \frac{8+3}{4} = \frac{11}{4}.$$

Now add.

$$3\frac{1}{2} + 2\frac{3}{4} = \frac{7}{2} + \frac{11}{4} = \frac{14}{4} + \frac{11}{4} = \frac{25}{4} \quad \text{or} \quad 6\frac{1}{4} \blacktriangleleft$$

Work Problem 8 at the side.

The *difference* of two numbers is found by subtracting the numbers. For example, $9 - 5 = 4$, so the difference of 9 and 5 is 4. Find the difference of two fractions as follows.

8. Add.

(a) $\dfrac{2}{3} + \dfrac{1}{12}$

(b) $\dfrac{3}{4} + \dfrac{1}{6}$

(c) $\dfrac{7}{30} + \dfrac{2}{45}$

ANSWERS

8. (a) $\dfrac{3}{4}$ (b) $\dfrac{11}{12}$ (c) $\dfrac{5}{18}$

1.1 Fractions

9. Subtract.

(a) $\dfrac{9}{11} - \dfrac{3}{11}$

(b) $\dfrac{2}{3} - \dfrac{1}{2}$

(c) $\dfrac{3}{10} - \dfrac{1}{4}$

(d) $2\dfrac{3}{8} - 1\dfrac{1}{2}$

Subtracting Fractions

The difference of two fractions with the same denominator is found by subtracting their numerators.

EXAMPLE 9 Subtract.

(a) $\dfrac{5}{8} - \dfrac{3}{8} = \dfrac{5-3}{8} = \dfrac{2}{8} = \dfrac{1}{4}$ (in lowest terms)

(b) $\dfrac{3}{4} - \dfrac{1}{3}$

The least common denominator is 12.

$$\dfrac{3}{4} - \dfrac{1}{3} = \dfrac{9}{12} - \dfrac{4}{12} = \dfrac{9-4}{12} = \dfrac{5}{12}$$

(c) $\dfrac{7}{9} - \dfrac{1}{6} = \dfrac{14}{18} - \dfrac{3}{18} = \dfrac{14-3}{18} = \dfrac{11}{18}$

Write each fraction with the least common denominator, 18.

(d) $2\dfrac{1}{2} - 1\dfrac{3}{4}$

First, change the mixed numbers $2\dfrac{1}{2}$ and $1\dfrac{3}{4}$ into improper fractions.

$$2\dfrac{1}{2} = 2 + \dfrac{1}{2} = \dfrac{4}{2} + \dfrac{1}{2} = \dfrac{5}{2}$$

$$1\dfrac{3}{4} = 1 + \dfrac{3}{4} = \dfrac{4}{4} + \dfrac{3}{4} = \dfrac{7}{4}$$

Now subtract.

$$2\dfrac{1}{2} - 1\dfrac{3}{4} = \dfrac{5}{2} - \dfrac{7}{4}$$

Multiply $\dfrac{5}{2}$ by $\dfrac{2}{2}$ to get a common denominator of 4.

$$\dfrac{5}{2} \cdot \dfrac{2}{2} = \dfrac{10}{4}$$

Finally,

$$\dfrac{10}{4} - \dfrac{7}{4} = \dfrac{3}{4}. \blacktriangleleft$$

Work Problem 9 at the side.

ANSWERS

9. (a) $\dfrac{6}{11}$ (b) $\dfrac{1}{6}$ (c) $\dfrac{1}{20}$ (d) $\dfrac{7}{8}$

name date hour

1.1 EXERCISES

Write each fraction in lowest terms. See Example 3.

1. $\dfrac{7}{14}$
2. $\dfrac{3}{9}$
3. $\dfrac{10}{12}$
4. $\dfrac{8}{10}$

5. $\dfrac{16}{18}$
6. $\dfrac{14}{20}$
7. $\dfrac{50}{75}$
8. $\dfrac{32}{48}$

9. $\dfrac{72}{108}$
10. $\dfrac{96}{120}$
11. $\dfrac{120}{144}$
12. $\dfrac{77}{132}$

Find the products or quotients. Write answers in lowest terms. See Examples 4 and 5.

13. $\dfrac{3}{4} \cdot \dfrac{3}{5}$
14. $\dfrac{3}{8} \cdot \dfrac{5}{7}$
15. $\dfrac{1}{10} \cdot \dfrac{6}{5}$
16. $\dfrac{6}{7} \cdot \dfrac{1}{3}$

17. $\dfrac{9}{4} \cdot \dfrac{8}{15}$
18. $\dfrac{3}{5} \cdot \dfrac{20}{15}$
19. $\dfrac{3}{8} \div \dfrac{5}{4}$
20. $\dfrac{9}{16} \div \dfrac{3}{8}$

21. $\dfrac{5}{12} \div \dfrac{15}{4}$
22. $\dfrac{21}{16} \div \dfrac{7}{8}$
23. $\dfrac{28}{3} \div \dfrac{7}{6}$
24. $\dfrac{121}{9} \div \dfrac{11}{18}$

25. $3\dfrac{15}{16} \div \dfrac{15}{4}$
26. $1\dfrac{15}{32} \div 2\dfrac{5}{8}$
27. $7\dfrac{4}{5} \div 2\dfrac{3}{10}$

28. $1\dfrac{5}{9} \cdot 3\dfrac{7}{10}$
29. $6\dfrac{7}{10} \cdot 4\dfrac{5}{7}$
30. $2\dfrac{3}{10} \cdot 5\dfrac{5}{3}$

Add or subtract. Write answers in lowest terms. See Examples 6–9. (Work from left to right in Exercises 49–54.)

31. $\dfrac{1}{12} + \dfrac{5}{12}$
32. $\dfrac{2}{5} + \dfrac{1}{5}$
33. $\dfrac{1}{10} + \dfrac{7}{10}$
34. $\dfrac{3}{8} + \dfrac{1}{8}$

35. $\dfrac{4}{9} + \dfrac{2}{3}$
36. $\dfrac{3}{5} + \dfrac{2}{15}$
37. $\dfrac{8}{11} + \dfrac{3}{22}$
38. $\dfrac{9}{10} + \dfrac{3}{5}$

39. $\dfrac{2}{3} - \dfrac{3}{5}$
40. $\dfrac{8}{12} - \dfrac{5}{9}$
41. $\dfrac{5}{6} - \dfrac{3}{10}$
42. $\dfrac{11}{4} - \dfrac{5}{8}$

43. $3\frac{1}{4} + \frac{1}{8}$

44. $5\frac{2}{3} + \frac{3}{4}$

45. $4\frac{1}{2} + 3\frac{2}{3}$

46. $7\frac{5}{8} + 3\frac{3}{4}$

47. $6\frac{1}{3} - 5\frac{1}{4}$

48. $8\frac{4}{9} - 7\frac{4}{5}$

49. $\frac{2}{5} + \frac{1}{3} + \frac{9}{10}$

50. $\frac{3}{8} + \frac{5}{6} + \frac{2}{3}$

51. $\frac{5}{7} + \frac{1}{4} - \frac{1}{2}$

52. $\frac{2}{3} + \frac{1}{6} - \frac{1}{2}$

53. $\frac{3}{4} + \frac{1}{8} - \frac{2}{3}$

54. $\frac{7}{10} + \frac{3}{5} - \frac{5}{8}$

Work each of the following word problems.

55. John Rizzo paid $\frac{1}{8}$ of a debt in January, $\frac{1}{3}$ in February, and $\frac{1}{4}$ in March. What portion of the debt was paid in these three months?

56. A rectangle is $\frac{5}{16}$ yard on each of two sides, and $\frac{7}{12}$ yard on each of the other two sides. Find the total distance around the rectangle.

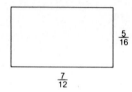

57. The Eastside Wholesale Market sold $3\frac{1}{4}$ tons of broccoli last month, $2\frac{3}{8}$ tons of spinach, $7\frac{1}{2}$ tons of corn, and $1\frac{5}{16}$ tons of turnips. Find the total number of tons of vegetables sold by the firm during the month.

58. Sharkey's Resort decided to expand by buying a piece of property next to the resort. The property has an irregular shape, with five sides. The lengths of the five sides are $146\frac{1}{2}$ feet, $98\frac{3}{4}$ feet, 196 feet, $76\frac{5}{8}$ feet, and $100\frac{7}{8}$ feet. Find the total distance around the piece of property.

59. Joann Kaufmann worked 40 hours during a certain week. She worked $8\frac{1}{4}$ hours on Monday, $6\frac{3}{8}$ hours on Tuesday, $7\frac{2}{3}$ hours on Wednesday, and $8\frac{3}{4}$ hours on Thursday. How many hours did she work on Friday?

60. A concrete truck is loaded with $9\frac{7}{8}$ cubic yards of concrete. The driver unloads $1\frac{1}{2}$ cubic yards at the first stop and $2\frac{3}{4}$ cubic yards at the second stop. At the third stop, $3\frac{5}{12}$ cubic yards are unloaded. How much concrete is left in the truck?

61. Rosario has 36 yards of material to make dresses to sell at a bazaar. Each dress needs $2\frac{1}{4}$ yards of material. How many dresses can be made?

62. Lindsay allows $1\frac{3}{5}$ bottles of beverage for each guest at a party. If he expects 35 guests, how many bottles of beverage will he need?

1.2 DECIMALS AND PERCENTS

Fractions are one way to represent parts of a whole. Another way is with decimals; a **decimal number** is a number written with a decimal point, such as 9.4. Each digit in a decimal number has a place value, as shown below.

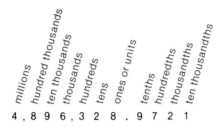

4,896,328.9721

Each successive place value is ten times larger than the place value to its right and is one-tenth as large as the place value to its left.

Prices are often written as decimals. The price $14.75 means 14 dollars and 75 cents, or 14 dollars and $\frac{75}{100}$ of a dollar.

▶ Find the place value for a digit in a number by writing the number in **expanded form.** For example, the expanded form for 1729 is

$$1729 = 1000 + 700 + 20 + 9.$$

This expanded form shows that 1 is in the thousands position, 7 is in the hundreds position, 2 is in the tens position, and 9 is in the ones or units position.

EXAMPLE 1   Write the following numbers in expanded form.

(a) $98.42 = 90 + 8 + .4 + .02$

(b) $.618 = .6 + .01 + .008$ ◀

Work Problem 1 at the side.

▶ Place value is used to write a decimal number as a fraction. For example, since the last digit of .67 is in the *hundredths* place,

$$.67 = \frac{67}{100}.$$

EXAMPLE 2   Write each decimal as a fraction in lowest terms.

(a) .9

This decimal is the same as $\frac{9}{10}$.

(b) $.6 = \frac{6}{10} = \frac{3}{5}$

(c) $.25 = \frac{25}{100} = \frac{1}{4}$

(d) $.295 = \frac{295}{1000} = \frac{59}{200}$ ◀

Work Problem 2 at the side.

Objectives

▶ Write a number in expanded form.

▶ Convert decimals to fractions.

▶ Add and subtract decimals.

▶ Multiply and divide decimals.

▶ Convert fractions to decimals.

▶ Convert percents to decimals and decimals to percents.

▶ Find percentages and percents.

1. Write in expanded form.

   (a) 478

   (b) 15,813

   (c) 12.46

   (d) .438

2. Convert to fractions. Write in lowest terms.

   (a) .5

   (b) .80

   (c) .15

   (d) .305

ANSWERS
1. (a) $400 + 70 + 8$
   (b) $10,000 + 5000 + 800 + 10 + 3$
   (c) $10 + 2 + .4 + .06$
   (d) $.4 + .03 + .008$
2. (a) $\frac{1}{2}$ (b) $\frac{4}{5}$ (c) $\frac{3}{20}$ (d) $\frac{61}{200}$

3. Add or subtract as indicated.

(a)   68.9
      42.72
    + 8.973

(b)   32.5
    − 21.72

(c) 42.83 + 71.629 + 3.074

(d) 351.8 − 2.706

ANSWERS
3. (a) 120.593   (b) 10.78
   (c) 117.533   (d) 349.094

---

3 The operations of addition and subtraction of decimals are explained in the next examples.

EXAMPLE 3 Add or subtract as indicated.

(a) 6.92 + 14.8 + 3.217

Place the numbers in a column, with decimal points lined up so that tenths are in one column, hundredths in another column, and so on.

$$\begin{array}{r} 6.92 \\ 14.8 \\ +\ 3.217 \\ \hline 24.937 \end{array}$$ Decimal points lined up

A good way to avoid errors is to attach zeros to make all the numbers the same length. For example,

$$\begin{array}{r} 6.92 \\ 14.8 \\ +\ 3.217 \end{array} \quad \text{becomes} \quad \begin{array}{r} 6.920 \\ 14.800 \\ +\ 3.217 \\ \hline 24.937. \end{array}$$ Attach zeros

(b) 47.6 − 32.509

Write the numbers in a column, attaching zeros to 47.6.

$$\begin{array}{r} 47.6 \\ -\ 32.509 \end{array} \quad \text{becomes} \quad \begin{array}{r} 47.600 \\ -\ 32.509 \\ \hline 15.091 \end{array}$$

(c) 3 − .253

A whole number is assumed to have the decimal at the right of the number. Write 3 as 3.000; then subtract.

$$\begin{array}{r} 3.000 \\ -\ .253 \\ \hline 2.747 \end{array}$$ ◀

Work Problem 3 at the side.

4 Multiplication with decimals is done as follows.

> **Multiplying Decimals**
> Ignore the decimal points and multiply as if the numbers were whole numbers. Then add together the number of **decimal places** (digits after the decimal point) in each number being multiplied. Locate the decimal point in the answer that many digits from the right.

EXAMPLE 4 Multiply.

(a) 29.3 × 4.52

Multiply as if the numbers were whole numbers.

$$\begin{array}{r} 29.3 \\ \times\ 4.52 \\ \hline 5\ 86 \\ 14\ 6\ 5 \\ 117\ 2 \\ \hline 132.4\ 36 \end{array}$$

1 decimal place in top number
2 decimal places in second number
1 + 2 = 3
3 decimal places in answer

Number Systems

(b) 7.003 × 55.8

$$\begin{array}{r} 7.003 \\ \times\ 55.8 \\ \hline 5\ 602\ 4 \\ 35\ 015 \\ 350\ 15 \\ \hline 390.767\ 4 \end{array}$$

3 decimal places
1 decimal place
3 + 1 = 4

4 decimal places

(c) 31.42 × 65

$$\begin{array}{r} 31.42 \\ \times\ \ \ 65 \\ \hline 157\ 10 \\ 1885\ 2 \\ \hline 2042.30 \end{array}$$

2 decimal places
0 decimal places
2 + 0 = 0

2 decimal places ◀

Work Problem 4 at the side.

Division of decimal numbers uses a slightly different process.

Dividing Decimals

Change the divisor (the number being divided by) into a whole number by moving the decimal point as many places as necessary to the right. Move the decimal point in the dividend (the number being divided) to the right by the same number of places. Then bring the decimal point straight up and divide as with whole numbers.

EXAMPLE 5   Divide.

(a) 279.45 ÷ 24.3

Step 1 Write the problem as follows.

$$24.3 \overline{)279.45}$$

Step 2 Move the decimal point in 24.3 one place to the right, to get the whole number 243. Then move the decimal point the same number of places in 279.45 to get 2794.5.

24.3.)279.4.5   Move one decimal place to the right

Step 3 Bring the decimal point straight up and divide as with whole numbers.

$$\begin{array}{r} 11.5 \\ 243\overline{)2794.5} \\ \underline{243} \\ 364 \\ \underline{243} \\ 121\ 5 \\ \underline{121\ 5} \\ 0 \end{array}$$

Move decimal point straight up

(b) $73.82 \overline{)1852.882}$

Move the decimal point two places to the right in 73.82, to get 7382. Do the same thing with 1852.882 to get 185288.2.

$$73.82.\overline{)1852.88.2}$$

---

4. Multiply.

(a) 2.13 × .05

(b) 69.32 × 1.04

(c) 397.12 × .06

(d) 42,980 × .012

ANSWERS
4. (a) .1065 **(b)** 72.0928
   (c) 23.8272 **(d)** 515.76

1.2 Decimals and Percents

5. Divide.

(a) $32.3 \overline{)481.27}$

(b) $.37 \overline{)5.476}$

(c) $375.125 \div 3.001$

Bring the decimal point straight up and divide as with whole numbers.

$$
\begin{array}{r}
25.1 \\
7382 \overline{)185288.2} \\
\underline{14764} \\
37648 \\
\underline{36910} \\
7382 \\
\underline{7382} \\
0
\end{array}
$$ ◀

Work Problem 5 at the side.

▶ A fraction is written in decimal form as follows.

> Convert a fraction to a decimal by dividing the denominator into the numerator.

EXAMPLE 6 Convert each fraction to a decimal.

(a) $\dfrac{1}{2}$

Divide the denominator, 2, into the numerator, 1. Attach zeros after the decimal point of the numerator as needed.

Decimal point comes straight up

$$
\begin{array}{r}
.5 \\
2 \overline{)1.0} \\
\underline{1\,0} \\
0
\end{array}
$$
Attach zero

$\dfrac{1}{2} = .5$

(b) $\dfrac{3}{8}$

$$
\begin{array}{r}
.375 \\
8 \overline{)3.000} \\
\underline{2\,4} \\
60 \\
\underline{56} \\
40 \\
\underline{40} \\
0
\end{array}
$$

$\dfrac{3}{8} = .375$

(c) $\dfrac{2}{3}$

$$
\begin{array}{r}
.6666 \\
3 \overline{)2.000\ldots} \\
\underline{1\,8} \\
20 \\
\underline{18} \\
20 \\
\underline{18} \\
20
\end{array}
$$

ANSWERS
5. (a) 14.9 (b) 14.8 (c) 125

The remainder in this division is never 0. This quotient is a **repeating decimal.** Round repeating decimal quotients to as many places as needed. For example, rounding to the nearest thousandth,

$$\frac{2}{3} = .667.$$

The last part of this example used the rule for rounding: if the first digit to be dropped is 5 or more, round the last digit to be kept to the next highest number. If the digit to be dropped is 4 or less, do not round up. For example, to the nearest thousandth,

.5555 = .556   and   .3333 = .333.
↑ 5 or more   ↑ 4 or less

Work Problem 6 at the side.

▶ An important application of decimals is in work with percents. The word **percent** means "per one hundred." Percent is written with the sign %. One percent means "one per one hundred" or "one one-hundredth."

$$1\% = .01 \quad \text{or} \quad 1\% = \frac{1}{100}$$

EXAMPLE 7

(a) Write 73% as a decimal.

Since 1% = .01,

$$73\% = 73 \times 1\% = 73 \times .01 = .73.$$

Also, 73% can be written as a decimal using the fraction form $1\% = \frac{1}{100}$.

$$73\% = 73 \times 1\% = 73 \times \left(\frac{1}{100}\right) = \frac{73}{100} = .73.$$

(b) Write 125% as a decimal.

$$125\% = 125 \times 1\% = 125 \times .01 = 1.25$$

(c) Write .32 as a percent.

Write .32 as 32 × .01. Then replace .01 with 1%.

$$.32 = 32 \times .01 = 32 \times 1\% = 32\%$$

(d) Write 2.63 as a percent.

$$2.63 = 263 \times .01 = 263 \times 1\% = 263\% \quad ◀$$

Work Problem 7 at the side.

▶ A **percentage** is part of a whole. For example, since 50% represents $\frac{50}{100} = \frac{1}{2}$ of a whole, 50% of 800 is half of 800, or 400. Percentages are found by multiplication, as in the next example.

---

6. Convert to decimals. Round to the nearest thousandth.

(a) $\frac{2}{9}$

(b) $\frac{17}{20}$

(c) $\frac{5}{8}$

7. Convert.

(a) 23% to a decimal

(b) 310% to a decimal

(c) .71 to a percent

(d) 1.32 to a percent

ANSWERS
6. (a) .222 **(b)** .85 **(c)** .625
7. (a) .23 **(b)** 3.10 **(c)** 71%
   (d) 132%

1.2 Decimals and Percents

8. Find the percentages.

   (a) 20% of 70

   (b) 36% of 500

   (c) Find the amount of discount on a television set with a regular price of $270 if the set is on sale at 25% off. Find the sale price of the set.

9. Find each of the following.

   (a) 90 is what percent of 270?

   (b) What percent of 70 is 14?

10. The interest in one year on deposits of $11,000 was $682. What percent interest was paid?

EXAMPLE 8 Find the percentages.

(a) 15% of 600

   In examples like this, the word *of* indicates multiplication, so that 15% of 600 is found by multiplying .15 and 600.

$$15\% \times 600 = .15 \times 600 = 90$$

(b) 125% of 80

$$125\% \times 80 = 1.25 \times 80 = 100$$

(c) A camera with a regular price of $18 is on sale this week at 22% off. Find the amount of the discount and the sale price of the camera.

   Find 22% of $18 by multiplying.

$$22\% \times \$18 = .22 \times \$18 = \$3.96$$

The discount is $3.96. The camera is on sale for $18 − $3.96 = $14.04. ◀

Work Problem 8 at the side.

To find the percentage of an amount, we multiplied by the percent. When the percentage and the amount are known, find the percent by *dividing* the percentage by the amount.

EXAMPLE 9 30 is what percent of 50?

The amount is 50 and the percentage is 30. To find the percent, divide 30 by 50.

$$\begin{array}{c}\text{Percentage} \rightarrow \\ \text{Amount} \rightarrow\end{array} \frac{30}{50} = .60$$

Since .60 = 60%, 30 is 60% of 50. ◀

EXAMPLE 10 The sales tax on an item selling for $150 is $9. What percent of the price was charged for sales tax?

Divide the percentage, $9, by the amount, $150.

$$\frac{9}{150} = .06 = 6\%$$

Sales tax of 6% was charged. ◀

Work Problems 9 and 10 at the side.

ANSWERS
8. (a) 14 (b) 180
   (c) $67.50; $202.50
9. (a) $33\frac{1}{3}$% (b) 20%
10. 6.2%

1.2 EXERCISES

Write in expanded form. See Example 1.

1. 86

2. 15

3. 694

4. 856

5. 5237

6. 4761

7. 36.81

8. 78.92

9. .567

Convert the decimals to fractions. Write in lowest terms. See Example 2.

10. .2

11. .8

12. .36

13. .72

14. .336

15. .215

16. .805

17. .625

Perform the indicated operations. See Examples 3–5.

18. 14.23 + 9.81 + 74.63 + 18.715

19. 89.416 + 21.32 + 478.91 + 298.213

20. 19.74 − 6.53

21. 27.96 − 8.39

22. 219.4 − 68

23. 283 − 12.42

24. 48.96
   37.421
 + 9.72

25. 9.71
   4.8
   3.6
   5.2
 + 8.17

26. 8.6
 − 3.751

27. 27.8
 − 13.582

28. 39.6 × 4.2

29. 18.7 × 2.3

30. 42.1 × 3.9

31. 19.63 × 4.08

32. .042 × 32

33. 571 × 2.9

34. 24.84 ÷ 6

35. 32.84 ÷ 4

36. 7.6266 ÷ 3.42 **37.** 14.9202 ÷ 2.43 **38.** 2496 ÷ .52 **39.** .56984 ÷ .034

Convert the following fractions to decimals. Round to the nearest thousandth. See Example 6.

40. $\frac{5}{8}$ **41.** $\frac{3}{8}$ **42.** $\frac{3}{4}$ **43.** $\frac{7}{16}$

44. $\frac{9}{16}$ **45.** $\frac{15}{16}$ **46.** $\frac{2}{3}$ **47.** $\frac{5}{6}$

***48.** $\frac{7}{13}$ **49.** $\frac{11}{15}$ **50.** $\frac{15}{17}$ **51.** $\frac{12}{19}$

Convert the following decimals to percents. See Examples 7(c) and 7(d).

52. .80 **53.** .75 **54.** .007 **55.** 1.4

56. .67 **57.** .003 **58.** .125 **59.** .983

Convert the following percents to decimals. See Examples 7(a) and 7(b).

60. 53% **61.** 38% **62.** 129% **63.** 174%

64. 96% **65.** 11% **66.** .9% **67.** .1%

Find each of the following. Round your answer to the nearest hundredth. See Examples 8 and 9.

68. What is 14% of 780? **69.** Find 12% of 350.

70. Find 22% of 1086. **71.** What is 20% of 1500?

---

*Calculator exercises have been included throughout this book and are identified with an exercise number printed in color.

72. 4 is what percent of 80?

73. 1300 is what percent of 2000?

74. What percent of 5820 is 6402?

75. What percent of 75 is 90?

76. 121 is what percent of 484?

77. What percent of 3200 is 64?

78. Find 118% of 125.8.

79. Find 3% of 128.

80. What is 91.72% of 8546.95?

81. Find 12.741% of 58.902.

82. What percent of 198.72 is 14.68?

83. 586.3 is what percent of 765.4?

Solve the word problems. See Examples 8 and 10.

84. A retailer has $23,000 invested in her business. She finds that she is earning 12% per year on this investment. How much money is she earning per year?

85. A family of four with a monthly income of $2000 spends 90% of its earnings and saves the rest. Find the *annual* savings of this family.

86. Harley Dabler recently bought a duplex for $144,000. He expects to earn $23,040 per year on the purchase price. What percent of the purchase price will he earn?

87. For a recent tour of the eastern United States, a travel agent figured that the trip totaled 2300 miles, with 805 miles of the trip by air. What percent of the trip was by air?

1.2 Exercises

88. Capitol Savings Bank pays 8.9% interest per year. What is the annual interest on an account of $3000?

89. Beth's Bargain Basement is having a sale this week. A purchase of $250 was discounted by $37.50. What percent was the discount?

90. When installing carpet, Delta Carpet Layers wasted 249 yards of a total of 4150 yards laid in April. What percent was wasted?

91. Scott must pay 6.5% sales tax on a new car. The cost of the car is $8600. Find the amount of the tax.

92. An ad for steel-belted radial tires promises 15% better mileage when using them. Alexandria's Escort now goes 420 miles on a tank of gas. If she switched to the new tires, how many extra miles could she drive on a tank of gas?

93. A home worth $77,000 at the beginning of the year has increased in value by $4620 over the last year. What percent of the value at the beginning of the year did the increase represent?

94. A small business takes in $274,600 per year and spends $30,755.20 for advertising. What percent of the income is spent on advertising?

95. A piece of property contains 126,000 square feet. Before the county will give a building permit, the owner of the property must donate 9.4% of the land for a park. How much land must be donated?

96. Self-employed people now must pay a Social Security tax of 11.7%. Find the tax due on earnings of $1756.

97. The Social Security tax on people who work for others is 7.15%. Find the tax on earnings of $2109.

1.3 SYMBOLS

So far only the symbols of arithmetic, such as $+$, $-$, $\times$ (or $\cdot$), and $\div$ have been used. Another common symbol is the one for equality, $=$, which says that two numbers are equal. This symbol with a slash through it, $\neq$, means "is *not* equal to." For example,

$$7 \neq 8$$

indicates that 7 is not equal to 8.

▶ If two numbers are not equal, then one of the numbers must be less than the other. The symbol $<$ represents "is less than," so that "7 is less than 8" is written

$$7 < 8.$$

Also, write "6 is less than 9" as $6 < 9$.

The symbol $>$ means "is greater than." Write "8 is greater than 2" as

$$8 > 2.$$

The statement "17 is greater than 11" becomes $17 > 11$.

Keep the symbols $<$ and $>$ straight by remembering that the symbol always points to the smaller number. For example, write "8 is less than 15" by pointing the symbol toward the 8:

$$8 < 15.$$

Work Problem 1 at the side.

▶ Word phrases must often be converted to symbols in algebra. The next example shows how to do this.

EXAMPLE 1 Write each word statement in symbols.

(a) Twelve equals ten plus two.
$$12 = 10 + 2$$

(b) Nine is less than ten.
$$9 < 10$$

(c) Fifteen is not equal to eighteen.
$$15 \neq 18$$

(d) Seven is greater than four.
$$7 > 4 \blacktriangleleft$$

Work Problem 2 at the side.

▶ Two other symbols, $\leq$ and $\geq$, also represent the idea of inequality. The symbol $\leq$ means "is less than or equal to," so that

$$5 \leq 9$$

means "5 is less than or equal to 9." This statement is true, since $5 < 9$ is true. If either the $<$ part or the $=$ part is true, then the inequality $\leq$ is true.

Objectives

▶ Know the meaning of $<$ and $>$.

▶ Translate word phrases to symbols.

▶ Know the meaning of $\leq$ and $\geq$.

▶ Write statements that change the direction of inequality symbols.

1. Write *true* or *false* for each statement.

(a) $7 < 5$

(b) $12 > 6$

(c) $4 \neq 10$

(d) $9 \neq 15 - 6$

(e) $28 = 4 \cdot 7$

2. Write in symbols.

(a) Nine equals eleven minus two.

(b) Seventeen is less than thirty.

(c) Eight is not equal to ten.

(d) Fourteen is greater than twelve.

(e) Six is less than eleven.

ANSWERS
1. (a) false (b) true (c) true
   (d) false (e) true
2. (a) $9 = 11 - 2$ (b) $17 < 30$
   (c) $8 \neq 10$ (d) $14 > 12$
   (e) $6 < 11$

1.3 Symbols

3. Tell whether each statement is true or false.

   (a) $30 \leq 40$

   (b) $25 \geq 10$

   (c) $40 \leq 10$

   (d) $21 \leq 21$

   (e) $3 \geq 3$

4. Write each statement with the inequality symbol reversed.

   (a) $8 < 10$

   (b) $3 > 1$

   (c) $9 \leq 15$

   (d) $6 \geq 2$

The symbol $\geq$ means "is greater than or equal to;"

$$9 \geq 5$$

is true because $9 > 5$ is true. Also, $8 \leq 8$ is true since $8 = 8$ is true. But $13 \leq 9$ is not true because neither $13 < 9$ nor $13 = 9$ is true.

EXAMPLE 2 Tell whether each statement is true or false.

(a) $15 \leq 20$

The statement $15 \leq 20$ is true since $15 < 20$.

(b) $25 \geq 30$

Both $25 > 30$ and $25 = 30$ are false. Because of this, $25 \geq 30$ is false.

(c) $12 \geq 12$

Since $12 = 12$, this statement is true. ◀

Work Problem 3 at the side.

▶ Any statement with $<$ can be converted to one with $>$, and any statement with $>$ can be converted to one with $<$. Do this by reversing both the order of the numbers and the direction of the symbol. For example, the statement $6 < 10$ can be written with $>$ as $10 > 6$.

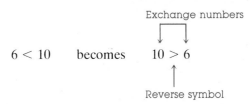

EXAMPLE 3 The following list shows the same statements written in two equally correct ways.

(a) $9 < 16;$ $16 > 9$

(b) $5 > 2;$ $2 < 5$

(c) $3 \leq 8;$ $8 \geq 3$

(d) $12 \geq 5;$ $5 \leq 12$ ◀

Work Problem 4 at the side.

Here is a summary of the symbols discussed in this section.

Symbols of Equality and Inequality

Symbol	Meaning
$=$	is equal to
$\neq$	is not equal to
$<$	is less than
$>$	is greater than
$\leq$	is less than or equal to
$\geq$	is greater than or equal to

ANSWERS
3. (a) true (b) true (c) false
   (d) true (e) true
4. (a) $10 > 8$ (b) $1 < 3$
   (c) $15 \geq 9$ (d) $2 \leq 6$

name   date   hour

1.3 EXERCISES

Insert < or > to make the following statements true.

1. 6   9   **2.** 5   3   **3.** 12   15   **4.** 8   10

5. 25   12   **6.** 17   9   **7.** 32   50   **8.** 41   72

9. $\frac{3}{4}$ 1   **10.** $\frac{2}{3}$ 0   **11.** $1\frac{5}{8}$ 1   **12.** $3\frac{7}{9}$ 2

Insert ≤ or ≥ to make the following statements true.

13. 12   17   **14.** 28   42   **15.** 16   14   **16.** 39   17

17. 8   28   **18.** 10   15   **19.** 35   42   **20.** 51   62

Which of the symbols <, >, ≤, and ≥ make the following statements true? Give all possible correct answers.

21. 6   9   **22.** 18   12   **23.** 51   50   **24.** 0   12

25. 5   5   **26.** 10   10   **27.** 48   0   **28.** 100   1000

29. 16   10   **30.** 5   3   **31.** $\frac{1}{4}$ $\frac{2}{5}$ **32.** $\frac{2}{3}$ $\frac{5}{8}$

33. .609   .61   **34.** .5   .499   **35.** $3\frac{1}{2}$ 4   **36.** $5\frac{7}{8}$ 6

Write the following word statements in symbols. See Example 1.

37. Seven equals five plus two.

38. Nine is greater than the product of four and two.

39. Three is less than the quotient of fifty and five.

40. Five equals ten minus five.

41. Twelve is not equal to five.

42. Fifteen does not equal sixteen.

43. Zero is greater than or equal to zero.

44. Six is less than or equal to six.

Tell whether each statement is true or false. See Example 2.

45. $8 + 2 = 10$

46. $8 \neq 9 - 1$

47. $12 \geq 10$

48. $45 < 45$

49. $0 < 15$

50. $16 \geq 10$

51. $9 + 12 = 21$

52. $9 < 12$

53. $25 \geq 19$

54. $18 < 5$

55. $9 < 0$

56. $15 \leq 32$

57. $6 \neq 5 + 1$

58. $15 < 21$

59. $11 < 11$

60. $29 \geq 30$

61. $8 \leq 0$

62. $26 \geq 50$

Rewrite the following statements so the inequality symbol points in the opposite direction. See Example 3.

63. $6 < 14$

64. $8 \leq 9$

65. $15 \geq 3$

66. $29 > 4$

67. $9 > 8$

68. $12 < 17$

69. $0 \leq 6$

70. $7 \leq 12$

71. $18 \geq 15$

72. $25 \geq 1$

73. $.481 \geq .439$

74. $.762 < .763$

1.4 EXPONENTS AND ORDER OF OPERATIONS

Objectives

▶ Use exponents.
▶ Use order of operations.
▶ Use brackets.
▶ Insert parentheses to make a statement true.

▶ It is common for a multiplication problem to have the same number (or **factor**) appearing several times. For example, in the product

$$3 \cdot 3 \cdot 3 \cdot 3 = 81$$

the factor 3 appears four times. To save space, repeated factors are written with an *exponent*. For example, in $3 \cdot 3 \cdot 3 \cdot 3$, the number 3 appears as a factor four times, so the product is written as 3^4.

$$3 \cdot 3 \cdot 3 \cdot 3 = 3^4$$

The number 4 is the **exponent** and 3 is the **base**. An exponent tells how many times the base is used as a factor in the multiplication problem.

EXAMPLE 1 Find the values of the following.

(a) 5^2

$$\underbrace{5 \cdot 5}_{} = 25$$
5 is used as a factor 2 times

Read 5^2 as "5 squared."

(b) 6^3

$$\underbrace{6 \cdot 6 \cdot 6}_{} = 216$$
6 is used as a factor 3 times

Read 6^3 as "6 cubed."

(c) 2^5

$$2 \cdot 2 \cdot 2 \cdot 2 \cdot 2 = 32 \qquad \text{2 is used as a factor 5 times}$$

Read 2^5 as "2 to the fifth power."

(d) 7^4

$$7 \cdot 7 \cdot 7 \cdot 7 = 2401 \qquad \text{7 is used as a factor 4 times}$$

Read 7^4 as "7 to the fourth power."

(e) $\left(\dfrac{2}{3}\right)^3$

$$\dfrac{2}{3} \cdot \dfrac{2}{3} \cdot \dfrac{2}{3} = \dfrac{8}{27} \qquad \tfrac{2}{3} \text{ is used as a factor 3 times} \blacktriangleleft$$

Work Problem 1 at the side.

▶ Many problems involve more than one symbol of arithmetic. For example, in finding the value of

$$5 + 2 \cdot 3,$$

which should be done first—multiplication or addition? The following **order of operations** has been agreed on as the most reasonable. (This is the order used by most calculators and computers.)

1. Find the value of each of the following.

  (a) 6^2

  (b) 3^5

  (c) 2^3

  (d) $\left(\dfrac{3}{4}\right)^2$

  (e) $\left(\dfrac{2}{9}\right)^3$

  (f) $\left(\dfrac{1}{2}\right)^4$

  (g) $(.4)^3$

ANSWERS

1. (a) 36 (b) 243 (c) 8 (d) $\dfrac{9}{16}$
   (e) $\dfrac{8}{729}$ (f) $\dfrac{1}{16}$ (g) .064

1.4 Exponents and Order of Operations   23

2. Find the following.

(a) $3 \cdot 8 + 7$

(b) $12 \cdot 6 + 9$

(c) $6 \cdot 11 + 4$

> **Order of Operations**
>
> *If parentheses or fraction bars are present,* simplify within parentheses, innermost first, and above and below fraction bars separately, in the following order.
>
> *Step 1* Apply all exponents.
>
> *Step 2* Do any multiplications or divisions in the order in which they occur, working from left to right.
>
> *Step 3* Do any additions or subtractions in the order in which they occur, working from left to right.
>
> *If no parentheses or fraction bars are present,* start with Step 1.

EXAMPLE 2 Find the value of $5 + 2 \cdot 3$.

Using the order of operations given above, first multiply 2 and 3, and then add 5.

$$5 + 2 \cdot 3 = 5 + 6 \quad \text{Multiply}$$
$$= 11 \quad \text{Add}$$

Therefore, $5 + 2 \cdot 3 = 11$. If the addition had been performed first, the result would have been $7 \cdot 3 = 21$, instead of 11. By the order of operations, only the first result, 11, is correct. ◀

Work Problem 2 at the side.

A dot has been used to show multiplication; another way to show multiplication is with parentheses. For example, 3(7) means $3 \cdot 7$ or 21. The next example shows the use of parentheses for multiplication.

EXAMPLE 3 Find the following.

(a) $9(6 + 11)$

Work first inside the parentheses.

$$9(6 + 11) = 9(17) \quad \text{Add inside parentheses}$$
$$= 153 \quad \text{Multiply}$$

(b) $2(5 + 6) + 7 \cdot 3 = 2(11) + 7 \cdot 3 \quad \text{Add inside parentheses}$
$$= 22 + 21 \quad \text{Multiply}$$
$$= 43 \quad \text{Add}$$

(c) $\dfrac{4(5 + 3) + 3}{2(3) - 1}$

Simplify the numerator and denominator separately.

$$\dfrac{4(5 + 3) + 3}{2(3) - 1} = \dfrac{4(8) + 3}{2(3) - 1} \quad \text{Add inside parentheses}$$
$$= \dfrac{32 + 3}{6 - 1} \quad \text{Multiply}$$
$$= \dfrac{35}{5} \quad \text{Add and subtract}$$
$$= 7 \quad \text{Divide}$$

ANSWERS

2. (a) 31 (b) 81 (c) 70

Number Systems

(d) $9 + 2^3 - 5$

Calculate 2^3 first.

$9 + 2^3 - 5 = 9 + 8 - 5$ Use the exponent
$ = 12$ Add, then subtract

Work Problem 3 at the side.

▶ An expression with double parentheses, such as $2(8 + 3(6 + 5))$, can be confusing. Avoid confusion by using square brackets, [ ], instead of one pair of parentheses.

EXAMPLE 4   Simplify $2[8 + 3(6 + 5)]$.

Work first within the parentheses, until a single number is found inside the brackets.

$2[8 + 3(6 + 5)] = 2[8 + 3(11)]$ Add
$ = 2[8 + 33]$ Multiply
$ = 2[41]$ Add
$ = 82$ Multiply

Work Problem 4 at the side.

▶ Some mathematical statements are not true as written, but they become true when parentheses are inserted in the proper place.

EXAMPLE 5   Insert parentheses so that the following statements are true.

(a) $9 - 3 - 2 = 8$

Use trial and error to see that this statement would be true if parentheses were inserted around $3 - 2$, since

$9 - (3 - 2) = 9 - 1 = 8.$

It is not true that $(9 - 3) - 2 = 8$, since $6 - 2 \neq 8$.

(b) $9 \cdot 2 - 4 \cdot 3 = 6$

Since $9 \cdot 2 - 4 \cdot 3 = 18 - 12 = 6$, no parentheses are needed here. If desired, parentheses may be placed as follows.

$(9 \cdot 2) - (4 \cdot 3) = 6$

Work Problem 5 at the side.

---

3. Find the value of these expressions.

(a) $2 \cdot 9 + 7 \cdot 3$

(b) $5 \cdot 8 + 6 \cdot 7$

(c) $7 \cdot 6 - 3(8 + 1)$

(d) $\dfrac{2(7 + 8) + 2}{3 \cdot 5 + 1}$

(e) $2 + 3^2 - 5$

4. Find the following.

(a) $4[7 + 3(6 + 1)]$

(b) $3[6 + (4 + 2)]$

(c) $9[(4 + 8) - 3]$

5. Insert parentheses as necessary to make each statement true.

(a) $14 - 3 - 1 = 12$

(b) $2 \cdot 5 + 3 \cdot 2 = 26$

(c) $3 + 4^2 \cdot 3 = 57$

ANSWERS
3. (a) 39 (b) 82 (c) 15 (d) 2
   (e) 6
4. (a) 112 (b) 36 (c) 81
5. (a) $14 - (3 - 1) = 14 - 2 = 12$
   (b) $(2 \cdot 5 + 3) \cdot 2 = 26$
   (c) $(3 + 4^2) \cdot 3 = 57$

1.4   Exponents and Order of Operations

1.4 EXERCISES

Find the values of the following. In Exercises 21–24, round to the nearest thousandth. See Example 1.

1. 6^2
2. 9^2
3. 8^2
4. 10^2

5. 17^2
6. 22^2
7. 5^3
8. 7^3

9. 6^4
10. 3^4
11. 2^5
12. 4^5

13. 3^6
14. 2^6
15. $\left(\dfrac{1}{2}\right)^2$
16. $\left(\dfrac{3}{4}\right)^2$

17. $\left(\dfrac{2}{5}\right)^3$
18. $\left(\dfrac{3}{7}\right)^3$
19. $\left(\dfrac{4}{5}\right)^3$
20. $\left(\dfrac{2}{3}\right)^5$

21. $(.83)^4$
22. $(.712)^2$
23. $(1.46)^3$
24. $(2.85)^4$

Find the values of the following expressions. See Examples 2–4.

25. $9 \cdot 3 - 11$
26. $6 \cdot 5 - 12$
27. $\dfrac{2}{5} \cdot \dfrac{11}{3} + \dfrac{2}{3} \cdot \dfrac{1}{4}$

28. $\dfrac{9}{4} \cdot \dfrac{2}{3} + \dfrac{4}{5} \cdot \dfrac{5}{3}$
29. $13 \cdot 2 - 15 \cdot 1$
30. $(6.1)(5.7) + (3.4)(12)$

31. $(5.4)(8.1) + (10.9)(4)$
32. $4[2 + 3(4)]$
33. $5[8 + (2 + 3)]$

34. $9[(14 + 5) - 10]$
35. $\dfrac{5(4 - 1) + 3}{2 \cdot 4 + 1}$
36. $\dfrac{7(3 + 1) - 2}{5 \cdot 2 + 3}$

37. $\dfrac{2(5 + 1) - 3(1 + 1)}{5(8 - 6) - 4 \cdot 2}$
38. $8^2 \div 2 \cdot 3$
39. $5^2 + 2^2 \div 4$

Tell whether each statement is true or false. (Hint: First simplify both sides of each statement.) See Examples 2–4.

40. $3 \cdot 8 - 4 \cdot 6 \leq 0$
41. $2 \cdot 20 - 8 \cdot 5 \geq 0$
42. $9 \cdot 2 - 6 \cdot 3 \geq 2$

43. $8 \cdot 3 - 4 \cdot 6 < 1$
44. $12 \cdot 3 - 6 \cdot 6 \leq 0$
45. $3[5(2) - 3] > 20$

1.4 Exercises 27

46. $2[2 + 3(2 + 5)] \leq 45$

47. $3[4 + 3(4 + 1)] \leq 55$

48. $\dfrac{2(5 + 3) + 2 \cdot 2}{2(4 - 1)} > 4$

49. $\dfrac{9(7 - 1) - 8 \cdot 2}{4(6 - 1)} > 2$

50. $\dfrac{3(8 - 3) + 2(4 - 1)}{9(6 - 2) - 8(5 - 2)} \geq 7$

51. $9 \leq 4^2 - 8$

52. $6^2 - 3^2 > 25$

53. $21.92 \leq 7.43^2 - 5.77^2$

54. $.479 > (.841)^3 - (.58)^4$

Insert parentheses in each expression so that the resulting statement is true. Some problems require no parentheses. See Example 5.

55. $10 - 7 - 3 = 6$

56. $16 - 4 - 3 = 15$

57. $3 \cdot 5 + 7 = 22$

58. $3 \cdot 5 + 7 = 36$

59. $3 \cdot 5 - 4 = 3$

60. $3 \cdot 5 - 4 = 11$

61. $3 \cdot 5 - 2 \cdot 4 = 36$

62. $100 \div 20 \div 5 = 1$

63. $360 \div 18 \div 4 = 5$

64. $2^2 + 4 \cdot 2 = 16$

65. $6 + 5 \cdot 3^2 = 99$

66. $3^3 - 2 \cdot 4 = 100$

67. $8 - 2^2 \cdot 2 = 8$

68. $3 \cdot \dfrac{2}{3} - \dfrac{1}{4} \cdot \dfrac{4}{5} = 1$

69. $\dfrac{1}{2} + \dfrac{5}{3} \cdot \dfrac{9}{7} - \dfrac{3}{2} = \dfrac{8}{7}$

Write the information given in the following problems using numerical symbols and parentheses.

70. Marjorie Jensen invested $600. After one year, her investment had tripled. She then took $150 and paid a car payment.

71. John Wilson had 5 decks of cards, each containing 52 cards. He removed all 4 aces from one deck.

72. A bus has 63 passengers. At one stop, 23 people get off and 17 new people get on.

73. An elevator has 5 passengers. At the first stop 6 get on and 1 gets off. At the next stop 3 get on and 2 get off.

1.5 VARIABLES AND EQUATIONS

▶ A **variable** is a symbol, usually a letter, such as x, y, or z, used to represent any unknown number. An **algebraic expression** is a collection of numbers, variables, symbols for operations, and symbols for grouping, such as parentheses or square brackets. For example,

$$x + 5, \quad 2m - 9, \quad \text{and} \quad 8p^2 + 6(p - 2)$$

are all algebraic expressions. In the algebraic expression $2m - 9$, the expression $2m$ means $2 \cdot m$, the product of 2 and m, and $8p^2$ shows the product of 8 and p^2. Also, $6(p - 2)$ means the product of 6 and $p - 2$.

▶ An algebraic expression takes on different numerical values as the variables take on different values.

EXAMPLE 1 Find the values of the following algebraic expressions when $m = 5$ and when $m = 9$.

(a) $8m$

Replace m with 5, to get

$$8m = 8 \cdot 5 \quad \text{Let } m = 5$$
$$= 40. \quad \text{Multiply}$$

If $m = 9$,

$$8m = 8 \cdot 9 \quad \text{Let } m = 9$$
$$= 72. \quad \text{Multiply}$$

(b) $3m^2$

For $m = 5$,

$$3m^2 = 3 \cdot 5^2 \quad \text{Let } m = 5$$
$$= 3 \cdot 25 \quad \text{Square}$$
$$= 75. \quad \text{Multiply}$$

For $m = 9$,

$$3m^2 = 3 \cdot 9^2$$
$$= 3 \cdot 81$$
$$= 243. \blacktriangleleft$$

In Example 1(b), it is important to notice that $3m^2$ means $3 \cdot m^2$; it *does not* mean $3m \cdot 3m$.

Work Problem 1 at the side.

EXAMPLE 2 Find the value of each expression when $x = 5$ and $y = 3$.

(a) $2x + 5y$

Replace x with 5 and y with 3. Do the multiplications first, and then add.

$$2x + 5y = 2 \cdot 5 + 5 \cdot 3 \quad \text{Let } x = 5 \text{ and } y = 3$$
$$= 10 + 15 \quad \text{Multiply}$$
$$= 25 \quad \text{Add}$$

Objectives

▶ Define *variable*.

▶ Find the value of algebraic expressions, given values for the variables.

▶ Convert statements from words to algebraic expressions.

▶ Identify solutions of equations.

▶ Define and use the domain of a set of numbers.

1. Find the value of each expression when $p = 3$.

   (a) $6p$

   (b) $p + 12$

   (c) $5p^2$

   (d) $2p^3$

   (e) p^4

ANSWERS
1. (a) 18 (b) 15 (c) 45 (d) 54
   (e) 81

1.5 Variables and Equations   29

2. Find the value of each expression when $x = 6$ and $y = 9$.

   (a) xy

   (b) $4x + 7y$

   (c) $\dfrac{4x - 2y}{x + 1}$

   (d) $x^2 + y^2$

   (e) $x^3 y$

3. Write as an algebraic expression. Use x as the variable.

   (a) The sum of 5 and a number

   (b) A number minus 4

   (c) The product of 6 and a number

   (d) 9 is multiplied by the sum of a number and 5

(b) $\dfrac{9x - 8y}{2x - y}$

Replace x with 5 and y with 3.

$$\dfrac{9x - 8y}{2x - y} = \dfrac{9 \cdot 5 - 8 \cdot 3}{2 \cdot 5 - 3} \qquad \text{Let } x = 5 \text{ and } y = 3$$

$$= \dfrac{45 - 24}{10 - 3} \qquad \text{Multiply}$$

$$= \dfrac{21}{7} \qquad \text{Subtract}$$

$$= 3 \qquad \text{Divide}$$

(c) $x^2 - 2y^2 = 5^2 - 2 \cdot 3^2 \qquad$ Let $x = 5$ and $y = 3$

$\qquad\qquad\quad = 25 - 2 \cdot 9 \qquad$ Use the exponents

$\qquad\qquad\quad = 25 - 18 \qquad$ Multiply

$\qquad\qquad\quad = 7 \qquad\qquad\;\;$ Subtract ◀

Work Problem 2 at the side.

▶ Variables are used in changing word phrases into algebraic expressions. The next example shows how to do this.

EXAMPLE 3 Change the following word phrases to algebraic expressions. Use x as the variable.

(a) The sum of a number and 9

"Sum" is the answer to an addition problem. This phrase translates as

$$x + 9 \quad \text{or} \quad 9 + x.$$

(b) 7 minus a number

"Minus" indicates subtraction, so the answer is

$$7 - x.$$

($x - 7$ would *not* be correct.)

(c) The product of 11 and a number

$$11 \cdot x \quad \text{or} \quad 11x$$

(d) 5 divided by a number

$$\dfrac{5}{x}$$

(e) The product of 2, and the sum of a number and 8

$$2(x + 8) \quad ◀$$

Work Problem 3 at the side.

ANSWERS

2. (a) 54 (b) 87 (c) $\dfrac{6}{7}$ (d) 117

   (e) 1944

3. (a) $5 + x$ (b) $x - 4$ (c) $6x$

   (d) $9(x + 5)$

▶ An **equation** is a statement that says two expressions are equal. Examples of equations are

$$x + 4 = 11, \quad 2y = 16, \quad \text{and} \quad 4p + 1 = 25 - p.$$

To **solve** an equation, find all values of the variable that make the equation true. The values of the variable that make the equation true are called the **solutions** of the equation.

EXAMPLE 4 Decide whether the given number is a solution of the equation.

(a) $5p + 1 = 36; \quad 7$

Replace p with 7.

$$5p + 1 = 36$$
$$5 \cdot 7 + 1 = 36 \quad \text{Let } p = 7$$
$$35 + 1 = 36$$
$$36 = 36 \quad \text{True}$$

The number 7 is a solution of the equation.

(b) $9m - 6 = 32; \quad 4$

$$9m - 6 = 32$$
$$9 \cdot 4 - 6 = 32 \quad \text{Let } m = 4$$
$$36 - 6 = 32$$
$$30 = 32 \quad \text{False}$$

The number 4 is not a solution of the equation. ◀

Work Problem 4 at the side.

▶ Sometimes the solutions of an equation must come from a certain list of numbers. This list of numbers is often written as a **set**, a collection of objects. For example, the set containing the numbers 1, 2, 3, 4, and 5 is written with **set braces, { }**, as

$$\{1, 2, 3, 4, 5\}.$$

For more information on sets, see Appendix A at the back of this book. The set of numbers from which the solutions of an equation must be chosen is called the **domain** of the equation.

EXAMPLE 5 Change each word statement to an equation. Use x as the variable. Then find all solutions for the equation from the domain

$$\{0, 2, 4, 6, 8, 10\}.$$

(a) The sum of a number and four is six.

The word "is" suggests "equals." Let x represent the unknown number and translate as follows.

The sum of
a number and four   is   six.
↓   ↓   ↓
$x + 4$ $=$ 6

4. Decide whether the given number is a solution of the equation

(a) $2x = 10; \quad 5$

(b) $p - 1 = 3; \quad 2$

(c) $2k + 3 = 15; \quad 7$

(d) $8p - 11 = 5; \quad 2$

(e) $9k - 8 = 19; \quad 4$

ANSWERS
4. (a) solution (b) not a solution
   (c) not a solution (d) solution
   (e) not a solution

1.5 Variables and Equations

5. Change each word statement to an equation. Find all solutions from the domain {0, 2, 4, 6, 8, 10}.

(a) The sum of a number and 13 is 19.

(b) Twice a number is added to 4, giving 20.

(c) Three times a number is subtracted from 21, giving 15.

(d) The quotient of a number and 2 is added to 5, giving 9.

Try each number from the given domain, {0, 2, 4, 6, 8, 10}, in turn.

x + 4	=	6	Given equation
0 + 4	=	6	False
2 + 4	=	6	True
4 + 4	=	6	False
6 + 4	=	6	False
8 + 4	=	6	False
10 + 4	=	6	False

The only solution of $x + 4 = 6$ is 2.

(b) Nine more than five times a number is 49.

Use x to represent the unknown number.

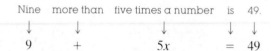

Try each number from the given domain, {0, 2, 4, 6, 8, 10}.

The solution is 8, since $9 + 5 \cdot 8 = 49$. ◀

Work Problem 5 at the side.

ANSWERS

5. (a) $x + 13 = 19$; 6
(b) $2x + 4 = 20$; 8
(c) $21 - 3x = 15$; 2
(d) $(x \div 2) + 5 = 9$
or $\frac{x}{2} + 5 = 9$; 8

name   date   hour

1.5 EXERCISES

Find the numerical values of the following when **(a)** $x = 3$ *and* **(b)** $x = 15$. *See Example 1.*

1. $x + 9$
   (a)   (b)

2. $x - 1$
   (a)   (b)

3. $5x$
   (a)   (b)

4. $7x$
   (a)   (b)

5. $\dfrac{2}{3}x + \dfrac{1}{3}$
   (a)   (b)

6. $\dfrac{9}{4}x - \dfrac{5}{3}$
   (a)   (b)

7. $\dfrac{x + 1}{3}$
   (a)   (b)

8. $\dfrac{x - 2}{5}$
   (a)   (b)

9. $\dfrac{3x - 5}{2x}$
   (a)   (b)

10. $\dfrac{x + 2}{x - 1}$
    (a)   (b)

11. $3x^2 + x$
    (a)   (b)

12. $2x + x^2$
    (a)   (b)

13. $6.459x$
    (a)   (b)

14. $.74x^2$
    (a)   (b)

15. $.0745(x^2 + 2)$
    (a)   (b)

16. $.204(3 + x)$
    (a)   (b)

1.5 Exercises

Find the numerical values of the following when (a) $x = 4$ and $y = 2$ and (b) $x = 1$ and $y = 5$. See Example 2. Round to the nearest thousandth in Exercises 37 and 38.

17. $8x + 3y + 5$
    (a)   (b)

18. $4x + 2y + 7$
    (a)   (b)

19. $3(x + 2y)$
    (a)   (b)

20. $2(2x + y)$
    (a)   (b)

21. $x + \dfrac{4}{y}$
    (a)   (b)

22. $y + \dfrac{8}{x}$
    (a)   (b)

23. $\dfrac{y}{3} + \dfrac{5}{y}$
    (a)   (b)

24. $\dfrac{x}{5} + \dfrac{y}{4}$
    (a)   (b)

25. $5\left(\dfrac{4}{3}x + \dfrac{7}{2}y\right)$
    (a)   (b)

26. $8\left(\dfrac{5}{2}x + \dfrac{9}{5}y\right)$
    (a)   (b)

27. $\dfrac{2x + 3y}{x + y + 1}$
    (a)   (b)

28. $\dfrac{5x + 3y + 1}{2x}$
    (a)   (b)

29. $\dfrac{2x + 4y - 6}{5y + 2}$
    (a)   (b)

30. $\dfrac{4x + 3y - 1}{x}$
    (a)   (b)

34   Number Systems

name   date   hour

31. $2y^2 + 5x$
(a)   (b)

32. $6x^2 + 4y$
(a)   (b)

33. $\dfrac{x^2 + y^2}{x + y}$
(a)   (b)

34. $\dfrac{9x^2 + 4y^2}{3x^2 + 2y}$
(a)   (b)

35. $\dfrac{3x + y^2}{2x + 3y}$
(a)   (b)

36. $\dfrac{x^2 + 1}{4x + 5y}$
(a)   (b)

37. $.841x^2 + .32y^2$
(a)   (b)

38. $\dfrac{3.4x + 2.59y}{0.8x + 0.3y^2}$
(a)   (b)

Change the word phrases to algebraic expressions. Use x to represent the variable. See Example 3.

39. Eight times a number

40. Fifteen times a number

41. Five times a number

42. Six added to a number

43. Four added to a number

44. A number subtracted from eight

45. Nine subtracted from a number

46. Eight subtracted from three times a number

47. Six added to two-thirds of a number

48. Three-fourths of a number, added to fifty-two

Decide whether the given number is a solution of the equation. See Example 4.

49. $p - 5 = 12$;   17

50. $x + 6 = 15$;   9

51. $5m + 2 = 7$;   2

1.5   Exercises   **35**

52. $3r + 5 = 8$;   2
53. $2y + 3(y - 2) = 14$;   4
54. $6a + 2(a + 3) = 14$;   1

55. $6p + 4p - 9 = 11$;   2
56. $2x + 3x + 8 = 38$;   6
57. $3r^2 - 2 = 46$;   4

58. $2x^2 + 1 = 19$;   3
59. $\dfrac{z + 4}{2 - z} = \dfrac{13}{5}$; $\dfrac{1}{3}$
60. $\dfrac{x + 6}{x - 2} = \dfrac{37}{5}$; $3\dfrac{1}{4}$

61. $9.54x + 3.811 = 0.4273x + 16.57718$;   1.4
62. $0.935(y + 6.1) + 0.0142 = 7.83y + .2017$;   .8

Change the word statements to equations. Use x as the variable. Find the solutions from the domain {0, 2, 4, 6, 8, 10}. *See Examples 3–5.*

63. The sum of a number and 8 is 12.
64. A number minus three equals seven.

65. Sixteen minus three-fourths of a number is ten.
66. The sum of six-fifths of a number and 6 is 18.

67. Five more than twice a number is 13.
68. The product of a number and 3 is 24.

69. Three times a number is equal to two more than twice the number.
70. Twelve divided by a number equals three times that number.

71. Twenty divided by five times a number is 2.
72. A number divided by 2 is 0.

name _____   date _____ hour _____

CHAPTER 1   REVIEW EXERCISES

For help with any of these exercises, look in the section given in brackets.

(1.1)   Write each fraction in lowest terms.

1. $\dfrac{3}{6}$
2. $\dfrac{5}{15}$
3. $\dfrac{20}{36}$
4. $\dfrac{27}{45}$

5. $\dfrac{18}{54}$
6. $\dfrac{60}{72}$
7. $\dfrac{114}{133}$
8. $\dfrac{204}{255}$

Find the products or quotients. Write the answers in lowest terms.

9. $\dfrac{5}{8} \cdot \dfrac{1}{3}$
10. $\dfrac{7}{10} \cdot \dfrac{1}{5}$
11. $\dfrac{3}{7} \cdot \dfrac{5}{6}$

12. $\dfrac{2}{9} \cdot \dfrac{6}{5}$
13. $\dfrac{3}{8} \div \dfrac{1}{4}$
14. $\dfrac{5}{12} \div \dfrac{7}{6}$

15. $\dfrac{8}{5} \div \dfrac{32}{15}$
16. $\dfrac{19}{115} \cdot \dfrac{46}{38}$
17. $\dfrac{56}{165} \div \dfrac{42}{33}$

Add or subtract. Write the answers in lowest terms.

18. $\dfrac{1}{3} + \dfrac{1}{3}$
19. $\dfrac{5}{7} + \dfrac{1}{7}$
20. $\dfrac{1}{10} + \dfrac{2}{5}$

21. $\dfrac{3}{8} + \dfrac{1}{2}$
22. $\dfrac{7}{10} - \dfrac{1}{4}$
23. $\dfrac{8}{9} - \dfrac{1}{6}$

24. $5\dfrac{1}{2} + 7\dfrac{2}{3}$
25. $273\dfrac{3}{4} + 198\dfrac{5}{12}$
26. $162\dfrac{1}{4} - 58\dfrac{5}{9}$

Work the following word problems.

27. John painted $\dfrac{1}{4}$ of a room on Monday and $\dfrac{1}{3}$ of the room on Tuesday. What portion was then painted?

28. A triangle has sides of length $2\dfrac{1}{4}$ feet, $1\dfrac{1}{2}$ feet, and $3\dfrac{1}{6}$ feet. Find the total distance around the triangle.

Chapter 1   Review Exercises   **37**

29. At a ceremony, each scout will get $\frac{3}{8}$ yard of ribbon. How many awards can be made from 15 yards of ribbon?

30. A contractor installs toolsheds. Each requires $1\frac{1}{4}$ cubic yards of concrete. How much concrete would be needed for 25 sheds?

(1.2) *Write in expanded form.*

31. 479

32. 8403

33. 15.4

34. 2.891

Convert the decimals to fractions. Write in lowest terms.

35. .4

36. .54

37. .85

38. .505

39. 2.345

40. 4.3758

Convert the fractions to decimals. Round to the nearest thousandth, if necessary.

41. $\frac{7}{8}$

42. $\frac{1}{4}$

43. $\frac{11}{16}$

44. $\frac{1}{6}$

45. $\frac{81}{95}$

46. $\frac{152}{233}$

Perform each operation.

47. 18.9
5.024
7.13
+ 256.9

48. 8.55
7.36
9.92
+ 5.47

49. 3.4
− 1.725

50. 8.
− .0546

51. (59.4)(3.7)

52. (124.9)(8.02)

53. 4.2282 ÷ .81

54. 2.0488 ÷ .26

Convert to percents.

55. .96

56. 1.42

57. .5

58. 5.136

Find the following. Round to the nearest hundredth.

59. 40% of 9000

60. 25% of 892

61. 124% of 176

62. What percent of 92 is 14?

63. 100.44 is what percent of 81?

Solve each word problem.

64. A fireman has $17,800 in the credit union. Each year the credit union gives a dividend of 8.4%. Find the dividend received by the fireman.

65. The total cost of a house is $79,500. The salesperson selling the house earns 3% of this amount. Find the fee earned by the salesperson.

66. A store is giving 20% off on all purchases. Find the amount paid for a chair with a regular price of $650.

67. A savings account of $2300 earned $138 interest in one year. Find the percent of interest earned.

(1.3) Which of the symbols $<$, $\leq$, $>$, and $\geq$ make the following statements true? Give all possible correct answers.

68. 3   4

69. 9   9

70. 22   15

71. $\dfrac{5}{8}$ $\dfrac{1}{2}$

72. .87   .865

73. .94   .904

74. $\dfrac{2}{3}$ .7

75. $\dfrac{3}{4}$ .8

Rewrite the following so that the inequality symbol points in the opposite direction.

76. $4 \leq 9$

77. $16 \geq 3$

78. $9 < 10$

79. $23 \leq 25$

(1.4) *Find the following.*

80. 4^3

81. 5^2

82. 2^5

83. 6^3

84. $\left(\dfrac{1}{2}\right)^4$

85. $\left(\dfrac{5}{8}\right)^2$

86. $\left(4\dfrac{2}{3}\right)^2$

87. $\left(2\dfrac{1}{3}\right)^3$

Chapter 1   Review Exercises   39

Work the problems and then decide whether each statement is true or false.

88. $5 \cdot 8 + 4 \geq 45$

89. $9 \cdot 5 - 8 \cdot 4 < 15$

90. $9 \cdot (6 + 2) - 4 > 70$

91. $5 \cdot (4 + 8) + 2 \cdot 3 \leq 68$

92. $5[6 - 2(4 - 1)] \leq 0$

93. $\dfrac{2(1 + 3)}{3(2 + 1)} < 1$

94. $\dfrac{3(7 + 4)}{2(1 + 5)} > 2$

95. $(9 - 3)^2 - 4^2 \geq 5$

96. $5^2 + (15 - 11)^2 \leq 40$

(1.5) Find the numerical value of the given expression when $x = 2$ and $y = 5$.

97. $3x$

98. $x + 3y$

99. $\dfrac{3x - y}{2x}$

100. $7x - y$

101. $9y + x$

102. $\dfrac{2x}{y + 1}$

103. $3(5x + 2y)$

104. $(x + 1)^2 + (y - 2)^2$

105. $\dfrac{3x^2 + 5y^2}{7x^2 - y^2}$

Write each word phrase or sentence using algebraic expressions. Use x as the variable.

106. 12 times a number

107. The sum of a number and nine

108. The difference of 4 and a number

109. Six subtracted from three times a number

110. The product of a number and one more than the number

111. The sum of a number and 4 is divided by 8.

112. Three-tenths of a number is subtracted from the sum of the number and 28.

113. The sum of a number and 16 is subtracted from ten-thirds of the number.

name _____ date _____ hour _____

CHAPTER 1 TEST

Write each fraction in lowest terms.

1. $\dfrac{15}{40}$

2. $\dfrac{84}{132}$

Perform each operation.

3. $\dfrac{5}{8} + \dfrac{9}{10}$

4. $\dfrac{3}{8} + \dfrac{7}{12} + \dfrac{11}{15}$

5. $21\dfrac{1}{4} - 7\dfrac{3}{8}$

6. $\dfrac{3}{2} \cdot \dfrac{4}{9}$

7. $\dfrac{6}{5} \div \dfrac{19}{15}$

8. Johnson Forest Products needed to raise some money quickly. To do so, the company sold $46\dfrac{1}{3}$ acres out of a $104\dfrac{7}{9}$-acre piece of land that it owned. How many acres of land were left?

9. Convert .625 to a fraction.

10. Convert $\dfrac{9}{16}$ to a decimal.

Perform the indicated operations.

11. 9.6 + 8.42 + 3.75

12. 123.4 − 98.7

13. 21.98 · (.72)

14. 252.008 ÷ 21.8

15. Convert .19 to a percent.

1. _____
2. _____
3. _____
4. _____
5. _____
6. _____
7. _____
8. _____
9. _____
10. _____
11. _____
12. _____
13. _____
14. _____
15. _____

Chapter 1 Test

16. Convert 76.2% to a decimal.

Find each of the following.

17. 8% of 170

18. 14.2% of 4600

19. A clock radio with a regular price of $48 is on sale this week at 20% off. Find the amount of the discount.

Find the value of each expression.

20. $4[5(1) - 3]$

21. $\dfrac{9(4) + 3(2)}{5 \cdot 4 + 1}$

22. $6^2 \cdot 2 - 24 - 5^2$

Find the numerical value of the given expression when $x = 3$ and $y = 7$.

23. $5x + 2y$

24. $x^2 + 3y$

25. $\dfrac{7x - y}{x + 4}$

Write each word phrase as an algebraic expression. Use x as the variable.

26. Four times a number

27. The sum of twice a number and 11

28. The quotient of 9 and the difference of a number and 8

Decide whether the given number is a solution for the equation.

29. $6m + m + 2 = 37$; 5

30. $8(y - 3) + 2y = 18$; 4

Number Systems

Operations with Real Numbers

2

2.1 REAL NUMBERS AND THE NUMBER LINE

Objectives

▶ Graphs can be a helpful way to picture sets of numbers.

Numbers are graphed on **number lines** like the one shown in Figure 1.

Figure 1

Draw a number line by locating any point on the line and calling it 0. Choose any point to the right of 0 and call it 1. The distance between 0 and 1 gives a unit of measure used to locate other points, as shown in Figure 1. The points labeled in Figure 1 correspond to the **whole numbers.**

Whole numbers $\{0, 1, 2, 3, 4, 5, \cdots\}$

The three dots show that the list of numbers continues in the same way indefinitely. Think of the graph as a picture of the set of numbers.

All the whole numbers starting with 1 are located to the right of 0 on the number line. But numbers also may be placed to the left of 0. These numbers, written $-1, -2, -3$, and so on, are shown in Figure 2. (The minus sign is used to show that the numbers are located to the *left* of 0.)

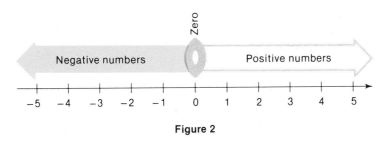

Figure 2

The numbers to the *left* of 0 are **negative numbers.** The numbers to the *right* of 0 are **positive numbers.** The number 0 itself is neither positive nor negative. Positive numbers and negative numbers are called **signed numbers.**

▶ Draw number lines.

▶ Identify whole numbers, integers, rational numbers, irrational numbers, and real numbers.

▶ Tell which of two real numbers is smaller.

▶ Find additive inverses.

▶ Find absolute values of real numbers.

1. Identify each number as positive, negative, or neither.

   (a) 15

   (b) −15

   (c) −6

   (d) $\frac{3}{4}$

   (e) $-\frac{5}{8}$

   (f) 0

There are many practical applications of negative numbers. For example, a temperature on a cold January day might be −10°, or 10 degrees below zero. A business that spends more than it takes in has a negative "profit."

Work Problem 1 at the side.

▶ The set of numbers marked on the number line in Figure 2 (including positive numbers, negative numbers, and zero) is called the set of **integers**.

> **Integers** {..., −3, −2, −1, 0, 1, 2, 3, ...}

Not all numbers are integers. For example, $\frac{1}{2}$ is not; it is a number halfway between the integers 0 and 1. Also, $3\frac{1}{4}$ is not an integer. Several numbers that are not integers are graphed in Figure 3.

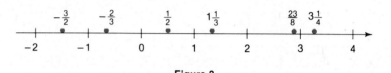

Figure 3

All the numbers in Figure 3 can be written as quotients of integers. These numbers are examples of **rational numbers**.

> **Rational numbers** {numbers that can be written as quotients of two integers, with denominator not 0}

Since any integer can be written as the quotient of itself and 1, all integers are also rational numbers. A decimal number such as .23 is also a rational number: $.23 = \frac{23}{100}$.

Although a great many numbers are rational, not all are. For example, a floor tile one foot on a side has a diagonal whose length is the square root of 2 (written $\sqrt{2}$). See Figure 4 below. It can be shown that $\sqrt{2}$ cannot be written as a quotient of two integers, so this number is not rational; it is **irrational**.

> **Irrational numbers** {numbers represented by points on the number line that are not rational}

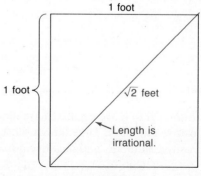

Figure 4

ANSWERS

1. (a) positive (b) negative
   (c) negative (d) positive
   (e) negative (f) neither

Operations with Real Numbers

Examples of irrational numbers include $\sqrt{3}$, $\sqrt{7}$, $-\sqrt{10}$, and π, which is the ratio of the distance around a circle to the distance across it.

Finally, *all* numbers that can be represented by points on the number line are called **real numbers.**

Real numbers	{all numbers that can be represented by points on the number line}

All the numbers mentioned above are real numbers. The relationships between the various types of numbers are shown in Figure 5.

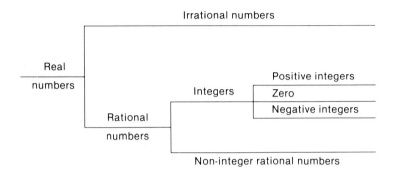

Figure 5

Work Problem 2 at the side.

3▶ If you are given any two whole numbers, you probably can tell which number is smaller. But what about two negative numbers, as in the set of integers? Moving from zero to the right along a number line, the positive numbers corresponding to the points on the number line *increase.* For example, $8 < 12$, and 8 is to the left of 12 on a number line. This ordering is extended to all real numbers by definition.

Ordering Real Numbers
The smaller of any two different real numbers is the one corresponding to the point that is to the left on the number line.

This means that any negative number is smaller than 0, and any negative number is smaller than any positive number. Also, 0 is smaller than any positive number.

EXAMPLE 1 Is it true that $-3 < -1$?

Find out by locating both numbers, -3 and -1, on a number line, as shown in Figure 6. Since -3 is to the left of -1 on the number line, -3 is smaller than -1. The statement $-3 < -1$ is true. ◀

Work Problem 3 at the side.

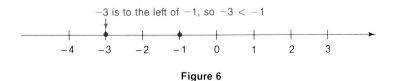

Figure 6

2. (a) Identify each real number in the set below as rational or irrational.

$$\left\{\frac{5}{8}, -7, -1\frac{3}{5}, 0, \sqrt{11}, -\pi\right\}$$

(b) Which of the numbers listed above are *integers*?

3. Indicate whether each statement is true or false.

(a) $-2 < 4$

(b) $6 > -3$

(c) $-9 < -12$

(d) $-4 \geq -1$

(e) $-6 \leq 0$

ANSWERS

2. (a) $\frac{5}{8}$, $-7 \left(\text{or } \frac{-7}{1}\right)$, $-1\frac{3}{5} \left(\text{or } \frac{-8}{5}\right)$, and $0 \left(\text{or } \frac{0}{1}\right)$ are rational; $\sqrt{11}$ and $-\pi$ are irrational.
(b) The only integers are -7 and 0.
3. (a) true (b) true (c) false
(d) false (e) true

2.1 Real Numbers and the Number Line

▶ By a property of the real numbers, for any real number *x* (except 0), there is exactly one number on the number line the same distance from 0 as *x* but on the opposite side of 0.

For example, Figure 7 shows that the numbers 3 and −3 are each the same distance from 0 but are on opposite sides of 0. The numbers 3 and −3 are called **additive inverses,** or **opposites,** of each other.

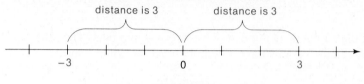

Figure 7

> The **additive inverse** of a number *a* is the number that is the same distance from 0 on the number line as *a*, but on the opposite side of 0.

The additive inverse of the number 0 is 0 itself. This makes 0 the only real number that is its own additive inverse. Other additive inverses occur in pairs. For example, 4 and −4, and 5 and −5, are additive inverses of each other. Several pairs of additive inverses are shown in Figure 8.

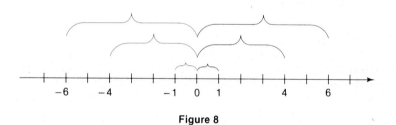

Figure 8

The additive inverse of a number can be indicated by writing the symbol − in front of the number. With this symbol, the additive inverse of 7 is written −7. The additive inverse of −4 can be written −(−4). It was suggested in Figure 8 that 4 is an additive inverse of −4. Since a number can have only one additive inverse, the symbols 4 and −(−4) must represent the same number, which means that

$$-(-4) = 4.$$

A generalization of this idea is given below.

> **Double Negative Rule**
>
> For any real number *a*,
>
> $$-(-a) = a.$$

Operations with Real Numbers

EXAMPLE 2 The following chart shows several numbers and their additive inverses.

Number	Additive inverse
−3	3
−4	−(−4), or 4
0	0
5	−5
19	−19

Example 2 suggests the following rule.

> Find the additive inverse of a number by changing the sign of the number.

Work Problem 4 at the side.

5 As mentioned above, additive inverses are numbers the same distance from 0 on the number line but on opposite sides of 0. This same idea can be expressed by saying that two numbers that are additive inverses have the same absolute value. The **absolute value** of a number is defined as the distance between 0 and the number on the number line. The symbol for the absolute value of the number a is $|a|$, read "the absolute value of a." For example, the distance between 2 and 0 on the number line is 2 units, so that

$$|2| = 2.$$

Also, the distance between −2 and 0 on the number line is 2, so that

$$|-2| = 2.$$

Since distance is a physical measurement, which is never negative,

> the absolute value of a number can never be negative.

For example,

$$|12| = 12 \quad \text{and} \quad |-12| = 12,$$

since both 12 and −12 lie at a distance of 12 units from 0 on the number line. Since 0 is a distance 0 units from 0, we have

$$|0| = 0.$$

EXAMPLE 3 Simplify by removing absolute value symbols.

(a) $|5| = 5$

(b) $|-5| = 5$

(c) $-|5| = -(5) = -5$

Find the absolute value first, then the additive inverse of the result.

4. Find the additive inverse.

   (a) 6

   (b) 15

   (c) −9

   (d) −12

   (e) 0

ANSWERS
4. (a) −6 (b) −15 (c) 9 (d) 12
   (e) 0

2.1 Real Numbers and the Number Line

5. Simplify by removing absolute value symbols.

(a) $|-6|$

(b) $|9|$

(c) $|-36|$

(d) $-|15|$

(e) $-|-9|$

(f) $-|32 - 2|$

(d) $-|-14| = -(14) = -14$

(e) $|8 - 2| = |6| = 6$ ◀

Part (e) in Example 3 shows that absolute value bars are also grouping symbols.

Work Problem 5 at the side.

ANSWERS

5. (a) 6 (b) 9 (c) 36 (d) -15
   (e) -9 (f) -30

Operations with Real Numbers

2.1 EXERCISES

Give the additive inverse of each number. For the exercises with absolute value, simplify first before deciding on the additive inverse. See Examples 2 and 3.

1. 8
2. 12
3. -9

4. -11
5. $-\dfrac{2}{3}$
6. -3.25

7. $|15|$
8. $|5|$
9. $|-8|$

Circle the smaller of the two given numbers. See Examples 1 and 3.

10. $-5,\ 5$
11. $9,\ -3$
12. $-12,\ -4$

13. $-9,\ -14$
14. $-8,\ -1$
15. $-\dfrac{4}{15},\ -\dfrac{1}{4}$

16. $\dfrac{3}{5},\ \left|-\dfrac{7}{9}\right|$
17. $5,\ |-2|$
18. $|-3|,\ |-4|$

19. $|-8|,\ |-9|$
20. $-|-6|,\ -|-4|$
21. $-|-2|,\ -|-3|$

Write true or false for each of the following statements. See Examples 1 and 3.

22. $6 < -3$
23. $-2 < -1$
24. $-8 < -4$

25. $-3 \geq -7$
26. $-9 \geq -12$
27. $-15 \leq -20$

28. $-21 \leq -27$
29. $-8 \leq -(-4)$
30. $-9 \leq -(-6)$

31. $0 \leq -(-4)$
32. $0 \geq -(-6)$
33. $-3.2 < -3.8$

2.1 Exercises

34. $-6.1 \geq -6.05$ 35. $6 > -(-2)$ 36. $-8 > -(-2)$

37. $-\dfrac{1}{2} < -\left(-\dfrac{2}{3}\right)$ 38. $-\dfrac{3}{5} > -\dfrac{5}{8}$ 39. $|-6| < |-9|$

40. $|-12| < |-20|$ 41. $-|8| > |-9|$ 42. $-|12| > |-15|$

43. $-|-5| \geq -|-9|$ 44. $-|-12| \leq -|-15|$ 45. $|-8| \geq |-2|$

Graph each group of numbers on the indicated number line. Simplify the expressions having absolute value bars before graphing them.

46. $0,\ 3,\ -5,\ -6$
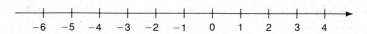

47. $2,\ 6,\ -2,\ -1$

48. $-2,\ -6,\ |-4|,\ 3,\ -|4|$

49. $-5,\ -3,\ -|-2|,\ 0,\ |-4|$
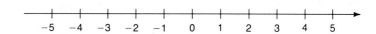

50. $|2|,\ -|4|,\ -|-3|,\ -|-5|$
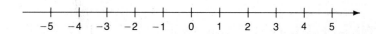

51. $\dfrac{1}{4},\ 2\dfrac{1}{2},\ -3\dfrac{4}{5},\ -4,\ -1\dfrac{5}{8}$
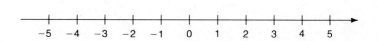

52. $5\dfrac{1}{4},\ 4\dfrac{5}{9},\ -2\dfrac{1}{3},\ 0,\ -3\dfrac{2}{5}$

50   Operations with Real Numbers

2.2 ADDITION OF REAL NUMBERS

▶ The number line can be used to give meaning to the addition of real numbers, as in the following examples.

EXAMPLE 1 Use the number line to find the sum 2 + 3.

Add the positive numbers 2 and 3 by starting at 0 and drawing an arrow two units to the *right*, as shown in Figure 9. This arrow represents the number 2 in the sum 2 + 3. Then, from the right end of this arrow draw another arrow three units to the right. The number below the end of this second arrow is 5, so 2 + 3 = 5. ◀

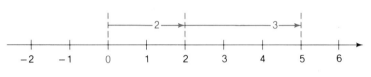

Figure 9

EXAMPLE 2 Use the number line to find the sum −2 + (−4). (Parentheses are placed around the −4 to avoid the confusing use of + and − next to each other.)

Add the negative numbers −2 and −4 on the number line by starting at 0 and drawing an arrow two units to the *left*, as shown in Figure 10. Draw the arrow to the left to represent the addition of a *negative* number. From the left end of this first arrow, draw a second arrow four units to the left. The number below the end of this second arrow is −6, so −2 + (−4) = −6. ◀

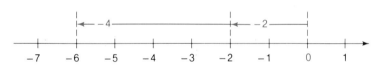

Figure 10

Work Problem 1 at the side.

In Example 2, the sum of the two negative numbers −2 and −4 is a negative number whose distance from 0 is the sum of the distance of −2 from 0 and the distance of −4 from 0. That is, *the sum of two negative numbers is the negative of the sum of their absolute values.*

$$-2 + (-4) = -(|-2| + |-4|) = -(2 + 4) = -6$$

> Add two numbers having the same signs by adding the absolute values of the numbers. Give the result the same sign as the numbers being added.

Objectives

▶ Add two numbers with the same sign on a number line.

▶ Add positive and negative numbers.

▶ Add mentally.

▶ Use the order of operations with real numbers.

1. Use the number lines to find the sums.

   (a) 1 + 4

   (b) −2 + (−5)

   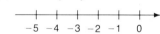

   (c) −3 + (−1)

ANSWERS

1. (a) 1 + 4 = 5

   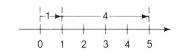

   (b) −2 + (−5) = −7

   (c) −3 + (−1) = −4

   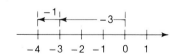

2.2 Addition of Real Numbers

2. Find the sums.

(a) $-7 + (-3)$

(b) $-12 + (-18)$

(c) $-15 + (-4)$

3. Use the number lines to find the sums.

(a) $6 + (-3)$

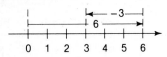

(b) $-5 + 1$

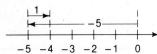

(c) $-7 + 2$

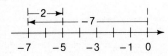

ANSWERS
2. (a) -10 (b) -30 (c) -19
3. (a) $6 + (-3) = 3$

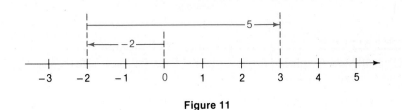

EXAMPLE 3 Find the sums.

(a) $-2 + (-9) = -11$ The sum of two negative numbers is negative
(b) $-8 + (-12) = -20$
(c) $-15 + (-3) = -18$ ◀

Work Problem 2 at the side.

▶ Use the number line again to give meaning to the sum of a positive number and a negative number.

EXAMPLE 4 Use the number line to find the sum $-2 + 5$.

Find the sum $-2 + 5$ on the number line by starting at 0 and drawing an arrow two units to the left. From the left end of this arrow, draw a second arrow five units to the right, as shown in Figure 11. The number below the end of this second arrow is 3, so $-2 + 5 = 3$. ◀

Figure 11

Work Problem 3 at the side.

Addition of numbers with different signs can also be defined using absolute value.

> Add numbers with different signs by first finding the difference of the absolute values of the numbers. Give the answer the same sign as the number with the larger absolute value.

For example, to add -12 and 5, find their absolute values: $|-12| = 12$ and $|5| = 5$. Then find the difference of these absolute values: $12 - 5 = 7$. Since $|-12| > |5|$, the sum will be negative, so that the final answer is $-12 + 5 = -7$.

▶ While a number line is useful in showing the rules for addition, it is important to be able to find sums mentally.

EXAMPLE 5 Check each answer, trying to work the addition mentally. If you get stuck, use a number line.

(a) $7 + (-4) = 3$

(b) $-8 + 12 = 4$

(c) $-\dfrac{1}{2} + \dfrac{1}{8} = -\dfrac{4}{8} + \dfrac{1}{8} = -\dfrac{3}{8}$ Remember to find a common denominator first

(d) $\dfrac{5}{6} + \left(-\dfrac{4}{3}\right) = -\dfrac{1}{2}$

(e) $-4.6 + 8.1 = 3.5$ ◀

Work Problem 4 at the side.

The rules for adding signed numbers are summarized below.

Adding Signed Numbers

Like signs   Add the absolute values of the numbers. Give the sum the same sign as the numbers being added.

Unlike signs   Find the difference of the larger absolute value and the smaller. Give the answer the sign of the number having the larger absolute value.

Sometimes a problem involves square brackets, [ ]. As mentioned earlier, brackets are treated just like parentheses. Do the calculations inside the brackets until a single number is obtained. Remember to use the order of operations given in Section 1.4 when adding more than two numbers.

EXAMPLE 6   Find the sums.

(a) $-3 + [4 + (-8)]$

First work inside the brackets. Follow the rules for the order of operations given in Section 1.4.

$$-3 + [4 + (-8)] = -3 + (-4) = -7$$

(b) $8 + [(-2 + 6) + (-3)] = 8 + [4 + (-3)] = 8 + 1 = 9$ ◀

Work Problem 5 at the side.

---

4. Check each answer, trying to work the addition in your head. If you get stuck, use a number line.

   (a) $-8 + 2 = -6$

   (b) $-15 + 4 = -11$

   (c) $17 + (-10) = 7$

   (d) $\dfrac{3}{4} + \left(-\dfrac{11}{8}\right) = -\dfrac{5}{8}$

   (e) $-9.5 + 3.8 = -5.7$

5. Find the sums.

   (a) $2 + [7 + (-3)]$

   (b) $6 + [(-2 + 5) + 7]$

   (c) $-9 + [-4 + (-8 + 6)]$

ANSWERS
4. All are correct.
5. (a) 6 (b) 16 (c) -15

2.2   Addition of Real Numbers

2.2 EXERCISES

Find the sums. See Examples 1–6.

1. $5 + (-3)$
2. $11 + (-8)$
3. $6 + (-8)$

4. $3 + (-7)$
5. $-6 + (-2)$
6. $-8 + (-3)$

7. $-9 + (-2)$
8. $-15 + (-6)$
9. $-3 + (-9)$

10. $-11.3 + (-5.8)$
11. $12.6 + (-8.42)$
12. $-10 + 2.531$

13. $4 + [13 + (-5)]$
14. $6 + [2 + (-13)]$
15. $8 + [-2 + (-1)]$

16. $12 + [-3 + (-4)]$
17. $-2 + [5 + (-1)]$
18. $-8 + [9 + (-2)]$

19. $-6 + [6 + (-9)]$
20. $-3 + [4 + (-8)]$
21. $[9 + (-2)] + 6$

22. $[8 + (-14)] + 10$
23. $[(-9) + (-14)] + 12$
24. $[(-8) + (-6)] + 10$

25. $-\dfrac{1}{6} + \dfrac{2}{3}$
26. $\dfrac{9}{10} + \left(-\dfrac{3}{5}\right)$
27. $\dfrac{5}{8} + \left(-\dfrac{17}{12}\right)$

28. $-\dfrac{6}{25} + \dfrac{19}{20}$
29. $2\dfrac{1}{2} + \left(-3\dfrac{1}{4}\right)$
30. $-4\dfrac{3}{8} + 6\dfrac{1}{2}$

31. $7.9 + (-8.4)$

32. $11.6 + (-15.4)$

33. $-6.1 + [3.2 + (-4.8)]$

34. $-9.4 + [-5.8 + (-1.4)]$

35. $[-3 + (-4)] + [5 + (-6)]$

36. $[-8 + (-3)] + [-7 + (-6)]$

37. $[-4 + (-3)] + [8 + (-1)]$

38. $[-5 + (-9)] + [16 + (-21)]$

39. $[-4 + (-6)] + [(-3) + (-8)] + [12 + (-11)]$

40. $[-2 + (-11)] + [12 + (-2)] + [18 + (-6)]$

41. $(-9.648 + 11.237) + [(-4.9123 + 1.8769) + 3.1589]$

42. $[-3.851 + (-2.4691)] + [11.809 + (-1.735)] + (-1.409)$

Write true or false for each statement.

43. $-4 + 0 = -4$

44. $-6 + 5 = -1$

45. $-8 + 12 = 8 + (-12)$

46. $15 + (-8) = 8 + (-15)$

47. $-9 + 5 + 6 = -2$

48. $-6 + (8 - 5) = -3$

49. $-\dfrac{3}{2} + \dfrac{5}{8} = \dfrac{5}{8} + \left(-\dfrac{3}{2}\right)$

50. $\dfrac{11}{5} + \dfrac{-6}{11} = \dfrac{-6}{11} + \dfrac{11}{5}$

51. $\left|\dfrac{-8}{13} + \dfrac{3}{4}\right| = \dfrac{8}{13} + \dfrac{3}{4}$

52. $|-4 + 2| = 4 + 2$

53. $|12 - 3| = 12 - 3$

54. $|-6 + 10| = 6 + 10$

55. $[4 + (-6)] + 6 = 4 + (-6 + 6)$

56. $[(-2) + (-3)] + (-6) = 12 + (-1)$

57. $-7 + [-5 + (-3)] = [(-7) + (-5)] + 3$

58. $6 + [-2 + (-5)] = [(-4) + (-2)] + 5$

59. $-5 + (-|-5|) = -10$

60. $|-3| + (-5) = -2$

Find all solutions for the following equations from the domain $\{-3, -2, -1, 0, 1, 2, 3\}$.

61. $x + 2 = 0$

62. $x + 3 = 0$

63. $x + 1 = -2$

64. $x + 2 = -1$

65. $14 + x = 12$

66. $x + 8 = 7$

67. $x + (-4) = -6$

68. $x + (-2) = -5$

69. $-8 + x = -6$

The word sum *indicates addition. Write a numerical expression for each statement and simplify.*

70. The sum of -9 and 2 and 6

71. The sum of 4 and -7 and -3

72. 12 added to the sum of -17 and -6

73. -3 added to the sum of 15 and -1

74. The sum of -11 and -4 increased by -5

75. The sum of -8 and -15 increased by -3

2.2 Exercises 57

Solve the following word problems.

76. Joann has $15. She then spends $6. How much is left?

77. An airplane is flying at an altitude of 6000 feet. It then descends 4000 feet. What is its final altitude?

78. Chuck is standing 15 feet below sea level in Death Valley. He then goes down another 120 feet. Find his final altitude.

79. Donna has $11 and spends $19. What is her final balance? (Write the answer with a negative number.)

80. One number of Nancy's blood pressure was 120, but then it changed by -30. Find her present blood pressure.

81. The temperature was $-14°$, but then it went down $12°$. Find the new temperature.

82. The temperature at 4 A.M. was $-22°$, but it went up $35°$ by noon. What was the temperature at noon?

83. A man owes $94 to a credit card company. He makes a payment of $60. What amount does he still owe?

84. Joann Post owes $983.72 on her Visa credit card. She returns items costing $74.18 and $12.53. She makes two purchases of $11.79 each and further purchases of $106.58, $29.81, and $73.24. She makes a payment of $186.50. Find the amount that she then owes.

85. A welder working with stainless steel must use precise measurements. Suppose a welder attaches two pieces of steel that are each 3.589 inches in length, then attaches an additional three pieces that are each 9.089 inches long, and finally cuts off a piece that is 7.612 inches long. Find the length of the welded piece of steel.

86. What number must be added to -9 to get 8?

87. What number must be added to -15 to get 3?

88. The sum of what number and 5 is -11?

89. The sum of what number and -6 is -9?

2.3 SUBTRACTION OF REAL NUMBERS

▶ As mentioned earlier, the answer to a subtraction problem is called a **difference.** Differences of signed numbers can be found by using a number line. Since *addition* of a positive number on the number line is shown by drawing an arrow to the *right*, *subtraction* of a positive number is shown by drawing an arrow to the *left*.

EXAMPLE 1 Use the number line to find the difference $7 - 4$.

To find the difference $7 - 4$ on the number line, begin at 0 and draw an arrow 7 units to the right. From the right end of this arrow, draw an arrow 4 units to the left, as shown in Figure 12. The number at the end of the second arrow shows that $7 - 4 = 3$. ◀

Figure 12

Work Problem 1 at the side.

▶ The procedure used in Example 1 to find $7 - 4$ is exactly the same procedure that would be used to find $7 + (-4)$, so that

$$7 - 4 = 7 + (-4).$$

It seems that *subtraction* of a positive number from a larger positive number is the same as *adding* the additive inverse of the smaller number to the larger. This result is extended as the definition of subtraction for all real numbers.

> **Definition of Subtraction**
> For any real numbers a and b,
> $$a - b = a + (-b).$$

That is, to **subtract** b from a, *add the additive inverse* (or opposite) of b to a. This definition leads to the following procedure for subtracting signed numbers.

> **Subtracting Signed Numbers**
> *Step 1* Change the subtraction symbol to addition.
> *Step 2* Change the sign of the number being subtracted.
> *Step 3* Add, as in the previous section.

EXAMPLE 2 Subtract.

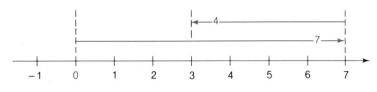

(a) $12 - 3 = 12 + (-3) = 9$

Objectives

▶ 1 Find a difference on the number line.

▶ 2 Use the definition of subtraction.

▶ 3 Work subtraction problems that involve brackets.

1. Use the number line to find the differences.

(a) $5 - 1$

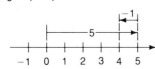

(b) $6 - 2$

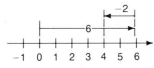

ANSWERS

1. (a) $5 - 1 = 4$

(b) $6 - 2 = 4$

2. Subtract.

(a) $6 - 10$

(b) $-2 - 4$

(c) $3 - (-5)$

(d) $-8 - (-12)$

3. Work each problem.

(a) $2 - [(-3) - (4 + 6)]$

(b) $[(5 - 7) + 3] - 8$

(c) $6 - [(-1 - 4) - 2]$

(d) $(-8 - 1) - [(-3 + 2) - (-4 + 1)]$

(b) $5 - 7 = 5 + (-7) = -2$

(c) $8 - 15 = 8 + (-15) = -7$

(d) $-3 - (-5) = -3 + (5) = 2$ (No change; Change $-$ to $+$; Additive inverse of -5)

(e) $-6 - (-9) = -6 + (9) = 3$

(f) $8 - (-5) = 8 + (5) = 13$ ◀

Work Problem 2 at the side.

Subtraction can be used to reverse the result of an addition problem. For example, if 4 is added to a number and then subtracted from the sum, the original number is the result.

The symbol $-$ has now been used for three purposes:

1. to represent subtraction, as in $9 - 5 = 4$;
2. to represent negative numbers, such as -10, -2, and -3;
3. to represent the additive inverse of a number, as in "the additive inverse of 8 is -8."

More than one use may appear in the same problem, such as $-6 - (-9)$, where -9 is subtracted from -6. The meaning of the symbol depends on its position in the algebraic expression.

▶ As before, with problems that have both parentheses and brackets, first do any operations inside the parentheses and brackets. Work from the inside out. Because subtraction is defined in terms of addition, the order of operations rules from Section 1.4 can be used.

EXAMPLE 3 Work each problem.

(a) $-6 - [2 - (8 + 3)] = -6 - [2 - 11]$
$= -6 - [2 + (-11)]$
$= -6 - (-9)$
$= -6 + (9)$
$= 3$

(b) $5 - [(-3 - 2) - (4 - 1)] = 5 - [(-3 + (-2)) - 3]$
$= 5 - [(-5) - 3]$
$= 5 - [(-5) + (-3)]$
$= 5 - (-8)$
$= 5 + 8$
$= 13$ ◀

Work Problem 3 at the side.

ANSWERS

2. (a) -4 (b) -6 (c) 8 (d) 4
3. (a) 15 (b) -7 (c) 13 (d) -11

2.3 EXERCISES

Find the differences. See Examples 1–3.

1. $3 - 6$

2. $7 - 12$

3. $5 - 9$

4. $8 - 13$

5. $-6 - 2$

6. $-11 - 4$

7. $-9 - 5$

8. $-12 - 15$

9. $6 - (-3)$

10. $8 - (-5)$

11. $5 - (-12)$

12. $12 - (-2)$

13. $-6 - (-2)$

14. $-7 - (-5)$

15. $2 - (3 - 5)$

16. $5 - (6 - 13)$

17. $-2 - (5 - 8)$

18. $-3 - (4 - 11)$

19. $\frac{1}{2} - \left(-\frac{1}{4}\right)$

20. $\frac{1}{3} - \left(-\frac{4}{3}\right)$

21. $-\frac{3}{4} - \frac{5}{8}$

22. $-\frac{5}{6} - \frac{1}{2}$

23. $\frac{5}{8} - \left(-\frac{1}{2} - \frac{3}{4}\right)$

24. $\frac{9}{10} - \left(\frac{1}{8} - \frac{3}{10}\right)$

25. $3.4 - (-8.2)$

26. $5.7 - (-11.6)$

27. $-6.4 - 3.5$

28. $-4.4 - 8.6$

29. $-4.1128 - (7.418 - 9.80632)$

30. $(-1.8142 - 3.7256) - (-9.8025)$

31. $[-7.6892 - (-3.2512)] - (-8.1243)$

32. $[(-2.1463 - 1.8374) - .28174] - [-3.258 - (-1.0926)]$

Work each problem. See Example 3.

33. $(4 - 6) + 12$ **34.** $(3 - 7) + 4$ **35.** $(8 - 1) - 12$ **36.** $(9 - 3) - 15$

37. $6 - (-8 + 3)$ **38.** $8 - (-9 + 5)$ **39.** $2 + (-4 - 8)$ **40.** $6 + (-9 - 2)$

41. $(-5 - 6) - (9 - 2)$ **42.** $(-4 - 8) - (6 - 1)$

43. $\left(-\dfrac{3}{8} - \dfrac{2}{3}\right) - \left(-\dfrac{9}{8} - 3\right)$ **44.** $\left(-\dfrac{3}{4} - \dfrac{5}{2}\right) - \left(-\dfrac{1}{8} - 1\right)$

45. $-9 - [(3 - 2) - (-4 - 2)]$ **46.** $-8 - [(-4 - 1) - (9 - 2)]$

47. $-3 + [(-5 - 8) - (-6 + 2)]$ **48.** $-4 + [(-12 + 1) - (-1 - 9)]$

49. $-9.1237 + [(-4.8099 - 3.2516) + 11.27903]$

50. $-7.6247 - [(-3.9928 + 1.42773) - (-2.80981)]$

62   Operations with Real Numbers

51. $[-12.1035 - (8.11725 + 3.83122)] - 17.40963$

52. $[-34.9122 + (6.45378 - 12.14273)] - 8.46922$

Write the given problem in symbols (no variables are needed). Then solve.

53. Subtract -6 from 12.

54. Subtract -8 from 15.

55. From -25, subtract -4.

56. What number is 6 less than -9?

57. $\frac{4}{27}$ is how much greater than $-\frac{5}{24}$?

58. How much greater is $\frac{8}{11}$ than $-\frac{5}{4}$?

59. How much greater is -7.3 than -8.4?

60. -12.4 is how much greater than -14.3?

The word difference *indicates subtraction. Write a numerical expression for each statement and simplify.*

61. Find the difference of 8 and -2.

62. Find the difference of 3 and -8.

63. Add -11 to the difference of -4 and 2.

64. Add 8 to the difference of 1 and -3.

65. From the sum of -12 and -3, subtract 4.

66. From the sum of 8 and -13, subtract -2.

Work the word problems.

67. The temperature dropped 10° below the previous temperature of −5°. Find the new temperature.

68. Bill owed his brother $10. He repaid $6 and later borrowed $7. What positive or negative number represents his present financial status?

69. The bottom of Death Valley is 282 feet below sea level. The top of Mt. Whitney has an altitude of 14,494 feet above sea level. Find the difference between these two elevations.

70. Tickets to the school play cost $15, and Joan is $12 in debt. How much must she earn before she can afford a ticket?

71. A chemist is running an experiment at a temperature of −174.6°. She then raises the temperature by 2.3°. Find the new temperature.

72. One year a company had a "profit" of −$25,000. The next year, the profit decreased by $7200. Find the profit the next year.

73. One company made a profit of $76,000, while another company lost $29,000. Find the difference between these "profits."

74. A first reading of a dial was 7.904. A second reading was −3.291. By how much had the reading gone down?

2.4 MULTIPLICATION OF REAL NUMBERS

We already know the rule for multiplying positive numbers:

> the product of two positive numbers is positive.

But what about multiplying other real numbers? Any rules for multiplication of real numbers should be consistent with the rules from arithmetic for multiplication. For example, the product of 0 and any real number (positive or negative) should be 0.

> For any number a,
> $a \cdot 0 = 0.$

▷ In order to define the product of a positive and a negative number so that the result is consistent with the multiplication of two positive numbers, look at the following pattern.

$$3 \cdot 5 = 15$$
$$3 \cdot 4 = 12$$
$$3 \cdot 3 = 9$$
$$3 \cdot 2 = 6$$
$$3 \cdot 1 = 3$$
$$3 \cdot 0 = 0$$
$$3 \cdot (-1) = ?$$

The numbers decrease by 3

What should $3(-1)$ equal? The product $3(-1)$ represents the sum

$$-1 + (-1) + (-1) = -3,$$

so the product should be -3. Also,

$$3(-2) = -2 + (-2) + (-2) = -6.$$

Work Problem 1 at the side.

The results from Problem 1 maintain the pattern in the list above, which suggests the following rule.

> The product of a positive number and a negative number is negative.

EXAMPLE 1 Find the products using the multiplication rule given above.

(a) $8(-5) = -(8 \cdot 5)$
$= -40$

(b) $5(-4) = -(5 \cdot 4)$
$= -20$

(c) $(-7)(2) = -(7 \cdot 2)$
$= -14$

Objectives

▷1 Find the product of a positive and a negative number.

▷2 Find the product of two negative numbers.

▷3 Use the order of operations.

▷4 Evaluate expressions involving variables.

1. Find each product by finding the sum of three numbers.

(a) $3(-3)$

(b) $3(-4)$

(c) $3(-5)$

ANSWERS
1. (a) -9 (b) -12 (c) -15

2.4 Multiplication of Real Numbers

2. Find the products.

   (a) $2(-6)$

   (b) $7(-8)$

   (c) $12(-15)$

   (d) $(-9)(2)$

   (e) $(-10)(3)$

   (f) $(-16)(19)$

3. Find the products.

   (a) $(-5)(-6)$

   (b) $(-7)(-3)$

   (c) $(-8)(-5)$

   (d) $(-11)(-2)$

   (e) $(-17)(-21)$

   (f) $(-82)(-13)$

ANSWERS
2. (a) -12 (b) -56 (c) -180
   (d) -18 (e) -30 (f) -304
3. (a) 30 (b) 21 (c) 40 (d) 22
   (e) 357 (f) 1066

(d) $(-9)\left(\dfrac{1}{3}\right) = -\left(9 \cdot \dfrac{1}{3}\right)$
$= -3$

(e) $(-6.2)(4.1) = -25.42$ ◀

Work Problem 2 at the side.

▶ The product of two positive numbers is positive, and the product of a positive number and a negative number is negative. What about the product of two negative numbers? Look at another pattern.

$(-5)(4) = -20$ The
$(-5)(3) = -15$ numbers
$(-5)(2) = -10$ increase
$(-5)(1) = -5$ by
$(-5)(0) = 0$ 5
$(-5)(-1) = ?$

The numbers on the left of the equals sign (in the boxes) decrease by 1 for each step down the list. The products on the right increase by 5 for each step down the list. To maintain this pattern, $(-5)(-1)$ should be 5 more than $(-5)(0)$, or 5 more than 0, so

$$(-5)(-1) = 5.$$

The pattern continues with

$(-5)(-2) = 10$
$(-5)(-3) = 15$
$(-5)(-4) = 20$
$(-5)(-5) = 25,$

and so on.

Multiply two negative numbers as follows.

> The product of two negative numbers is positive.

EXAMPLE 2 Find the products using the multiplication rule given above.

(a) $(-9)(-2) = 9 \cdot 2$
$= 18$

(b) $(-6)(-12) = 6 \cdot 12$
$= 72$

(c) $(-8)(-1) = 8 \cdot 1$
$= 8$

(d) $(-15)(-2) = 15 \cdot 2$
$= 30$ ◀

Work Problem 3 at the side.

A summary of the results for multiplying positive and negative numbers is given here.

> **Multiplying Signed Numbers**
> The product of two numbers having the *same* signs is *positive*, and the product of two numbers having *different* signs is *negative*.

▶ The next example shows the order of operations discussed in Chapter 1 used with the multiplication of positive and negative numbers.

EXAMPLE 3 Simplify.

(a) $(-9)(2) - (-3)(2)$

First find all products, working from left to right.

$$(-9)(2) - (-3)(2) = -18 - (-6)$$

Now perform the subtraction.

$$-18 - (-6) = -18 + 6$$
$$= -12$$

(b) $(-6)(-2) - (3)(-4) = 12 - (-12)$
$$= 12 + 12$$
$$= 24$$

(c) $-5(-2 - 3) = -5(-5) = 25$ ◀

Work Problem 4 at the side.

▶ The last two examples show how numbers may be substituted for variables.

EXAMPLE 4 Evaluate the expression

$$(3x + 4y)(-2m)$$

for each set of values.

(a) $x = -1$, $y = -2$, $m = -3$

First substitute the given values for the variables. Then find the value of the expression. Put parentheses around the number for each variable.

$(3x + 4y)(-2m)$
$= [3(-1) + 4(-2)][-2(-3)]$ Find the products
$= [-3 + (-8)][6]$ Use order of operations
$= (-11)(6)$
$= -66$

4. Simplify.

(a) $(-3)(4) - (2)(6)$

(b) $(-5)(-6) - (8)(-3)$

(c) $-7(-2 - 5)$

(d) $-4(-7 - 9)$

(e) $-8[-1 - (-4)(-5)]$

ANSWERS
4. (a) -24 (b) 54 (c) 49 (d) 64
   (e) 168

2.4 Multiplication of Real Numbers

5. Evaluate the following expressions.

   (a) $2x - 7(y + 1)$, if $x = -4$ and $y = 3$

   (b) $(-3x)(4x - 2y)$, if $x = 2$ and $y = -1$

   (c) $2x^2 - 4y^2$, if $x = -2$ and $y = -3$

(b) $x = 7, \quad y = -9, \quad m = 5$

Substitute. Put parentheses around -9.

$$(3x + 4y)(-2m)$$
$$= [3 \cdot 7 + 4(-9)](-2 \cdot 5)$$
$$= [21 + (-36)](-10) \quad \text{Find the products}$$
$$= (-15)(-10)$$
$$= 150 \blacktriangleleft$$

EXAMPLE 5 Evaluate $2x^2 - 3y^2$ for $x = -3$ and $y = -4$. Use parentheses as shown.

$$2(-3)^2 - 3(-4)^2 = 2(9) - 3(16) \quad \text{Square } -3 \text{ and } -4$$
$$= 18 - 48 \quad \text{Multiply}$$
$$= -30 \quad \text{Subtract} \blacktriangleleft$$

Work Problem 5 at the side.

ANSWERS

5. (a) -36 (b) -60 (c) -28

2.4 EXERCISES

Find the products. See Examples 1 and 2. In Exercises 22–24, round to the nearest thousandth.

1. $(-3)(-4)$ **2.** $(-3)(4)$ **3.** $3(-4)$ **4.** $-2(-8)$

5. $(-1)(-5)$ **6.** $(-9)(-5)$ **7.** $(-4)(-11)$ **8.** $(-5)(7)$

9. $(-10)(-12)$ **10.** $9(-5)$ **11.** $(8)(-6)$ **12.** $(13)(-2)$

13. $(-6)(5)$ **14.** $(-9)(0)$ **15.** $0(-11)$ **16.** $(15)(-11)$

17. $\left(-\dfrac{7}{3}\right)\left(\dfrac{8}{21}\right)$ **18.** $\left(-\dfrac{3}{8}\right)\left(-\dfrac{10}{9}\right)$ **19.** $\left(-\dfrac{5}{4}\right)\left(\dfrac{6}{15}\right)$

20. $(-5.1)(.02)$ **21.** $(-3.7)(-2.1)$ **22.** $(-12.804)(4.12)$

23. $(3.871)(-5.463)$ **24.** $(-6.4972)(-13.8015)$

Simplify. See Example 3.

25. $6 - 4 \cdot 5$ **26.** $3 - 2 \cdot 9$ **27.** $-9 - (-2) \cdot 3$

28. $-11 - (-7) \cdot 4$ **29.** $9(6 - 10)$ **30.** $5(12 - 15)$

31. $-6(2 - 4)$ **32.** $-9(5 - 8)$ **33.** $(4 - 9)(2 - 3)$

34. $(6 - 11)(3 - 6)$ **35.** $(2 - 5)(3 - 7)$ **36.** $(5 - 12)(2 - 6)$

37. $(-4 - 3)(-2) + 4$ **38.** $(-5 - 2)(-3) + 6$ **39.** $3(-4) - (-2)$

40. $5(-2) - (-9)$ **41.** $(-8 - 2)(-4) - (-5)$ **42.** $(-9 - 1)(-2) - (-6)$

43. $|-4(-2)| + |-4|$ **44.** $|8(-5)| + |-2|$ **45.** $|2|(-4) + |6| \cdot |-4|$

Evaluate the following expressions, given $x = -2$, $y = 3$, and $a = -4$. See Example 4.

46. $5x - 2y + 3a$

47. $6x - 5y + 4a$

48. $(2x + y)(3a)$

49. $(5x - 2y)(-2a)$

50. $\left(\dfrac{1}{3}x - \dfrac{4}{5}y\right)\left(-\dfrac{1}{5}a\right)$

51. $\left(\dfrac{5}{6}x + \dfrac{3}{2}y\right)\left(-\dfrac{1}{3}a\right)$

52. $(-5 + x)(-3 + y)(2 - a)$

53. $(6 - x)(5 + y)(3 + a)$

54. $-2y^2 + 3a$

55. $5x - 4a^2$

56. $3a^2 - x^2$

57. $4y^2 - 2x^2$

Find the solution for each of the following equations from the domain $\{-3, -2, -1, 0, 1, 2, 3\}$.

58. $2x = -4$

59. $3k = -6$

60. $-4m = 0$

61. $-9y = 0$

62. $-8p = 16$

63. $-9r = 27$

64. $2x + 1 = -3$

65. $3w + 3 = -3$

66. $-4a + 2 = 10$

67. $-5t + 6 = 11$

68. $\dfrac{3}{5}m + \dfrac{2}{3} = -\dfrac{17}{15}$

69. $-\dfrac{10}{9}r + \dfrac{4}{3} = \dfrac{2}{9}$

The word product indicates multiplication. Write a numerical expression for each statement and simplify.

70. The product of -9 and 2 is added to 6.

71. The product of 4 and -7 is added to -9.

72. After the product of -1 and 6 is found, the result is subtracted from -9.

73. Twice the product of -8 and 2 is subtracted from -4.

74. Nine is subtracted from the product of 7 and -6.

75. Three is subtracted from the product of -2 and 3.

2.5 DIVISION OF REAL NUMBERS

Objectives

▶ Find the reciprocal, or multiplicative inverse, of a number.

▶ Divide with signed numbers.

▶ Simplify numerical expressions.

▶ Factor integers.

▶ The difference of two numbers is found by adding the additive inverse of the second number to the first. Division is related to multiplication in a similar way. The *quotient* of two numbers is found by *multiplying* by the *multiplicative inverse*. By definition, since

$$8 \cdot \frac{1}{8} = \frac{8}{8} = 1 \quad \text{and} \quad \frac{5}{4} \cdot \frac{4}{5} = \frac{20}{20} = 1,$$

the multiplicative inverse of 8 is $\frac{1}{8}$, and of $\frac{5}{4}$ is $\frac{4}{5}$.

> Pairs of numbers whose product is 1 are called **multiplicative inverses,** or **reciprocals,** of each other.

EXAMPLE 1 The following chart shows several numbers and the multiplicative inverse (if it exists) of each number.

Number	Multiplicative inverse (reciprocal)
4	$\frac{1}{4}$
-5	$\frac{1}{-5}$ or $-\frac{1}{5}$
$\frac{3}{4}$	$\frac{4}{3}$
$-\frac{5}{8}$	$-\frac{8}{5}$
0	None

Why is there no multiplicative inverse for the number 0? Suppose that k is to be the multiplicative inverse of 0. Then $k \cdot 0$ should equal 1. But $k \cdot 0 = 0$ for any number k. Since there is no value of k that is a solution of the equation $k \cdot 0 = 1$,

> 0 has no multiplicative inverse.

Work Problem 1 at the side.

▶ In a way similar to that used for subtraction, the *quotient* of a and b is defined to be the product of a and the multiplicative inverse of b.

> **Definition of Division**
> For any real numbers a and b, with $b \neq 0$,
> $$\frac{a}{b} = a \cdot \frac{1}{b}.$$

1. Complete the chart.

Number	Multiplicative inverse
(a) 6	
(b) -2	
(c) $\frac{2}{3}$	
(d) $-\frac{1}{4}$	
(e) 0	

ANSWERS

1. (a) $\frac{1}{6}$ (b) $-\frac{1}{2}$ (c) $\frac{3}{2}$ (d) -4
   (e) none

2.5 Division of Real Numbers

2. Find the quotients.

(a) $\dfrac{42}{7}$

(b) $\dfrac{-36}{6}$

(c) $\dfrac{-12}{-4}$

(d) $\dfrac{18}{-9}$

(e) $\dfrac{-3}{0}$

3. Find the quotients.

(a) $\dfrac{-8}{-2}$

(b) $\dfrac{-16}{2}$

(c) $\dfrac{1}{4} \div \left(-\dfrac{2}{3}\right)$

ANSWERS
2. (a) 6 (b) −6 (c) 3 (d) −2
   (e) meaningless
3. (a) 4 (b) 8 (c) $\dfrac{3}{8}$

The definition above indicates that b, the number to divide by, cannot be 0. The reason is that 0 has no multiplicative inverse, so $\tfrac{1}{0}$ is not a number. For this reason,

> division by 0 is meaningless

and is never permitted. If a division problem turns out to involve division by 0, write "meaningless."

Since division is defined in terms of multiplication, all the rules for multiplication of signed numbers also apply to division.

EXAMPLE 2 Write each quotient as a product and evaluate.

(a) $\dfrac{12}{3} = 12 \cdot \dfrac{1}{3} = 4$

(b) $\dfrac{-10}{2} = -10 \cdot \dfrac{1}{2} = -5$

(c) $\dfrac{8}{-4} = 8 \cdot \left(\dfrac{1}{-4}\right) = -2$

(d) $\dfrac{-14}{-7} = -14\left(\dfrac{1}{-7}\right) = 2$

(e) $\dfrac{-100}{-20} = -100\left(\dfrac{1}{-20}\right) = 5$

(f) $\dfrac{-10}{0}$ Meaningless ◀

Work Problem 2 at the side.

The following rule for division with signed numbers follows from the definition of division and the rules for multiplication with signed numbers.

> **Dividing Signed Numbers**
>
> The quotient of two numbers having the *same* sign is *positive*; the quotient of two numbers having *different* signs is *negative*.

EXAMPLE 3 Find the quotients.

(a) $\dfrac{8}{-2} = -4$

(b) $\dfrac{-45}{-9} = 5$

(c) $-\dfrac{1}{8} \div \left(-\dfrac{3}{4}\right) = -\dfrac{1}{8} \cdot \left(-\dfrac{4}{3}\right) = \dfrac{1}{6}$ ◀

Work Problem 3 at the side.

From the definitions of multiplication and division of real numbers,

$$\dfrac{-40}{8} = -40 \cdot \dfrac{1}{8} = -5,$$

and
$$\frac{40}{-8} = 40\left(\frac{1}{-8}\right)$$
$$= -5,$$
so that
$$\frac{-40}{8} = \frac{40}{-8}.$$

Based on this example, the quotient of a positive and a negative number can be expressed in any of the following three forms.

> For any positive real numbers a and b, with $b \neq 0$,
> $$\frac{-a}{b} = \frac{a}{-b} = -\frac{a}{b}.$$

The form $\frac{a}{-b}$ is seldom used.

The quotient of two negative numbers can be expressed as the quotient of two positive numbers.

> For any positive real numbers a and b, with $b \neq 0$,
> $$\frac{-a}{-b} = \frac{a}{b}.$$

▶ The next example shows how to simplify numerical expressions involving quotients.

EXAMPLE 4 Simplify each expression.

(a) $\dfrac{5(-2) - (3)(4)}{2(1 - 6)}$

Simplify the numerator and denominator separately. Then divide or write in lowest terms.

$$\frac{5(-2) - (3)(4)}{2(1 - 6)} = \frac{-10 - 12}{2(-5)} \quad \text{Multiply in numerator}$$
$$\text{Subtract in denominator}$$
$$= \frac{-22}{-10} \quad \text{Subtract in numerator}$$
$$\text{Multiply in denominator}$$
$$= \frac{11}{5} \quad \text{Lowest terms}$$

(b) $\dfrac{4^2 - 6^2}{5(-3 + 2)}$

$$\frac{4^2 - 6^2}{5(-3 + 2)} = \frac{16 - 36}{5(-1)} \quad \text{Square 4 and 6}$$
$$\text{Add } -3 \text{ and } 2$$
$$= \frac{-20}{-5} \quad \text{Subtract in numerator}$$
$$\text{Multiply in denominator}$$
$$= 4 \quad \text{Divide} \blacktriangleleft$$

2.5 Division of Real Numbers

4. Simplify.

(a) $\dfrac{5(-4)}{-2-8}$

(b) $\dfrac{6(-4)-2(5)}{3(2-7)}$

(c) $\dfrac{-6(-8)+(-3)9}{(-2)[4-(-3)]}$

(d) $\dfrac{5^2+3^2}{3(-4)-5}$

5. Find all factors of each number.

(a) 24

(b) 30

(c) 19

(d) 37

Work Problem 4 at the side.

The rules for operations with signed numbers are summarized here.

Operations with Signed Numbers
Addition
  Like signs Add the absolute values of the numbers. The result is given the same sign as the numbers.
  Unlike signs Subtract the smaller absolute value from the larger absolute value. Give the result the sign of the number having the larger absolute value.
Subtraction
  Add the additive inverse, or opposite, of the second number.
Multiplication and Division
  Like signs The product or quotient of two numbers with like signs is positive.
  Unlike signs The product or quotient of two numbers with unlike signs is negative.
  Division by 0 is meaningless.

In Section 1.1 the definition of a *factor* was given for whole numbers. (For example, since $9 \cdot 5 = 45$, both 9 and 5 are factors of 45.) The definition can now be extended to integers.

If the product of two integers is a third integer, then each of the two integers is a **factor** of the third. For example, $(-3)(-4) = 12$, so -3 and -4 are both factors of 12. The factors of 12 are the numbers -12, -6, -4, -3, -2, -1, 1, 2, 3, 4, 6, and 12.

EXAMPLE 5 The following chart shows several integers and the factors of those integers.

Integer	Factors
18	$-18, -9, -6, -3, -2, -1, 1, 2, 3, 6, 9, 18$
20	$-20, -10, -5, -4, -2, -1, 1, 2, 4, 5, 10, 20$
15	$-15, -5, -3, -1, 1, 3, 5, 15$
7	$-7, -1, 1, 7$
1	$-1, 1$

Work Problem 5 at the side.

ANSWERS

4. (a) 2 (b) $\dfrac{34}{15}$ (c) $-\dfrac{3}{2}$ (d) -2

5. (a) $-24, -12, -8, -6, -4, -3,$
   $-2, -1, 1, 2, 3, 4, 6, 8, 12, 24$
   (b) $-30, -15, -10, -6, -5, -3,$
   $-2, -1, 1, 2, 3, 5, 6, 10, 15, 30$
   (c) $-19, -1, 1, 19$
   (d) $-37, -1, 1, 37$

2.5 EXERCISES

Find the multiplicative inverse (if one exists) for each number. Round to the nearest thousandth in Exercises 11 and 12. See Example 1.

1. 9
2. 8
3. -4
4. -10

5. $\dfrac{2}{3}$
6. $\dfrac{3}{4}$
7. $\dfrac{-9}{10}$
8. $\dfrac{-4}{5}$

9. 0
10. $\dfrac{0}{5}$
11. .8697
12. 1.4385

Find the quotients. See Examples 2 and 3.

13. $\dfrac{-10}{5}$
14. $\dfrac{-12}{3}$
15. $\dfrac{-15}{5}$
16. $\dfrac{-20}{2}$

17. $\dfrac{18}{-3}$
18. $\dfrac{24}{-6}$
19. $\dfrac{100}{-20}$
20. $\dfrac{250}{-25}$

21. $\dfrac{-12}{-6}$
22. $\dfrac{-25}{-5}$
23. $\dfrac{-150}{-10}$
24. $\dfrac{-280}{-20}$

25. $\dfrac{-180}{-5}$
26. $\dfrac{-350}{-7}$
27. $\dfrac{0}{-2}$
28. $\dfrac{0}{12}$

29. $-\dfrac{1}{2} \div \left(-\dfrac{3}{4}\right)$
30. $-\dfrac{5}{8} \div \left(-\dfrac{3}{16}\right)$
31. $(-4.2) \div (-2)$
32. $(-9.8) \div (-7)$

33. $\dfrac{4}{-.8}$ 34. $\dfrac{-6}{.3}$ 35. $\dfrac{12}{2-5}$ 36. $\dfrac{15}{3-8}$

37. $\dfrac{50}{2-7}$ 38. $\dfrac{30}{5-8}$ 39. $\dfrac{-30}{2-8}$ 40. $\dfrac{-50}{6-11}$

41. $\dfrac{-40}{8-(-2)}$ 42. $\dfrac{-72}{6-(-2)}$ 43. $\dfrac{-120}{-3-(-5)}$ 44. $\dfrac{-200}{-6-(-4)}$

45. $\dfrac{-15-3}{3}$ 46. $\dfrac{16-(-2)}{-6}$ 47. $\dfrac{-30-(-8)}{-11}$

48. $\dfrac{-17-(-12)}{5}$ 49. $\dfrac{-6.42-(-3.891)}{-.05}$ 50. $\dfrac{11.096-(-8.1151)}{.8-.05}$

Simplify the numerators and denominators separately. Then find the quotients. Round to the nearest thousandth in Exercises 67 and 68. See Example 4.

51. $\dfrac{-8(-2)}{3-(-1)}$ 52. $\dfrac{-12(-3)}{-15-(-3)}$ 53. $\dfrac{-15(2)}{-7-3}$

54. $\dfrac{-20(6)}{-5-1}$ 55. $\dfrac{-2(6)+3}{2-(-1)}$ 56. $\dfrac{3(-8)+4}{-6+1}$

76 Operations with Real Numbers

name date hour

57. $\dfrac{-5(2) + 3(-2)}{-3 - (-1)}$

58. $\dfrac{4(-1) + 3(-2)}{-2 - 3}$

59. $\dfrac{-9(-2) - (-4)(-2)}{-2(3) - 2(2)}$

60. $\dfrac{5(-2) - 3(4)}{-2[3 - (-2)] - 1}$

61. $\dfrac{4(-2) - 5(-3)}{2[-1 + (-3)] - (-8)}$

62. $\dfrac{5(-3) - (-2)(-4)}{5[-4 + (-2)] + 3(10)}$

63. $\dfrac{4^2 - 5^2}{3(6 - 9 + 2)}$

64. $\dfrac{6^2 + 4^2}{5(2 + 13)}$

65. $\dfrac{3^2 + 5^2}{4^2 + 1^2}$

66. $\dfrac{10^2 - 5^2}{8^2 + 3^2 + 2}$

67. $\dfrac{(.86)^2 + (2.5)^2}{(-1.43)^3 - (-3.76)}$

68. $\dfrac{(-.49)^2 - (.21)^2}{3.58 - (-1.12)^3}$

Find the solution of each equation from the domain $\{-8, -6, -4, -2, 0, 2, 4, 6, 8\}$.

69. $\dfrac{x}{4} = -2$

70. $\dfrac{x}{2} = -1$

71. $\dfrac{n}{-2} = 3$

72. $\dfrac{t}{-2} = -2$

73. $\dfrac{q}{-3} = 0$

74. $\dfrac{p}{5} = 0$

75. $\dfrac{m}{-2} = -4$

76. $\dfrac{y}{-1} = 2$

2.5 Exercises

Find all integer factors of each number. See Example 5.

77. 36

78. 32

79. 25

80. 14

81. 40

82. 50

83. 17

84. 13

85. 29

Write the following in symbols, using x as the variable, and find the solution. All solutions come from the list of integers between −12 and 12, inclusive.

86. Six times a number is −42.

87. Four times a number is −32.

88. When a number is divided by 5, the result is 2.

89. When a number is divided by 4, the result is −2.

90. When a number is divided by 3, the result is −3.

91. When a number is divided by −3, the result is −4.

92. The quotient of a number and 2 is −6. (Write the quotient as $\frac{x}{2}$.)

93. The quotient of a number and −1 is 2.

94. The quotient of 6 and one more than a number is 3.

95. When the square of a number is divided by 3, the result is 12.

78   Operations with Real Numbers

2.6 PROPERTIES OF ADDITION AND MULTIPLICATION

The basic properties of addition and multiplication of real numbers are discussed in this section. In the following statements, a, b, and c represent real numbers.

▶ **Commutative properties** By the commutative properties, two numbers added, or multiplied, in any order give the same result.

$$a + b = b + a$$
$$ab = ba$$

EXAMPLE 1 Use a commutative property to complete each statement.

(a) $-8 + 5 = 5 +$ _____

By the commutative property for addition, the missing number is -8, since $-8 + 5 = 5 + (-8)$.

(b) $(-2)(7) =$ _____ (-2)

By the commutative property for multiplication, the missing number is 7, since $(-2)(7) = (7)(-2)$. ◀

Work Problem 1 at the side.

▶ **Associative properties** By the associative properties, when adding or multiplying three numbers, the first two may be grouped together or the last two may be grouped together without affecting the answer.

$$(a + b) + c = a + (b + c)$$
$$(ab)c = a(bc)$$

EXAMPLE 2 Use an associative property to complete each statement.

(a) $8 + (-1 + 4) = (8 +$ _____ $) + 4$

The missing number is -1.

(b) $[2 \cdot (-7)] \cdot 6 = 2 \cdot$ _____

The completed expression on the right should be $2 \cdot [(-7) \cdot 6]$. ◀

Work Problem 2 at the side.

By the associative property of addition, the sum of three numbers will be the same no matter which way the numbers are "associated" in groups. For this reason, parentheses can be left out in many addition problems. For example, both

$$(-1 + 2) + 3 \quad \text{and} \quad -1 + (2 + 3)$$

can be written as

$$-1 + 2 + 3.$$

In the same way, parentheses also can be left out of many multiplication problems.

Objectives

Identify the use of the following properties:

▶ commutative,
▶ associative,
▶ identity,
▶ inverse,
▶ distributive.

1. Complete each statement. Use a commutative property.

   (a) $x + 9 = 9 +$ _____

   (b) $(-12)(4) =$ _____ (-12)

   (c) $9(-11) = (-11)$ _____

   (d) $5x = x \cdot$ _____

2. Complete each statement. Use an associative property.

   (a) $(9 + 10) + (-3)$
   $= 9 + [$ _____ $+ (-3)]$

   (b) $-5 + (2 + 8)$
   $= ($ _____ $) + 8$

   (c) $10 \cdot [(-8) \cdot (-3)]$
   $=$ _____

ANSWERS
1. (a) x (b) 4 (c) 9 (d) 5
2. (a) 10 (b) $-5 + 2$
   (c) $[10 \cdot (-8)] \cdot (-3)$

3. Decide whether each statement is an example of the commutative property, the associative property, or both.

   (a) $2(4 \cdot 6) = (2 \cdot 4)6$

   (b) $(2 \cdot 4)6 = (4 \cdot 2)6$

   (c) $(2 + 4) + 6 = 4 + (2 + 6)$

4. Use the identity property to complete each statement.

   (a) $9 + 0 = $ ____

   (b) ____ $+ (-7) = -7$

   (c) $8 \cdot$ ____ $= 8$

   (d) ____ $\cdot 1 = 5$

ANSWERS
3. (a) associative (b) commutative (c) both
4. (a) 9 (b) 0 (c) 1 (d) 5

EXAMPLE 3 (a) Is $(2 + 4) + 5 = 2 + (4 + 5)$ an example of the associative property?

The order of the three numbers is the same on both sides of the equals sign. The only change is in the grouping, or association, of the numbers. Therefore, this is an example of the associative property.

(b) Is $6(3 \cdot 10) = 6(10 \cdot 3)$ an example of the associative property or the commutative property?

The same numbers, 3 and 10, are grouped on each side. On the left, however, the 3 appears first in $(3 \cdot 10)$. On the right, the 10 appears first. Since the only change involves the order of the numbers, this statement is an example of the commutative property.

(c) Is $(8 + 1) + 7 = 8 + (7 + 1)$ an example of the associative property or the commutative property?

In the statement, both the order and the grouping are changed. On the left the order of the three numbers is 8, 1, and 7. On the right it is 8, 7, and 1. On the left the 8 and 1 are grouped, and on the right the 7 and 1 are grouped. Therefore, both the associative and the commutative properties are used. ◂

Work Problem 3 at the side.

Identity properties The identity properties say that the sum of 0 and any number equals that number, and the product of 1 and any number equals that number.

$$a + 0 = a \quad \text{and} \quad 0 + a = a$$
$$a \cdot 1 = a \quad \text{and} \quad 1 \cdot a = a$$

The number 0 leaves the identity, or value, of any real number unchanged by addition. For this reason, 0 is called the **identity element for addition.** Since multiplication by 1 leaves any real number unchanged, 1 is the **identity element for multiplication.**

EXAMPLE 4 These statements are examples of the identity properties.

(a) $-3 + 0 = -3$

(b) $0 + \frac{1}{2} = \frac{1}{2}$

(c) $-\frac{3}{4} \cdot 1 = -\frac{3}{4}$

(e) $1 \cdot 25 = 25$ ◂

Work Problem 4 at the side.

Inverse properties By the inverse properties, the sum of the numbers a and $-a$ is 0, and the product of the nonzero numbers a and $\frac{1}{a}$ is 1.

$$a + (-a) = 0 \quad \text{and} \quad -a + a = 0$$
$$a \cdot \frac{1}{a} = 1 \quad \text{and} \quad \frac{1}{a} \cdot a = 1 \quad (a \neq 0)$$

Recall that $-a$ is the **additive inverse** of a and $\frac{1}{a}$ is the **multiplicative inverse** of the nonzero number a.

Operations with Real Numbers

EXAMPLE 5 These statements are examples of the inverse properties.

(a) $\dfrac{2}{3} \cdot \dfrac{3}{2} = 1$

(b) $(-5)\left(-\dfrac{1}{5}\right) = 1$

(c) $-\dfrac{1}{2} + \dfrac{1}{2} = 0$

(d) $4 + (-4) = 0$ ◀

Work Problem 5 at the side.

▶ 5 Look at the following statements.
$$2(5 + 8) = 2(13) = 26$$
$$2(5) + 2(8) = 10 + 16 = 26$$

Since both expressions equal 26,
$$2(5 + 8) = 2(5) + 2(8).$$

This result is an example of the *distributive property,* the only property involving *both* addition and multiplication. With this property, a product can be changed to a sum or difference.

Distributive property By the distributive property, multiplying a number *a* by a sum of numbers $b + c$ gives the same result as multiplying *a* by *b* and *a* by *c* and then adding the two products.

$$\overset{\frown}{a(b + c)} = ab + ac \qquad \text{and} \qquad (b + c)a = ba + ca$$

As the arrows show, the *a* outside the parentheses is "distributed" over the *b* and *c* inside. Another form of the distributive property is valid for subtraction.

$$a(b - c) = ab - ac \qquad \text{and} \qquad (b - c)a = ba - ca$$

The distributive property also can be extended to more than two numbers.

$$a(b + c + d) = ab + ac + ad$$

EXAMPLE 6 Use the distributive property to rewrite each expression.

(a) $5(9 + 6) = 5 \cdot 9 + 5 \cdot 6$ Multiply both terms by 5
$ = 45 + 30$
$ = 75$

(b) $4(x + 5 + y) = 4x + 4 \cdot 5 + 4y$
$ = 4x + 20 + 4y$

(c) $-2(x + 3) = -2x + (-2)(3)$
$ = -2x - 6$

(d) $3(k - 9) = 3k - 3 \cdot 9$
$ = 3k - 27$

(e) $6 \cdot 8 + 6 \cdot 2 = 6(8 + 2)$ Distributive property
$ = 6(10) = 60$

5. Complete the statements so that they are examples of either an identity property or an inverse property. Tell which property.

(a) $-6 + \underline{} = 0$

(b) $\dfrac{4}{3} \cdot \underline{} = 1$

(c) $\dfrac{-1}{9} \cdot \underline{} = 1$

(d) $275 + \underline{} = 275$

ANSWERS

5. (a) 6; inverse (b) $\dfrac{3}{4}$; inverse
   (c) -9; inverse (d) 0; identity

6. Use the distributive property to rewrite each expression.

(a) $2(p + 5)$

(b) $9(x + 2)$

(c) $-4(y + 7)$

(d) $5(m - 4)$

(e) $9 \cdot k + 9 \cdot 5$

(f) $3a - 3b$

(g) $7(2y + 7k - 9m)$

7. Write without parentheses.

(a) $-(3k - 5)$

(b) $-(2 - r)$

(c) $-(-5y + 8)$

(d) $-(-z + 4)$

ANSWERS
6. (a) $2p + 10$ (b) $9x + 18$
   (c) $-4y - 28$ (d) $5m - 20$
   (e) $9(k + 5)$ (f) $3(a - b)$
   (g) $14y + 49k - 63m$
7. (a) $-3k + 5$ (b) $-2 + r$
   (c) $5y - 8$ (d) $z - 4$

(f) $4x - 4m = 4(x - m)$

(g) $8(3r + 11t + 5z) = 8(3r) + 8(11t) + 8(5z)$
$= (8 \cdot 3)r + (8 \cdot 11)t + (8 \cdot 5)z$ Associative property
$= 24r + 88t + 40z$

Work Problem 6 at the side.

The distributive property is used to remove the parentheses from expressions such as $-(2y + 3)$. Do this by first writing $-(2y + 3)$ as $-1 \cdot (2y + 3)$.

$-(2y + 3) = -1 \cdot (2y + 3)$
$= -1 \cdot (2y) + (-1) \cdot (3)$ Distributive property
$= -2y - 3$ Multiply

EXAMPLE 7 Write without parentheses.

(a) $-(7r - 8) = -1(7r) + (-1)(-8)$ Distributive property
$= -7r + 8$

(b) $-(-9w + 2) = 9w - 2$

Work Problem 7 at the side.

The properties of addition and multiplication of the real numbers are summarized below.

Properties of Addition and Multiplication

For any real numbers a, b, and c, the following properties hold.

Commutative properties $a + b = b + a$
$ab = ba$

Associative properties $(a + b) + c = a + (b + c)$
$(ab)c = a(bc)$

Identity properties There is a real number 0 such that
$a + 0 = a$ and $0 + a = a$.
There is a real number 1 such that
$a \cdot 1 = a$ and $1 \cdot a = a$.

Inverse properties For each real number a, there is a single real number $-a$ such that
$a + (-a) = 0$ and $(-a) + a = 0$.
For each nonzero real number a, there is a single real number $\frac{1}{a}$ such that
$a \cdot \frac{1}{a} = 1$ and $\frac{1}{a} \cdot a = 1$.

Distributive property $a(b + c) = ab + ac$

Operations with Real Numbers

2.6 EXERCISES

Label each statement as an example of the commutative, associative, identity, inverse, or distributive property. See Examples 1–6.

1. $6 + 15 = 15 + 6$

2. $9 + (11 + 4) = (9 + 11) + 4$

3. $5(15 \cdot 8) = (5 \cdot 15)8$

4. $(23)(9) = (9)(23)$

5. $12(-8 \cdot 3) = (-8)(12 \cdot 3)$

6. $(-9)[6(-2)] = [-9(6)](-2)$

7. $2 + (p + r) = (p + r) + 2$

8. $(m + n) + 4 = 4 + (m + n)$

9. $-\dfrac{6}{5} + \dfrac{5}{12} = \dfrac{5}{12} + \left(-\dfrac{6}{5}\right)$

10. $\left(-\dfrac{9}{5}\right)\left(-\dfrac{3}{11}\right) = \left(-\dfrac{3}{11}\right)\left(-\dfrac{9}{5}\right)$

11. $6 + (-6) = 0$

12. $-8 + 8 = 0$

13. $-4 + 0 = -4$

14. $0 + (-9) = -9$

15. $3\left(\dfrac{1}{3}\right) = 1$

16. $-7\left(-\dfrac{1}{7}\right) = 1$

17. $\dfrac{2}{3} \cdot 1 = \dfrac{2}{3}$

18. $-\dfrac{9}{4} \cdot 1 = -\dfrac{9}{4}$

19. $6(5 - 2x) = 6 \cdot 5 - 6(2x)$

20. $5(2m) + 5(7n) = 5(2m + 7n)$

Use the indicated property to write a new expression that is equal to the given expression. Simplify the new expression if possible. See Examples 1, 2, 4, and 5.

21. $9 + k$; commutative

22. $z + 5$; commutative

23. $m + 0$; identity

24. $(-9) + 0$; identity

25. $3(r + m)$; distributive

26. $11(k + z)$; distributive

27. $8 \cdot \dfrac{1}{8}$; inverse

28. $\dfrac{1}{6} \cdot 6$; inverse

29. $12 + (-12)$; inverse

30. $-8 + 8$; inverse

31. $5 + (-5)$; commutative

32. $-9 + 9$; commutative

33. $-3(r + 2)$; distributive

34. $4(k - 5)$; distributive

35. $9 \cdot 1$; identity

36. $1(-4)$; identity

name date hour

37. $(k + 5) + (-6)$; associative

38. $(m + 4) + (-2)$; associative

39. $(4z + 2r) + 3k$; associative

40. $(6m + 2n) + 5r$; associative

Use the distributive property to rewrite each expression. Simplify if possible. See Examples 6 and 7.

41. $5(m + 2)$

42. $6(k + 5)$

43. $-4(r + 2)$

44. $-3(m + 5)$

45. $-8(k - 2)$

46. $-4(z - 5)$

47. $-\dfrac{2}{3}(a + 9)$

48. $-\dfrac{3}{7}(p + 14)$

49. $\left(r + \dfrac{8}{3}\right)\dfrac{3}{4}$

50. $\left(m + \dfrac{12}{7}\right)\dfrac{3}{4}$

51. $(8 - k)(-2)$

52. $(9 - r)(-3)$

53. $2(5r + 6m)$

54. $5(2a + 4b)$

55. $-4(3x - 4y)$

56. $-9(5k - 12m)$

57. $5 \cdot 8 + 5 \cdot 9$

58. $(4.6)(3.54) + (4.6)(8.46)$

2.6 Exercises **85**

59. $(7.12)(2.3) + (7.12)(7.7)$ **60.** $6x + 6m$ **61.** $9p + 9q$

62. $8(2x) + 8(3y)$ **63.** $5(7z) + 5(8w)$ **64.** $11(2r) + 11(3s)$

Use the distributive property to write each of the following without parentheses. See Example 7.

65. $-(3k + 5)$ **66.** $-(2z + 12)$

67. $-(4y - 8)$ **68.** $-(3r - 15)$

69. $-(-4 + p)$ **70.** $-(-12 + 3a)$

71. $-(-1 - 15r)$ **72.** $-(-14 - 6y)$

Tell whether or not the events in Exercises 73–76 are commutative.

73. Getting out of bed and taking a shower.   **74.** Putting on your right shoe or your left shoe first.

75. Taking English or taking history.   **76.** Putting on your shoe or putting on your sock.

77. Evaluate $25 - (6 - 2)$ and evaluate $(25 - 6) - 2$. Do you think subtraction is associative?

78. Evaluate $180 \div (15 \div 3)$ and evaluate $(180 \div 15) \div 3$. Do you think division is associative?

2.7 SIMPLIFYING EXPRESSIONS

▶ The properties of addition and multiplication introduced in the previous section are used to simplify algebraic expressions.

EXAMPLE 1 Simplify the following expressions.

(a) $4x + 8 + 9$

Since $8 + 9 = 17$,
$$4x + 8 + 9 = 4x + 17.$$

(b) $4(3m - 2n)$

Use the distributive property.
$$4(3m - 2n) = 4(3m) - 4(2n)$$
$$= 12m - 8n$$

(c) $6 + 3(4k + 5) = 6 + 3(4k) + 3(5)$ — Distributive property
$= 6 + 12k + 15$ — Multiply
$= 21 + 12k$ — Add

(d) $5 - (2y - 8) = 5 - 1 \cdot (2y - 8)$ — Write $(2y - 8)$ as $1(2y - 8)$
$= 5 - 2y + 8$ — Distributive property
$= 13 - 2y$ — Add ◀

Work Problem 1 at the side.

▶ A **term** is a single number, or a product of a number and one or more variables raised to powers. Examples of terms include
$$-9x^2, \quad 15y, \quad -3, \quad 8m^2n, \quad \text{and} \quad k.$$

The **numerical coefficient** of the term $9m$ is 9, the numerical coefficient of $-15x^3y^2$ is -15, the numerical coefficient of x is 1, and the numerical coefficient of 8 is 8.

EXAMPLE 2 Give the numerical coefficient of the following terms.

Term	Numerical coefficient
$-7y$	-7
$34r^3$	34
$-26x^5yz^4$	-26
$-k$	-1
r	1 ◀

Work Problem 2 at the side.

▶ Terms with exactly the same variables (including the same exponents) are called **like terms.** For example, $9m$ and $4m$ have the same variables and are like terms. Also, $6x^3$ and $-5x^3$ are like terms. The terms $-4y^3$ and $4y^2$ have different exponents and are **unlike terms.**

The sum or difference of like terms may be expressed as one term by using the distributive property. For example,
$$3x + 5x = (3 + 5)x = 8x.$$

Objectives

▶ Simplify expressions.
▶ Identify terms and numerical coefficients.
▶ Identify like terms.
▶ Combine like terms.
▶ Simplify expressions from word problems.

1. Simplify each expression.

   (a) $9k + 12 - 5$

   (b) $7(3p + 2q)$

   (c) $2 + 5(3z - 1)$

   (d) $-3 - (2 + 5y)$

   (e) $-(7 - 6k) + 9$

2. Give the numerical coefficient of each term.

   (a) $15q$

   (b) $-2m^3$

   (c) $-18m^7q^4$

   (d) $-r$

ANSWERS
1. (a) $9k + 7$ (b) $21p + 14q$
   (c) $15z - 3$ (d) $-5 - 5y$
   (e) $2 + 6k$
2. (a) 15 (b) -2 (c) -18 (d) -1

3. Combine terms.

   (a) $4k + 7k$

   (b) $4r - r$

   (c) $5z + 9z - 4z$

   (d) $8p + 8p^2$

4. Simplify.

   (a) $10p + 3(5 + 2p)$

   (b) $7z - 2 - 4(1 + z)$

   (c) $-(3 + 5k) + 7k$

5. Write the following statement as a mathematical expression and simplify:

   Three times a number is subtracted from the sum of the number and 8.

ANSWERS
3. (a) $11k$ (b) $3r$ (c) $10z$
   (d) cannot be simplified
4. (a) $16p + 15$ (b) $3z - 6$
   (c) $2k - 3$
5. $8 - 2x$

▶ This process is called **combining terms.** Remember that

only *like terms* may be combined.

EXAMPLE 3 Combine terms in the following expressions.

(a) $6r + 3r + 2r$

Use the distributive property to combine like terms.
$$6r + 3r + 2r = (6 + 3 + 2)r = 11r$$

(b) $4x + x = 4x + 1x = 5x$ Note: $x = 1x$

(c) $16y - 9y = (16 - 9)y = 7y$

(d) $32y + 10y^2$ cannot be simplified because $32y$ and $10y^2$ are unlike terms. ◀

Work Problem 3 at the side.

EXAMPLE 4 Simplify the following expressions.

(a) $14y + 2(6 + 3y) = 14y + 2(6) + 2(3y)$ Distributive property
$ = 14y + 12 + 6y$ Multiply
$ = 20y + 12$ Combine like terms

(b) $9k - 6 - 3(2 - 5k) = 9k - 6 - 3(2) - 3(-5k)$ Distributive property
$ = 9k - 6 - 6 + 15k$ Multiply
$ = 24k - 12$ Combine like terms

(c) $-(2 - r) + 10r = -1(2 - r) + 10r$
$ = -1(2) - 1(-r) + 10r$
$ = -2 + r + 10r$
$ = -2 + 11r$

(d) $5(2a - 6) - 3(4a - 9) = 10a - 30 - 12a + 27$
$ = -2a - 3$ ◀

Work Problem 4 at the side.

▶ The next example shows how to simplify the result of converting a word phrase to a mathematical expression.

EXAMPLE 5 Five times a number, four times a number, and six times a number are added to 9.

$5x + 4x + 6x + 9$ Write with symbols
$ = 15x + 9$ Combine terms ◀

Work Problem 5 at the side.

88 Operations with Real Numbers

name   date   hour

2.7 EXERCISES

Give the numerical coefficient of each of the following terms. See Example 2.

1. $15y$
2. $7z$
3. $-22m^4$

4. $-2k^7$
5. $35a^4b^2$
6. $12m^5n^4$

7. -9
8. 21
9. y^2

10. x^4
11. $-r$
12. $-z$

Write like *or* unlike *for the following groups of terms.*

13. $6m, \; -14m$
14. $-2a, \; 5a$
15. $7z^3, \; 7z^2$

16. $10m^5, \; 10m^6$
17. $25y, \; -14y, \; 8y$
18. $-11x, \; 5x, \; 7x$

19. $2, \; 5, \; -2$
20. $-8, \; 3, \; 9$
21. $p, \; -5p, \; 12p$

Simplify the following expressions by combining terms. See Examples 1 and 3.

22. $9y + 8y$
23. $15m + 12m$

24. $-4a - 2a$
25. $2k + 9 + 5k + 6$

26. $2 + 17z + 1 + 2z$
27. $m + 1 - m + 2 + m - 4$

2.7 Exercises   89

28. $12 - 13x - 27 + 2x - x$

29. $-2x + 3 + 4x - 17 + 20$

30. $r - 6 - 12r - 4 + 6r$

31. $16 - 5m - 4m - 2 + 2m$

32. $6 - 3z - 2z - 5 + z - 3z$

33. $-\dfrac{10}{3} + x + \dfrac{1}{4}x - 6 - \dfrac{5}{2}x$

34. $-p + \dfrac{1}{5}p - \dfrac{3}{5}p - 4 - \dfrac{1}{3}p$

35. $1.9 + 7.2 + 11x - 1.9 + .56x$

36. $-4.2r + 2.8 - r + .3 + 1.9r$

37. $6y^2 + 11y^2 - 8y^2$

38. $-9m^3 + 3m^3 - 7m^3$

39. $2p^2 + 3p^2 - 8p^3 - 6p^3$

name date hour

40. $5y^3 + 6y^3 - 3y^2 - 4y^2$

41. $-7.913q^2 + 2.804q - 11.723 + 5.069q^2 - 8.124q - 6.977$

42. $8.271m^3 - 3.722m^2 + 5.006 - 3.994m^3 + 4.129m^2 - 1.728$

Use the distributive property and combine terms to simplify the following expressions. See Example 4.

43. $6(5t + 11)$ **44.** $2(3x + 4)$

45. $-3(n + 5)$ **46.** $-4(y - 8)$

47. $5(-2 + t) + 4t$ **48.** $6t - (3t + 2)$

49. $2a + 3(a - 2) - 1$ **50.** $4 + 2(b - 5) - 3b$

51. $-3(2r - 3) + 2(5r + 3)$ **52.** $-4(5y - 7) + 3(2y - 5)$

2.7 Exercises 91

53. $8(2k - 1) - (4k - 3)$

54. $6(3p - 2) - (5p + 1)$

55. $6 - 4(x + 2) - (4x + 3)$

56. $2z + (-3z - 1) - 4(1 - z)$

57. $-2(-3k + 2) - (5k - 6) - 3k - 5$

58. $-2(3r - 4) - (6 - r) + 2r - 5$

59. $-7.916(3y - 2.8) - 4.72(9.1 - 5y)$

60. $4.873(8.2q - 7.3) + 1.29(3.9 - .42q)$

Convert the following statements into mathematical expressions. Use x as the variable. Combine terms when possible. See Example 5.

61. Two times a number is subtracted from the sum of the number and 2.

62. Four times a number is added to the sum of the number and -15.

63. Three times a number is subtracted from twice the number. This result is subtracted from 9 times the number.

64. A number is subtracted from 4 times the number, with this result subtracted from the sum of 6 and five times the number.

65. Nine is multiplied by the sum of five times a number and 4, with the result subtracted from the difference of 4 and twice the number.

66. Seven times a number is added to -9. This result is subtracted from four times the sum of three times the number and 5.

92   Operations with Real Numbers

CHAPTER 2   REVIEW EXERCISES

(2.1) *Circle the smaller number in each pair.*

1. $-9,\ 4$
2. $3,\ -5$
3. $-8,\ -7$
4. $-\dfrac{3}{4},\ -\dfrac{7}{8}$

5. $\dfrac{5}{3},\ \left|-\dfrac{3}{2}\right|$
6. $9,\ |-7|$
7. $-|-2|,\ -|-9|$
8. $-|-7|,\ -|-4|$

Write true or false for each statement.

9. $-9 < 9$
10. $7 < -7$
11. $0 \le -2$
12. $-5 \ge -5$

13. $3 \le -(-5)$
14. $9 \ge -(-10)$
15. $-3.25 > -2.25$
16. $-5.493 < -4.875$

17. $-|7| \le -|-2|$
18. $-|4| > -|-3|$
19. $2|-(-2)| > -|-2| \cdot 2$
20. $|-5| \cdot \dot{0} < |5 - 5|$

Graph each group of numbers on the indicated number line.

21. $5,\ -4,\ 3,\ -2,\ 0$
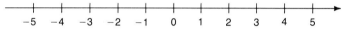

22. $-1,\ -3,\ |-4|,\ |-1|$
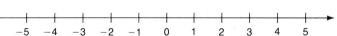

23. $3\dfrac{1}{4},\ -2\dfrac{4}{5},\ -1\dfrac{1}{8},\ \dfrac{2}{3}$
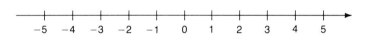

24. $|-2|,\ -|-5|,\ -|3|,\ -|0|$

(2.2) *Find the sums.*

25. $-9 + 3$
26. $12 + (-15)$
27. $-7 + (-8)$

28. $\dfrac{7}{8} + \left(-\dfrac{3}{10}\right)$
29. $\dfrac{7}{12} + \left(-\dfrac{2}{9}\right)$
30. $-11.3 + (-2.9)$

31. $1.64 + (-2.97)$
32. $-7 + (-2 + 8)$
33. $(-9 + 6) + (-10)$

34. $[-3 + (-5)] + (-8)$

35. $[(-2) + (-11)] + [7 + (-12)]$

36. $[(-4) + 6 + (-9)] + [-3 + (-5)]$

37. $[(-6) + (-7) + 8] + [8 + (-15)]$

38. Tom has $9. He spends $11. Find his new balance.

39. The temperature is 15°. It goes down 22°. Find the new temperature.

40. One year a company spent $28,000 on advertising. The next year it changed the amount spent on advertising by −$7000. How much was spent on advertising the second year?

41. On a cold day, the temperature was −17°. It then increased by 19°. Find the new temperature.

(2.3) *Find the differences.*

42. $-6 - (-4)$

43. $-2 - (-11)$

44. $6 - (-10)$

45. $15 - (-3)$

46. $-8 - 9$

47. $-12 - 27$

48. $\dfrac{3}{4} - \left(-\dfrac{2}{3}\right)$

49. $-\dfrac{1}{5} - \left(-\dfrac{7}{10}\right)$

50. $-12.8 - (-15.4)$

51. $-46.9 - (-21.8)$

52. $(-9 + 6) - (-3)$

53. $(-15 - 7) - (-9)$

54. $(-8 - 2) - 4\left[\dfrac{(-1 - 4) - (-2)}{-1 + (-2)}\right]$

55. $-3 + 8\left[\dfrac{(-12 + 15) - (-4 - 5)}{-3 - (2 - 1)}\right]$

56. What number is 6 less than -15?

57. How much greater is 7 than -15?

(2.4) *Simplify.*

58. $(-11)(-3)$

59. $17(-5)$

60. $(-9)(12)$

61. $-\dfrac{4}{5}\left(-\dfrac{10}{7}\right)$

62. $-\dfrac{3}{8}\left(-\dfrac{16}{15}\right)$

63. $(-11.3)(2.5)$

64. $(-9.4)(-2.8)$

65. $4(3 - 7)$

66. $(6 - 4)(9 - 11)$

67. $(5 - 1)(3 - 8)$

68. $8(-9) - (6)(-2)$

69. $3(-7) - (-9)$

70. $-11(-4) - (3)(-7)$

71. $|6(-8)| - |-3|$

72. $-|(-9)(4)| - |-2|$

Evaluate the following expressions, given $x = -5$, $y = 4$, and $z = -3$.

73. $5x - 4z$

74. $2y + 7x$

75. $5z + 11y - x$

76. $y^2 - 2z^2$

77. $(2x - 8y)(z^2)$

78. $3z^2 - 4x^2$

(2.5) *Find the quotients.*

79. $\dfrac{-25}{-5}$

80. $\dfrac{280}{-7}$

81. $-\dfrac{2}{3} \div \dfrac{1}{2}$

82. $44.8 \div (-4)$

83. $\dfrac{36}{9 + (-3)}$

84. $\dfrac{-50}{-4 - 1}$

85. $\dfrac{8 - 4(-2)}{-5(3) - 1}$

86. $\dfrac{5(-3) - 8(3)}{(-5)(-4) + (-7)}$

87. $\dfrac{7^2 - 3^2}{2^2 + 4^2}$

88. $\dfrac{11^2 - 5^2 + 2}{5^2 + 5^2 - 1^2}$

89. $(18 - 2) - 4\left[\dfrac{(-1 - 4) - (-2)}{-1 + (-2)}\right]$

90. $-3 + 8\left[\dfrac{(-12 + 15) - (-4 - 5)}{-3 - (2 - 1)}\right]$

(2.6) *Label each statement as an example of the commutative, associative, identity, inverse, or distributive property.*

91. $8.974 \cdot 1 = 8.974$

92. $-\dfrac{2}{3} + \dfrac{2}{3} = 0$

93. $7 + 4m = 4m + 7$

94. $8(4 \cdot 3) = (8 \cdot 4) \cdot 3$

95. $\dfrac{5}{8} \cdot \dfrac{8}{5} = 1$

96. $9p + 0 = 9p$

Match the property in Column I with all examples of it from Column II.

Column I

97. Commutative ____

98. Associative ____

99. Identity ____

100. Inverse ____

101. Distributive ____

Column II

A. $-2 + 2 = 0$

B. $3 + (7 + x) = (3 + 7) + x$

C. $8 + 0 = 8$

D. $17 \cdot 1 = 17$

E. $3(x + y) = 3x + 3y$

F. $8 + m = m + 8$

G. $-5\left(\dfrac{1}{-5}\right) = 1$

H. $mn = nm$

(2.7) *Combine terms whenever possible.*

102. $2m + 9m$

103. $15p^2 - 7p^2 + 8p^2$

104. $5p^2 - 4p + 6p + 11p^2$

105. $-2(3k - 5) + 2(k + 1)$

106. $7(2m + 3) - 2(8m - 4)$

107. $-(2k + 8) - (3k - 7)$

Operations with Real Numbers

CHAPTER 2   TEST

Graph each set of numbers on the indicated number line.

1. $-4,\ 4,\ -3,\ 0,\ 2\frac{1}{2},\ -1\frac{7}{8}$

1. [number line from -5 to 5]

2. $|-2|,\ -|3|,\ -2\frac{3}{8},\ -|-1|$

2. [number line from -5 to 5]

Select the smaller number from each pair.

3. $-.742,\ \ -.705$

3. _____

4. $6,\ \ -|-8|$

4. _____

Write each word phrase as an algebraic expression with x as the variable.

5. Twice a number subtracted from 11

5. _____

6. The quotient of 9 and the difference of a number and 8

6. _____

Perform the indicated operations whenever possible.

7. $-9 - (4 - 11) + (-5)$

7. _____

8. $-2\frac{1}{5} + 5\frac{1}{4}$

8. _____

9. $-6 - [-5 + (8 - 9)]$

9. _____

10. $3^2 + (-7) - (2^3 - 5)$

10. _____

11. $|-6| \cdot (-5) + 2 \cdot |8|$

11. _____

12. $\dfrac{-7 - (-5 + 1)}{-4 - (-3)}$

12. _____

13. $\dfrac{-6[5 - (-1 + 4)]}{-9[2 - (-1)] - 6(-4)}$

13. _____

Chapter 2   Test

14. $\dfrac{15(-4-2)}{16(-2)+(-7-1)(-3-1)}$

Find the solution for each equation. Choose solutions from the domain $\{-9, -5, -4, -2, -1, 3, 9\}$.

15. $\dfrac{t}{-3} = 3$

16. $2x + 1 = -7$

17. $-4x - 2 = 6$

Evaluate the following expressions, given $m = -2$ *and* $p = 6$.

18. $4m - 3p^2$

19. $\dfrac{6m + 5p}{p - 3}$

Match the property in Column I with all examples of it from Column II.

Column I

20. Commutative
21. Associative
22. Identity
23. Inverse
24. Distributive

Column II

A. $12 + (-12) = -12 + 12$
B. $\dfrac{6}{5}\left(\dfrac{5}{6}\right) = 1$
C. $1(-9) = -9$
D. $2(5) + 2(9) = 2(5 + 9)$
E. $8x = x \cdot 8$
F. $-(2 - p) = -2 + p$
G. $-7 + 7 = 0$
H. $9 + (2 + y) = (9 + 2) + y$
I. $-12 + 0 = -12$

Simplify by combining like terms.

25. $2x + 5 + 5x - 3x - 3$

26. $4(2m + 1) - (m + 5)$

Operations with Real Numbers

CUMULATIVE REVIEW EXERCISES

Write these fractions in lowest terms.

1. $\dfrac{15}{40}$
2. $\dfrac{27}{45}$
3. $\dfrac{108}{144}$

Work the following problems.

4. $\dfrac{3}{4} + \dfrac{7}{8}$
5. $\dfrac{5}{6} + \dfrac{1}{4} + \dfrac{7}{15}$
6. $16\dfrac{7}{8} - 3\dfrac{1}{10}$

7. $\dfrac{9}{8} \cdot \dfrac{16}{3}$
8. $\dfrac{3}{4} \div \dfrac{5}{8}$
9. $\dfrac{4}{15} \cdot \dfrac{5}{9} \div \dfrac{10}{27}$

10. One dog weighs $8\dfrac{1}{3}$ pounds, and another dog weighs $12\dfrac{5}{8}$ pounds. Find the total weight of both dogs.

11. In making dresses, EarthWorks uses $\dfrac{5}{8}$ yard of trim per dress. How many yards of trim would be used to make 56 dresses?

12. A cook wants to increase a recipe that serves 6 to make enough for 20 people. The recipe calls for $1\dfrac{1}{4}$ cups of cheese. How much cheese will be needed to serve 20?

13. A dog handler wants to build three pens of equal length in a space that is $25\dfrac{7}{8}$ meters* long. How long should each pen be?

Work the following problems.

14. Convert .375 to a fraction.
15. Convert $\dfrac{13}{16}$ to a decimal.
16. $4.8 + 12.5 + 16.73$

*A meter is a unit of length in the metric system. No knowledge of the metric system is required for this exercise.

Cumulative Review Exercises

17. 56.3 − 28.99

18. 67.8(.45)

19. 42.56 ÷ 3.2

20. 236.46 ÷ 4.2

21. (39.8)(.41)

22. Convert .45 to a percent.

23. Convert 39.2% to a decimal.

24. Find 12% of 180.

25. Find 11.6% of 1500.

26. What percent is 24 of 64?

27. 65 is what percent of 40?

28. What percent of 72 is 18?

29. A purchasing agent bought 3 desks at $211.40 each and 3 chairs for $195, $189.95, and $168.50. What was the final bill (without tax)?

30. The purchasing agent in Exercise 29 paid a sales tax of $6\frac{1}{4}\%$ on his purchase. What was the final bill with tax?

31. This year only 306 of 450 graduates went to graduation. What percent attended graduation?

32. A car has a price of $5000. For trading in her old car, Rosalie will get 25% off. Find the price of the car with the trade-in.

Tell whether each of the following is true or false.

33. $5[3 + 8(2 + 1)] \leq 130$

34. $7[5(2 + 5) + 3] \leq 260$

35. $\dfrac{8(7) - 5(6 + 2)}{3 \cdot 5 + 1} \geq 1$

36. $\dfrac{4(9+3)-8(4)}{2+3-3} \geq 2$

37. $\dfrac{3^3 + 2(5+4^2)}{5^2 - 8} < 3$

38. $\dfrac{(9-2)^2 - 5 \cdot 3^2}{6^2 - 4 \cdot 8} > 2$

Perform the indicated operations.

39. $-11 + 20 + (-2)$

40. $13 + (-19) + 7$

41. $9 - (-4)$

42. $2 - (-13)$

43. $-2(-5)(-4)$

44. $(-8)(-7)(-9)$

45. $\dfrac{4 \cdot 9}{-3}$

46. $\dfrac{8}{7-7}$

47. $(-5 + 8) + (-2 - 7)$

48. $(-2 + 7) + (-3 - 9)$

49. $(-7-1)(-4) + (-4)$

50. $-2 - (-3)(7) + (-5)$

51. $\dfrac{-3 - (-5)}{1 - (-1)}$

52. $\dfrac{-2 - (-8)}{-5 - (-4)}$

53. $\dfrac{6(9) + 3}{6(-4) + 8(2) - 4(-2)}$

54. $\dfrac{6(11) + 8}{-9(3) + (-8)(-3) + 3}$

55. $\dfrac{-2(5^3) - 6}{4^2 + 2(-5) + (-2)}$

56. $\dfrac{(-3)^2 - (-4)(2^4)}{5 \cdot 2 - (-2)^3}$

Cumulative Review Exercises

Circle the smaller number in each pair.

57. −7, −11

58. −15, −9

59. −8.23, −|−7|

60. 3, −|−7|

61. |7.23|, −3

62. |−11|, −10

Find the value of each expression when $x = -2, y = -4,$ and $z = 3$.

63. xyz

64. $(x + 4y)3z$

65. $2x^2 - y^2$

66. $xz^3 - 5y^2$

67. $\dfrac{3x - y^3}{-4z}$

68. $\dfrac{x^3 - 3y}{3(x - y)}$

Find the solution for each equation from the domain $\{-3, -2, -1, 0, 1, 2, 3\}$.

69. $-5 + x = -2$

70. $2x - 3 = -7$

71. $-4x - 7 = -3$

Write an equation using x as the variable. Then find the solution of the equation from the domain $\{-3, -2, -1, 0, 1, 2, 3\}$.

72. Half of a number increased by 3 is 2.

73. When the sum of a number and 6 is divided by the square of the number, the result is 1.

74. The additive inverse of a number plus 7 is 4.

75. The square of a number multiplied by 4 equals 0.

Name the property illustrated by each of the following examples.

76. $5 + 11 = 11 + 5$

77. $18 \cdot 1 = 18$

78. $7(k + m) = 7k + 7m$

79. $3 + (5 + 2) = 3 + (2 + 5)$

80. $7 + (-7) = 0$

81. $\dfrac{6}{5} \cdot \dfrac{5}{6} = 1$

Solving Equations and Inequalities

3

3.1 THE ADDITION PROPERTY OF EQUALITY

▶ Methods of solving *linear equations* will be introduced in this section.

> **A linear equation** can be written in the form
> $$ax + b = c,$$
> for real numbers a, b, and c, with $a \neq 0$.

Linear equations are solved by using a series of steps to produce a simpler equation of the form

$$x = \text{a number}.$$

▶ According to the equation

$$x - 5 = 2,$$

both $x - 5$ and 2 represent the same number, since this is the meaning of the equals sign. Solve the equation by changing the left side from $x - 5$ to just x. This is done by adding 5 to $x - 5$. Keep the two sides equal by also adding 5 on the right side.

$$x - 5 = 2 \qquad \text{Given equation}$$
$$x - 5 + 5 = 2 + 5 \qquad \text{Add 5 on both sides}$$

Here 5 was added on both sides of the equation. Now simplify each side separately to get

$$x = 7.$$

The solution of the given equation is 7. Check by replacing x with 7 in the given equation.

$$x - 5 = 2 \qquad \text{Given equation}$$
$$7 - 5 = 2 \qquad \text{Let } x = 7$$
$$2 = 2 \qquad \text{True}$$

Since the final statement is true, 7 checks as the solution.

The equation above was solved by adding the same number to both sides, as justified by the **addition property of equality.**

Objectives

▶ Identify linear equations.

▶ Use the addition property of equality.

▶ Simplify equations, and then use the addition property of equality.

1. Complete each step in solving the following equations.

   (a) $r + 11 = 20$
   $r + 11 + \underline{} = 20 + \underline{}$
   $r = \underline{}$

   (b) $p + 2 = 8$
   $p + 2 + \underline{} = 8 + \underline{}$
   $p = \underline{}$

   (c) $z + 9 = 3$
   $z + 9 + \underline{} = 3 + \underline{}$
   $z = \underline{}$

2. Use the addition property of equality to solve the following equations.

   (a) $m - 2 = 6$

   (b) $y + \dfrac{3}{4} = \dfrac{19}{4}$

   (c) $a + 2 = -3$

   (d) $p + 6 = 2$

ANSWERS
1. (a) $-11, -11, 9$ (b) $-2, -2, 6$
   (c) $-9, -9, -6$
2. (a) 8 (b) 4 (c) -5 (d) -4

Addition Property of Equality

If A, B, and C are algebraic expressions, then the equations

$$A = B$$

and

$$A + C = B + C$$

have exactly the same solutions. In other words, the same expression may be added to both sides of an equation.

The addition property of equality applies to any equation, not just linear equations.

Work Problem 1 at the side.

EXAMPLE 1 Solve the equation $x - 16 = 7$.

If the left side of this equation were just x, the solution could be found. Get x alone by using the addition property of equality and adding 16 on both sides.

$$x - 16 = 7$$
$$x - 16 + 16 = 7 + 16$$
$$x = 23$$

Check by substituting 23 for x in the original equation.

$x - 16 = 7$ Given equation
$23 - 16 = 7$ Let $x = 23$
$7 = 7$ True

Since the check results in a true statement, 23 is the solution. ◀

In this example, why was 16 added to both sides of the equation $x - 16 = 7$? The equation would be solved if it could be rewritten so that one side contained only the variable and the other side contained only a number. Since $x - 16 + 16 = x + 0 = x$, adding 16 on the left side simplifies that side to just x, the variable, as desired.

The addition property of equality says that the same number may be *added* to both sides of an equation. As was shown in Chapter 2, subtraction is defined in terms of addition. Because of the way subtraction is defined, the addition property also permits *subtracting* the same number on both sides of an equation.

Work Problem 2 at the side.

EXAMPLE 2 Solve the equation $3k + 17 = 4k$.

As a first step, get all terms that contain variables on the same side of the equation. One way to do this is to subtract $3k$ from each side.

$$3k + 17 = 4k$$
$$3k + 17 - 3k = 4k - 3k \quad \text{Subtract } 3k$$
$$17 = k$$

The solution is 17.

Solving Equations and Inequalities

The equation $3k + 17 = 4k$ also could be solved by first subtracting $4k$ from each side, as follows.

$$3k + 17 = 4k$$
$$3k + 17 - 4k = 4k - 4k \quad \text{Subtract } 4k$$
$$17 - k = 0$$

Now subtract 17 from both sides.

$$17 - k - 17 = 0 - 17 \quad \text{Subtract 17}$$
$$-k = -17$$

This result gives the value of $-k$, but not of k itself. However, this result does say that the additive inverse of k is -17, which means that k must be 17.

$$-k = -17$$
$$k = 17$$

This solution agrees with the first one. Check the solution by replacing k with 17 in the given equation. ◄

Work Problem 3 at the side.

3 Sometimes an equation must be simplified as a first step in its solution.

EXAMPLE 3 Solve the equation $4r + 5r - 3 + 8 - 3r - 5r = 12 + 8$.

First, simplify the equation by combining terms.

$$4r + 5r - 3r - 5r - 3 + 8 = 12 + 8$$
$$r + 5 = 20$$

Subtract 5 from both sides of this equation.

$$r + 5 - 5 = 20 - 5 \quad \text{Subtract 5}$$
$$r = 15$$

The solution of the given equation is 15. (Check this.) ◄

Work Problem 4 at the side.

EXAMPLE 4 Solve the equation $3(2 + 5x) - (1 + 14x) = 6$.

Use the distributive property to simplify the equation.

$$3(2 + 5x) - (1 + 14x) = 6$$
$$3(2) + 3(5x) - 1(1) - 1(14x) = 6 \quad \text{Distributive property}$$
$$6 + 15x - 1 - 14x = 6 \quad \text{Multiply}$$
$$x + 5 = 6 \quad \text{Combine terms}$$

Subtract 5 on both sides of the equation to get the variable term alone on one side.

$$x = 1$$

Check by substituting 1 for x in the original equation. ◄

Work Problem 5 at the side.

3. Solve each equation.

(a) $5m = 4m + 6$

(b) $3y = 2y - 9$

(c) $2k - 8 = 3k$

(d) $\dfrac{7}{2}m + 1 = \dfrac{9}{2}m$

4. Solve each equation.

(a) $7p + 2p - 8p + 5 = 9 + 1$

(b) $11k - 6 - 4 - 10k = -5 + 5$

(c) $-4 + 3 - 2m + 3m = 10 - 5 - 7$

5. Solve each equation.

(a) $2(a + 4) - (3 + a) = 8$

(b) $-(5 - 3r) + 4(-r + 1) = 1$

(c) $-3(m - 4) + 2(5 + 2m) = 29$

ANSWERS
3. (a) 6 (b) -9 (c) -8 (d) 1
4. (a) 5 (b) 10 (c) -1
5. (a) 3 (b) -2 (c) 7

3.1 The Addition Property of Equality

name   date   hour

3.1 EXERCISES

Solve each equation by using the addition property of equality. Check each solution. See Examples 1 and 2.

1. $x - 3 = 7$
2. $x + 5 = 13$
3. $7 + k = 5$
4. $9 + m = 4$

5. $3r = 2r + 10$
6. $2p = p + 3$
7. $7z = -8 + 6z$
8. $4y = 3y - 5$

9. $2p + 6 = 10 + p$
10. $5r + 2 = -1 + 4r$
11. $2k + 2 = -3 + k$
12. $6 + 7x = 6x + 3$

13. $x - 5 = 2x + 6$
14. $-3r + 7 = -4r - 19$
15. $6z + 3 = 5z - 3$
16. $6t + 5 = 5t + 7$

17. $2p = p + \frac{1}{2}$
18. $5m = 4m + \frac{2}{3}$
19. $\frac{4}{3}z = \frac{1}{3}z - 5$
20. $\frac{9}{5}m = \frac{4}{5}m + 6$

21. $\frac{11}{4}r - \frac{1}{2} = \frac{7}{4}r + \frac{2}{3}$
22. $\frac{9}{2}y - \frac{3}{4} = \frac{7}{2}y + \frac{5}{8}$
23. $2.7a + 5 = 1.7a$
24. $4.7p - 3 = 3.7p$

Solve the following equations. First simplify each side of the equation as much as possible. Check each solution. See Example 3.

25. $4x + 3 + 2x - 5x = 2 + 8$
26. $3x + 2x - 6 + x - 5x = 9 + 4$

27. $9r + 4r + 6 - 8 = 10r + 6 + 2r$
28. $-3t + 5t - 6t + 4 - 3 = -3t + 2$

29. $11z + 2 + 4z - 3z = 5z - 8 + 6z$
30. $2k + 8k + 6k - 4k - 8 + 2 = 3k + 2 + 10k$

31. $.15y - .4y + 8 - 2.3 + 7.6 = y - .25y$

32. $4.12m + 8m - .9m + 2.3 - 5 = 10.22m + 6.8$

Solve the following equations. Check each solution. See Example 4.

33. $(5y + 6) - (3 + 4y) = 9$

34. $(8p - 3) - (7p + 1) = -2$

35. $2(r + 5) - (9 + r) = -1$

36. $4(y - 6) - (3y + 2) = 8$

37. $-6(2a + 1) + (13a - 7) = 4$

38. $-5(3k - 3) + (1 + 16k) = 2$

39. $4(7x - 1) + 3(2 - 5x) = 4(3x + 5)$

40. $9(2m - 3) - 4(5 + 3m) = 5(4 + m)$

41. $-2(8p + 7) - 3(4 - 7p) = 2(3 + 2p) - 6$

42. $-5(8 - 2z) + 4(7 - z) = 7(8 + z) - 3$

Write an equation using the following information. Then solve the equation and check the solution. Use x as the variable.

43. Three times a number is 17 more than twice the number. Find the number.

44. If six times a number is subtracted from seven times a number, the result is -9. Find the number.

45. If five times a number is added to three times the number, the result is the sum of seven times the number and 9. Find the number.

46. If nine times a number is subtracted from eleven times the number, the result is 4 less than three times the number. Find the number.

47. The sum of twice a number and 5 is multiplied by 6. The result is 8 less than 13 times the number. Find the number.

48. Four times a number is subtracted from 7. The result is multiplied by 5, giving 3 more than -19 times the number. Find the number.

3.2 THE MULTIPLICATION PROPERTY OF EQUALITY

Objectives

1. Use the multiplication property of equality.
2. Simplify equations, and then use the multiplication property of equality.
3. Solve equations such as $-r = 4$.
4. Use the multiplication property of equality to solve equations with decimals.

The addition property of equality by itself is not enough to solve an equation like $3x + 2 = 17$.

$$3x + 2 = 17$$
$$3x + 2 - 2 = 17 - 2 \quad \text{Subtract 2 from both sides}$$
$$3x = 15 \quad \text{Simplify}$$

Instead of just x on the left side, the equation has $3x$. Another property is needed to change $3x = 15$ to $x = $ a number.

▶ If $3x = 15$, then $3x$ and 15 both represent the same number. Multiplying both $3x$ and 15 by the same number will also result in an equality. The **multiplication property of equality** states that both sides of an equation can be multiplied by the same quantity.

Multiplication Property of Equality

If A, B, and C represent algebraic expressions not equal to zero, the equations

$$A = B$$

and

$$AC = BC$$

have exactly the same solutions.

In other words, both sides of an equation may be multiplied by the same nonzero expression.

This property can be used to solve $3x = 15$. The $3x$ on the left must be changed to $1x$, or x, instead of $3x$. Get x by multiplying both sides of the equation by $\frac{1}{3}$. Use $\frac{1}{3}$ because $\frac{1}{3} \cdot 3 = \frac{3}{3} = 1$, since $\frac{1}{3}$ is the reciprocal of 3.

$$3x = 15$$
$$\frac{1}{3}(3x) = \frac{1}{3} \cdot 15 \quad \text{Multiply both sides by } \frac{1}{3}$$
$$\left(\frac{1}{3} \cdot 3\right)x = \frac{1}{3} \cdot 15 \quad \text{Associative property}$$
$$1x = 5$$
$$x = 5 \quad \text{Simplify}$$

The solution of the equation is 5. This result can be checked in the original equation.

Work Problem 1 at the side.

Just as the addition property of equality permits *subtracting* the same number from both sides of an equation, the multiplication property of equality permits *dividing* both sides of an equation by the same nonzero number. For example, the equation $3x = 15$, solved above by

---

1. Check that 5 is the solution of $3x = 15$.

ANSWER
1. The solution is 5.

2. Solve each equation.

   (a) $7m = 56$

   (b) $3r = -12$

   (c) $8y = 108$

   (d) $-2m = 16$

   (e) $-6p = -14$

multiplication, also could be solved by dividing both sides by 3, as follows.

$$3x = 15$$
$$\frac{3x}{3} = \frac{15}{3} \quad \text{Divide by 3}$$
$$x = 5 \quad \text{Simplify}$$

EXAMPLE 1 Solve the equation $25p = 30$.

Get p on the left (instead of $25p$) by using the multiplication property of equality to divide both sides of the equation by 25, the coefficient of p.

$$25p = 30$$
$$\frac{25p}{25} = \frac{30}{25} \quad \text{Divide by 25}$$
$$p = \frac{30}{25} = \frac{6}{5} \quad \text{Simplify}$$

To check, substitute $\frac{6}{5}$ for p in the given equation.

$$25p = 30$$
$$\frac{25}{1}\left(\frac{6}{5}\right) = 30 \quad \text{Let } p = \frac{6}{5}$$
$$30 = 30 \quad \text{True}$$

The solution is $\frac{6}{5}$. ◀

Work Problem 2 at the side.

In the next two examples, multiplication produces the solution more quickly than division would.

EXAMPLE 2 Solve the equation $\frac{a}{4} = 3$.

Replace $\frac{a}{4}$ by $\frac{1}{4}a$, since division by 4 is the same as multiplication by $\frac{1}{4}$. To get a alone on the left, multiply both sides by 4, the reciprocal of the coefficient of a.

$$\frac{a}{4} = 3$$
$$\frac{1}{4}a = 3 \quad \text{Change } \frac{a}{4} \text{ to } \frac{1}{4}a$$
$$4 \cdot \frac{1}{4}a = 4 \cdot 3 \quad \text{Multiply by 4}$$
$$1a = 12$$
$$a = 12$$

Check the answer.

$$\frac{a}{4} = 3 \quad \text{Given equation}$$
$$\frac{12}{4} = 3 \quad \text{Let } a = 12$$
$$3 = 3 \quad \text{True}$$

The solution 12 is correct. ◀

ANSWERS

2. (a) 8 (b) -4 (c) $\frac{27}{2}$ or $13\frac{1}{2}$

   (d) -8 (e) $\frac{7}{3}$

Solving Equations and Inequalities

Work Problem 3 at the side.

EXAMPLE 3 Solve the equation $\frac{3}{4}h = 6$.

Get h alone on the left by multiplying both sides of the equation by $\frac{4}{3}$. Use $\frac{4}{3}$ because $\frac{4}{3} \cdot \frac{3}{4}h = 1 \cdot h = h$.

$$\frac{3}{4}h = 6$$
$$\frac{4}{3}\left(\frac{3}{4}h\right) = \frac{4}{3} \cdot 6 \quad \text{Multiply by } \frac{4}{3}$$
$$1 \cdot h = \frac{4}{3} \cdot \frac{6}{1}$$
$$h = 8$$

The solution is 8. Check the answer by substitution in the given equation. ◀

Work Problem 4 at the side.

▷ In the next example, it is necessary to simplify the equation before using the multiplication property of equality.

EXAMPLE 4 Solve the equation $5m + 6m = 33$.

Simplify the equation by using the distributive property to combine terms.

$$5m + 6m = 33$$
$$11m = 33 \quad \text{Distributive property}$$

Now divide both sides by 11.

$$\frac{11m}{11} = \frac{33}{11} \quad \text{Divide by 11}$$
$$1m = 3$$
$$m = 3 \quad \text{Simplify}$$

The solution is 3. Check this solution. ◀

Work Problem 5 at the side.

▷ The following example shows how to use the multiplication property of equality to solve equations such as $-r = 4$.

EXAMPLE 5 Solve the equation $-r = 4$.

On the left side, change $-r$ to r by first writing $-r$ as $-1 \cdot r$.

$$-r = 4$$
$$-1 \cdot r = 4 \quad -r = -1 \cdot r$$

Now multiply both sides of this last equation by -1.

$$-1(-1 \cdot r) = -1 \cdot 4 \quad \text{Multiply by } -1$$
$$(-1)(-1) \cdot r = -4 \quad \text{Associative property}$$
$$1 \cdot r = -4$$
$$r = -4 \quad \text{Identity property}$$

3.2 The Multiplication Property of Equality

---

3. Solve each equation.

(a) $\dfrac{m}{2} = 6$

(b) $\dfrac{y}{5} = 5$

(c) $\dfrac{p}{4} = -6$

(d) $\dfrac{a}{-2} = 8$

4. Solve each equation.

(a) $\dfrac{2}{3}m = 8$

(b) $\dfrac{7}{8}m = 28$

(c) $\dfrac{3}{4}k = -21$

(d) $\dfrac{1}{2}p = \dfrac{7}{4}$

5. Solve each equation.

(a) $5p + 2p = 28$

(b) $9k - k = -56$

(c) $7m - 5m = -12$

(d) $4r - 9r = 20$

ANSWERS
3. (a) 12 (b) 25 (c) -24 (d) -16
4. (a) 12 (b) 32 (c) -28 (d) $\dfrac{7}{2}$
5. (a) 4 (b) -7 (c) -6 (d) -4

6. Solve each equation.

   (a) $-m = 2$

   (b) $-p = -7$

7. Solve each equation.

   (a) $-1.5p = 4.5$

   (b) $12.5k = -63.75$

   (c) $-.7m = -5.04$

Check this solution.

$$-r = 4 \quad \text{Given equation}$$
$$-(-4) = 4 \quad \text{Let } r = -4$$
$$4 = 4$$

The solution, -4, checks. ◀

Work Problem 6 at the side.

▶ The final example shows how to solve equations with decimals.

EXAMPLE 6 Solve the equation $2.1x = 6.09$.

Divide both sides by 2.1.

$$\frac{2.1x}{2.1} = \frac{6.09}{2.1}$$
$$1x = 2.9$$
$$x = 2.9$$

Check that the solution is 2.9. ◀

Work Problem 7 at the side.

ANSWERS
6. (a) -2 (b) 7
7. (a) -3 (b) -5.1 (c) 7.2

name   date   hour

3.2 EXERCISES

Solve each equation and check your solution. See Examples 1–6.

1. $5x = 25$
2. $7x = 28$
3. $2m = 50$
4. $6y = 72$

5. $3a = -24$
6. $5k = -60$
7. $8s = -56$
8. $10t = -36$

9. $-6x = 16$
10. $-6x = 24$
11. $-18z = 108$
12. $-11p = 77$

13. $5r = 0$
14. $2x = 0$
15. $-y = 6$
16. $-m = 2$

17. $-n = -4$
18. $-p = -8$
19. $2x + 3x = 20$
20. $3k + 4k = 14$

21. $5m + 6m - 2m = 72$
22. $11r - 5r + 6r = 84$
23. $k + k + 2k = 80$

24. $4z + z + 2z = 28$
25. $3r - 5r = 6$
26. $9p - 13p = 12$

27. $7r - 13r = -24$
28. $12a - 18a = -36$
29. $-7y + 8y - 9y = -56$

30. $-11b + 7b + 2b = -100$
31. $\dfrac{m}{2} = 16$
32. $\dfrac{p}{5} = 3$

3.2   Exercises   **113**

33. $\dfrac{x}{7} = 7$

34. $\dfrac{k}{8} = 2$

35. $\dfrac{2}{3}t = 6$

36. $\dfrac{4}{3}m = 18$

37. $\dfrac{15}{2}z = 20$

38. $\dfrac{12}{5}r = 18$

39. $\dfrac{3}{4}p = -60$

40. $\dfrac{5}{8}z = -40$

41. $\dfrac{2}{3}k = 5$

42. $\dfrac{5}{3}m = 6$

43. $-\dfrac{2}{7}p = -7$

44. $-\dfrac{3}{11}y = -2$

45. $1.7p = 5.1$

46. $2.3k = 11.04$

47. $-4.2m = 25.62$

48. $-3.9a = -15.6$

49. $8.974z = 6.2818$

50. $11.506y = 3.4518$

51. $-9.273k = 10.2003$

52. $-1.565r = 3.9125$

53. $.9123p = -5.20011$

54. $.8761a = -3.32918$

Write an equation for each problem. Then solve the equation and check the solution.

55. When a number is multiplied by 4, the result is 6. Find the number.

56. When a number is multiplied by -5, the result is 2. Find the number.

57. Chuck decided to divide a sum of money equally among four relatives, Dennis, Mike, Ed, and Joyce. Each relative received $62. Find the amount that was originally divided.

58. If twice a number is divided by 5, the result is 4. Find the number.

59. Twice a number is divided by 1.74, producing -8.38 as a quotient. Find the number.

60. A number is multiplied by 12 and then divided by 54.96. The result is -3.1. Find the number.

Solving Equations and Inequalities

3.3 SOLVING LINEAR EQUATIONS

▶ This section shows how to use both the addition property and the multiplication property to solve more complicated equations.

Solving Linear Equations

Step 1 Combine like terms to simplify. Use the commutative, associative, and distributive properties as needed.

Step 2 If necessary, use the addition property of equality to simplify further, so that the variable term is on one side of the equals sign and the number is on the other.

Step 3 If necessary, use the multiplication property of equality to simplify further. This gives an equation of the form $x =$ a number.

Step 4 Check the solution by substituting into the original equation. (Do *not* substitute into an intermediate step.)

The check is used only to catch errors in carrying out the steps of the solution.

EXAMPLE 1 Solve the equation $2x + 3x + 3 = 38$.

Follow the four steps summarized above.

Step 1 Combine like terms.

$$2x + 3x + 3 = 38$$
$$5x + 3 = 38 \quad \text{Combine like terms.}$$

Step 2 Use the addition property of equality. Subtract 3 from both sides.

$$5x + 3 - 3 = 38 - 3 \quad \text{Subtract 3}$$
$$5x = 35$$

Step 3 Use the multiplication property of equality. Divide both sides by 5.

$$\frac{5x}{5} = \frac{35}{5} \quad \text{Divide by 5}$$
$$x = 7$$

Step 4 Check the solution. Substitute 7 for x in the given equation.

$$2x + 3x + 3 = 38$$
$$2(7) + 3(7) + 3 = 38 \quad \text{Let } x = 7$$
$$14 + 21 + 3 = 38$$
$$38 = 38 \quad \text{True}$$

Since the final statement is true, 7 is the solution. ◀

Work Problem 1 at the side.

EXAMPLE 2 Solve the equation $3r + 4 - 2r - 7 = 4r + 3$.

Use the four steps again.

Step 1

$$3r + 4 - 2r - 7 = 4r + 3$$
$$r - 3 = 4r + 3 \quad \text{Combine like terms}$$

Objectives

▶ Learn the four steps in solving a linear equation and how to use them.

▶ Write word phrases as mathematical phrases.

1. Solve each equation.

   (a) $3k + 2k + 7 = 17$

   (b) $7m + 9m + 5 = 43$

   (c) $9p - 5p + p - 8 = -18$

ANSWERS

1. (a) 2 (b) $\frac{19}{8}$ (c) -2

3.3 Solving Linear Equations   115

2. Solve each equation.

(a) $7 + 4p - 3p + 8$
$= 9p + 7$

(b) $5y - 7y + 6y - 9$
$= 3 + 2y$

(c) $-3k - 5k - 6 + 11$
$= 2k - 5$

(d) $2y + 5y - 6 + 8$
$= 9y - 1$

3. Solve each equation.

(a) $7(p - 2) + p = 2p + 4$

(b) $11 + 3(a + 1)$
$= 5a + 16$

(c) $3(m + 5) - 5 + 2m$
$= 2(m - 10)$

ANSWERS

2. (a) 1 (b) 6 (c) 1 (d) $\frac{3}{2}$
3. (a) 3 (b) -1 (c) -10

Step 2
$$r - 3 + 3 = 4r + 3 + 3 \quad \text{Add 3}$$
$$r = 4r + 6$$
$$r - 4r = 4r + 6 - 4r \quad \text{Subtract } 4r$$
$$-3r = 6$$

Step 3
$$\frac{-3r}{-3} = \frac{6}{-3} \quad \text{Divide by } -3$$
$$r = -2$$

Step 4 Substitute -2 for r in the original equation.
$$3r + 4 - 2r - 7 = 4r + 3$$
$$3(-2) + 4 - 2(-2) - 7 = 4(-2) + 3$$
$$-6 + 4 + 4 - 7 = -8 + 3$$
$$-5 = -5 \quad \text{True}$$

The solution of the given equation is -2. ◀

In Step 2 of Example 2, the terms were added and subtracted in such a way that the variable term ended up on the left side of the equation. Choosing differently would lead to the variable term on the right side of the equation. Usually there is no real advantage either way.

Work Problem 2 at the side.

EXAMPLE 3 Solve the equation $4(k - 3) - k = k - 6$.

Step 1 Before combining like terms, use the distributive property to simplify $4(k - 3)$.
$$4(k - 3) = 4k - 4 \cdot 3 = 4k - 12$$

Now combine like terms.
$$4k - 12 - k = k - 6$$
$$3k - 12 = k - 6$$

Step 2
$$3k - 12 + 12 = k - 6 + 12 \quad \text{Add 12}$$
$$3k = k + 6$$
$$3k - k = k + 6 - k \quad \text{Subtract } k$$
$$2k = 6$$

Step 3
$$\frac{2k}{2} = \frac{6}{2} \quad \text{Divide by 2}$$
$$k = 3$$

Step 4 Check this answer by substituting 3 for k in the given equation. Remember to do all the work inside the parentheses first.
$$4(k - 3) - k = k - 6$$
$$4(3 - 3) - 3 = 3 - 6 \quad \text{Let } k = 3$$
$$4(0) - 3 = 3 - 6$$
$$0 - 3 = 3 - 6$$
$$-3 = -3 \quad \text{True}$$

The solution of the equation is 3. ◀

Work Problem 3 at the side.

EXAMPLE 4 Solve the equation $8a - (3 + 2a) = 3a + 1$.

Step 1 Simplify.

$$8a - (3 + 2a) = 3a + 1$$
$$8a - 1 \cdot (3 + 2a) = 3a + 1$$
$$8a - 3 - 2a = 3a + 1 \quad \text{Distributive property}$$
$$6a - 3 = 3a + 1$$

Step 2 First, add 3 to both sides; then subtract $3a$.

$$6a - 3 \boxed{+ 3} = 3a + 1 \boxed{+ 3} \quad \text{Add 3}$$
$$6a = 3a + 4$$
$$6a \boxed{- 3a} = 3a + 4 \boxed{- 3a} \quad \text{Subtract } 3a$$
$$3a = 4$$

Step 3
$$\frac{3a}{\boxed{3}} = \frac{4}{\boxed{3}} \quad \text{Divide by 3}$$
$$a = \frac{4}{3}$$

Check that the solution is $\frac{4}{3}$. ◀

Work Problem 4 at the side.

EXAMPLE 5 Solve the equation $4(8 - 3t) = 32 - 8(t + 2)$.

Step 1 Simplify.

$$4(8 - 3t) = 32 - 8(t + 2)$$
$$32 - 12t = 32 - 8t - 16$$
$$32 - 12t = 16 - 8t$$

Step 2 Subtract 32 from both sides, and then add $8t$ to both sides.

$$32 - 12t \boxed{- 32} = 16 - 8t \boxed{- 32} \quad \text{Subtract 32}$$
$$-12t = -16 - 8t$$
$$-12t \boxed{+ 8t} = -16 - 8t \boxed{+ 8t} \quad \text{Add } 8t$$
$$-4t = -16$$

Step 3 Divide both sides by -4.

$$\frac{-4t}{\boxed{-4}} = \frac{-16}{\boxed{-4}} \quad \text{Divide by } -4$$
$$t = 4$$

Step 4 Check this solution in the given equation.

$$4(8 - 3\boxed{t}) = 32 - 8(\boxed{t} + 2)$$
$$4(8 - 3 \cdot \boxed{4}) = 32 - 8(\boxed{4} + 2) \quad \text{Let } t = 4$$
$$4(8 - 12) = 32 - 8(6)$$
$$4(-4) = 32 - 48$$
$$-16 = -16 \quad \text{True}$$

The solution, 4, checks. ◀

Work Problem 5 at the side.

▶ The next section includes a detailed discussion of the methods of solving word problems. One of the main steps in solving word problems is converting the phrases in the word problems into mathematical

3.3 Solving Linear Equations

4. Solve each equation.

(a) $4y - (y + 7) = 9$

(b) $7m - (2m - 9) = 39$

(c) $5(m - 2) - (3m - 6) = 3m - 1$

5. Solve each equation.

(a) $2(4 + 3r) = 3(r + 1) + 11$

(b) $-3(m + 2) = 4(2m + 1) + 1$

(c) $2 - 3(2 + 6z) = 4(z + 1) + 18$

ANSWERS

4. (a) $\frac{16}{3}$ (b) 6 (c) -3

5. (a) 2 (b) -1 (c) $-\frac{13}{11}$

117

6. Write each phrase as a mathematical expression. Use x as the variable.

   (a) 8 more than a number

   (b) The sum of a number and 9

   (c) 2 less than a number

7. Write each of the following as a mathematical expression. Use x as the variable.

   (a) The product of a number and 5

   (b) -3 times a number

   (c) Five-eighths of a number

   (d) The quotient of a number and 10

8. Write each phrase as a mathematical expression. Use x as the variable.

   (a) 10 added to twice a number

   (b) The product of 5 and 2 less than a number

   (c) The quotient of 8 plus a number and 3 times the number

   (d) Twice a number added to the reciprocal of 5

ANSWERS
6. (a) $8 + x$ or $x + 8$
   (b) $9 + x$ or $x + 9$ (c) $x - 2$
7. (a) $5x$ (b) $-3x$ (c) $\frac{5}{8}x$ (d) $\frac{x}{10}$
8. (a) $10 + 2x$ (b) $5(x - 2)$
   (c) $\frac{8 + x}{3x}$ (d) $2x + \frac{1}{5}$

expressions. The next few examples show how to translate phrases that occur frequently in word problems.

EXAMPLE 6 Write the following phrases as mathematical expressions. Use x to represent the unknown quantity. (Other letters could be used to represent the unknown.)

(a) The sum of a number and 12

"The **sum** of a number and 12" is written as $x + 12$ or $12 + x$.

(b) 7 more than a number

The phrase "7 **more than** a number" is translated as $7 + x$ or $x + 7$. Notice the difference between the wording of part (b) and "7 *is* more than a number" which translates as $7 > x$.

(c) 3 less than a number

The words *less than* here indicate subtraction, so "3 **less than** a number" is written $x - 3$. ($3 - x$ would *not* be correct.) Compare this example with "3 *is* less than a number", written $3 < x$.

(d) A number decreased by 14

"A number **decreased by** 14" is $x - 14$.

(e) Ten fewer than x

"Ten **fewer than** x" translates as $x - 10$. ◀

Work Problem 6 at the side.

EXAMPLE 7 Write the following phrases as mathematical expressions. Use x as the variable.

(a) "The product of a number and 3" is written as $3x$, since *product* indicates multiplication.

(b) "Three times a number" is also $3x$.

(c) "Two-thirds of a number" is $\frac{2}{3}x$.

(d) "The quotient of a number and 2" is $\frac{x}{2}$. (The word *quotient* indicates division; use a fraction bar instead of $\div$.)

(e) "The reciprocal of a nonzero number" is $\frac{1}{x}$. ◀

Work Problem 7 at the side.

Some word problems involve a combination of symbols. The next example shows how to translate such phrases.

EXAMPLE 8 Write the following phrases as mathematical expressions. Use x as the variable.

(a) "Seven less than 4 times a number" is written $4x - 7$.

(b) "A nonzero number plus its reciprocal" is $x + \frac{1}{x}$.

(c) "Five times the sum of a number and 2" is written as $5(x + 2)$.

(d) "A number divided by the sum of 4 and the number" is $\frac{x}{4 + x}$. ◀

Work Problem 8 at the side.

Solving Equations and Inequalities

name _____ date _____ hour _____

3.3 EXERCISES

Solve each equation and check your solution. See Examples 1–5.

1. $4h + 8 = 16$

2. $3x - 15 = 9$

3. $6k + 12 = -12 + 7k$

4. $2m - 6 = 6 + 3m$

5. $12p + 18 = 14p$

6. $10m - 15 = 7m$

7. $3x + 9 = -3(2x + 2)$

8. $4z + 2 = -2(z + 3)$

9. $2(2r - 1) = -3(r + 3)$

10. $3(3k + 5) = 2(5k + 5)$

11. $\frac{3}{2}\left(\frac{1}{3}x + 4\right) = 6\left(\frac{1}{4} + x\right)$

12. $\frac{4}{3}p + \frac{4}{3} = \frac{2}{3}p - \frac{1}{3}$

13. $3(5 + 1.4x) = 3x$

14. $2(-3 + 2.1x) = 2x + x$

15. $7.492y - 3.86 = 5.562y$

16. $4.813q + 1.769 = 5.25525q$

3.3 Exercises 119

17. $.291z + 3.715 = -.874z + 1.9675$

18. $1.043k - 2.816 = .359k - 4.2524$

Combine terms as necessary; then solve the equations. Check your answers. See Examples 3–5.

19. $-5 - 3(2x + 1) = 12$

20. $10 - 2(3x - 4) = 2x$

21. $-5k - 8 = 2(k + 6) + 1$

22. $4a - 7 = 3(2a + 5) - 2$

23. $5(2m - 1) = 4(2m + 1) + 7$

24. $3(3k - 5) = 4(3k - 1) - 17$

25. $5(4t + 3) = 6(3t + 2) - 1$

26. $7(2y + 6) = 9(y + 3) + 5$

27. $5(x - 3) + 2 = 5(2x - 8) - 3$

28. $6(2v - 1) - 5 = 7(3v - 2) - 24$

29. $-2(3s + 9) - 6 = -3(4s + 11) - 6$

30. $-3(5z + 24) + 2 = 2(3 - 2z) - 10$

31. $6(2p - 8) + 24 = 3(5p - 6) - 6$

32. $2(5x + 3) - 3 = 6(2x - 3) + 15$

33. $-(4m + 2) - (-3m - 5) = 3$

34. $-(6k - 5) - (-5k + 8) = -4$

35. $2(4x - 1) - 3(-2x + 5) = 3$

36. $4(3x - 3) - 3(-x - 4) = 20$

37. $3(4x + 2) - 2(5x - 1) = 0$

38. $5(3 - x) + 3(2x - 2) = 2$

39. $5(y - 2) + 6y = 4(y + 3) - 1$

40. $2(3x + 1) - x = 4 - (2x + 3) - 6$

41. $1.2(x + 5) = 3(2x - 8) + 23.28$

42. $4.7(m - 1) = 2(3m + 5) - 17.69$

43. $5.1(p + 6) - 3.8p = 4.9(p + 3) + 34.62$

44. $2.7(k - 3) + 1.1k = 3.4(k + 1) - 10.54$

Write each of the following as a mathematical expression. Use x as the variable. See Examples 6–8.

45. A number added to -6

46. -1 added to a number

47. The sum of a number and 12

48. A quantity increased by -18

49. 5 less than a number

50. A number decreased by 6

51. Subtract 9 from a number.

52. 16 fewer than a number

53. The product of a number and 9

54. Double a number

55. Triple a number

56. Three-fifths of a number

57. The quotient of a number and 6

58. The quotient of -9 and a number

59. A number divided by -4

60. 7 divided by a number

61. The product of 8 and the sum of a number and 3

62. A nonzero number subtracted from its reciprocal

63. Three times the quotient of a number and 2

64. Eight times the difference of a number and 8

3.4 FROM WORD PROBLEMS TO EQUATIONS

As mentioned in the last section, one of the main skills involved in solving word problems is translating words and phrases into mathematical expressions. To get the solution to a word problem, first read the problem carefully and determine what facts are given and what must be found. Next, use the five steps given below.

Solving Word Problems
Step 1 Choose a variable to represent the numerical value that you are asked to find—the unknown number.
Step 2 Translate the problem into an equation.
Step 3 Solve the equation.
Step 4 Answer the question asked in the problem.
Step 5 Check your solution by using the original words of the problem.

The second step is often the hardest. Start to translate the problem into an equation by writing the given phrases as mathematical expressions.

▶ Since equal mathematical expressions are names for the same number, translate any words that mean *equal* or *same* as $=$. The $=$ sign leads to an equation to be solved.

EXAMPLE 1 Translate "the result of decreasing a number by 7, then multiplying the difference by 4, is 100" into an equation. Use x as the variable. Solve the equation.
Translate as follows.

$$(x - 7) \cdot 4 = 100$$

Solve the equation using the steps given in the previous section.

$(x - 7) \cdot 4 = 100$
$4x - 28 = 100$ Distributive property
$4x = 128$ Add 28 to both sides
$x = 32$ Divide by 4

Now check the answer by substituting 32 for x into the original problem. Decrease the number, 32, by 7: $32 - 7 = 25$. Multiply the result by 4: $25 \cdot 4 = 100$. The solution, 32, gives a correct result. ◀

Work Problem 1 at the side.

Notice how the five steps for solving word problems are used to solve the next few problems.

EXAMPLE 2 If three times the sum of a number and 4 is decreased by twice the number, the result is -6. Find the number.

Let x represent the unknown number. "Three times the sum of a number and 4" translates into symbols as $3(x + 4)$. "Twice the

Objective

▶ Solve word problems.

1. Write equations for each of the following and then solve.

(a) When 3 times a number is added to 9, the answer is 12.

(b) If you add 10 to a number, the result is 20.

(c) Twice a number, increased by 3, equals 17.

ANSWERS
1. (a) $3x + 9 = 12$; 1
   (b) $10 + x = 20$; 10
   (c) $2x + 3 = 17$; 7

2. Solve each word problem.

   (a) If 2 is added to the product of 7 and a number, the result is 8 more than the number. Find the number.

   (b) If a number is multiplied by 4, with the result added to 9, the sum is 3 more than the number. Find the number.

---

number" is $2x$. Now write an equation using the information given in the problem.

Three times the sum of a number and 4 — decreased by — twice the number — is — -6.

$$3(x + 4) - 2x = -6$$

Solve the equation.

$$3(x + 4) - 2x = -6$$
$$3x + 12 - 2x = -6 \quad \text{Distributive property}$$
$$x + 12 = -6 \quad \text{Combine terms}$$
$$x = -18 \quad \text{Subtract 12}$$

Check that -18 is the correct answer by substituting this result into the words of the original problem. Three times the sum of -18 and 4 is $3(-18 + 4) = 3(-14) = -42$. Twice -18 is -36; subtract -36 from -42 to get $-42 - (-36) = -6$ as required. ◀

Work Problem 2 at the side.

EXAMPLE 3 The audience at a concert included 25 more women than men. The total number of people at the concert was 139. Find the number of men.

Let x represent the number of men. The number of women is given by $x + 25$. Now write the equation.

The total — is — the number of men — plus — the number of women.

$$139 = x + x + 25$$

Solve the equation.

$$139 = 2x + 25$$
$$139 - 25 = 2x + 25 - 25 \quad \text{Subtract 25}$$
$$114 = 2x \quad \text{Simplify}$$
$$57 = x \quad \text{Divide by 2}$$

Check the solution in the words of the problem. If the number of men was 57, then the number of women was $25 + 57 = 82$. The total number was $57 + 82 = 139$, as required. There were 57 men at the concert. ◀

EXAMPLE 4 The owner of a small cafe found one day that the number of orders for tea was $\frac{1}{3}$ the number of orders for coffee. If the total number of orders for the two drinks was 76, how many orders were placed for coffee.

Let x equal the number of orders for coffee and $\frac{1}{3}x$ equal the number of orders for tea. Use the fact that the total number of orders was 76 to write an equation.

The total — is — orders for coffee — plus — orders for tea.

$$76 = x + \frac{1}{3}x$$

ANSWERS

2. (a) 1 (b) -2

Solving Equations and Inequalities

Now solve the equation.

$$76 = \frac{4}{3}x \quad \text{Combine terms}$$

$$\frac{3}{4}(76) = \frac{3}{4}\left(\frac{4}{3}x\right) \quad \text{Multiply by } \frac{3}{4}$$

$$57 = x$$

There were 57 orders for coffee and $(\frac{1}{3})(57) = 19$ orders for tea, giving a total of $57 + 19 = 76$ orders, as required. ◀

Work Problem 3 at the side.

The next example uses *consecutive odd integers*, which are odd integers in a row, such as 5 and 7, or 31 and 33. When you work with consecutive integers, the following information may be helpful.

For a pair of consecutive	use these variables:
integers,	x, $x + 1$;
odd integers,	x, $x + 2$;
even integers,	x, $x + 2$.

EXAMPLE 5 If the smaller of two consecutive odd integers is doubled, the result is 7 more than the larger of the two integers. Find the two integers.

Let x be the smaller integer. Since the two numbers are consecutive *odd* integers, $x + 2$ is the larger. Now write an equation from the words of the problem.

If the smaller is doubled	the result is	7	more than	the larger.
↓	↓	↓	↓	↓
$2x$	$=$	7	$+$	$x + 2$

Solve the equation.

$$2x = 7 + x + 2$$
$$2x = 9 + x \quad \text{Combine terms}$$
$$2x - x = 9 + x - x \quad \text{Subtract } x$$
$$x = 9 \quad \text{Combine terms}$$

The smaller integer is 9. The larger integer would be $9 + 2 = 11$. Check that twice the smaller integer is 7 more than the larger. ◀

EXAMPLE 6 Ten less than three times the smallest of three consecutive integers is nine more than the sum of the other two.

If x is the smallest integer, then the next consecutive integer is $x + 1$ and the one after that is $(x + 1) + 1 = x + 2$. Use the statement of the problem to write an equation.

Three times smallest	less 10	is	9	more than	the sum of the other two.
↓	↓	↓	↓	↓	↓
$3x$	$- 10$	$=$	9	$+$	$(x + 1) + (x + 2)$

3. Solve each word problem.

(a) A farmer has 28 more hens than roosters, with 170 chickens in all. Find the number of roosters the farmer has.

(b) In a given amount of time, Larry drove twice as far as Rick. Altogether they drove 90 miles. Find the number of miles driven by each.

ANSWERS
3. **(a)** 71 roosters **(b)** 60 for Larry, 30 for Rick

3.4 From Word Problems to Equations

4. Solve each word problem.

   (a) The larger of two consecutive integers is added to 14, giving a result twice the smaller. Find the integers.

   (b) If the largest of three consecutive integers is tripled, the result is twice the sum of the other two integers. Find the integers.

Solve this equation.
$$3x - 10 = 9 + x + 1 + x + 2$$
$$3x - 10 = 2x + 12 \quad \text{Combine terms}$$
$$x - 10 = 12 \quad \text{Subtract } 2x$$
$$x = 22 \quad \text{Add 10}$$

The smallest integer is 22, the next is 22 + 1 = 23, and the largest is 22 + 2 = 24. Check these numbers in the words of the original problem. ◀

Work Problem 4 at the side.

ANSWERS
4. (a) 15, 16 (b) 4, 5, 6

Solving Equations and Inequalities

3.4 EXERCISES

Solve the following word problems. Follow Steps 1–5. See Examples 1 and 2.

Step 1   Choose a variable to represent the unknown quantity.
Step 2   Translate the problem into an equation.
Step 3   Solve the equation.
Step 4   Answer the question asked in the problem.
Step 5   Check the solution by using the original words of the problem.

1. If three times a number is decreased by 2, the result is 22. Find the number.

2. When 6 is added to four times a number, the result is 42. Find the number.

3. The sum of a number and 3 is multiplied by 4, giving 36 as a result. Find the number.

4. The sum of a number and 8 is multiplied by 5, giving 60 as the answer. Find the number.

5. Twice a number is added to the number, giving 90. Find the number.

6. If the sum of a number and 8 is multiplied by -2, the result is -8. Find the number.

7. When 6 is subtracted from a number, the result is seven times the number. Find the number.

8. If 4 is subtracted from twice a number, the result is 4 less than the number. Find the number.

9. When five times a number is added to twice the number, the result is 10. Find the number.

10. If seven times a number is subtracted from eleven times a number, the result is 9. Find the number.

Solve the following word problems. Check the answers in the words of the original problem. See Examples 3 and 4.

11. Tony has a board 44 inches long. He wishes to cut it into two pieces so that one piece will be 6 inches longer than the other. How long should the shorter piece be?

12. Nevarez and Smith were opposing candidates in the school board election. Nevarez received 30 more votes than did Smith, with 516 total votes cast. How many votes did Smith receive?

13. On an algebra test, the highest grade was 42 points more than the lowest grade. The sum of the two grades was 138. Find the lowest grade.

14. In a physical fitness test, Rolfe did 25 more push-ups than Chuck did. The total number of push-ups for both men was 173. Find the number of push-ups that Chuck did.

15. A pharmacist found that at the end of the day she had $\frac{4}{3}$ as many prescriptions for antibiotics as she had for tranquilizers. She had 84 prescriptions altogether for these two types of drugs. How many did she have for tranquilizers?

16. Mark White gives rides in his glass-bottomed boat in the Bahama Islands. One day he noticed that the boat contained 17 more men (counting himself) than women. The number of men was 57 less than twice the number of women. How many women were on the boat?

17. Joann McKillip runs a dairy farm. Last year her cow Bessie gave 238 fewer gallons of milk than one of her other cows, Louise. The amount Louise gave was 375 gallons less than twice the amount produced by Bessie. How many gallons of milk did Bessie give?

18. Rick is 8 years older than Steve. The sum of their ages is 46 years. How old is each now?

Solve the following word problems. Check the answers in the words of the original problem. See Examples 5 and 6.

19. When the smaller of two consecutive integers is added to three times the larger, the result is 43. Find the integers.

20. If five times the smaller of two consecutive integers is added to three times the larger, the result is 59. Find the integers.

21. Find two consecutive odd integers such that when twice the larger is added to the smaller, the result is 169.

22. Find two consecutive even integers such that when the smaller is multiplied by 5, the result is 110 more than the larger.

23. The smallest of three consecutive integers is added to twice the largest, producing a result 15 less than four times the middle integer. Find the integers.

24. If the middle of three consecutive integers is added to 100, the result is 1 less than the sum of the third integer and twice the smallest. Find the integers.

3.4 Exercises 129

25. If 9 is added to the largest of three consecutive odd integers, the answer equals the sum of the first and second integers. Find the integers.

26. If the first and third of three consecutive even integers are added, the result is 20 more than the second integer. Find the integers.

The following word problems are real "head-scratchers."

27. Kevin is three times as old as Bob. Three years ago the sum of their ages was 22 years. How old is each now? (*Hint:* First write an expression for the age of each now, then for the age of each three years ago.)

28. A store has 39 quarts of milk, some in pint cartons and some in quart cartons. There are six times as many quart cartons as pint cartons. How many quart cartons are there? (*Hint:* 1 quart = 2 pints.)

29. A table is three times as long as it is wide. If it were 3 feet shorter and 3 feet wider, it would be square (with all sides equal). How long and how wide is the table?

30. Kevin works for $6 an hour. A total of 25% of his salary is deducted for taxes and insurance. How many hours must he work to take home $450?

31. Paula received a paycheck for $585 for her weekly wages less 10% deductions. How much was she paid before the deductions were made?

32. At the end of a day, the owner of a gift shop had $2394 in the cash register. This included sales tax of 5% on all sales. Find the amount of the sales.

3.5 FORMULAS

Many word problems can be solved with a formula. Formulas exist for geometric figures such as squares and circles, for distance, for money earned on bank savings, and for converting English measurements to metric measurements, for example. A list of the formulas used in this book is given inside the front and back covers.

▶ Given the values of all but one of the variables in a formula, the value of the remaining variable can be found by using the methods introduced in this chapter for solving equations.

Objectives

1▶ Solve a formula for one variable given the values of the other variables.

2▶ Use a formula to solve a word problem.

3▶ Solve a formula for a specified variable.

EXAMPLE 1 Find the value of the remaining variable in each of the following.

(a) $A = LW$; $A = 64, L = 10$

As shown in Figure 1, this formula gives the area of a rectangle with length L and width W. Substitute the given values into the formula and then solve for W.

$A = LW$
$64 = 10 W$ Let $A = 64, L = 10$
$6.4 = W$ Divide by 10

Check that the width of the rectangle is 6.4.

Rectangle
$A = LW$

Figure 1

(b) $A = \frac{1}{2}(b + B)h$; $A = 210, B = 27, h = 10$

This formula gives the area of a trapezoid with parallel sides of length b and B and distance h between the parallel sides. See Figure 2.

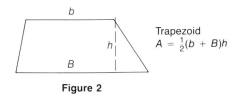

Trapezoid
$A = \frac{1}{2}(b + B)h$

Figure 2

Again, begin by substituting the given values into the formula.

$A = \frac{1}{2}(b + B)h$

$210 = \frac{1}{2}(b + 27)(10)$ $A = 210, B = 27, h = 10$

3.5 Formulas   131

1. Find the value of the remaining variable in each of the following.

   (a) $I = prt$; $I = \$246$, $r = .06, t = 2$

   (b) $P = 2L + 2W$; $P = 126$, $W = 25$

2. The perimeter of a triangle is 48 centimeters. One of the sides is 10 centimeters long, and the other two sides are equal in length. Find the length of each equal side.

Now solve for b.

$$210 = \frac{1}{2}(10)(b + 27) \quad \text{Commutative property}$$
$$210 = 5(b + 27)$$
$$210 = 5b + 135 \quad \text{Distributive property}$$
$$210 - 135 = 5b + 135 - 135 \quad \text{Subtract 135}$$
$$75 = 5b$$
$$\frac{75}{5} = \frac{5b}{5} \quad \text{Divide by 5}$$
$$15 = b$$

Check that the length of the shorter parallel side, b, is 15. ◀

Work Problem 1 at the side.

▶ As the next examples show, formulas are often used to solve word problems.

EXAMPLE 2 The perimeter of a square is 96 inches. Find the length of a side.

The list of formulas inside the front cover gives the formula $P = 4s$ for the **perimeter,** or the distance around a square, where s is the length of a side of a square. Here, the perimeter is given as 96 inches, so that $P = 96$. Substitute 96 for P in the formula.

$$P = 4s$$
$$96 = 4s \quad P = 96$$
$$24 = s \quad \text{Divide by 4}$$

Check this solution to see that each side of the square is 24 inches long. ◀

Work Problem 2 at the side.

EXAMPLE 3 The perimeter of a rectangle is 80 meters, and the length is 25 meters. (See Figure 1.) Find the width of the rectangle.

The distance around a rectangle is called the perimeter of the rectangle. The formula for the perimeter of a rectangle is found by adding the lengths of the four sides.

$$P = L + L + W + W = 2L + 2W$$

Find the width by substituting 80 for P and 25 for L in the formula $P = 2L + 2W$.

$$P = 2L + 2W$$
$$80 = 2(25) + 2W \quad P = 80,\; L = 25$$
$$80 = 50 + 2W \quad \text{Multiply}$$
$$30 = 2W \quad \text{Subtract 50}$$
$$15 = W \quad \text{Divide by 2}$$

Check this result. If W is 15 and L is 25, the perimeter will be

$$2(15) + 2(25) = 30 + 50 = 80,$$

as required. The width of the rectangle is 15 meters. ◀

ANSWERS
1. (a) $2050 (b) 38
2. 19 centimeters

Work Problem 3 at the side.

EXAMPLE 4 The area of a triangle is 126 square meters. The base of the triangle is 21 meters. Find the height.

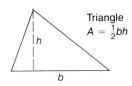

Figure 3

The formula for the area of a triangle is $A = \frac{1}{2}bh$, where A is area, b is the base, and h is the height. See Figure 3. Substitute 126 for A and 21 for b in the formula.

$$A = \frac{1}{2}bh$$

$$126 = \frac{1}{2}(21)h \qquad A = 126, b = 21$$

Simplify the problem by eliminating the fraction $\frac{1}{2}$. Multiply both sides of the equation by 2.

$$2(126) = 2\left(\frac{1}{2}\right)(21)h \qquad \text{Multiply by 2}$$

$$252 = 21h$$

$$12 = h \qquad \text{Divide by 21}$$

Check in the words of the given problem that the height of the triangle is 12 meters. ◀

Work Problem 4 at the side.

3 Sometimes it is necessary to solve a large number of problems that use the same formula. For example, a surveying class might need to solve several problems that involve the formula for the area of a rectangle, $A = LW$. Suppose that in each problem the area (A) and the length (L) of a rectangle are given and the width (W) must be found. Rather than solving for W each time the formula is used, it would be simpler to rewrite the *formula* so that it is solved for W. This process is called **solving for a specified variable.** As the following examples will show, solving a formula for a specified variable requires the same steps used earlier to solve equations with just one variable.

EXAMPLE 5 Solve $A = LW$ for W.

Think of undoing what has been done to W. Since W is multiplied by L, undo the multiplication by dividing both sides of $A = LW$ by L.

$$A = LW$$

$$\frac{A}{L} = \frac{LW}{L} \qquad \text{Divide by } L$$

$$\frac{A}{L} = W$$

The formula is now solved for W. ◀

3. Work the following word problems.

   (a) The width of a swimming pool is 50 feet, and the length is 75 feet. Find the perimeter of the pool.

   (b) A farmer has 1000 meters of fencing material to enclose a rectangular field. The width of the field will be 200 meters. Find the length.

4. Work the following word problems.

   (a) The base of a triangle is 35 centimeters. The height is 40 centimeters. Find the area of the triangle.

   (b) The area of a triangle is 54 square meters. The height is 18 meters. Find the length of the base of the triangle.

ANSWERS
3. (a) 250 feet   (b) 300 meters
4. (a) 700 square centimeters
   (b) 6 meters

3.5 Formulas

5. Solve each formula for the specified variable.

(a) $d = rt$; for t

(b) $I = prt$; for t

(c) $P = a + b + c$; for a

6. Solve each equation for x.

(a) $6x + 2y = 9$

(b) $y = \frac{2}{3}x - b$

(c) $y = \frac{1}{2}(x + h)$

7. Solve each formula for the specified variable.

(a) $A = p + prt$; for t

(b) $y = mx + b$; for x

ANSWERS

5. (a) $t = \dfrac{d}{r}$ (b) $t = \dfrac{I}{pr}$
   (c) $a = P - b - c$
6. (a) $x = \dfrac{9 - 2y}{6}$
   (b) $x = \dfrac{3y + 3b}{2}$ (c) $x = 2y - h$
7. (a) $\dfrac{A - p}{pr}$ (b) $\dfrac{y - b}{m}$

Work Problem 5 at the side.

The formula for converting temperatures given in degrees Celsius to degrees Fahrenheit is

$$F = \frac{9}{5}C + 32.$$

The next example shows how to solve this formula for C.

EXAMPLE 6 Solve $F = \frac{9}{5}C + 32$ for C.

First, undo the addition of 32 to $\frac{9}{5}C$ by subtracting 32 from both sides.

$$F = \frac{9}{5}C + 32$$

$$F - 32 = \frac{9}{5}C + 32 - 32 \quad \text{Subtract 32}$$

$$F - 32 = \frac{9}{5}C$$

Now multiply both sides by $\frac{5}{9}$. Use parentheses on the left.

$$\frac{5}{9}(F - 32) = \frac{5}{9} \cdot \frac{9}{5}C \quad \text{Multiply by } \frac{5}{9}$$

$$\frac{5}{9}(F - 32) = C$$

This last result is the formula for converting temperatures from Fahrenheit to Celsius. ◀

Work Problem 6 at the side.

EXAMPLE 7 Solve $P = 2L + 2W$ for L.

Begin by subtracting $2W$ on both sides.

$$P = 2L + 2W$$

$$P - 2W = 2L + 2W - 2W \quad \text{Subtract } 2W$$

$$P - 2W = 2L$$

$$\frac{P - 2W}{2} = \frac{2L}{2} \quad \text{Divide by 2}$$

$$\frac{P - 2W}{2} = L$$

The last step gives the formula solved for L, as required. ◀

Work Problem 7 at the side.

name date hour

3.5 EXERCISES

In the following exercises a formula is given, along with the values of all but one of the variables in the formula. Find the value of the variable that is not given. See Example 1.

1. $P = 4s$; $s = 32$

2. $P = 2L + 2W$; $L = 5, W = 3$

3. $A = \frac{1}{2}bh$; $b = 6, h = 12$

4. $A = \frac{1}{2}bh$; $b = 9, h = 24$

5. $d = rt$; $d = 8, r = 2$

6. $d = rt$; $d = 100, t = 5$

7. $A = \frac{1}{2}bh$; $A = 20, b = 5$

8. $A = \frac{1}{2}bh$; $A = 30, b = 6$

9. $P = 2L + 2W$; $P = 40, W = 6$

10. $P = 2L + 2W$; $P = 180, L = 50$

11. $V = \frac{1}{3}Bh$; $V = 80, B = 24$

12. $V = \frac{1}{3}Bh$; $V = 52, h = 13$

13. $C = 2\pi r$; $C = 9.42, \pi = 3.14$*

14. $C = 2\pi r$; $C = 25.12, \pi = 3.14$

15. $A = \pi r^2$; $r = 9, \pi = 3.14$

16. $A = \pi r^2$; $r = 15, \pi = 3.14$

17. $I = prt$; $I = 100, p = 500, r = .10$

18. $I = prt$; $I = 60, p = 150, r = .08$

---

*Actually, π is approximately equal to 3.14, not *exactly* equal to 3.14.

19. $V = LWH$; $V = 150, L = 10, W = 5$

20. $V = LWH$; $V = 800, L = 40, W = 10$

21. $A = \frac{1}{2}(b + B)h$; $A = 42, b = 5, B = 7$

22. $A = \frac{1}{2}(b + B)h$; $A = 70, b = 15, B = 20$

Use a formula to write an equation for each word problem; then solve it. Check the solution in the words of the original problem. The necessary formulas are given inside the front cover of the book. See Examples 2–4. Use 3.14 as an approximation for π.

23. The area of a rectangle is 60 square meters, and the width is 6 meters. Find the length.

24. The perimeter of a square is 80 centimeters. Find the length of a side.

25. The radius of a circle is 6 feet. Find the circumference.

26. The length of a rectangle is 15 inches, and the perimeter is 50 inches. Find the width.

27. The perimeter of a triangle is 72 meters. One side is 16 meters, and another side is 32 meters. Find the third side.

28. The shorter base of a trapezoid is 16 centimeters, and the longer base is 20 centimeters. The height is 6 centimeters. Find the area.

29. Find the radius of a circle with an area of 78.5 square inches.

30. The area of a triangle is 30 square centimeters. The base is 12 centimeters. Find the height.

31. The area of a trapezoid is 33 square meters. One base has a length of 8 meters and the height is 6 meters. Find the length of the other base.

32. The volume of a right circular cylinder with a radius of 4 centimeters is 502.4 cubic centimeters. Find the height of the cylinder.

33. Find the area of the base of a right pyramid with a height of 12 feet and a volume of 144 cubic feet.

34. Find the height of a right circular cylinder with a radius of 3 inches and a surface area of 339.12 square inches.

Solve the given formula for the indicated variable. See Examples 5–7.

35. $A = LW$; for L

36. $d = rt$; for r

37. $d = rt$; for t

3.5 Exercises

38. $V = LWH$; for W

39. $V = LWH$; for H

40. $I = prt$; for p

41. $I = prt$; for t

42. $C = 2\pi r$; for r

43. $c^2 = a^2 + b^2$; for a^2

44. $a + b + c = P$; for b

45. $A = \frac{1}{2}bh$; for b

46. $A = \frac{1}{2}bh$; for h

47. $V = \pi r^2 h$; for r^2

48. $V = \frac{1}{3}\pi r^2 h$; for r^2

49. $P = 2L + 2W$; for W

50. $A = p + prt$; for r

51. $P = A - Art$; for t

52. $y = ax + b$; for a

53. $d = gt^2 + vt$; for v

54. $S = 2\pi rh + 2\pi r^2$; for h

55. $A = \frac{r}{2l}$; for l

56. $R = \frac{E}{I}$; for I

57. $A = \frac{1}{2}(b + B)h$; for b

58. $C = \frac{5}{9}(F - 32)$; for F

3.6 RATIOS AND PROPORTIONS

▶ Ratios provide a way of comparing two numbers or quantities. A **ratio** is a quotient of two quantities with the same units.

> The ratio of the number a to the number b is written
>
> a to b, $a:b$,   or $\dfrac{a}{b}$.

This last way of writing a ratio is most common in algebra.

EXAMPLE 1   Write a ratio for each word phrase.

(a) The ratio of 5 hours to 3 hours is

$$\frac{5}{3}.$$

(b) To find the ratio of 6 hours to 3 days, first convert 3 days to hours.

$$3 \text{ days} = 3 \cdot 24$$
$$= 72 \text{ hours}$$

The ratio of 6 hours to 3 days is thus

$$\frac{6}{72} = \frac{1}{12}. \blacktriangleleft$$

Work Problem 1 at the side.

▶ A ratio is used to compare two numbers or amounts. A **proportion** says that two ratios are equal. For example,

$$\frac{3}{4} = \frac{15}{20}$$

is a proportion that says that the ratios $\frac{3}{4}$ and $\frac{15}{20}$ are equal. In the proportion

$$\frac{a}{b} = \frac{c}{d},$$

a, b, c, and d are the **terms** of the proportion. Beginning with the proportion

$$\frac{a}{b} = \frac{c}{d}$$

and multiplying both sides by the common denominator, bd, gives

$$bd \cdot \frac{a}{b} = bd \cdot \frac{c}{d}$$
$$\frac{b}{b}(d \cdot a) = \frac{d}{d}(b \cdot c) \quad \text{Associative and commutative properties}$$
$$ad = bc. \quad \text{Commutative property}$$

Objectives

▶ Write ratios.

▶ Decide whether proportions are true.

▶ Solve proportions.

▶ Solve word problems using proportions.

1. Write each ratio.

   (a) 9 women to 5 women

   (b) 4 inches to 1 foot

   (c) 3 days to 2 weeks

   (d) $6 to 10 quarters

ANSWERS

1. (a) $\dfrac{9}{5}$ (b) $\dfrac{4}{12}$ or $\dfrac{1}{3}$ (c) $\dfrac{3}{14}$
   (d) $\dfrac{24}{10}$ or $\dfrac{12}{5}$

2. Decide whether each proportion is true or false.

   (a) $\dfrac{7}{12} = \dfrac{35}{60}$

   (b) $\dfrac{21}{15} = \dfrac{62}{45}$

   (c) $\dfrac{58}{25} = \dfrac{638}{275}$

The products ad and bc are found by multiplying diagonally, as shown below.

$$\dfrac{a}{b} = \dfrac{c}{d}$$

with cross products bc and ad.

For this reason, ad and bc are called **cross products.**

> If $\dfrac{a}{b} = \dfrac{c}{d}$ then the cross products ad and bc are equal.
>
> Also, if $ad = bc$, then $\dfrac{a}{b} = \dfrac{c}{d}$.

From the rule given above, if $\dfrac{a}{b} = \dfrac{c}{d}$ then $ad = bc$. However, if $\dfrac{a}{c} = \dfrac{b}{d}$, then $ad = cb$, or $ad = bc$. This means that the two proportions are equivalent, and

the proportion $\dfrac{a}{b} = \dfrac{c}{d}$ can always be written as $\dfrac{a}{c} = \dfrac{b}{d}$.

Sometimes one form is more convenient to work with than the other.

EXAMPLE 2 Decide whether the following proportions are true or false.

(a) $\dfrac{3}{4} = \dfrac{15}{20}$

Check to see whether the cross products are equal.

$$4 \cdot 15 = 60$$
$$\dfrac{3}{4} = \dfrac{15}{20}$$
$$3 \cdot 20 = 60$$

The cross products are equal, so the proportion is true.

(b) $\dfrac{6}{7} = \dfrac{30}{32}$

The cross products are $6 \cdot 32 = 192$ and $7 \cdot 30 = 210$. The cross products are not equal, so the proportion is false. ◀

Work Problem 2 at the side.

▶ Four numbers are used in a proportion. If any three of these numbers are known, the fourth can be found.

EXAMPLE 3 (a) Find x in the proportion

$$\dfrac{5}{9} = \dfrac{x}{63}.$$

ANSWERS
2. (a) true (b) false (c) true

Solving Equations and Inequalities

The cross products must be equal,

$$5 \cdot 63 = 9 \cdot x$$
$$315 = 9x$$
$$35 = x. \quad \text{Divide by 9}$$

(b) Solve for r in the proportion

$$\frac{8}{5} = \frac{12}{r}.$$

Set the cross products equal to each other.

$$8r = 5 \cdot 12$$
$$8r = 60$$
$$r = \frac{60}{8} = \frac{15}{2} \blacktriangleleft$$

Work Problem 3 at the side.

EXAMPLE 4 Solve the equation

$$\frac{m-2}{m+1} = \frac{5}{3}.$$

Find the cross products.

$$3(m - 2) = 5(m + 1)$$
$$3m - 6 = 5m + 5$$
$$3m = 5m + 11 \quad \text{Add 6}$$
$$-2m = 11 \quad \text{Subtract } 5m$$
$$m = -\frac{11}{2} \quad \text{Divide by } -2 \blacktriangleleft$$

Work Problem 4 at the side.

▶ Proportions occur in many practical applications, as the next example shows.

EXAMPLE 5 A hospital charges a patient $7.80 for 12 capsules. How much should it charge for 18 capsules?

Let x be the cost of 18 capsules. Set up a proportion; one ratio in the proportion can involve the number of capsules, and the other ratio can use the costs. Make sure that corresponding numbers appear in the numerator and the denominator.

$$\begin{array}{cc} \text{Cost} & \text{Number} \\ \dfrac{\text{Cost of 18}}{\text{Cost of 12}} = \dfrac{18}{12} \\ \dfrac{x}{7.80} = \dfrac{18}{12} \end{array}$$

3. Solve each equation.

(a) $\dfrac{25}{11} = \dfrac{125}{k}$

(b) $\dfrac{y}{6} = \dfrac{35}{42}$

(c) $\dfrac{24}{a} = \dfrac{16}{15}$

4. Solve each equation.

(a) $\dfrac{z}{z+1} = \dfrac{2}{3}$

(b) $\dfrac{p+3}{p-5} = \dfrac{3}{4}$

(c) $\dfrac{a+6}{2a} = \dfrac{1}{5}$

ANSWERS

3. (a) 55 (b) 5 (c) $\dfrac{45}{2}$

4. (a) 2 (b) -27 (c) -10

3.6 Ratios and Proportions

5. Solve each word problem.

(a) On a map, 12 inches represents 500 miles. How many miles would be represented by 30 inches?

(b) If 7 shirts cost $87.50, find the cost of 11 shirts.

As shown above, this proportion could also be written as

$$\frac{18}{\text{Cost of 18}} = \frac{12}{\text{Cost of 12}}$$

$$\frac{18}{x} = \frac{12}{7.80}.$$

Solve either equation. Find the cross products, and set them equal.

$$12x = 18(7.80)$$
$$12x = 140.40$$
$$x = 11.70 \quad \text{Divide by 12}$$

The 18 capsules should cost $11.70. ◀

Work Problem 5 at the side.

ANSWERS
5. (a) 1250 miles **(b)** $137.50

3.6 EXERCISES

Write the following ratios. Write each ratio in lowest terms. See Example 1.

1. 30 miles to 20 miles
2. 50 feet to 90 feet
3. 72 dollars to 110 dollars

4. 120 people to 80 people
5. 6 feet to 5 yards
6. 10 yards to 8 feet

7. 30 inches to 4 feet
8. 100 inches to 5 yards
9. 12 minutes to 2 hours

10. 8 quarts to 5 pints
11. 4 dollars to 10 quarters
12. 35 dimes to 6 dollars

13. 20 hours to 5 days
14. 6 days to 9 hours
15. 80¢ to $3

Decide whether the following proportions are true. See Example 2.

16. $\dfrac{4}{7} = \dfrac{12}{21}$
17. $\dfrac{9}{10} = \dfrac{18}{20}$
18. $\dfrac{6}{8} = \dfrac{15}{20}$
19. $\dfrac{12}{18} = \dfrac{8}{12}$

20. $\dfrac{5}{8} = \dfrac{35}{56}$
21. $\dfrac{12}{7} = \dfrac{36}{20}$
22. $\dfrac{7}{10} = \dfrac{82}{120}$
23. $\dfrac{18}{20} = \dfrac{56}{60}$

24. $\dfrac{16}{40} = \dfrac{22}{55}$
25. $\dfrac{19}{30} = \dfrac{57}{90}$
26. $\dfrac{110}{18} = \dfrac{160}{27}$
27. $\dfrac{420}{600} = \dfrac{14}{20}$

28. $\dfrac{12.39}{8.91} = \dfrac{4.13}{2.97}$ 29. $\dfrac{.612}{1.05} = \dfrac{1.0404}{1.785}$ 30. $\dfrac{6.354}{7.823} = \dfrac{3.987}{4.096}$ 31. $\dfrac{3.827}{2.4513} = \dfrac{8.0946}{5.7619}$

Solve each of the following proportions. See Examples 3 and 4.

32. $\dfrac{35}{4} = \dfrac{k}{20}$ 33. $\dfrac{z}{56} = \dfrac{7}{8}$ 34. $\dfrac{m}{32} = \dfrac{3}{24}$

35. $\dfrac{6}{x} = \dfrac{4}{18}$ 36. $\dfrac{z-2}{80} = \dfrac{20}{100}$ 37. $\dfrac{25}{100} = \dfrac{8}{m+6}$

38. $\dfrac{2}{3} = \dfrac{y+2}{7}$ 39. $\dfrac{5}{8} = \dfrac{m-1}{5}$ 40. $\dfrac{5}{9} = \dfrac{2z+2}{15}$

41. $\dfrac{3}{4} = \dfrac{3n-1}{10}$ 42. $\dfrac{6}{2k+3} = \dfrac{8}{5}$ 43. $\dfrac{5}{4p-5} = \dfrac{3}{4}$

44. $\dfrac{m}{m-3} = \dfrac{5}{3}$ 45. $\dfrac{r+1}{r} = \dfrac{1}{3}$ 46. $\dfrac{3k-1}{k} = \dfrac{6}{7}$

47. $\dfrac{y}{6y-5} = \dfrac{5}{11}$ 48. $\dfrac{p+7}{p-1} = \dfrac{3}{4}$ 49. $\dfrac{r+8}{r-9} = \dfrac{7}{3}$

name   date   hour

Solve the following applications involving proportions. See Example 5.

50. A bus ride of 80 miles costs $5. Find the charge to ride 180 miles.

51. A certain lawn mower uses 3 tanks of gas to cut 10 acres of lawn. How many tanks of gas would be needed for 30 acres?

52. A Hershey bar contains 200 calories. How many bars would you need to eat to get 500 calories?

53. José can assemble 12 car parts in 40 minutes. How many minutes would he need to assemble 15 car parts?

54. If 2 pounds of fertilizer will cover 50 square feet of garden, how many pounds would be needed for 225 square feet?

55. The tax on a $20 item is $1. Find the tax on a $110 item.

56. On a road map, 3 inches represents 8 miles. How many inches would represent a distance of 24 miles?

57. A garden service charges $30 to install 50 square feet of sod. Find the charge to install 125 square feet.

58. If 9 pairs of jeans cost $121.50, find the cost of 5 pairs.

59. Suppose that 7 sacks of fertilizer cover 3325 square feet of lawn. Find the number of sacks needed for 7125 square feet.

60. The distance between two cities on a road map is 11 inches. Actually, the cities are 308 miles apart. The distance between two other cities is 15 inches. How far apart are these cities?

61. Twelve yards of material are needed for 5 dresses. How much material would be needed for 8 dresses?

62. The charge to move a load of freight 750 miles is $90. Find the charge to move the freight 1200 miles.

63. If 8 ounces of medicine must be mixed with 20 ounces of water, how many ounces of medicine must be mixed with 50 ounces of water?

3.7 THE ADDITION PROPERTY OF INEQUALITY

Inequalities are statements with algebraic expressions related by

$<$ "is less than"
$\leq$ "is less than or equal to"
$>$ "is greater than"
$\geq$ "is greater than or equal to."

An inequality is solved by finding all real number solutions for it. For example, the solution of $x \leq 2$ includes all *real numbers* that are less than or equal to 2, and not just the *integers* less than or equal to 2.

▶ Graphing is a good way to show the solution of an inequality. To graph all real numbers satisfying $x \leq 2$, place a dot at 2 on a number line and draw an arrow extending from the dot to the left (to represent the fact that all numbers less than 2 are also part of the graph.) The graph is shown in Figure 4.

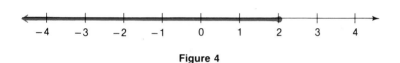

Figure 4

EXAMPLE 1   Graph $x > -5$.

The statement $x > -5$ says that x can take any value greater than -5, but x cannot equal -5 itself. Show this on a graph by placing an open circle at -5 and drawing an arrow to the right, as in Figure 5. The open circle at -5 shows that -5 is not part of the graph. ◀

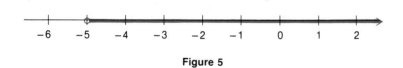

Figure 5

EXAMPLE 2   Graph $3 > x$.

The statement $3 > x$ means the same as $x < 3$. The graph of $x < 3$ is shown in Figure 6. ◀

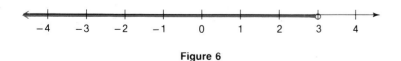

Figure 6

Work Problem 1 at the side.

Objectives

1. Graph intervals on a number line.
2. Learn the addition property of inequality.
3. Use the addition property to solve inequalities.
4. Write phrases from word problems as inequalities.
5. Use the addition property to solve inequalities involving numbers between other numbers.

1. Graph each of the following.

(a) $x \leq 3$

(b) $x > -4$

(c) $-4 \geq x$

(d) $0 < x$

ANSWERS

1.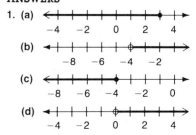

3.7 The Addition Property of Inequality   **147**

2. Graph each inequality.

(a) $4 < x \leq 8$

(b) $-7 < x < -2$

(c) $-6 < x \leq -4$

3. Solve each of the following inequalities. Graph each solution.

(a) $x - 1 < 6$

(b) $2m \geq m - 4$

(c) $-1 + 8r < 7r + 2$

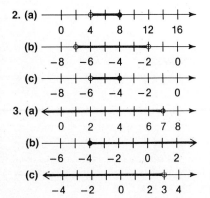

EXAMPLE 3 Graph $-3 \leq x < 2$.

The statement $-3 \leq x < 2$ is read "-3 is less than or equal to x and x is less than 2." Graph this inequality by placing a solid dot at -3 (because -3 is part of the graph) and an open circle at 2 (because 2 is not part of the graph). Then draw a line segment between the two circles, as in Figure 7. ◀

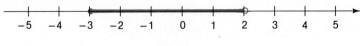

Figure 7

Work Problem 2 at the side.

Inequalities such as $x + 4 \leq 9$ are solved in much the same way as equations. First use the **addition property of inequality,** which states that the same term can be added to both sides of an inequality.

Addition Property of Inequality

For any expressions A, B, and C, the inequalities

$$A < B \quad \text{and} \quad A + C < B + C$$

have exactly the same solutions.

In other words, the same expression may be added to both sides of an inequality.

The addition property of inequality also works with $>$, $\leq$, or $\geq$. Just as with the addition property of equality, the same expression may be *subtracted* on both sides of an inequality.

The following examples show how the addition property is used to solve inequalities.

EXAMPLE 4 Solve the inequality $7 + 3k > 2k - 5$.

Use the addition property of inequality twice, once to get the terms containing k on one side of the inequality and a second time to get the integers together on the other side.

$$7 + 3k > 2k - 5$$
$$7 + 3k - 2k > 2k - 5 - 2k \quad \text{Subtract } 2k$$
$$7 + k > -5 \quad \text{Simplify}$$
$$7 + k - 7 > -5 - 7 \quad \text{Subtract } 7$$
$$k > -12$$

The graph of the solution, $k > -12$, is shown in Figure 8. ◀

Figure 8

Work Problem 3 at the side.

148   Solving Equations and Inequalities

EXAMPLE 5 Solve $6 + 3y \geq 4y - 5$.

First subtract $3y$ from both sides; then add 5 to both sides.

$6 + 3y - 3y \geq 4y - 5 - 3y$ Subtract $3y$
$6 \geq y - 5$
$6 + 5 \geq y - 5 + 5$ Add 5
$11 \geq y$

This solution is correct, but it is customary to write the solution to an inequality with the variable on the left. The statement $11 \geq y$ says that 11 is greater than or equal to y. This is the same as saying that y is less than or equal to 11, written

$$y \leq 11.$$

The graph of the solution $y \leq 11$ is shown in Figure 9. ◀

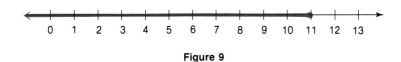

Figure 9

The inequality $6 + 3y \geq 4y - 5$ from Example 5 also can be solved by first subtracting $4y$ from both sides.

$6 + 3y - 4y \geq 4y - 5 - 4y$ Subtract $4y$
$6 - y \geq -5$
$6 - y - 6 \geq -5 - 6$ Subtract 6
$-y \geq -11$

Complete the solution by adding y and 11 on both sides.

$-y \geq -11$
$-y + y \geq -11 + y$ Add y
$0 \geq -11 + y$
$0 + 11 \geq -11 + y + 11$ Add 11
$11 \geq y$

As above, this solution can be rewritten as $y \leq 11$.

Work Problem 4 at the side.

▶ The next example shows how inequalities can be used in the solution of a word problem.

EXAMPLE 6 If 2 is added to five times a number, the result is greater than or equal to 5 more than four times the number. Find the number.

First translate this word problem into an inequality. Let x represent the unknown number. Then "2 is added to five times a number" is expressed as $5x + 2$, and "5 more than four times the number" is $4x + 5$. The two expressions are related by "is greater than or equal to."

$$5x + 2 \geq 4x + 5$$

4. Solve each of the following inequalities. Graph each solution.

(a) $3m \geq 2m + 1$

―――――――――

(b) $12y > 13y - 2$

―――――――――

(c) $1 - 6y \leq -5y$

―――――――――

(d) $2 + 9a \geq 10a - 8$

―――――――――

ANSWERS
4.
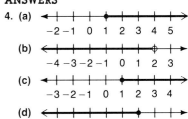

3.7 The Addition Property of Inequality   149

5. Solve the following word problems.

   (a) Twice a number is less than or equal to the sum of the number and 2. Find all possible values for the number.

   (b) If nine times a number is subtracted from 8, the result is less than the product of -10 and the number. Find all possible values for the number.

6. Solve and graph.

   (a) $5 \leq x - 3 \leq 8$

   (b) $-12 < a + 1 \leq -9$

   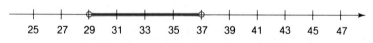

Begin the solution by subtracting $4x$ from both sides; then subtract 2 from both sides.

$$5x + 2 - 4x \geq 4x + 5 - 4x \quad \text{Subtract } 4x$$
$$x + 2 \geq 5$$
$$x + 2 - 2 \geq 5 - 2 \quad \text{Subtract } 2$$
$$x \geq 3$$

The solution is any number greater than or equal to 3. ◄

Work Problem 5 at the side.

Inequalities that say that one number is *between* two other numbers also can be solved by using the addition property of inequality.

EXAMPLE 7   Solve each inequality.

(a) $4 \leq x + 5 \leq 15$

Subtract 5 from each part of the inequality.

$$4 \leq x + 5 \leq 15$$
$$4 - 5 \leq x + 5 - 5 \leq 15 - 5 \quad \text{Subtract } 5$$
$$-1 \leq x \leq 10$$

A graph of this result is shown in Figure 10.

Figure 10

(b) $24 < y - 5 < 32$

Add 5 throughout.

$$24 + 5 < y - 5 + 5 < 32 + 5 \quad \text{Add } 5$$
$$29 < y < 37$$

The graph is shown in Figure 11. ◄

Figure 11

Work Problem 6 at the side.

ANSWERS
5. (a) $x \leq 2$ (b) $x < -8$
6. (a) $8 \leq x \leq 11$

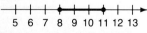

(b) $-13 < a \leq -10$

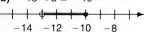

3.7 EXERCISES

Graph each inequality on the given number line. See Examples 1–3.

1. $x \leq 4$

2. $x \leq -3$

3. $a < 3$

4. $p > 4$

5. $-2 \leq x \leq 5$

6. $8 \leq m \leq 10$

7. $3 \leq y < 5$

8. $0 < y \leq 10$

Solve each inequality. See Examples 4 and 5.

9. $a + 6 < 8$

10. $k - 4 < 2$

11. $z - 3 \geq -2$

12. $p + 2 \geq -6$

13. $4x < 3x + 6$

14. $5x \leq 4x - 8$

15. $\dfrac{4}{3}k \leq \dfrac{1}{3}k - 9$

16. $\dfrac{5}{6}p > 5 - \dfrac{1}{6}p$

Solve each inequality and then graph your solutions. See Examples 4 and 5.

17. $3n + 5 \leq 2n - 6$

18. $5x - 2 < 4x - 5$

19. $\dfrac{7}{5}(y - 5) + \dfrac{2}{5} < \dfrac{2}{5}(y - 4)$

20. $\dfrac{4}{3}(x + 6) + \dfrac{25}{3} > \dfrac{1}{3}(x + 1)$

21. $-6(k + 2) + 3 \geq -7(k - 5)$

22. $-3(m - 5) + 8 < -4(m + 2)$

3.7 Exercises

23. $5(2k + 3) - 2(k - 8) > 3(2k + 4) + k - 2$

24. $2(3z - 5) + 4(z + 6) \geq 2(3z + 2) + 3(z - 5)$

Solve each inequality and then graph your solution. See Example 7.

25. $8 \leq p + 2 \leq 15$

26. $4 \leq k - 1 \leq 10$

27. $-3 < y - 8 < 4$

28. $-6 < r - 1 < -2$

In the following exercises, write an inequality using the information given in the problem and then solve it. See Example 6.

29. If four times a number is added to 8, the result is less than three times the number added to 5. Find all possible values of the number.

30. The product of 7 and a number is added to 4, giving a result that is greater than or equal to six times the number. Find all possible values of the number.

31. If the length of a rectangle is to be twice the width and the difference between the two dimensions is to be less than or equal to 7 meters, what is the largest possible value for the width?

32. The perimeter of a triangle must be no more than 55 centimeters. One side of the triangle is 18 centimeters, and a second side is 13 centimeters. Find the largest possible length for the third side.

33. The length of a rectangle is four times the width. If three times the width is subtracted from the length, the result is at least 15. Find the smallest possible values for the length and width of the rectangle.

34. The perimeter of a triangle cannot be more than 25 meters. One side is 7 meters. A second side is 9 meters. Find the longest possible length of the third side.

152 Solving Equations and Inequalities

3.8 THE MULTIPLICATION PROPERTY OF INEQUALITY

Objectives

1. Learn the multiplication property of inequality.
2. Use the multiplication property to solve inequalities.
3. Use the multiplication property to solve inequalities about numbers between other numbers.

▶ The addition property of inequality alone cannot be used to solve inequalities such as $4y \geq 28$. These inequalities require the *multiplication property of inequality*. To see how this property works, it will be helpful to look at some examples.

First start with the inequality $3 < 7$ and multiply both sides by the positive number 2.

$$3 < 7$$
$$2(3) < 2(7) \quad \text{Multiply both sides by 2}$$
$$6 < 14 \quad \text{True}$$

Now multiply both sides of $3 < 7$ by the negative number -5.

$$3 < 7$$
$$-5(3) < -5(7) \quad \text{Multiply both sides by } -5$$
$$-15 < -35 \quad \text{False}$$

Getting a true statement when multiplying both sides by -5 requires reversing the direction of the inequality symbol.

$$3 < 7$$
$$-5(3) > -5(7) \quad \text{Multiply by } -5\text{; reverse the symbol}$$
$$-15 > -35 \quad \text{True}$$

Take the inequality $-6 < 2$ as another example. Multiply both sides by the positive number 4.

$$-6 < 2$$
$$4(-6) < 4(2) \quad \text{Multiply by 4}$$
$$-24 < 8 \quad \text{True}$$

Multiplying both sides of $-6 < 2$ by -5 *and at the same time reversing the direction of the inequality symbol* gives

$$-6 < 2$$
$$(-5)(-6) > (-5)(2) \quad \text{Multiply by } -5$$
$$30 > -10. \quad \text{True}$$

Work Problem 1 at the side.

1. (a) Multiply both sides of $-2 < 8$ by 6 and then by -5.

   (b) Multiply both sides of $-4 > -9$ by 2 and then by -8.

In summary, the two parts of the multiplication property of inequality are as stated here.

Multiplication Property of Inequality

For any expressions A, B, and C $(C \neq 0)$,

(1) if C is *positive*, then the inequalities

$$A < B \quad \text{and} \quad AC < BC$$

have exactly the same solutions;

(2) if C is *negative*, then the inequalities

$$A < B \quad \text{and} \quad AC > BC$$

have exactly the same solutions.

In other words, both sides of an inequality may be multiplied by the same nonzero number. If the number is negative, reverse the inequality symbol.

ANSWERS
1. (a) $-12 < 48$; $10 > -40$
   (b) $-8 > -18$; $32 < 72$

2. Solve each inequality. Graph the solution.

(a) $5x \geq 30$

(b) $9y < -18$

(c) $-4p \leq 32$

(d) $-2r > -12$

(e) $-5p \leq 0$

The multiplication property of inequality also works with $>$, $\leq$, or $\geq$.

It is important to remember the differences in the multiplication property for positive and negative numbers.

1. When both sides of an inequality are multiplied or divided by a positive number, the direction of the inequality symbol *does not change*. Adding or subtracting terms on both sides also does not change the symbol.
2. When both sides of an inequality are multiplied or divided by a negative number, the direction of the symbol *does change*. Reverse the symbol of inequality only when multiplying or dividing by a negative number.

▶ The next examples show how to use the multiplication property to solve inequalities.

EXAMPLE 1 Solve the inequality $3r < 18$.

Use the multiplication property of inequality and divide both sides by 3. Since 3 is a positive number, the direction of the inequality symbol *does not* change.

$$3r < 18$$
$$\frac{3r}{3} < \frac{18}{3} \quad \text{Divide by 3}$$
$$r < 6$$

The graph of this solution is shown in Figure 12. ◀

Figure 12

EXAMPLE 2 Solve the inequality $-4t \geq 8$.

Here both sides of the inequality must be divided by -4, a negative number, which *does* change the direction of the inequality symbol.

$$-4t \geq 8$$
$$\frac{-4t}{-4} \leq \frac{8}{-4} \quad \text{Divide by } -4; \text{ symbol reversed}$$
$$t \leq -2$$

The solution is graphed in Figure 13. ◀

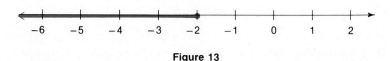

Figure 13

Work Problem 2 at the side.

EXAMPLE 3 Solve the inequality $-y \leq -11$.

In the previous section this inequality was solved by using the addition property. It can be solved more quickly by using the multiplication

ANSWERS
2. (a) $x \geq 6$

(b) $y < -2$
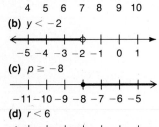
(c) $p \geq -8$

(d) $r < 6$

(e) $p \geq 0$

property of inequality; multiply both sides by -1. Since -1 is negative, change the direction of the inequality symbol.

$$-y \leq -11$$
$$(-1)(-y) \geq (-1)(-11) \quad \text{Multiply by } -1$$
$$y \geq 11$$

The solution is graphed in Figure 14. ◀

Figure 14

Work Problem 3 at the side.

The steps in solving an inequality are summarized below. (Remember that $<$ can be replaced with $>$, $\leq$, or $\geq$ in this summary.)

> **Solving Inequalities**
> *Step 1* Use the associative, commutative, and distributive properties to combine like terms on both sides of the inequality.
> *Step 2* Use the addition property of inequality to simplify the inequality to one of the form $ax < b$, where a and b are real numbers.
> *Step 3* Use the multiplication property of inequality to simplify further to an inequality of the form $x < c$ or $x > c$, where c is a real number.

Notice how these steps are used in the next example.

EXAMPLE 4 Solve the inequality $3z + 2 - 5 > -z + 7 + 2z$.

Step 1 Simplify and combine terms.
$$3z + 2 - 5 > -z + 7 + 2z$$
$$3z - 3 > z + 7$$

Step 2 Use the addition property of inequality.
$$3z - 3 \boxed{+3} > z + 7 \boxed{+3} \quad \text{Add 3}$$
$$3z > z + 10$$
$$3z \boxed{-z} > z + 10 \boxed{-z} \quad \text{Subtract } z$$
$$2z > 10$$

Step 3 Use the multiplication property of inequality.
$$\frac{2z}{\boxed{2}} > \frac{10}{\boxed{2}} \quad \text{Divide by 2, which is positive}$$
$$z > 5$$

Since 2 is positive, the direction of the inequality symbol was not changed in the third step. A graph of the solution is shown in Figure 15. ◀

Figure 15

3.8 The Multiplication Property of Inequality

3. Solve each inequality. Graph the solution.

(a) $-p > 4$

(b) $-r \leq -2$

ANSWERS
3. (a)

4. Solve. Graph each solution.

   (a) $2m - 4 \geq 3m - 1$

   (b) $5r - r + 2 < 7r - 5$

5. Solve $4(y - 1) - 3y > -15 - (2y + 1)$. Graph the solution.

6. Solve $2 \leq 3p - 1 \leq 8$. Graph the solution.

ANSWERS

4. (a) $m \leq -3$

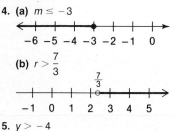

5. $y > -4$

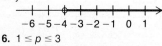

6. $1 \leq p \leq 3$

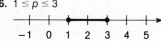

Work Problem 4 at the side.

EXAMPLE 5 Solve $5(k - 3) - 7k \geq 4(k - 3) + 9$.

Step 1 Use the distributive property; then combine like terms.

$$5(k - 3) - 7k \geq 4(k - 3) + 9$$
$$5k - 15 - 7k \geq 4k - 12 + 9 \quad \text{Distributive property}$$
$$-2k - 15 \geq 4k - 3 \quad \text{Combine like terms}$$

Step 2 Use the addition property.

$$-2k - 15 - 4k \geq 4k - 3 - 4k \quad \text{Subtract } 4k$$
$$-6k - 15 \geq -3$$
$$-6k - 15 + 15 \geq -3 + 15 \quad \text{Add 15}$$
$$-6k \geq 12$$

Step 3 Divide both sides by -6, a negative number. Change the direction of the inequality symbol.

$$\frac{-6k}{-6} \leq \frac{12}{-6} \quad \text{Divide by } -6$$
$$k \leq -2$$

A graph of the solution is shown in Figure 16.

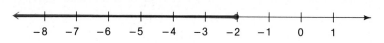

Figure 16

Work Problem 5 at the side.

▶ The next example shows how to solve inequalities in which one expression is between two other ones.

EXAMPLE 6 Solve $4 \leq 3x - 5 < 6$.

First add 5 to each part.

$$4 \leq 3x - 5 < 6$$
$$4 + 5 \leq 3x - 5 + 5 < 6 + 5 \quad \text{Add 5}$$
$$9 \leq 3x < 11$$

Now divide each part by the positive number 3.

$$\frac{9}{3} \leq \frac{3x}{3} < \frac{11}{3} \quad \text{Divide by 3}$$
$$3 \leq x < \frac{11}{3}$$

A graph of the solution is shown in Figure 17.

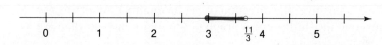

Figure 17

Work Problem 6 at the side.

name _____ date _____ hour _____

3.8 EXERCISES

Solve each inequality, and graph the solution. See Examples 1–5.

1. $3x < 27$

2. $5h \geq 20$

3. $4r \geq -12$

4. $6a < -18$

5. $-2k \leq 12$

6. $-3v > 6$

7. $4k + 1 \geq 2k - 9$

8. $5y + 3 < 2y + 12$

9. $4q + 1 - 5 < 8q + 4$

10. $5x - 2 \leq 2x + 6 - x$

11. $3 - 2z + 2 > 4 - z$

12. $3 - 5t < 8 + 2t - 2$

13. $12 - w \leq 4 + 4w - 7$

14. $10 - 4k + 8 \geq 6 - 2k + 10$

15. $-k + 4 + 7k \leq -1 + 3k + 5$

16. $6y - 2y - 4 + 7y > 3y - 4 + 7y$

17. $2(x - 5) + 3x < 4(x - 6) + 3$

18. $5(t + 3) - 6t \leq 3(2t + 1) - 4t$

19. $-3.762p > 3.68676$

20. $.428k \leq -.53928$

Solve each inequality, and graph the solution. See Example 6.

21. $-5 \leq 2x - 3 \leq 9$

22. $-7 \leq 3x - 4 \leq 8$

23. $-12 \leq \dfrac{1}{2}z + 1 \leq 4$

24. $-6 \leq \dfrac{1}{3}a - 3 \leq 5$

Solve each word problem.

25. One side of a triangle is twice as long as a second side. The third side of the triangle is 17 feet long. The perimeter of the triangle cannot be more than 50 feet. Find the longest possible values for the other two sides of the triangle.

26. The perimeter of a rectangle must be no greater than 120 meters. The width of the rectangle must be 22 meters. Find the greatest possible value for the length of the rectangle.

27. One side of a rectangle is 8 meters long. The area of the rectangle must be at least 240 square meters. Find the shortest possible length for the rectangle.

28. A triangle has a height of 20 meters. The area of the triangle must be less than or equal to 40 square meters. Find the greatest possible length for the base of the triangle.

29. In order to qualify for a company pension plan, an employee must average at least $1000 per month in earnings. During the first four months of the year, an employee made $900, $1200, $1040, and $760. What amount of earnings during the fifth month will qualify the employee?

30. When four times a number is subtracted from 8, the result is less than 15. Find all numbers that satisfy this condition.

31. If half a number is added to 5, the result is greater than or equal to -3. Find all such numbers.

32. If two-thirds of a number is added to -3, the result is no more than 2. Find all such numbers.

158 Solving Equations and Inequalities

CHAPTER 3   REVIEW EXERCISES

(3.1–3.3)   *Solve each equation. Check each solution.*

1. $m - 5 = 1$
2. $y + 8 = -4$
3. $3k + 1 = 2k + 8$

4. $5k = 4k + \dfrac{2}{3}$
5. $(4r - 2) - (3r + 1) = 8$
6. $3(2y - 5) = 2 + 5y$

7. $7k = 35$
8. $12r = -48$
9. $2p - 7p + 8p = 15$

10. $\dfrac{m}{12} = -1$
11. $\dfrac{5}{8}k = 8$
12. $2k - 5 = 4k + 7$

13. $2 - 3(y - 5) = 4 + y$
14. $2(5m - 1) = 4(m + 3) - 2$
15. $3 - 3(6 - r) = -2(r + 5)$

(3.4)   *Solve the word problems. Check your answers.*

16. If 4 is subtracted from twice a number, the result is 8. Find the number.

17. The sum of a number and 5 is multiplied by 6, giving 72 as a result. Find the number.

18. If three times the smaller of two consecutive odd integers is added to twice the larger, the result is 99. Find the integers.

19. On a test in geometry, the highest grade was 35 points more than the lowest. The sum of the highest and lowest grades was 157. Find the lowest and the highest scores.

20. In a marathon, Susan ran $\frac{2}{3}$ as far as Linda. In all, the two people ran 30 miles. How many miles did Susan run?

21. Find three consecutive integers such that the sum of the smaller two equals the larger less 4.

(3.5) *A formula is given in the following exercises, along with the values of some of the variables. Find the value of the variable that is not given.*

22. $A = \frac{1}{2}bh$; $A = 22$; $b = 4$

23. $A = \frac{1}{2}(b + B)h$; $b = 9, B = 12, h = 8$

24. $C = 2\pi r$; $C = 12.56, \pi = 3.14$

25. $V = \frac{4}{3}\pi r^3$; $\pi = 3.14, r = 1$

Solve each formula for the specified variable.

26. $A = LW$; for W

27. $I = prt$; for r

28. $m + a = x^2 - m$; for m

29. $A = \frac{1}{2}(b + B)h$; for h

Solve the word problems. Check your answers.

30. The area of a triangle is 25 square meters. The base is 10 meters in length. Find the height.

31. The shorter base of a trapezoid is 42 centimeters long, and the longer base is 48 centimeters long. The height of the trapezoid is 8 centimeters. Find the area of the trapezoid.

(3.6) *Write the following ratios in lowest terms.*

32. 50 centimeters to 30 centimeters

33. 6 days to 1 week

34. 45 inches to 5 feet

35. 2 months to 3 years

Decide whether or not the following proportions are true.

36. $\dfrac{15}{18} = \dfrac{45}{54}$

37. $\dfrac{11}{19} = \dfrac{55}{105}$

Solve each equation. Check your answers.

38. $\dfrac{p}{5} = \dfrac{21}{30}$

39. $\dfrac{3}{5} = \dfrac{k}{12}$

40. $\dfrac{y}{y-5} = \dfrac{7}{2}$

41. $\dfrac{2+m}{2-m} = \dfrac{3}{4}$

Solve the word problems. Check your answers.

42. If 1 quart of oil must be mixed with 24 quarts of gasoline, how much oil would be needed for 192 quarts of gasoline?

43. Lupe can paint 3 walls in 7 hours. How long would it take her to paint 27 walls?

44. If 15 yards of cloth are needed for 18 nurses' smocks, how much cloth would be needed for 90 smocks?

45. The distance between two cities on a road map is 16 centimeters. The two cities are actually 150 kilometers apart. The distance on the map between two other cities is 40 centimeters. How far apart are these cities?

(3.7–3.8) *Graph each inequality on a number line.*

46. $m \geq -2$

47. $-5 \leq p < 6$

Chapter 3 Review Exercises

Solve each inequality. Graph the solution.

48. $y + 5 \geq 2$

49. $5y > 4y + 8$

50. $9(k - 5) - (3 + 8k) \geq 5$

51. $3(2z + 5) + 4(8 + 3z) \leq 5(3z + 2) + 2z$

52. $-6 \leq x + 2 \leq 0$

53. $3 < y - 4 < 5$

54. $6k \geq -18$

55. $-11y < 22$

56. $-z > -4$

57. $5m - 9 \geq 7m + 3$

58. $6p - 5p > 2 - 4p + 7p + 8$

59. $-(y + 2) + 3(2y - 7) \leq 4 - 5y$

60. $-3 \leq 2m + 1 \leq 4$

61. $9 < 3m + 5 \leq 20$

162 Solving Equations and Inequalities

Solve each problem and check your answer in the words of the original problem.

62. The perimeter of a square cannot be greater than 200 meters. Find the greatest possible value for the length of a side.

63. One side of a triangle is 3 centimeters longer than the short side. The third side is twice as long as the short side. If the perimeter of the triangle cannot exceed 39 centimeters, find the greatest possible length for the short side.

Here are some additional word problems. Solve each problem and check your answer in the words of the original problem.

64. Two-thirds of a number added to the number is 10. What is the number?

65. If three-fourths of a number is subtracted from twice the number, the result is 15. Find the number.

66. If the larger of two consecutive even integers is subtracted from twice the smaller, the result is 88. What is the smaller integer?

67. The sum of three consecutive integers is 144. What is the first of the three numbers.

Chapter 3   Review Exercises

68. The perimeter of a rectangle is 288 feet. The length is 4 feet longer than the width. Find the width.

69. The perimeter of a triangle is 84 meters. One side is twice as long as another, and the third side is 18 meters long. What is the length of the longest side?

70. The area of a triangle is 182 square inches. The height is 14 inches. Find the length of the base.

71. The perimeter of a rectangle is 86 inches. The width is 17 inches. What is the length?

72. The Celsius temperature is 22°C. What is the corresponding Fahrenheit temperature?

73. Kay bought a car for $6988. She made a $2000 down payment and paid off the car with a single payment including simple interest of 7.7% for 4 years on the balance. How much interest did she pay?

name _____ date _____ hour _____

74. On a recent diet, Joe lost 18 more pounds than Sue. Their total weight loss was 42 pounds. How much did Sue lose?

75. Rick and Steve drove from different towns to a family reunion. Rick drove 43 miles farther than Steve. The two men drove a total of 293 miles. How far did Steve drive to the reunion?

76. A teacher noted that one week he had graded 32 more algebra tests than geometry tests. He had graded 102 tests altogether. How many geometry tests did he grade that week?

77. A parking lot attendant parked 63 cars one day. He parked 11 fewer large cars than small cars. How many small cars did he park?

78. A bird-watcher in the North Atlantic counted 23 more fulmars than shearwaters. Altogether she counted 53 birds of the two species. How many were shearwaters?

79. A recipe calls for $1\frac{1}{3}$ cups of milk. How much milk would be needed to make half the recipe?

Chapter 3 Review Exercises   **165**

80. On a certain map, 2 inches represents 100 kilometers. How many kilometers would 5 inches represent?

81. The tax on an $18 item is $1.17. How much tax would be charged on a $215 item?

82. It costs $4.50 to plant 2 square yards with a certain ground cover. How much would it cost to plant $6\frac{1}{2}$ square yards?

83. If $7\frac{3}{4}$ yards of fabric are needed to make 2 dresses, how many yards would be needed to make 9 dresses?

name _____   date _____ hour _____

CHAPTER 3 TEST

Solve each equation and check the solution.

1. $2m - 5 = 3$

2. $6v + 3 = 8v - 7$

3. $3(a + 12) = 1 - 2(a - 5)$

4. $4k - 6k + 8(k - 3) = -2(k + 12)$

5. $\dfrac{m}{5} = 2$

6. $6x - 4 = 12$

7. $4 - (3 - m) = 12 + 3m$

8. $-(r + 4) = 2 + r$

9. If 8 is added to four times a number, the result is -12. Find the number.

10. Vern paid $57 more to tune up his Bronco than his Oldsmobile. He paid $257 in all. How much did he pay for the tune-up on the Oldsmobile?

11. Find two consecutive odd integers such that 13 subtracted from twice the smaller gives the larger integer.

12. _____

13. _____

14. _____

15. _____

16. _____

17. _____

18. _____

19. _____

20. _____→

21. _____→

22. _____→

23. _____→

24. _____→

12. Solve the formula $I = prt$ for p.

13. Solve the formula $m = an + bp$ for n.

14. Solve the formula $A = \frac{1}{2}(b + B)h$ for h.

15. The perimeter of a rectangle is 48 meters. The width is 10 meters. Find the length.

16. Is the proportion $\frac{15}{79} = \frac{465}{2449}$ true or not?

Solve.

17. $\frac{z}{16} = \frac{3}{48}$

18. $\frac{y + 5}{y - 2} = \frac{1}{4}$

19. If 11 hamburgers cost $6.05, find the cost of 32 hamburgers.

Solve the following inequalities. Graph each solution.

20. $x - 1 \leq 3$

21. $-2m < -14$

22. $5(k - 2) + 3 \leq 2(k - 3) + 2k$

23. $-4r + 2(r - 3) \geq 5r - (3 + 6r) + 1 - 8$

24. $-8 < 3k - 2 \leq 12$

Solving Equations and Inequalities

Exponents and Polynomials 4

4.1 EXPONENTS

In Chapter 1 exponents were used to write repeated products:

$$5^2 = 5 \cdot 5 = 25,$$
$$4^3 = 4 \cdot 4 \cdot 4 = 64$$
$$9^1 = 9,$$

and so on. In the expression 5^2, the number 5 is called the **base** and 2 is called the **exponent**. The expression 5^2 is called an **exponential expression**.

EXAMPLE 1 Write $3 \cdot 3 \cdot 3 \cdot 3 \cdot 3$ in exponential form and evaulate the exponential expression.

Since 3 occurs as a factor five times, the base is 3 and the exponent is 5. The exponential expression is 3^5. The value is

$$3^5 = 3 \cdot 3 \cdot 3 \cdot 3 \cdot 3 = 243.$$ ◀

Work Problem 1 at the side.

EXAMPLE 2 Evaluate each exponential expression. Name the base and the exponent.

	Base	Exponent
(a) $5^4 = 5 \cdot 5 \cdot 5 \cdot 5 = 625$	5	4
(b) $-5^4 = -1 \cdot 5^4 = -1 \cdot (5 \cdot 5 \cdot 5 \cdot 5) = -625$	5	4
(c) $(-5)^4 = (-5)(-5)(-5)(-5) = 625$	-5	4 ◀

It is important to understand the differences between parts (b) and (c) of Example 2. In -5^4 the lack of parentheses shows that the exponent 4 refers only to the base 5, and not -5; in $(-5)^4$ the parentheses show that the exponent 4 refers to the base -5. In summary,

$-a^n$ and $(-a)^n$ are not necessarily the same.

Expression	Base	Exponent	Example
$-a^n$	a	n	$-3^2 = -(3 \cdot 3) = -9$
$(-a)^n$	$-a$	n	$(-3)^2 = (-3)(-3) = 9$

Objectives

▶ Use exponents.
▶ Learn the product rule for exponents.
▶ Use zero as an exponent.
▶ Use negative numbers as exponents.
▶ Learn the quotient rule for exponents.

1. Write each product in exponential form and evaluate.

   (a) $2 \cdot 2 \cdot 2 \cdot 2$

   (b) $7 \cdot 7 \cdot 7$

ANSWERS
1. (a) $2^4 = 16$ (b) $7^3 = 343$

4.1 Exponents

2. Evaluate each exponential expression. Name the base and the exponent.

   (a) 6^3

   (b) $(-2)^5$

   (c) -2^5

   (d) -3^2

   (e) $(-3)^2$

3. Find each product by the product rule, if possible.

   (a) $8^2 \cdot 8^5$

   (b) $3^5 \cdot 2^6$

   (c) $(-7)^5 \cdot (-7)^3$

   (d) $y^3 \cdot y$

   (e) $k^4 \cdot k^5$

4. Multiply.

   (a) $5m^2 \cdot 2m^6$

   (b) $3p^5 \cdot 9p^4$

   (c) $-7p^5 \cdot (3p^8)$

ANSWERS
2. (a) 216; 6; 3 (b) -32; -2; 5
   (c) -32; 2; 5 (d) -9; 3; 2
   (e) 9; -3; 2
3. (a) 8^7 (b) cannot use product rule (c) $(-7)^8$ (d) y^4 (e) k^9
4. (a) $10m^8$ (b) $27p^9$ (c) $-21p^{13}$

Work Problem 2 at the side.

▶ One reason for the importance of exponents is that several useful rules can be developed as shortcuts for working many problems. For example, by the definition of exponential expressions,

$$2^4 \cdot 2^3 = (2 \cdot 2 \cdot 2 \cdot 2)(2 \cdot 2 \cdot 2)$$
$$= 2 \cdot 2 \cdot 2 \cdot 2 \cdot 2 \cdot 2 \cdot 2$$
$$= 2^7.$$

Generalizing from this example, $2^4 \cdot 2^3 = 2^{4+3} = 2^7$, suggests the **product rule for exponents.**

> For any positive integers m and n, $a^m \cdot a^n = a^{m+n}$.

Be careful: The bases must be the same before the product rule for exponents can be applied.

EXAMPLE 3 Use the product rule for exponents to find each product.

(a) $6^3 \cdot 6^5 = 6^{3+5} = 6^8$ by the product rule.

(b) $(-4)^5(-4)^3 = (-4)^{5+3} = (-4)^8$ by the product rule.

(c) The product rule does not apply to the product $2^3 \cdot 3^2$, since the bases are different.

(d) $x^2 \cdot x = x^2 \cdot x^1 = x^{2+1} = x^3$

(e) $m^4 \cdot m^3 = m^{4+3} = m^7$ ◀

Work Problem 3 at the side.

EXAMPLE 4 Multiply $2x^3$ and $3x^7$.

Since $2x^3$ means $2 \cdot x^3$ and $3x^7$ means $3 \cdot x^7$, use the associative and commutative properties and the product rule to get

$$2x^3 \cdot 3x^7 = 2 \cdot 3 \cdot x^3 \cdot x^7 = 6x^{10}. \;◀$$

Be sure you understand the difference between *adding* and *multiplying* terms. For example,

$$8x^3 + 5x^3 = 13x^3,$$

but

$$(8x^3)(5x^3) = 8 \cdot 5 \cdot x^{3+3} = 40x^6.$$

Work Problem 4 at the side.

▶ The rule for division with exponents is similar to the product rule for exponents. For example,

$$\frac{6^5}{6^2} = \frac{6 \cdot 6 \cdot 6 \cdot 6 \cdot 6}{6 \cdot 6} = 6 \cdot 6 \cdot 6 = 6^3.$$

The difference of the exponents, $5 - 2$, gives the exponent in the quotient, 3.

If the exponents in the numerator and denominator are equal, then, for example,

$$\frac{6^5}{6^5} = \frac{6 \cdot 6 \cdot 6 \cdot 6 \cdot 6}{6 \cdot 6 \cdot 6 \cdot 6 \cdot 6} = 1.$$

If, however, the exponents are subtracted as above,

$$\frac{6^5}{6^5} = 6^{5-5} = 6^0.$$

This means that $6^0 = 1$. Based on this,

for any nonzero real number a, $\quad a^0 = 1 \quad (a \neq 0)$.

EXAMPLE 5 Evaluate each exponential expression.

(a) $60^0 = 1$

(b) $(-60)^0 = 1$

(c) $-60^0 = -(1) = -1$

(d) $y^0 = 1$, if $y \neq 0$

(e) $-r^0 = -1$, if $r \neq 0$ ◀

Notice the difference between parts (b) and (c) of Example 5. In Example 5(b) the base is -60 and the exponent is 0. Any nonzero base raised to a zero exponent is 1. But in Example 5(c), the base is 60. Then $60^0 = 1$, and $-60^0 = -1$.

Work Problem 5 at the side.

▶ In the discussion above, we found that

$$\frac{6^5}{6^2} = 6^3,$$

where the bottom exponent was subtracted from the top exponent. If the bottom exponent were larger than the top exponent, subtracting would result in a negative exponent.

For example,

$$\frac{6^2}{6^5} = 6^{2-5} = 6^{-3}.$$

On the other hand,

$$\frac{6^2}{6^5} = \frac{6 \cdot 6}{6 \cdot 6 \cdot 6 \cdot 6 \cdot 6} = \frac{1}{6 \cdot 6 \cdot 6} = \frac{1}{6^3},$$

so that

$$6^{-3} = \frac{1}{6^3}.$$

This example suggests that **negative exponents** be defined as follows.

5. Evaluate.

(a) 28^0

(b) $(-16)^0$

(c) -7^0

(d) $m^0, \quad m \neq 0$

(e) $-p^0, \quad p \neq 0$

ANSWERS
5. (a) 1 (b) 1 (c) -1 (d) 1
   (e) -1

4.1 Exponents

6. Simplify by using the definition of negative exponents.

   (a) 4^{-3}

   (b) 6^{-2}

   (c) $\left(\dfrac{2}{3}\right)^{-2}$

   (d) $2^{-1} + 5^{-1}$

   (e) $m^{-5}, \quad m \neq 0$

   (f) $\dfrac{1}{z^{-4}}, \quad z \neq 0$

For any nonzero real number a and any integer n,
$$a^{-n} = \frac{1}{a^n} \quad (a \neq 0).$$

By definition, a^{-n} and a^n are reciprocals, since
$$a^n \cdot a^{-n} = a^n \cdot \frac{1}{a^n} = 1.$$

The definition of a^{-n} also can be written as
$$a^{-n} = \frac{1}{a^n} = \left(\frac{1}{a}\right)^n.$$

For example, using the last result above,
$$6^{-3} = \left(\frac{1}{6}\right)^3 \quad \text{and} \quad \left(\frac{1}{3}\right)^{-2} = 3^2.$$

EXAMPLE 6 Simplify by using the definition of negative exponents.

(a) $3^{-2} = \dfrac{1}{3^2} = \dfrac{1}{9}$

(b) $5^{-3} = \dfrac{1}{5^3} = \dfrac{1}{125}$

(c) $\left(\dfrac{1}{2}\right)^{-3} = 2^3 = 8$

(d) $\left(\dfrac{2}{5}\right)^{-4} = \left(\dfrac{5}{2}\right)^4$

(e) $4^{-1} - 2^{-1} = \dfrac{1}{4} - \dfrac{1}{2} = \dfrac{1}{4} - \dfrac{2}{4} = -\dfrac{1}{4}$

(f) $p^{-2} = \dfrac{1}{p^2}, \quad p \neq 0$

(g) $\dfrac{1}{x^{-4}} = \left(\dfrac{1}{x}\right)^{-4} = x^4, \quad x \neq 0$ ◀

Be careful:

a negative exponent does not indicate a negative number; negative exponents lead to reciprocals.

Expression	Example	
a^{-n}	$3^{-2} = \dfrac{1}{3^2} = \dfrac{1}{9}$	Not negative
$-a^{-n}$	$-3^{-2} = -\dfrac{1}{3^2} = -\dfrac{1}{9}$	Negative

Work Problem 6 at the side.

ANSWERS

6. (a) $\dfrac{1}{4^3}$ (b) $\dfrac{1}{6^2}$ (c) $\left(\dfrac{3}{2}\right)^2$ (d) $\dfrac{7}{10}$

   (e) $\dfrac{1}{m^5}$ (f) z^4

Exponents and Polynomials

The definition of negative exponents allows some factors in a fraction to be moved by changing the sign of the exponent. For example,

$$\frac{2^{-3}}{3^{-4}} = \frac{\frac{1}{2^3}}{\frac{1}{3^4}} = \frac{1}{2^3} \cdot \frac{3^4}{1} = \frac{3^4}{2^3},$$

so that

$$\frac{2^{-3}}{3^{-4}} = \frac{3^4}{2^3}.$$

EXAMPLE 7 Write with only positive exponents. Assume all variables represent nonzero real numbers.

(a) $\dfrac{4^{-2}}{5^{-3}} = \dfrac{5^3}{4^2}$

(b) $\dfrac{m^{-5}}{p^{-1}} = \dfrac{p^1}{m^5} = \dfrac{p}{m^5}$

(c) $\dfrac{a^{-2}b}{3d^{-3}} = \dfrac{bd^3}{3a^2}$

(d) $x^3 y^{-4} = \dfrac{x^3}{y^4}$ ◀

Work Problem 7 at the side.

5 Now that zero and negative exponents have been defined, we can state the **quotient rule for exponents.**

For any nonzero real number a and any integers m and n,

$$\frac{a^m}{a^n} = a^{m-n} \quad (a \neq 0).$$

EXAMPLE 8 Simplify, using the quotient rule for exponents. Write answers with positive exponents.

(a) $\dfrac{5^8}{5^6} = 5^{8-6} = 5^2$

(b) $\dfrac{4^2}{4^9} = 4^{2-9} = 4^{-7} = \dfrac{1}{4^7}$

(c) $\dfrac{5^{-3}}{5^{-7}} = 5^{-3-(-7)} = 5^4$

(d) $\dfrac{q^5}{q^{-3}} = q^{5-(-3)} = q^8, \quad q \neq 0$

(e) $\dfrac{3^2 x^5}{3^4 x^3} = \dfrac{3^2}{3^4} \cdot \dfrac{x^5}{x^3} = 3^{2-4} \cdot x^{5-3}$

$\qquad = 3^{-2} x^2 = \dfrac{x^2}{3^2}, \quad x \neq 0$

7. Write with only positive exponents. Assume all variables represent nonzero real numbers.

(a) $\dfrac{7^{-1}}{5^{-4}}$

(b) $\dfrac{x^{-3}}{y^{-2}}$

(c) $\dfrac{4h^{-5}}{m^{-2}k}$

(d) $p^2 q^{-5}$

ANSWERS

7. (a) $\dfrac{5^4}{7}$ (b) $\dfrac{y^2}{x^3}$ (c) $\dfrac{4m^2}{h^5 k}$ (d) $\dfrac{p^2}{q^5}$

4.1 Exponents

8. Simplify. Write answers with positive exponents.

(a) $\dfrac{5^{11}}{5^8}$

(b) $\dfrac{4^7}{4^{10}}$

(c) $\dfrac{6^{-5}}{6^{-2}}$

(d) $\dfrac{8^4 \cdot m^9}{8^5 \cdot m^{10}}$

9. Simplify. Assume all variables represent nonzero real numbers.

(a) $12^5 \cdot 12^{-7} \cdot 12^6$

(b) $y^{-2} \cdot y^5 \cdot y^{-8}$

(c) $\dfrac{6x^{-1}}{3x^2}$

(d) $\dfrac{3^9 \cdot x^2 y^{-2}}{3^3 \cdot x^{-4} y}$

Sometimes numerical expressions with small exponents, such as 3^2, are evaluated. Doing that would give the result as $\dfrac{x^2}{9}$. ◀

Work Problem 8 at the side.

Since exponential expressions with negative exponents can be written with positive exponents, the product rule for exponents is also true for negative exponents. The rules for exponents discussed in this section are summarized below.

Definitions and Rules for Exponents

For any integers m and n, and real numbers a,

Product rule	$a^m \cdot a^n = a^{m+n}$	
Zero exponent	$a^0 = 1$	$(a \neq 0)$
Negative exponent	$a^{-n} = \dfrac{1}{a^n}$	$(a \neq 0)$
Quotient rule	$\dfrac{a^m}{a^n} = a^{m-n}$	$(a \neq 0)$.

EXAMPLE 9 Simplify by using the rules for exponents. Assume all variables represent nonzero real numbers.

(a) $5^{-4} \cdot 5^7 = 5^{-4+7} = 5^3$

(b) $x^{-8} \cdot x^7 \cdot x^{-3} = x^{-8+7+(-3)} = x^{-4} = \dfrac{1}{x^4}$

(c) $\dfrac{m^{-2}}{m^{-5}} = \dfrac{m^5}{m^2} = m^3$ ◀

Work Problem 9 at the side.

ANSWERS

8. (a) 5^3 (b) $\dfrac{1}{4^3}$ (c) $\dfrac{1}{6^3}$ (d) $\dfrac{1}{8m}$

9. (a) 12^4 (b) $\dfrac{1}{y^5}$ (c) $\dfrac{2}{x^3}$ (d) $\dfrac{3^6 x^6}{y^3}$

name   date   hour

4.1 EXERCISES

Identify the base and exponent for each exponential expression. See Example 2.

1. 5^{12}
2. a^6
3. $(3m)^4$
4. -2^4

5. -125^3
6. $(-1)^8$
7. $(-2x)^2$
8. $-(-p)^5$

9. $3m^2$
10. $5y^3$
11. $-r^5$
12. $(-y)^5$

Write each expression using exponents. See Example 1.

13. $3 \cdot 3 \cdot 3 \cdot 3 \cdot 3$
14. $4 \cdot 4 \cdot 4$
15. $(-2)(-2)(-2)(-2)(-2)$

16. $(-1)(-1)(-1)(-1)$
17. $p \cdot p \cdot p \cdot p \cdot p$
18. $\dfrac{1}{4 \cdot 4 \cdot 4 \cdot 4 \cdot 4}$

19. $\dfrac{1}{(-2)(-2)(-2)}$
20. $\dfrac{1}{3 \cdot 3 \cdot 3 \cdot 3}$
21. $\dfrac{1}{2 \cdot 2 \cdot 2 \cdot 2 \cdot 2}$

22. $\dfrac{1}{y \cdot y \cdot y \cdot y}$
23. $(-2z)(-2z)(-2z)$
24. $(-3m)(-3m)(-3m)(-3m)$

Evaluate each expression. For example, $5^2 + 5^3 = 25 + 125 = 150$. In Exercises 55–58, round to the nearest thousandth. See Examples 5 and 6.

25. $3^2 + 3^4$
26. $2^8 - 2^6$
27. $4^2 + 4^3$
28. $3^3 + 3^4$

29. $2^2 + 2^5$
30. $4^2 + 4^1$
31. $4^0 + 5^0$
32. $3^0 + 8^0$

33. $(-9)^0 + 9^0$
34. $8^1 - 8^0$
35. 3^{-3}
36. 4^{-2}

37. 5^{-2}
38. 2^{-5}
39. 9^{-1}
40. $(-12)^{-1}$

41. $(-6)^{-2}$
42. 8^{-3}
43. 7^{-1}
44. 12^{-2}

45. $\left(\dfrac{1}{2}\right)^{-5}$
46. $\left(\dfrac{1}{5}\right)^{-2}$
47. $\left(\dfrac{1}{2}\right)^{-1}$
48. $\left(\dfrac{3}{4}\right)^{-1}$

49. $\left(\dfrac{2}{3}\right)^{-3}$

50. $\left(\dfrac{5}{4}\right)^{-2}$

51. $2^{-1} + 3^{-1}$

52. $3^{-1} - 4^{-1}$

53. $4^{-2} + 4^{-3}$

54. $2^{-4} - 2^{-3}$

55. $(.98)^{-2}$

56. $(1.76)^{-2}$

57. $(3.918)^{-3}$

58. $(.162)^{-3}$

Use the product rule to simplify each expression. Write each answer in exponential form with only positive exponents. See Example 3.

59. $4^2 \cdot 4^3$

60. $3^5 \cdot 3^4$

61. $9^5 \cdot 9^3$

62. $a^6 \cdot a^4$

63. $k^4 \cdot k^{-7}$

64. $p^{-5} \cdot p^{15}$

65. $4^3 \cdot 4^5 \cdot 4^{-10}$

66. $2^3 \cdot 2^4 \cdot 2^{-6}$

67. $(-3)^3(-3)^2$

68. $(-t)^5(-t)^3$

69. $(-z)^3(-z)^6$

70. $(-m)^4(-m)^6$

Use the quotient rule to simplify each expression. Leave each answer in exponential form with only positive exponents. See Examples 7–9.

71. $\dfrac{4^7}{4^2}$

72. $\dfrac{11^5}{11^3}$

73. $\dfrac{4^2}{4^4}$

74. $\dfrac{14^{11}}{14^{15}}$

75. $\dfrac{8^3}{8^9}$

76. $\dfrac{5^4}{5^{10}}$

77. $\dfrac{q^{-4}}{q^2}$

78. $\dfrac{c^{-5}}{c^3}$

79. $\dfrac{d^{-2}}{d^{-5}}$

80. $\dfrac{x^6}{x^{-9}}$

81. $\dfrac{y^2}{y^5}$

82. $\dfrac{z}{z^{-1}}$

83. $\dfrac{r^{-1}}{r}$

84. $\dfrac{4k^{-3}m^5}{4^{-1}k^{-7}m^{-3}}$

85. $\dfrac{6^{-1}y^{-2}z^5}{6^2y^{-1}z^{-2}}$

In Exercises 86–93, first add the given terms; then start over and multiply them. See Example 4.

86. $4m^3,\ 9m^3$

87. $8y^2,\ 7y^2$

88. $-12p,\ 11p$

89. $3q^4,\ 5q^4$

90. $7r,\ 3r,\ 5r$

91. $9a^3,\ 2a^3,\ 3a^3$

92. $-5a^{-2},\ 3a^{-2}$

93. $6r^{-4},\ -8r^{-4}$

4.2 POWER RULES FOR EXPONENTS

Additional rules for exponents are developed in this section.

▶ Simplify an expression such as $(8^3)^2$ with the product rule for exponents, as follows.

$$(8^3)^2 = (8^3)(8^3) = 8^{3+3} = 8^6$$

The exponents in $(8^3)^2$ are multiplied to give the exponent in 8^6: $3 \cdot 2 = 6$. This example suggests the first **power rule for exponents.**

> For any integers m and n, $(a^m)^n = a^{mn}$.

EXAMPLE 1 Use the power rule for exponents to simplify each expression.

(a) $(2^5)^3 = 2^{5 \cdot 3} = 2^{15}$

(b) $(5^7)^{-2} = 5^{7(-2)} = 5^{-14} = \dfrac{1}{5^{14}}$

(c) $(x^{-2})^{-5} = x^{(-2)(-5)} = x^{10}, \quad x \neq 0$

(d) $(n^{-3})^2 = n^{(-3)(2)} = n^{-6} = \dfrac{1}{n^6}, \quad n \neq 0$ ◀

Work Problem 1 at the side.

▶ The properties studied in Chapter 2 can be used to develop two more rules for exponents. Using the definition of an exponential expression and the commutative and associative properties, the expression $(4 \cdot 8)^3$ can be evaluated as shown below.

$$(4 \cdot 8)^3 = (4 \cdot 8)(4 \cdot 8)(4 \cdot 8)$$
$$= 4 \cdot 4 \cdot 4 \cdot 8 \cdot 8 \cdot 8 \quad \text{Commutative and associative properties}$$
$$= 4^3 \cdot 8^3$$

This example suggests the following rule, another of the **power rules for exponents.**

> For any integer m, $(ab)^m = a^m b^m$.

EXAMPLE 2 Use the rule given above to simplify each expression.

(a) $(3xy)^2 = 3^2 x^2 y^2$
$= 9x^2 y^2$

(b) $9(pq)^2 = 9(p^2 q^2)$
$= 9p^2 q^2$

(c) $(2m^2 p^3)^4 = 2^4 (m^2)^4 (p^3)^4$
$= 2^4 m^8 p^{12}$

Objectives

▶ Use $(a^m)^n = a^{mn}$.

▶ Use $(ab)^m = a^m b^m$.

▶ Use $\left(\dfrac{a}{b}\right)^m = \dfrac{a^m}{b^m}$.

▶ Use combinations of rules.

1. Simplify each expression and write with positive exponents.

(a) $(5^3)^4$

(b) $(6^{-2})^5$

(c) $(3^{-2})^{-1}$

(d) $(a^{-6})^{-5}, \quad a \neq 0$

ANSWERS

1. (a) 5^{12} (b) $\dfrac{1}{6^{10}}$ (c) 3^2 (d) a^{30}

4.2 Power Rules for Exponents

2. Simplify and write with positive exponents.

   (a) $5(mn)^3$

   (b) $(3a^2b^4)^5$

   (c) $(5m^{-1})^3$

   (d) $(6a^{-3})^{-2}$

3. Simplify and write with positive exponents. Assume all variables represent nonzero real numbers.

   (a) $\left(\dfrac{5}{2}\right)^4$

   (b) $\left(\dfrac{p}{q}\right)^2$

   (c) $\left(\dfrac{9}{10}\right)^{-2}$

   (d) $\left(\dfrac{r}{t}\right)^{-3}$

(d) $(3k^{-2})^{-3} = 3^{-3}k^{(-2)(-3)}$

$= 3^{-3}k^6$

$= \dfrac{k^6}{3^3}$ ◀

Work Problem 2 at the side.

Since the quotient $\dfrac{a}{b}$ can be written as $a \cdot \dfrac{1}{b}$, the rule above, together with some of the properties of real numbers, gives the final power rule for exponents.

For any integer m, $\left(\dfrac{a}{b}\right)^m = \dfrac{a^m}{b^m}$ $(b \neq 0)$.

EXAMPLE 3 Simplify each expression.

(a) $\left(\dfrac{2}{3}\right)^5 = \dfrac{2^5}{3^5}$

(b) $\left(\dfrac{4}{5}\right)^{-2}$

With negative exponents, it is simpler to use the reciprocals first.

$$\left(\dfrac{4}{5}\right)^{-2} = \left(\dfrac{5}{4}\right)^2 = \dfrac{5^2}{4^2}$$

(c) $\left(\dfrac{a}{b}\right)^4 = \dfrac{a^4}{b^4},\ b \neq 0$ ◀

Work Problem 3 at the side.

The next example shows that more than one rule may be needed to simplify an expression.

EXAMPLE 4 Use a combination of the rules for exponents to simplify each expression. Write answers with positive exponents. Assume all variables represent nonzero real numbers.

(a) $\dfrac{(4^2)^3}{4^5}$

Use a power rule and then the quotient rule.

$\dfrac{(4^2)^3}{4^5} = \dfrac{4^6}{4^5}$ Power rule

$= 4^1$ Quotient rule

$= 4$

(b) $(2x)^3(2x)^2$

Use the product rule first, then a power rule.

$(2x)^3(2x)^2 = (2x)^5$

$= 2^5x^5$ or $32x^5$

ANSWERS

2. (a) $5m^3n^3$ (b) $3^5a^{10}b^{20}$ (c) $\dfrac{5^3}{m^3}$

   (d) $\dfrac{a^6}{6^2}$

3. (a) $\dfrac{5^4}{2^4}$ (b) $\dfrac{p^2}{q^2}$ (c) $\dfrac{10^2}{9^2}$ (d) $\dfrac{t^3}{r^3}$

(c) $\left(\dfrac{2a^3}{5}\right)^4$

By the power rules,

$$\left(\dfrac{2a^3}{5}\right)^4 = \dfrac{2^4 a^{12}}{5^4} \quad \text{or} \quad \dfrac{16a^{12}}{625}.$$

(d) $\left(\dfrac{3x^{-2}}{4^{-1}y^3}\right)^{-3} = \dfrac{3^{-3}x^6}{4^3 y^{-9}}$ Power rule

Move factors with negative exponents.

$$= \dfrac{x^6 y^9}{3^3 \cdot 4^3}$$

(e) $\dfrac{(5m^2)^{-1}(3m^{-4})^{-2}}{(3m)^{-3}} = \dfrac{5^{-1}m^{-2} \cdot 3^{-2}m^8}{3^{-3}m^{-3}}$

$$= \dfrac{3^3 m^3 m^8}{5 \cdot 3^2 m^2}$$

$$= \dfrac{1}{5} \cdot \dfrac{3^3}{3^2} \cdot \dfrac{m^{11}}{m^2}$$

$$= \dfrac{1}{5} \cdot 3 \cdot m^9$$

$$= \dfrac{3m^9}{5} \blacktriangleleft$$

Work Problem 4 at the side.

We again summarize the rules for exponents, adding the three rules from this section.

Definitions and Rules for Exponents

For any integers m and n:

Product rule $a^m \cdot a^n = a^{m+n}$

Zero exponent $a^0 = 1$ $(a \neq 0)$

Negative exponent $a^{-n} = \dfrac{1}{a^n}$ $(a \neq 0)$

Quotient rule $\dfrac{a^m}{a^n} = a^{m-n}$ $(a \neq 0)$

Power rules $(a^m)^n = a^{mn}$
$(ab)^m = a^m b^m$
$\left(\dfrac{a}{b}\right)^m = \dfrac{a^m}{b^m}$ $(b \neq 0)$.

---

4. Simplify. Write answers with positive exponents. Assume all variables represent nonzero real numbers.

(a) $\dfrac{(3^4)^2}{3^5}$

(b) $(2m)^{-1}(2m)^4$

(c) $\left(\dfrac{5k^{-1}}{8}\right)^{-2}$

(d) $\left(\dfrac{9x^{-1}}{2y^{-4}}\right)^{-3}$

(e) $\left(\dfrac{r^{-2}s^4}{r^{-3}s^6}\right)^{-1}$

(f) $\dfrac{(3k^4)^{-2}(4k^{-1})^3}{(5k)^{-2}}$

ANSWERS

4. (a) 3^3 (b) $2^3 m^3$ (c) $\dfrac{8^2 k^2}{5^2}$
(d) $\dfrac{2^3 x^3}{9^3 y^{12}}$ (e) $\dfrac{s^2}{r}$ (f) $\dfrac{5^2 \cdot 4^3}{3^2 k^9}$

4.2 Power Rules for Exponents

name _____ date _____ hour _____

4.2 EXERCISES

Use the rules for exponents to simplify each expression. Assume all variables represent nonzero numbers. Write each answer in exponential form (using only positive exponents). See Examples 1–4.

1. $(6^3)^2$
2. $(8^4)^6$
3. $(9^{-3})^2$
4. $(2^{-3})^4$

5. $(3^{-5})^{-2}$
6. $(8^{-4})^{-1}$
7. $\dfrac{(y^3)^3}{(y^2)^2}$
8. $\dfrac{(r^2)^4}{(r^3)^2}$

9. $\dfrac{(k^2)^4}{(k^6)^2}$
10. $\dfrac{(w^4)^2}{(w^7)^3}$
11. $\dfrac{a^6 \cdot a^5}{(a^2)^4}$
12. $\dfrac{s^7 \cdot s^9}{(s^5)^2}$

13. $\dfrac{4^3 \cdot 4^{-5}}{4^7}$
14. $\dfrac{2^5 \cdot 2^{-4}}{2^{-1}}$
15. $\dfrac{5^{-3} \cdot 5^{-2}}{5^4}$
16. $\dfrac{8^{-2} \cdot 8^5}{8^6}$

17. $\dfrac{m^4 \cdot m^{-5}}{m^{-6}}$
18. $\dfrac{p^3 \cdot p^{-5}}{p^5}$
19. $\dfrac{a^{11} \cdot a^{-7}}{a^5}$
20. $\dfrac{z^3 \cdot z^{-5}}{z^{10}}$

21. $\dfrac{r^5 \cdot r^{-8}}{r^{-6} \cdot r^4}$
22. $\dfrac{x^3 \cdot x^{-1}}{x^8 \cdot x^{-2}}$
23. $\dfrac{a^6 \cdot a^{-3}}{a^{-5} \cdot a}$
24. $\dfrac{b^{10} \cdot b^{-2}}{b^{-8} \cdot b^6}$

25. $(5m)^3$
26. $(2xy)^4$
27. $(3mn)^4$
28. $(-2ab)^5$

29. $(-3x^5)^2$
30. $(4m^3n^2)^4$
31. $(5p^2q)^3$
32. $(2p^2a^4)^5$

4.2 Exercises 181

33. $(3x^{-5})^2$ 34. $(5p^{-4})^{-2}$ 35. $(9^{-1}y^5)^{-2}$ 36. $(4^{-2}m^{-3})^{-2}$

37. $\left(\dfrac{a}{5}\right)^3$ 38. $\left(\dfrac{9}{x}\right)^2$ 39. $\left(\dfrac{3mn}{2}\right)^5$ 40. $\left(\dfrac{2x}{5y}\right)^4$

41. $\left(\dfrac{a}{bc}\right)^{-1}$ 42. $\left(\dfrac{2a}{3d}\right)^{-2}$ 43. $\left(\dfrac{5m^{-2}}{m^{-1}}\right)^2$ 44. $\left(\dfrac{4x^3}{3^{-1}}\right)^{-1}$

45. $\dfrac{(x^3)^2}{x^9 x^7}$ 46. $\dfrac{(m^2)^4}{(m^9)^3}$ 47. $\dfrac{(b^2)^4}{b^3(b^2)^6}$

48. $\dfrac{(8r^2)^3}{(8r^3)^4}$ 49. $\dfrac{(3x^2)^{-2}(5x^{-1})^3}{3x^{-5}}$ 50. $\dfrac{(2k^3)^{-1}(3k^{-2})^2}{2k^{-4}}$

51. $\dfrac{(4b^3)^{-2}(2b^{-1})^3}{(a^3b)^{-4}}$ 52. $\dfrac{(3n)^{-2}(5n^{-2})^3}{n^{-2}}$ 53. $\dfrac{(2y^{-1}z^2)^2(3y^{-2}z^{-3})^3}{(y^3z^2)^{-1}}$

54. $\dfrac{(3p^{-2}q^3)^2(5p^{-1}q^{-4})^{-1}}{(p^2q^{-2})^{-3}}$ 55. $\dfrac{(9^{-1}z^{-2}x)^{-1}(4z^2x^4)^{-2}}{(5z^{-2}x^{-3})^2}$ 56. $\dfrac{(4^{-1}a^{-1}b^{-2})^{-2}(5a^{-3}b^4)^{-2}}{(3a^{-3}b^{-5})^2}$

4.3 AN APPLICATION OF EXPONENTS: SCIENTIFIC NOTATION

Objectives
1. Express numbers in scientific notation.
2. Convert numbers in scientific notation to numbers without exponents.
3. Use scientific notation in calculations.

▶ One example of the use of exponents comes from science. The numbers occurring in science are often extremely large (such as the distance from the earth to the sun, 93,000,000 miles) or extremely small (the wavelength of yellow-green light is approximately .0000006 meters). Because of the difficulty of working with many zeros, scientists often express such numbers with exponents. Each number is written as $a \times 10^n$, where $1 \leq |a| < 10$ and n is an integer. This form is called **scientific notation.** There is always one nonzero digit before the decimal point. For example, 35 is written 3.5×10^1, or 3.5×10; 56,200 is written 5.62×10^4, since

$$56,200 = 5.62 \times 10,000 = 5.62 \times 10^4.$$

The steps involved in writing a number in scientific notation are given below.

Writing a Number in Scientific Notation

Step 1 Place a caret, ∧, to the right of the first nonzero digit.

Step 2 Count the number of places from the caret to the decimal point.

Step 3 The number of places in Step 2 is the absolute value of the exponent on 10.

Step 4 The exponent on 10 is positive if you counted from left to right; the exponent is negative if you counted from right to left.

EXAMPLE 1 Write each number in scientific notation.

(a) 93,000,000

Place a caret after the first nonzero digit.

9∧3,000,000 ← Caret to right of first nonzero digit

Count from the caret to the decimal point.

9∧3,000,000   7 places

Since the counting was from left to right, the exponent on 10 is positive, and $93,000,000 = 9.3 \times 10^7$.

(b) $4∧63,000,000,000,000 = 4.63 \times 10^{14}$ 14 places

(c) $302,100 = 3.021 \times 10^5$

(d) .00462

Place a caret to the right of the first nonzero digit.

.004∧62

Count from the caret to the decimal point.

.004∧62   3 places

1. Write each number in scientific notation.

   (a) 63,000

   (b) 5,870,000

   (c) .0571

   (d) .000062

2. Write without exponents.

   (a) 4.2×10^3

   (b) 8.7×10^5

   (c) 6.42×10^{-3}

3. Express without exponents.

   (a) $(2.6 \times 10^4)(2 \times 10^{-6})$

   (b) $\dfrac{4.8 \times 10^2}{2.4 \times 10^{-3}}$

Since we are counting from right to left, the exponent will be negative.
$$.00462 = 4.62 \times 10^{-3}$$

(e) $.0000762 = .00007{\scriptstyle\wedge}62 = 7.62 \times 10^{-5}$ ◀

Work Problem 1 at the side.

▶ 2 To convert a number written in scientific notation to a number without exponents, use the steps given below.

Converting from Scientific Notation

Step 1 Count from the decimal point the same number of places as the exponent on 10, attaching additional zeros as necessary.

Step 2 Move to the right if the exponent on 10 is positive; move to the left if the exponent is negative.

EXAMPLE 2 Write each number without exponents.

(a) 6.2×10^3

   Moving the decimal point 3 places to the right gives
$$6.2 \times 10^3 = 6.200 = 6200.$$

(b) $4.283 \times 10^5 = 4.28300 = 428,300$

(c) $9.73 \times 10^{-2} = 09.73 = .0973$ ◀

As these examples show, the exponent tells the number of places that the decimal point is moved.

Work Problem 2 at the side.

▶ 3 The next example shows how scientific notation can be used with products and quotients.

EXAMPLE 3 Write each number without exponents.

(a) $(6 \times 10^3)(5 \times 10^{-4})$

$$\begin{aligned}
(6 \times 10^3)(5 \times 10^{-4}) & \\
= (6 \times 5)(10^3 \times 10^{-4}) &\quad \text{Commutative and associative properties} \\
= 30 \times 10^{-1} &\quad \text{Product rule for exponents} \\
= 3.0 &\quad \text{Write without exponents}
\end{aligned}$$

(b) $\dfrac{6 \times 10^{-5}}{2 \times 10^3} = \dfrac{6}{2} \times \dfrac{10^{-5}}{10^3} = 3 \times 10^{-8} = .00000003$ ◀

Work Problem 3 at the side.

ANSWERS
1. (a) 6.3×10^4 (b) 5.87×10^6
   (c) 5.71×10^{-2} (d) 6.2×10^{-5}
2. (a) 4200   (b) 870,000   (c) .00642
3. (a) .052   (b) 200,000

4.3 EXERCISES

Write each number in scientific notation. See Example 1.

1. 6,835,000,000
2. 321,000,000,000,000
3. 8,360,000,000,000
4. 6850
5. 25,000
6. 110,000,000
7. .0101
8. .0000006
9. .000012
10. .000000982
11. .834
12. .0069

Write each number without exponents. See Examples 2 and 3.

13. 8.1×10^9
14. 3.5×10^2
15. 9.132×10^6
16. 3.24×10^8
17. 3.2×10^{-4}
18. 5.76×10^{-5}
19. $(2 \times 10^8) \times (4 \times 10^{-3})$
20. $(5 \times 10^4) \times (3 \times 10^{-2})$
21. $(4 \times 10^{-1}) \times (1 \times 10^{-5})$
22. $(6 \times 10^{-5}) \times (2 \times 10^4)$
23. $(7 \times 10^3) \times (2 \times 10^2) \times (3 \times 10^{-4})$
24. $(3 \times 10^{-5}) \times (3 \times 10^2) \times (5 \times 10^{-2})$
25. $(1.2 \times 10^2) \times (5 \times 10^{-3}) \times (2.4 \times 10^3)$
26. $(4.6 \times 10^{-3}) \times (2 \times 10^{-1}) \times (4 \times 10^5)$

27. $\dfrac{9 \times 10^5}{3 \times 10^{-1}}$

28. $\dfrac{12 \times 10^{-4}}{4 \times 10^4}$

29. $\dfrac{8 \times 10^{-3}}{2 \times 10^{-2}}$

30. $\dfrac{5 \times 10^{-1}}{1 \times 10^{-5}}$

31. $\dfrac{2.6 \times 10^5}{2 \times 10^2}$

32. $\dfrac{9.5 \times 10^{-1}}{5 \times 10^3}$

33. $\dfrac{7.2 \times 10^{-3} \times 1.6 \times 10^5}{4 \times 10^{-2} \times 3.6 \times 10^9}$

34. $\dfrac{8.7 \times 10^{-2} \times 1.2 \times 10^{-6}}{3 \times 10^{-4} \times 2.9 \times 10^{11}}$

Write the numbers in each statement in scientific notation. See Example 1.

35. Light visible to the human eye has a wavelength between .0004 millimeters and .0008 millimeters.

36. In the ocean, the amount of oxygen per cubic mile of water is 4,037,000,000 tons, and the amount of radium is .0003 tons.

37. Each tide in the Bay of Fundy carries more than 3,680,000,000,000,000 cubic feet of water into the bay.

38. The mean (average) diameter of the sun is about 865,000 miles.

Write the numbers in each statement without exponents. See Example 2.

39. There are 1×10^3 cubic millimeters in 6.102×10^{-2} cubic inches.

40. In the food chain that links the largest sea creature (the whale) to the smallest (the diatom), 4×10^{14} diatoms sustain a medium-sized whale for only a few hours.

41. Many ocean trenches have a depth of 3.5×10^4 feet.

42. The average life span of a human is 1×10^9 seconds.

4.4 POLYNOMIALS

▶ Recall that in an expression such as

$$4x^3 + 6x^2 + 5x$$

the quantities that are added, $4x^3$, $6x^2$, and $5x$, are called **terms.** In the term $4x^3$, the number 4 is called the **numerical coefficient,** or simply the **coefficient,** of x^3. In the same way, 6 is the coefficient of x^2 in the term $6x^2$, and 5 is the coefficient of x in the term $5x$.

EXAMPLE 1 Name the coefficient of each term in these expressions.

(a) $4x^3$

The coefficient is 4.

(b) $x - 6x^4$

The coefficient of x is 1 because $x = 1 \cdot x$. The coefficient of x^4 is -6, since $x - 6x^4$ can be written as the sum $x + (-6x^4)$.

(c) $5 - v^3$

The coefficient of the term 5 is 5 since $5 = 5v^0$. By writing $5 - v^3$ as a sum, $5 + (-v^3)$, or $5 + (-1v^3)$, the coefficient of v^3 can be identified as -1. ◀

Work Problem 1 at the side.

▶ Recall that **like terms** have exactly the same combination of variables with the same exponent. Only the coefficients may be different. Examples of like terms are

$$19m^5 \text{ and } 14m^5;$$
$$6y^9, -37y^9, \text{ and } y^9;$$
$$3pq, -2pq, \text{ and } 4pq.$$

Add like terms with the distributive property.

EXAMPLE 2 Simplify each expression by adding like terms.

(a) $-4x^3 + 6x^3 = (-4 + 6)x^3$ Distributive property
$= 2x^3$

(b) $3x^4 + 5x^4 = (3 + 5)x^4 = 8x^4$

(c) $9x^6 - 14x^6 + x^6 = (9 - 14 + 1)x^6 = -4x^6$

(d) $12m^2 + 5m + 4m^2 = (12 + 4)m^2 + 5m$
$= 16m^2 + 5m$

(e) $3x^2y + 4x^2y - x^2y = (3 + 4 - 1)x^2y = 6x^2y$ ◀

Example 2(d) shows that it is not possible to add $16m^2$ and $5m$. These two terms are unlike because the exponents on the variables are different. **Unlike terms** have different variables or different exponents on the same variables.

Work Problem 2 at the side.

Objectives

▶ Identify terms and coefficients.
▶ Combine like terms.
▶ Know the vocabulary for polynomials.
▶ Add polynomials.
▶ Subtract polynomials.

1. Name the coefficient of each term in these expressions.

   (a) $3m^2$

   (b) $2x^3 - x$

   (c) $x + 8$

2. Add like terms.

   (a) $5x^4 + 7x^4$

   (b) $9pq + 3pq - 2pq$

   (c) $r^2 + 3r + 5r^2$

   (d) $8t + 6w$

ANSWERS
1. (a) 3 (b) 2; -1 (c) 1; 8
2. (a) $12x^4$ (b) $10pq$ (c) $6r^2 + 3r$
   (d) cannot be added—unlike terms

4.4 Polynomials

3. Choose one or more of the following descriptions for each of the expressions in parts (a)–(d).

   (1) Polynomial
   (2) Polynomial written in descending order
   (3) Not a polynomial

   (a) $3m^3 + 5m^2 - 2m + 1$

   (b) $2p^4 + p^6$

   (c) $\dfrac{1}{x} + 2x^2 + 3$

   (d) $x - 3$

4. For each polynomial, first simplify if possible. Then give the degree and tell whether the polynomial is a monomial, binomial, trinomial, or none of these.

   (a) $3x^2 + 2x - 4$

   (b) $x^3 + 4x^3$

   (c) $x^8 - x^7 + 2x^8$

ANSWERS
3. (a) 1 and 2 (b) 1 (c) 3
   (d) 1 and 2
4. (a) degree 2; trinomial
   (b) degree 3; monomial (simplify to $5x^3$)
   (c) degree 8; binomial (simplify to $3x^8 - x^7$)

▶ *Polynomials* are basic to algebra. A **polynomial in x** is the sum of a finite number of terms of the form ax^n, for any real number a and any whole number n. For example,

$$16x^8 - 7x^6 + 5x^4 - 3x^2 + 4$$

is a polynomial in x (here 4 can be written as $4x^0$). This polynomial is written in **descending powers** of the variable, since the exponents on x decrease from left to right. On the other hand,

$$2x^3 - x^2 + \frac{4}{x}$$

is not a polynomial in x, since $\frac{4}{x} = 4x^{-1}$ is not a *product*, ax^n, for a whole number n. (Of course, a *polynomial* could be defined using any variable, or variables, and not just x.)

Work Problem 3 at the side.

The **degree** of a term with one variable is the exponent on the variable. For example, $3x^4$ has degree 4, $6x^{17}$ has degree 17, $5x$ has degree 1, and -7 has degree 0 (since -7 can be written as $-7x^0$). The **degree of a polynomial** in one variable is the highest exponent found in any nonzero term of the polynomial. For example, $3x^4 - 5x^2 + 6$ is of degree 4, while $5x$ is of degree 1, and 3 (or $3x^0$) is of degree 0.

Three types of polynomials are very common and are given special names. A polynomial with exactly three terms is called a **trinomial**. (*Tri-* means "three," as in *tri*angle.) Examples are

$$9m^3 - 4m^2 + 6, \quad 19y^2 + 8y + 5, \quad \text{and} \quad -3m^5 - 9m^2 + 2.$$

A polynomial with exactly two terms is called a **binomial**. (*Bi-* means "two," as in *bi*cycle.) Examples are

$$-9x^4 + 9x^3, \quad 8m^2 + 6m, \quad \text{and} \quad 3m^5 - 9m^2.$$

A polynomial with only one term is called a **monomial**. (*Mon(o)-* means "one," as in *mono*rail.) Examples are

$$9m, \quad -6y^5, \quad a^2, \quad \text{and} \quad 6.$$

EXAMPLE 3 For each polynomial, first simplify if possible by combining like terms. Then give the degree and tell whether the polynomial is a monomial, a binomial, a trinomial, or none of these.

(a) $2x^3 + 5$

The polynomial cannot be simplified. The degree is 3. The polynomial is a binomial.

(b) $4x - 5x + 2x$

Add like terms to simplify: $4x - 5x + 2x = x$. The degree is 1. The polynomial is a monomial. ◀

Work Problem 4 at the side.

▶ Polynomials may be added, subtracted, multiplied, and divided. Polynomial addition and subtraction are explained in the rest of this section.

Exponents and Polynomials

Adding Polynomials
Add two polynomials by adding like terms.

EXAMPLE 4 Add $6x^3 - 4x^2 + 3$ and $-2x^3 + 7x^2 - 5$.

Write like terms in columns.
$$6x^3 - 4x^2 + 3$$
$$-2x^3 + 7x^2 - 5$$

Now add, column by column.

$$\begin{array}{ccc} 6x^3 & -4x^2 & 3 \\ -2x^3 & 7x^2 & -5 \\ \hline 4x^3 & 3x^2 & -2 \end{array}$$

Add the three sums together.
$$4x^3 + 3x^2 + (-2) = 4x^3 + 3x^2 - 2$$

Work Problem 5 at the side.

The polynomials in Example 4 also could be added horizontally as shown in the next example.

EXAMPLE 5 Add $6x^3 - 4x^2 + 3$ and $-2x^3 + 7x^2 - 5$.

Write the sum.
$$(6x^3 - 4x^2 + 3) + (-2x^3 + 7x^2 - 5)$$

Rewrite this sum with the parentheses removed and with subtractions changed to addition of inverses.
$$6x^3 + (-4x^2) + 3 + (-2x^3) + 7x^2 + (-5)$$

Place like terms together.
$$6x^3 + (-2x^3) + (-4x^2) + 7x^2 + 3 + (-5)$$

Combine like terms to get
$$4x^3 + 3x^2 + (-2) \quad \text{or just} \quad 4x^3 + 3x^2 - 2,$$

the same answer found in Example 4.

Work Problem 6 at the side.

Earlier, the difference $x - y$ was defined as $x + (-y)$. (Find the difference $x - y$ by adding x and the opposite of y.) For example,
$$7 - 2 = 7 + (-2)$$
$$= 5$$

and
$$-8 - (-2) = -8 + 2 = -6.$$

A similar method is used to subtract polynomials.

---

5. Add each pair of polynomials.

(a) $4x^3 - 3x^2 + 2x$
    $6x^3 + 2x^2 - 3x$

(b) $x^2 - 2x + 5$
    $4x^2 + 3x - 2$

6. Find each sum.

(a) $(2x^4 - 6x^2 + 7)$
    $+ (-3x^4 + 5x^2 + 2)$

(b) $(3x^3 + 4x + 2)$
    $+ (6x^3 - 5x - 7)$

ANSWERS
5. (a) $10x^3 - x^2 - x$ (b) $5x^2 + x + 3$
6. (a) $-x^4 - x^2 + 9$ (b) $9x^3 - x - 5$

4.4 Polynomials

7. Subtract, and check your answers by addition.

   (a) $(14y^3 - 6y^2 + 2y - 5)$
       $-(2y^3 - 7y^2 - 4y + 6)$

   (b) $(7y^2 - 11y + 8)$
       $-(-3y^2 + 4y + 6)$

8. Use the method of subtracting by columns to solve each problem.

   (a) $(14y^3 - 6y^2 + 2y)$
       $-(2y^3 - 7y^2 + 6)$

   (b) $(6p^4 - 8p^3 + 2p - 1)$
       $-(-7p^4 + 6p^2 - 12)$

ANSWERS
7. (a) $12y^3 + y^2 + 6y - 11$
   (b) $10y^2 - 15y + 2$
8. (a) $12y^3 + y^2 + 2y - 6$
   (b) $13p^4 - 8p^3 - 6p^2 + 2p + 11$

EXAMPLE 6 Subtract: $(5x - 2) - (3x - 8)$.

By the definition of subtraction,

$$(5x - 2) - (3x - 8) = (5x - 2) \boxed{+} [\boxed{-}(3x - 8)].$$

From Chapter 2,

$$\boxed{-}(3x - 8) = \boxed{-1}(3x - 8)$$
$$= -3x + 8.$$

Then

$$(5x - 2) - (3x - 8) = (5x - 2) + (-3x + 8)$$
$$= 2x + 6. \blacktriangleleft$$

In summary, polynomials are subtracted as follows.

Subtracting Polynomials
Subtract two polynomials by changing all the signs of the second polynomial and adding the result to the first polynomial.

EXAMPLE 7 Subtract $6x^3 - 4x^2 + 2$ from $11x^3 + 2x^2 - 8$.

Start with

$$(11x^3 + 2x^2 - 8) - (6x^3 - 4x^2 + 2).$$

Change all the signs on the second polynomial and add like terms.

$$(11x^3 + 2x^2 - 8) \boxed{+} (\boxed{-}6x^3 \boxed{+} 4x^2 \boxed{-} 2)$$
$$= 5x^3 + 6x^2 - 10$$

Check a subtraction problem such as this by using the fact that if $a - b = c$, then $a = b + c$. For example, $6 - 2 = 4$. Check by writing $6 = 2 + 4$, which is correct. Check the polynomial subtraction above by adding $6x^3 - 4x^2 + 2$ and $5x^3 + 6x^2 - 10$. Since the sum is $11x^3 + 2x^2 - 8$, the subtraction was performed correctly. $\blacktriangleleft$

Work Problem 7 at the side.

Subtraction also can be done in columns.

EXAMPLE 8 Use the method of subtracting by columns to find $(14y^3 - 6y^2 + 2y - 5) - (2y^3 - 7y^2 - 4y + 6)$.

Step 1 Arrange like terms in columns.

$$\begin{array}{r} 14y^3 - 6y^2 + 2y - 5 \\ 2y^3 - 7y^2 - 4y + 6 \end{array}$$

Step 2 Change all signs in the second row, and then add (*Step 3*).

$$\begin{array}{r} 14y^3 - 6y^2 + 2y - 5 \\ \boxed{-}2y^3 \boxed{+} 7y^2 \boxed{+} 4y \boxed{-} 6 \\ \hline 12y^3 + y^2 + 6y - 11 \end{array} \quad \text{All signs changed} \\ \text{Add} \blacktriangleleft$$

Either the horizontal or the vertical method may be used for adding and subtracting polynomials.

Work Problem 8 at the side.

Exponents and Polynomials

name　　　　　　　　　　　　　　　　　　　　　　　　　　date　　　　　　hour

4.4　EXERCISES

In each polynomial, combine terms whenever possible. Write the results in descending powers of the variable. See Examples 2 and 3.

1. $3m^5 + 5m^5$
2. $-4y^3 + 3y^3$
3. $2r^5 + (-3r^5)$

4. $-19y^2 + 9y^2$
5. $2m^5 - 5m^2$
6. $-9y + 9y^2$

7. $3x^5 + 2x^5 - 4x^5$
8. $6x^3 - 8x^7 - 9x^3$
9. $-4p^7 + 8p^7 - 5p^7$

10. $-3a^8 + 4a^8 - 3a^8 + 2a^8$
11. $4y^2 + 3y^2 - 2y^2 + y^2$
12. $3r^5 - 8r^5 + r^5 - 2r^5$

13. $-5p^5 + 8p^5 - 2p^5 - p^5$
14. $6k^3 - 9k^3 + 8k^3 - 2k^3$

15. $y^4 + 8y^4 - 9y^2 + 6y^2 + 10y^2$
16. $11a^2 - 10a^2 + 2a^2 - a^6 + 2a^6$

17. $4z^5 - 9z^3 + 8z^2 + 10z^5$
18. $-9m^3 + 2m^3 - 11m^3 + 15m^2 - 9m$

19. $2p^7 - 8p^6 + 5p^4 - 9p$
20. $7y^3 - 8y^2 + 6y + 2$

21. $-.823q^2 + 1.725q - .374 + 1.994q^2 - .324q + .122$

22. $5.893r^3 - 2.776r^2 + 5.409r - 6.783r^3 + 1.437r - r^2$

For each polynomial, first simplify, if possible; then give the degree of the polynomial and tell whether it is a monomial, a binomial, a trinomial, or none of these. See Example 3.

                                                       Simplified form   *Degree*   *Kind of polynomial*

23. $5x^4 - 8x$

24. $4y - 8y$

25. $23x^9 - \frac{1}{2}x^2 + x$

26. $2m^7 - 3m^6 + 2m^5 + m$

27. $x^8 + 3x^7 - 5x^4$

28. $2x - 2x^2$

29. $\frac{3}{5}x^5 + \frac{2}{5}x^5$

30. $\frac{9}{11}x^2$

31. -8

32. $2m^8 - 5m^9$

Tell whether each statement is true always, sometimes, *or* never.

33. A binomial is a polynomial.

34. A polynomial is a trinomial.

35. A trinomial is a binomial.

36. A monomial has no coefficient.

37. A binomial is a trinomial.

38. A polynomial of degree 4 has 4 terms.

name   date   hour

Add or subtract as indicated. See Examples 4 and 8.

39. Add.

$3m^2 + 5m$
$\underline{2m^2 - 2m}$

40. Add.

$4a^3 - 4a^2$
$\underline{6a^3 + 5a^2}$

41. Subtract.

$12x^4 - x^2$
$\underline{8x^4 + 3x^2}$

42. Subtract.

$2a + 5d$
$\underline{3a - 6d}$

43. Subtract.

$2n^5 - 5n^3 + 6$
$\underline{3n^5 + 7n^3 + 8}$

44. Subtract.

$3r^2 - 4r + 2$
$\underline{7r^2 + 2r - 3}$

45. Add.

$9m^3 - 5m^2 + 4m - 8$
$\underline{3m^3 + 6m^2 + 8m - 6}$

46. Add.

$12r^5 + 11r^4 - 7r^3 - 2r^2 - 5r - 3$
$\underline{-8r^5 - 10r^4 + 3r^3 + 2r^2 - 5r + 7}$

47. Add.

$12m^2 - 8m + 6$
$\underline{3m^2 + 5m - 2}$

48. Subtract.

$5a^4 - 3a^3 + 2a^2$
$\underline{a^3 - a^2 + a - 1}$

49. Add.

$5b^2 + 6b + 2$
$\underline{3b^2 - 4b + 5}$

50. Add.

$3w^2 - 5w + 2$
$4w^2 + 6w - 5$
$\underline{8w^2 + 7w - 2}$

Perform the indicated operations. See Examples 5–7.

51. $(2r^2 + 3r) - (3r^2 + 5r)$

52. $(3r^2 + 5r - 6) + (2r - 5r^2)$

53. $(8m^2 - 7m) - (3m^2 + 7m)$

54. $(x^2 + x) - (3x^2 + 2x - 1)$

55. $8 - (6s^2 - 5s + 7)$

56. $2 - [3 - (4 + s)]$

57. $(8s - 3s^2) + (-4s + 5s^2)$

58. $(3x^2 + 2x + 5) + (8x^2 - 5x - 4)$

4.4 Exercises

59. $(16x^3 - x^2 + 3x) + (-12x^3 + 3x^2 + 2x)$

60. $(-2b^6 + 3b^4 - b^2) - (b^6 + 2b^4 + 2b^2)$

61. $(7y^4 + 3y^2 + 2y) - (18y^4 - 5y^2 - y)$

62. $(3x^2 + 2x + 5) + (-7x^2 - 8x + 2) + (3x^2 - 4x + 7)$

63. $(9a^4 - 3a^2 + 2) + (4a^4 - 4a^2 + 2) + (-12a^4 + 6a^2 - 3)$

64. $(4m^2 - 3m + 2) + (5m^2 + 13m - 4) - (16m^2 + 4m - 3)$

65. $[(8m^2 + 4m - 7) - (2m^2 - 5m + 2)] - (m^2 + m + 1)$

66. $[(9b^3 - 4b^2 + 3b + 2) - (-2b^3 - 3b^2 + b)] - (8b^3 + 6b + 4)$

67. $(3.127m^2 - 5.148m - 3.947) - (-.259m^2 + 7.125m - 8.9)$

68. $(-4.009k^2 + 3.176k + 4.1) - (1.795k^2 - .165k - .9935)$

Write each statement as an equation or an inequality. Do not solve.

69. When $4 + x^2$ is added to $-9x + 2$, the result is larger than 8.

70. When $6 + 3x$ is subtracted from $5 + 2x$, the difference is larger than $8x + x^2$.

71. The sum of $5 + x^2$ and $3 - 2x$ is not equal to 5.

72. The sum of $3 - 2x + x^2$ and $8 - 9x + 3x^2$ is negative.

4.5 MULTIPLICATION OF POLYNOMIALS

Objectives

▶ Multiply a monomial and a polynomial.
▶ Multiply two polynomials.
▶ Multiply vertically.

▶ As shown earlier, the product of two monomials is found by using the rules for exponents and the commutative and associative properties. For example,

$$(6x^3)(4x^4) = 6 \cdot 4 \cdot x^3 \cdot x^4 = 24x^7.$$

Also,

$$(-8m^6)(-9n^6) = (-8)(-9)(m^6)(n^6) = 72m^6n^6.$$

Find the product of a monomial and a polynomial with more than one term by first using the distributive property and then the method shown above.

EXAMPLE 1 Use the distributive property to find each product.

(a) $4x^2(3x + 5)$

$$\begin{aligned} 4x^2(3x + 5) &= (4x^2)(3x) + (4x^2)(5) \quad \text{Distributive property} \\ &= 12x^3 + 20x^2 \quad \text{Multiply} \end{aligned}$$

(b) $-8m^3(4m^3 + 3m^2 + 2m - 1)$

$$\begin{aligned} &= (-8m^3)(4m^3) + (-8m^3)(3m^2) \\ &\quad + (-8m^3)(2m) + (-8m^3)(-1) \\ &= -32m^6 - 24m^5 - 16m^4 + 8m^3 \end{aligned}$$

Work Problem 1 at the side.

1. Find each product.

(a) $5m^3(2m + 7)$

(b) $2x^4(3x^2 + 2x - 5)$

(c) $-4y^2(3y^3 + 2y^2 - 4y + 8)$

▶ The distributive property is also used to find the product of any two polynomials. For example, to find the product of the polynomials $x + 1$ and $x - 4$, think of $x + 1$ as a single quantity and use the distributive property as follows.

$$(x + 1)(x - 4) = (x + 1)x + (x + 1)(-4)$$

Now use the distributive property to find $(x + 1)x$ and $(x + 1)(-4)$.

$$\begin{aligned} (x + 1)x + (x + 1)(-4) &= x(x) + 1(x) + x(-4) + 1(-4) \\ &= x^2 + x + (-4x) + (-4) \\ &= x^2 - 3x - 4 \end{aligned}$$

EXAMPLE 2 Find the product $(2x + 1)(3x + 5)$.

$$\begin{aligned} (2x + 1)(3x + 5) &= (2x + 1)(3x) + (2x + 1)(5) \\ &= (2x)(3x) + (1)(3x) + (2x)(5) + (1)(5) \\ &= 6x^2 + 3x + 10x + 5 \\ &= 6x^2 + 13x + 5 \end{aligned}$$

Work Problem 2 at the side.

2. Find each product.

(a) $(4x + 3)(2x + 1)$

(b) $(3k - 2)(2k + 1)$

(c) $(m + 5)(3m - 4)$

▶ When at least one of the factors in a product of polynomials has three or more terms, the multiplication can be simplified by writing one polynomial above the other.

ANSWERS
1. (a) $10m^4 + 35m^3$
   (b) $6x^6 + 4x^5 - 10x^4$
   (c) $-12y^5 - 8y^4 + 16y^3 - 32y^2$
2. (a) $8x^2 + 10x + 3$
   (b) $6k^2 - k - 2$
   (c) $3m^2 + 11m - 20$

3. Find each product.

   (a) $4k - 6$
        $2k + 5$

   (b) $3x^2 + 4x - 5$
                $x + 4$

4. Find each product.

   (a) $2m + 3p$
        $5m - 4p$

   (b) $k^3 - k^2 + k + 1$
                      $k + 1$

   (c) $a^3 + 3a - 4$
        $2a^2 + 6a + 5$

EXAMPLE 3 Multiply $2x^2 + 4x + 1$ by $3x + 5$.

Start with
$$2x^2 + 4x + 1$$
$$3x + 5.$$

It is not necessary to line up terms in columns, because any terms may be multiplied (not just like terms). Begin by multiplying each of the terms in the top row by 5.

Step 1

$$2x^2 + 4x + 1$$
$$3x + 5$$
$$\overline{10x^2 + 20x + 5} \quad \leftarrow \text{Product of 5 and } 2x^2 + 4x + 1$$

Notice how this process is similar to multiplication of whole numbers. Now multiply each term in the top row by $3x$. Be careful to place the like terms in columns, since the final step will involve addition (as in multiplying two whole numbers).

Step 2

$$2x^2 + 4x + 1$$
$$3x + 5$$
$$\overline{10x^2 + 20x + 5}$$
$$6x^3 + 12x^2 + 3x \quad \leftarrow \text{Product of } 3x \text{ and } 2x^2 + 4x + 1$$

Step 3 Add like terms.

$$2x^2 + 4x + 1$$
$$3x + 5$$
$$\overline{10x^2 + 20x + 5}$$
$$6x^3 + 12x^2 + 3x$$
$$\overline{6x^3 + 22x^2 + 23x + 5}$$

The product is $6x^3 + 22x^2 + 23x + 5$. ◀

Work Problem 3 at the side.

EXAMPLE 4 Find the product of $3p - 5q$ and $2p + 7q$.

$$3p - 5q$$
$$2p + 7q$$
$$\overline{21pq - 35q^2} \quad \leftarrow 7q(3p - 5q)$$
$$6p^2 - 10pq \quad \leftarrow 2p(3p - 5q)$$
$$\overline{6p^2 + 11pq - 35q^2} \blacktriangleleft$$

EXAMPLE 5 Find the product of $4m^3 - 2m^2 + 4m$ and $m^2 + 5$.

$$4m^3 - 2m^2 + 4m$$
$$m^2 + 5$$
$$\overline{20m^3 - 10m^2 + 20m} \quad \text{Terms of top row multiplied by 5}$$
$$4m^5 - 2m^4 + 4m^3 \quad \text{Terms of top row multiplied by } m^2 \blacktriangleleft$$
$$\overline{4m^5 - 2m^4 + 24m^3 - 10m^2 + 20m}$$

Work Problem 4 at the side.

ANSWERS
3. (a) $8k^2 + 8k - 30$
   (b) $3x^3 + 16x^2 + 11x - 20$
4. (a) $10m^2 + 7mp - 12p^2$
   (b) $k^4 + 2k + 1$
   (c) $2a^5 + 6a^4 + 11a^3 + 10a^2 - 9a - 20$

EXAMPLE 6 Find $(3n^2 + 5n - 1)^2$.

By definition, $(3n^2 + 5n - 1)^2 = (3n^2 + 5n - 1)(3n^2 + 5n - 1)$. Use the vertical method of multiplication.

$$\begin{array}{r}
3n^2 + 5n - 1 \\
3n^2 + 5n - 1 \\
\hline
-3n^2 - 5n + 1 \\
15n^3 + 25n^2 - 5n \\
9n^4 + 15n^3 - 3n^2 \\
\hline
9n^4 + 30n^3 + 19n^2 - 10n + 1
\end{array}$$ ◀

EXAMPLE 7 Find $(x + 5)^3$.

Since $(x + 5)^3 = (x + 5)(x + 5)(x + 5)$, the first step is to find the product $(x + 5)(x + 5)$.

$$\begin{array}{r}
x + 5 \\
x + 5 \\
\hline
5x + 25 \\
x^2 + 5x \\
\hline
x^2 + 10x + 25
\end{array}$$

Now multiply this result by $x + 5$.

$$\begin{array}{r}
x^2 + 10x + 25 \\
x + 5 \\
\hline
5x^2 + 50x + 125 \\
x^3 + 10x^2 + 25x \\
\hline
x^3 + 15x^2 + 75x + 125
\end{array}$$ ◀

Work Problem 5 at the side.

---

5. Find each product.

   (a) $(3k - 2)^2$

   (b) $(2x^2 - 3x + 4)^2$

   (c) $(m + 1)^3$

ANSWERS
5. (a) $9k^2 - 12k + 4$
   (b) $4x^4 - 12x^3 + 25x^2 - 24x + 16$
   (c) $m^3 + 3m^2 + 3m + 1$

4.5 EXERCISES

Find each product. See Examples 1 and 2.

1. $(-4x^5)(8x^2)$

2. $(-3x^7)(2x^5)$

3. $(5y^4)(3y^7)$

4. $(10p^2)(5p^3)$

5. $(15a^4)(2a^5)$

6. $(-3m^6)(-5m^4)$

7. $2m(3m + 2)$

8. $-5p(6 - 3p)$

9. $3p(-2p^3 + 4p^2)$

10. $4x(3 + 2x + 5x^3)$

11. $-8z(2z + 3z^2 + 3z^3)$

12. $7y(3 + 5y^2 - 2y^3)$

13. $2y(3 + 2y + 5y^4)$

14. $-2m^4(3m^2 + 5m + 6)$

15. $4z^3(8z^2 + 5xz - 3x^2)$

Find each binomial product. See Example 4.

16. $(n - 1)(n + 4)$

17. $(x + 5)(x - 5)$

18. $(y + 8)(y - 8)$

19. $(3r - 1)(r - 4)$

20. $(2k + 5)(k + 2)$

21. $(6p + 5)(p - 1)$

22. $(2x + 3)(6x - 4)$

23. $(4m + 3)(4m + 3)$

24. $(3x - 2)(3x - 2)$

25. $(b + 8)(6b - 2)$

26. $(5a + 1)(2a + 7)$

27. $(8b - 3a)(2b + a)$

28. $(6p - 5m)(2p + 3m)$

29. $(-4h + k)(2h - k)$

30. $(5y - 3x)(4y + x)$

Find each product. See Examples 3 and 5.

31. $(6x + 1)(2x^2 + 4x + 1)$

32. $(9y - 2)(8y^2 - 6y + 1)$

33. $(9a + 2)(9a^2 + a + 1)$

34. $(2r - 1)(3r^2 + 4r - 4)$

35. $(4m + 3)(5m^3 - 4m^2 + m - 5)$

36. $(y + 4)(3y^4 - 2y^2 + 1)$

37. $(2x - 1)(3x^5 - 2x^3 + x^2 - 2x + 3)$

38. $(2a + 3)(a^4 - a^3 + a^2 - a + 1)$

39. $(5x^2 + 2x + 1)(x^2 - 3x + 5)$

40. $(2m^2 + m - 3)(m^2 - 4m + 5)$

Find each product. See Examples 6 and 7.

41. $(x + 7)^2$

42. $(m + 6)^2$

43. $(a - 4)^2$

44. $(b - 10)^2$

45. $(2p - 5)^2$

46. $(3m + 1)^2$

47. $(5k + 8)^2$

48. $(8m - 3)^2$

49. $(m - 5)^3$

50. $(p + 3)^3$

51. $(2a + 1)^3$

52. $(3m - 1)^3$

53. $(k + 1)^4$

54. $(r - 1)^4$

55. $(3r - 2s)^3$

56. $(2z + 5y)^3$

4.6 PRODUCTS OF BINOMIALS

Objectives

▶ Multiply binomials by the FOIL method.
▶ Square binomials.
▶ Find the product of the sum and difference of two terms.

▶ The methods introduced in the last section can be used to find the product of any two polynomials. They are the only practical methods for multiplying polynomials with three or more terms. However, many of the polynomials to be multiplied are binomials, with only two terms, so this section discusses a shortcut that eliminates the need to write all the steps. To develop this shortcut, let us first multiply $x + 3$ and $x + 5$ using the distributive property.

$$(x + 3)(x + 5) = (x + 3)x + (x + 3)5$$
$$= (x)(x) + (3)(x) + (x)(5) + (3)(5)$$
$$= x^2 + 3x + 5x + 15$$
$$= x^2 + 8x + 15$$

The first term in the second line, $(x)(x)$, is the product of the first terms of the two binomials.

$(x + 3)(x + 5)$ Multiply the first terms: $(x)(x)$

The term $(x)(5)$ is the product of the first term of the first binomial and the last term of the second binomial. This is the **outer product.**

$(x + 3)(x + 5)$ Multiply the outer terms: $(x)(5)$

The term $(3)(x)$ is the product of the last term of the first binomial and the first term of the second binomial. The product of these middle terms is called the **inner product.**

$(x + 3)(x + 5)$ Multiply the inner terms: $(3)(x)$

Finally, $(3)(5)$ is the product of the last terms of the two binomials.

$(x + 3)(x + 5)$ Multiply the last terms: $(3)(5)$

In the third step of the multiplication above, the inner product and the outer product are added. This step should be performed mentally, so that the three terms of the answer can be written without extra steps as

$$(x + 3)(x + 5) = x^2 + 8x + 15.$$

Work Problem 1 at the side.

A summary of these steps is given below. This procedure is sometimes called the **FOIL method,** which comes from the abbreviation for *First, Outer, Inner, Last.*

1. For the product $(2p - 5)(3p + 7)$, find the following.

   (a) Product of first terms

   (b) Outer product

   (c) Inner product

   (d) Product of last terms

Multiplying Binomials by the FOIL Method

Step 1 Multiply the two first terms of the binomials to get the first term of the answer.

Step 2 Find the outer product and the inner product and mentally add them, when possible, to get the middle term of the answer.

Step 3 Multiply the two last terms of the binomials to get the last term of the answer.

ANSWERS
1. (a) $2p(3p) = 6p^2$
   (b) $2p(7) = 14p$
   (c) $-5(3p) = -15p$
   (d) $-5(7) = -35$

4.6 Products of Binomials   201

2. Use the FOIL method to find each product.

(a) $(m + 4)(m - 3)$

(b) $(y + 7)(y + 2)$

(c) $(r - 8)(r - 5)$

3. Find each product.

(a) $(4k - 1)(2k + 3)$

(b) $(6m + 5)(m - 4)$

(c) $(8y + 3)(2y + 1)$

(d) $(3r + 2t)(3r + 4t)$

EXAMPLE 1 Use the FOIL method to find the product $(x + 8)(x - 6)$.

Step 1 F Multiply the first terms.
$$x(x) = x^2$$

Step 2 O Find the product of the outer terms.
$$x(-6) = -6x$$

I Find the product of the inner terms.
$$8(x) = 8x$$

Add the outer and inner products mentally.
$$-6x + 8x = 2x$$

Step 3 L Multiply the last terms.
$$8(-6) = -48$$

The product of $x + 8$ and $x - 6$ is found by adding the terms found in the three steps above, so
$$(x + 8)(x - 6) = x^2 + 2x - 48.$$

As a shortcut, this product can be found in the following manner.

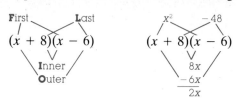

Work Problem 2 at the side.

EXAMPLE 2 Multiply $9x - 2$ and $3x + 1$.

First $(9x - 2)(3x + 1)$ $27x^2$
Outer $(9x - 2)(3x + 1)$ $9x$
Inner $(9x - 2)(3x + 1)$ $-6x$
Last $(9x - 2)(3x + 1)$ -2

$$\begin{aligned} & \quad\quad\quad\quad\quad\quad F \quad O \quad I \quad L \\ (9x - 2)(3x + 1) &= 27x^2 + 9x - 6x - 2 \\ &= 27x^2 + 3x - 2 \end{aligned}$$

EXAMPLE 3 Find the following products.

$$\begin{aligned} & \quad\quad\quad\quad\quad\quad\quad F \quad\quad\quad\quad O \quad\quad\quad\quad I \quad\quad\quad\quad L \\ \text{(a) } (2k + 5y)(k + 3y) &= (2k)(k) + (2k)(3y) + (5y)(k) + (5y)(3y) \\ &= 2k^2 + 6ky + 5ky + 15y^2 \\ &= 2k^2 + 11ky + 15y^2 \end{aligned}$$

(b) $(7p + 2q)(3p - q) = 21p^2 - pq - 2q^2$

Work Problem 3 at the side.

▶ Certain special types of binomial products occur so often that the form of the answers should be memorized. For example, to find the square of a binomial quickly, use the method shown in Example 4.

ANSWERS

2. (a) $m^2 + m - 12$
   (b) $y^2 + 9y + 14$
   (c) $r^2 - 13r + 40$
3. (a) $8k^2 + 10k - 3$
   (b) $6m^2 - 19m - 20$
   (c) $16y^2 + 14y + 3$
   (d) $9r^2 + 18rt + 8t^2$

EXAMPLE 4 Find $(2m + 3)^2$.

Squaring $2m + 3$ by the FOIL method gives
$$(2m + 3)(2m + 3) = 4m^2 + 12m + 9.$$

The result has the square of both the first and the last terms of the binomial:
$$4m^2 = \boxed{(2m)^2} \quad \text{and} \quad 9 = \boxed{3^2}.$$

The middle term is twice the product of the two terms of the binomial, that is,
$$12m = \boxed{2(2m)(3)}. \blacktriangleleft$$

This example suggests the following rule.

Square of a Binomial

The square of a binomial is a trinomial made up of the square of the first term, plus twice the product of the two terms, plus the square of the last term of the binomial. For a and b,
$$(a + b)^2 = a^2 + 2ab + b^2.$$
Also,
$$(a - b)^2 = a^2 - 2ab + b^2.$$

EXAMPLE 5 Use the formula to square each binomial.

(a) $(5z - 1)^2 = \boxed{(5z)^2 - 2(5z)(1) + (1)^2}$
$= 25z^2 - 10z + 1$

Recall that $(5z)^2 = 5^2 z^2 = 25z^2$.

(b) $(3b + 5r)^2 = (3b)^2 + 2(3b)(5r) + (5r)^2$
$= 9b^2 + 30br + 25r^2$

(c) $(2a - 9x)^2 = 4a^2 - 36ax + 81x^2$

(d) $\left(4m + \dfrac{1}{2}\right)^2 = (4m)^2 + 2(4m)\left(\dfrac{1}{2}\right) + \left(\dfrac{1}{2}\right)^2$
$= 16m^2 + 4m + \dfrac{1}{4} \blacktriangleleft$

Work Problem 4 at the side.

3 Binomial products of the form $(a + b)(a - b)$ also occur frequently. In these products, one binomial is the sum of two terms, and the other is the difference of the same two terms. As an example, the product of $x + 2$ and $x - 2$ is
$$(x + 2)(x - 2) = x^2 - 2x + 2x - 4$$
$$= x^2 - 4.$$

As shown with the FOIL method, the product of $a + b$ and $a - b$ will be the difference of two squares.

Product of the Sum and Difference of Two Terms

$$(a + b)(a - b) = a^2 - b^2$$

4.6 Products of Binomials

4. Find each square.

(a) $(t + u)^2$

(b) $(2m - p)^2$

(c) $(4p + 3q)^2$

(d) $(5r - 6s)^2$

(e) $\left(3k - \dfrac{1}{2}\right)^2$

ANSWERS
4. (a) $t^2 + 2tu + u^2$
   (b) $4m^2 - 4mp + p^2$
   (c) $16p^2 + 24pq + 9q^2$
   (d) $25r^2 - 60rs + 36s^2$
   (e) $9k^2 - 3k + \dfrac{1}{4}$

5. Find each product by using the pattern for the sum and difference of two terms.

   (a) $(6a + 3)(6a - 3)$

   (b) $(10m + 7)(10m - 7)$

   (c) $(7p + 2q)(7p - 2q)$

   (d) $(8r + 5s)(8r - 5s)$

   (e) $\left(3r - \frac{1}{2}\right)\left(3r + \frac{1}{2}\right)$

EXAMPLE 6 Find each product.

(a) $(5m + 3)(5m - 3)$

Use the pattern for the sum and difference of two terms.
$$(5m + 3)(5m - 3) = (5m)^2 - 3^2$$
$$= 25m^2 - 9$$

(b) $(4x + y)(4x - y) = (4x)^2 - y^2$
$$= 16x^2 - y^2$$

(c) $\left(z - \frac{1}{4}\right)\left(z + \frac{1}{4}\right) = z^2 - \frac{1}{16}$ ◀

Work Problem 5 at the side.

The product formulas of this section will be very useful in later work, particularly in Chapter 5. Therefore, it is important to memorize these formulas and practice using them.

ANSWERS

5. (a) $36a^2 - 9$ (b) $100m^2 - 49$
   (c) $49p^2 - 4q^2$ (d) $64r^2 - 25s^2$
   (e) $9r^2 - \frac{1}{4}$

name   date   hour

4.6 EXERCISES

Find each product. See Examples 1–3.

1. $(r - 1)(r + 3)$
2. $(x + 2)(x - 5)$
3. $(x - 7)(x - 3)$

4. $(r + 3)(r + 6)$
5. $(2x - 1)(3x + 2)$
6. $(4y - 5)(2y + 1)$

7. $(6z + 5)(z - 3)$
8. $(8a + 3)(6a + 1)$
9. $(a + 4)(2a + 1)$

10. $(3x - 1)(2x + 3)$
11. $(2r - 1)(4r + 3)$
12. $(5m + 2)(3m - 4)$

13. $(2a + 4)(3a - 2)$
14. $(11m - 10)(10m + 11)$
15. $(4 + 5x)(5 - 4x)$

16. $(8 + 3x)(2 - x)$
17. $(-3 + 2r)(4 + r)$
18. $(-5 + 6z)(2 - z)$

19. $(-3 + a)(-5 - 2a)$
20. $(-6 - 3y)(1 - 4y)$
21. $(p + 3q)(p + q)$

22. $(2r - 3s)(3r + s)$
23. $(5y + z)(2y - z)$
24. $(9m + 4k)(2m - 3k)$

25. $(8y - 9z)(y + 5z)$
26. $(3a + 7b)(-4a + b)$
27. $(4r + 9s)(-2r + 5s)$

28. $(7m + 11n)(3m - 8n)$
29. $(2.13y + 4.06)(1.92y - 3.9)$
30. $(8.17m - 2.4)(3.5m + 1.8)$

Find each square. See Examples 4 and 5.

31. $(m + 2)^2$
32. $(x + 8)^2$
33. $(x - 2y)^2$

34. $(3m - n)^2$
35. $\left(2z - \dfrac{5}{2}x\right)^2$
36. $\left(6a - \dfrac{3}{2}b\right)^2$

37. $(5p + 2q)^2$
38. $(8a - 3b)^2$
39. $\left(4a - \dfrac{5}{4}b\right)^2$

40. $\left(9y + \dfrac{z}{3}\right)^2$
41. $(.85r + .23s)^2$
42. $(.67m - .17k)^2$

Find the following products. See Example 6.

43. $(p + 2)(p - 2)$
44. $(a + 8)(a - 8)$
45. $(2b + 5)(2b - 5)$

46. $(3x + 4)(3x - 4)$
47. $(6a - p)(6a + p)$
48. $(5y + 3x)(5y - 3x)$

49. $\left(2m - \dfrac{5}{3}\right)\left(2m + \dfrac{5}{3}\right)$
50. $\left(3a - \dfrac{4}{5}\right)\left(3a + \dfrac{4}{5}\right)$
51. $(7y^2 + 10z)(7y^2 - 10z)$

52. $(6x + 5y^2)(6x - 5y^2)$
53. $(.48q + .37r)(.48q - .37r)$
54. $(.26a + .15b)(.26a - .15b)$

Write each statement as an equation or an inequality, using x to represent the unknown number. Do not solve.

55. The square of 3 more than a number is 5.

56. The square of the sum of a number and 6 is less than 3.

57. When 3 plus a number is multiplied by the number less 4, the result is greater than 7.

58. Twice a number plus 4, multiplied by 6 times the number, less 5, gives 8.

Exponents and Polynomials

4.7 DIVIDING A POLYNOMIAL BY A MONOMIAL

Objective

▶ Divide a polynomial by a monomial.

▶ The quotient rule for exponents is used to divide a monomial by another monomial. For example,

$$\frac{12x^2}{6x} = 2x, \quad \frac{25m^5}{5m^2} = 5m^3, \quad \text{and} \quad \frac{30a^2b^8}{15a^3b^3} = \frac{2b^5}{a}.$$

Dividing a Polynomial by a Monomial

Divide a polynomial by a monomial by dividing each term of the polynomial by the monomial:

$$\frac{a+b}{c} = \frac{a}{c} + \frac{b}{c} \quad (c \neq 0).$$

EXAMPLE 1 Divide $5m^5 - 10m^3$ by $5m^2$.

Use the rule above, with $+$ replaced by $-$.

$$\frac{5m^5 - 10m^3}{5m^2} = \frac{5m^5}{5m^2} - \frac{10m^3}{5m^2} = m^3 - 2m$$

Check by multiplication.

$$5m^2(m^3 - 2m) = 5m^5 - 10m^3$$

Since division by 0 is meaningless, the quotient

$$\frac{5m^5 - 10m^3}{5m^2}$$

has no value if $m = 0$. In the rest of the chapter, we assume that no denominators are 0. ◀

Work Problem 1 at the side.

EXAMPLE 2 Divide: $\dfrac{16a^5 - 12a^4 + 8a^2}{4a^3}$.

Divide each term of $16a^5 - 12a^4 + 8a^2$ by $4a^3$.

$$\frac{16a^5 - 12a^4 + 8a^2}{4a^3} = \frac{16a^5}{4a^3} - \frac{12a^4}{4a^3} + \frac{8a^2}{4a^3}$$

$$= 4a^2 - 3a + \frac{2}{a}$$

The result is not a polynomial because of the expression $\frac{2}{a}$, which has a variable in the denominator. While the sum, difference, and product of two polynomials are always polynomials, the quotient of two polynomials may not be.

Again, check by multiplying.

$$4a^3\left(4a^2 - 3a + \frac{2}{a}\right) = 4a^3(4a^2) - 4a^3(3a) + 4a^3\left(\frac{2}{a}\right)$$

$$= 16a^5 - 12a^4 + 8a^2. \blacktriangleleft$$

Work Problem 2 at the side.

1. Divide.

    (a) $\dfrac{6p^4 + 18p^7}{3p^2}$

    (b) $\dfrac{12m^6 + 18m^5 + 30m^4}{6m^2}$

    (c) $(18r^7 - 9r^2) \div (3r)$

2. Divide.

    (a) $\dfrac{20x^4 - 25x^3 + 5x}{5x^2}$

    (b) $\dfrac{50m^4 - 30m^3 + 20m}{10m^3}$

ANSWERS

1. (a) $2p^2 + 6p^5$
   (b) $2m^4 + 3m^3 + 5m^2$
   (c) $6r^6 - 3r$

2. (a) $4x^2 - 5x + \dfrac{1}{x}$
   (b) $5m - 3 + \dfrac{2}{m^2}$

4.7 Dividing a Polynomial by a Monomial

3. Divide.

(a) $\dfrac{8y^7 - 9y^6 - 11y - 4}{y^2}$

(b) $\dfrac{12p^5 + 8p^4 + 6p^3 - 5p^2}{3p^3}$

EXAMPLE 3 Divide.

$$\frac{12x^4 - 7x^3 + x - 4}{4x} = \frac{12x^4}{4x} - \frac{7x^3}{4x} + \frac{x}{4x} - \frac{4}{4x}$$
$$= 3x^3 - \frac{7x^2}{4} + \frac{1}{4} - \frac{1}{x}$$

Check by multiplication. ◀

EXAMPLE 4 Divide the polynomial
$$180y^{10} - 150y^8 + 120y^6 - 90y^4 + 100y$$
by the monomial $-30y^2$.

Using the methods of this section,
$$\frac{180y^{10} - 150y^8 + 120y^6 - 90y^4 + 100y}{-30y^2}$$
$$= \frac{180y^{10}}{-30y^2} - \frac{150y^8}{-30y^2} + \frac{120y^6}{-30y^2} - \frac{90y^4}{-30y^2} + \frac{100y}{-30y^2}$$
$$= -6y^8 + 5y^6 - 4y^4 + 3y^2 - \frac{10}{3y}.$$

To check, multiply this result and $-30y^2$. ◀

Work Problem 3 at the side.

ANSWERS

3. (a) $8y^5 - 9y^4 - \dfrac{11}{y} - \dfrac{4}{y^2}$

(b) $4p^2 + \dfrac{8p}{3} + 2 - \dfrac{5}{3p}$

Exponents and Polynomials

name　　　　　　　　　　　　　　　　　　date　　　　　　hour

4.7 EXERCISES

Divide.

1. $\dfrac{4x^2}{2x}$
2. $\dfrac{8m^7}{2m}$
3. $\dfrac{10a^3}{5a}$

4. $\dfrac{36p^8}{4p^3}$
5. $\dfrac{27k^4m^5}{3km^6}$
6. $\dfrac{18x^5y^6}{3x^2y^2}$

Divide each polynomial by 2m. See Examples 1–4.

7. $60m^4 - 20m^2$
8. $120m^6 - 60m^3 + 80m^2$
9. $10m^5 - 16m^2 + 8m^3$

10. $6m^5 - 4m^3 + 2m^2$
11. $8m^5 - 4m^3 + 4m^2$
12. $8m^3 - 4m^2 + 6m$

13. $2m^5 - 4m^2 + 8m$
14. $m^2 + m + 1$
15. $2m^2 - 2m + 5$

Divide each polynomial by $3x^2$. See Examples 1–4.

16. $15x^2 - 9x^3$
17. $12x^4 - 3x^3 + 3x^2$
18. $45x^3 + 15x^2 - 9x^5$

19. $27x^3 - 9x^4 + 18x^5$
20. $-18x^6 + 6x^5 + 3x^4 - 9x^3$
21. $36x + 24x^2 + 3x^3$

22. $4x^4 - 3x^3 + 2x$
23. $x^3 + 6x^2 - x$
24. $6x^5 - 3x^4 + 9x^2 + 27$

4.7　Exercises　　209

Perform each division. See Examples 1–4.

25. $\dfrac{8k^4 - 12k^3 - 2k^2 + 7k - 3}{2k}$

26. $\dfrac{27r^4 - 36r^3 - 6r^2 + 26r - 2}{3r}$

27. $\dfrac{100p^5 - 50p^4 + 30p^3 - 30p}{-10p^2}$

28. $\dfrac{2m^5 - 6m^4 + 8m^2}{-2m^3}$

29. $\dfrac{8x + 16x^2 + 10x^3}{4x^4}$

30. $\dfrac{36m^5 - 24m^4 + 16m^3 - 8m^2}{4m^3}$

31. $(16y^5 - 8y^2 + 12y) \div (4y^2)$

32. $(20a^4 - 15a^5 + 25a^3) \div (15a^4)$

33. $(120x^{11} - 60x^{10} + 140x^9 - 100x^8) \div (10x^{12})$

34. $(5 + x + 6x^2 + 8x^3) \div (3x^4)$

Solve each problem.

35. What polynomial, when divided by $3x^2$, yields $4x^3 + 3x^2 - 4x + 2$ as a quotient?

36. What polynomial, when divided by $4m^3$, yields $-6m^2 + 4m$ as a quotient?

37. The quotient of a certain polynomial and $-7y^2$ is $9y^2 + 3y + 5 - \dfrac{2}{y}$. Find the polynomial.

38. The quotient of a certain polynomial and a is $2a^2 + 3a + 5$. Find the polynomial.

4.8 THE QUOTIENT OF TWO POLYNOMIALS

Objective

▶ Divide a polynomial by a polynomial.

▶ A method of "long division" is used to divide a polynomial by a polynomial (other than a monomial). This method is similar to the method of long division used for two whole numbers. For comparison, the division of whole numbers is shown alongside the division of polynomials.

Step 1

Divide 27 into 6696.

$$27\overline{)6696}$$

Divide $2x + 3$ into $8x^3 - 4x^2 - 14x + 15$.

$$2x + 3\overline{)8x^3 - 4x^2 - 14x + 15}$$

Step 2

27 divides into 66 **2** times; $2 \cdot 27 = 54$.

$$\begin{array}{r} 2 \\ 27\overline{)6696} \\ 54 \end{array}$$

$2x$ divides into $8x^3$ **$4x^2$** times; $4x^2(2x + 3) = 8x^3 + 12x^2$.

$$\begin{array}{r} 4x^2 \\ 2x + 3\overline{)8x^3 - 4x^2 - 14x + 15} \\ 8x^3 + 12x^2 \end{array}$$

Step 3

Subtract; then bring down the next term.

$$\begin{array}{r} 2 \\ 27\overline{)6696} \\ 54\downarrow \\ \overline{12\mathbf{9}} \end{array}$$

Subtract; then bring down the next term.

$$\begin{array}{r} 4x^2 \\ 2x + 3\overline{)8x^3 - 4x^2 - 14x + 15} \\ 8x^3 + 12x^2 \\ \overline{-16x^2 - \mathbf{14x}} \end{array}$$

(To subtract two polynomials, change the sign of the second and then add.)

Step 4

27 divides into 129 **4** times; $4 \cdot 27 = 108$.

$$\begin{array}{r} 24 \\ 27\overline{)6696} \\ 54 \\ \overline{129} \\ 108 \end{array}$$

$2x$ divides into $-16x^2$ **$-8x$** times; $-8x(2x + 3) = -16x^2 - 24x$.

$$\begin{array}{r} 4x^2 - \mathbf{8x} \\ 2x + 3\overline{)8x^3 - 4x^2 - 14x + 15} \\ 8x^3 + 12x^2 \\ \overline{-16x^2 - 14x} \\ -16x^2 - 24x \end{array}$$

Step 5

Subtract; then bring down the next term.

$$\begin{array}{r} 24 \\ 27\overline{)6696} \\ 54\downarrow \\ \overline{129} \\ 108\downarrow \\ \overline{21\mathbf{6}} \end{array}$$

Subtract; then bring down the next term.

$$\begin{array}{r} 4x^2 - 8x \\ 2x + 3\overline{)8x^3 - 4x^2 - 14x + 15} \\ 8x^3 + 12x^2 \\ \overline{-16x^2 - 14x} \\ -16x^2 - 24x \downarrow \\ \overline{10x + \mathbf{15}} \end{array}$$

4.8 The Quotient of Two Polynomials

1. Divide.

(a) $(2y^2 - y - 21) \div (y + 3)$

(b) $(x^3 + x^2 + 4x - 6) \div (x - 1)$

(c) $\dfrac{p^3 - 2p^2 - 5p + 9}{p + 2}$

(d) $\dfrac{2m^4 - 11m^3 + 16m^2 - 5m + 6}{m - 3}$

Step 6

27 divides into 216 **8** times; $8 \cdot 27 = 216$.

$$\begin{array}{r} 248 \\ 27\overline{)6696} \\ \underline{54} \\ 129 \\ \underline{108} \\ 216 \\ \underline{216} \\ \end{array}$$

6696 divided by 27 is 248. There is no remainder.

2x divides into $10x$ **5** times; $5(2x + 3) = 10x + 15$.

$$\begin{array}{r} 4x^2 - 8x + 5 \\ 2x + 3\overline{)8x^3 - 4x^2 - 14x + 15} \\ \underline{8x^3 + 12x^2} \\ -16x^2 - 14x \\ \underline{-16x^2 - 24x} \\ 10x + 15 \\ \underline{10x + 15} \\ \end{array}$$

$8x^3 - 4x^2 - 14x + 15$ divided by $2x + 3$ is $4x^2 - 8x + 5$. There is no remainder.

Step 7

Check by multiplication.

$27 \cdot 248 = 6696$

Check by multiplication.

$(2x + 3)(4x^2 - 8x + 5)$
$= 8x^3 - 4x^2 - 14x + 15$

EXAMPLE 1 Divide $4x^3 - 4x^2 + 5x - 8$ by $2x - 1$.

$$\begin{array}{r} 2x^2 - x + 2 \\ 2x - 1\overline{)4x^3 - 4x^2 + 5x - 8} \\ \underline{4x^3 - 2x^2} \\ -2x^2 + 5x \\ \underline{-2x^2 + x} \\ 4x - 8 \\ \underline{4x - 2} \\ -6 \\ \end{array}$$

Step 1 2x divides into $4x^3$ **$2x^2$** times; $2x^2(2x - 1) = 4x^3 - 2x^2$.

Step 2 Subtract; bring down the next term.

Step 3 2x divides into $-2x^2$ **$-x$** times; $-x(2x - 1) = -2x^2 + x$.

Step 4 Subtract; bring down the next term.

Step 5 2x divides into $4x$ **2** times; $2(2x - 1) = 4x - 2$.

Step 6 Subtract. The remainder is -6. Thus $2x - 1$ divides into $4x^3 - 4x^2 + 5x - 8$ with a quotient of $2x^2 - x + 2$ and a remainder of -6. Write the remainder as a fraction. The result is not a polynomial because of the remainder.

$$\frac{4x^3 - 4x^2 + 5x - 8}{2x - 1} = 2x^2 - x + 2 + \frac{-6}{2x - 1}$$

Step 7 Check by multiplication.

$$(2x - 1)\left(2x^2 - x + 2 + \frac{-6}{2x - 1}\right)$$
$$= 4x^3 - 4x^2 + 5x - 8 \quad \blacktriangleleft$$

Work Problem 1 at the side.

ANSWERS

1. (a) $2y - 7$ (b) $x^2 + 2x + 6$

(c) $p^2 - 4p + 3 + \dfrac{3}{p + 2}$

(d) $2m^3 - 5m^2 + m - 2$

Exponents and Polynomials

EXAMPLE 2 Divide $x^3 - 1$ by $x - 1$.

Here the polynomial $x^3 - 1$ is missing the x^2 term and the x term. When terms are missing, use 0 as the coefficient for the missing terms.

$$x^3 - 1 = x^3 + \boxed{0x^2} + \boxed{0x} - 1$$

Now divide.

$$\begin{array}{r}
x^2 + x + 1 \\
x - 1 \overline{\smash{)}x^3 + 0x^2 + 0x - 1} \\
\underline{x^3 - x^2} \\
x^2 + 0x \\
\underline{x^2 - x} \\
x - 1 \\
\underline{x - 1} \\
0
\end{array}$$

The remainder is 0. The quotient is $x^2 + x + 1$. Check by multiplication.

$$(x^2 + x + 1)(x - 1) = x^3 - 1 \blacktriangleleft$$

Work Problem 2 at the side.

EXAMPLE 3 Divide $x^4 + 2x^3 + 2x^2 - x - 1$ by $x^2 + 1$.

Since $x^2 + 1$ has a missing x term, write it as $x^2 + 0x + 1$. Then proceed through the division process as follows.

$$\begin{array}{r}
x^2 + 2x + 1 \\
x^2 \boxed{+ 0x} + 1 \overline{\smash{)}x^4 + 2x^3 + 2x^2 - x - 1} \\
\underline{x^4 + 0x^3 + x^2} \\
2x^3 + x^2 - x \\
\underline{2x^3 + 0x^2 + 2x} \\
x^2 - 3x - 1 \\
\underline{x^2 + 0x + 1} \\
-3x - 2
\end{array}$$

When the result of subtracting ($-3x - 2$, in this case) is a polynomial of smaller degree than the divisor ($x^2 + 0x + 1$), that polynomial is the remainder. Write the result as

$$x^2 + 2x + 1 + \frac{-3x - 2}{x^2 + 1}. \blacktriangleleft$$

Work Problem 3 at the side.

2. Divide.

(a) $\dfrac{r^2 - 5}{r + 4}$

(b) $(x^3 - 8) \div (x - 2)$

3. Divide.

(a) $(2x^4 + 3x^3 - x^2 + 6x + 5) \div (x^2 - 1)$

(b) $\dfrac{2m^5 + m^4 + 6m^3 - 3m^2 - 18}{m^2 + 3}$

ANSWERS

2. (a) $r - 4 + \dfrac{11}{r + 4}$
   (b) $x^2 + 2x + 4$
3. (a) $2x^2 + 3x + 1 + \dfrac{9x + 6}{x^2 - 1}$
   (b) $2m^3 + m^2 - 6$

4.8 The Quotient of Two Polynomials

4.8 EXERCISES

Perform each division. See Example 1.

1. $\dfrac{x^2 - x - 6}{x - 3}$

2. $\dfrac{m^2 - 2m - 24}{m + 4}$

3. $\dfrac{2y^2 + 9y - 35}{y + 7}$

4. $\dfrac{y^2 + 2y + 1}{y + 1}$

5. $\dfrac{p^2 + 2p - 20}{p + 6}$

6. $\dfrac{x^2 + 11x + 16}{x + 8}$

7. $\dfrac{r^2 - 8r + 15}{r - 3}$

8. $\dfrac{t^2 - 3t - 10}{t - 5}$

9. $\dfrac{12m^2 - 20m + 3}{2m - 3}$

10. $\dfrac{2y^2 - 5y - 3}{2y + 1}$

11. $\dfrac{2a^2 - 11a + 16}{2a + 3}$

12. $\dfrac{9w^2 + 6w + 10}{3w - 2}$

13. $\dfrac{2x^2 + 5x + 3}{2x + 1}$

14. $\dfrac{4m^2 - 4m + 5}{2m - 1}$

15. $\dfrac{14k^2 + 19k - 30}{7k - 8}$

16. $\dfrac{15m^2 + 34m + 28}{5m + 3}$

17. $\dfrac{2x^3 - x^2 + 3x + 2}{2x + 1}$

18. $\dfrac{12t^3 - 11t^2 + 9t + 18}{4t + 3}$

19. $\dfrac{8k^4 - 12k^3 - 2k^2 + 7k - 6}{2k - 3}$

20. $\dfrac{27r^4 - 36r^3 - 6r^2 + 26r - 24}{3r - 4}$

Perform each division. See Examples 2 and 3.

21. $\dfrac{3y^3 + y^2 + 3y + 1}{y^2 + 1}$

22. $\dfrac{2r^3 - 5r^2 - 6r + 15}{r^2 - 3}$

23. $\dfrac{3k^3 - 4k^2 - 6k + 10}{k^2 - 2}$

24. $\dfrac{5z^3 - z^2 + 10z + 2}{z^2 + 2}$

25. $\dfrac{x^4 - x^2 - 6x}{x^2 - 2}$

26. $\dfrac{r^4 - 2r^2 + 5}{r^2 - 1}$

27. $\dfrac{6p^4 - 15p^3 + 14p^2 - 5p + 10}{3p^2 + 1}$

28. $\dfrac{6r^4 - 10r^3 - r^2 + 15r - 8}{2r^2 - 3}$

29. $\dfrac{4m^5 - 8m^4 - 3m^3 + 22m^2 - 15}{4m^2 - 3}$

30. $\dfrac{2x^5 + 6x^4 - x^3 + 3x^2 - x + 5}{2x^2 + 1}$

31. $\dfrac{y^3 + 1}{y + 1}$

32. $\dfrac{x^4 - 1}{x^2 - 1}$

33. $\dfrac{a^4 - 1}{a^2 + 1}$

34. $\dfrac{p^5 - 2}{p^2 - 1}$

216 Exponents and Polynomials

name   date   hour

CHAPTER 4   REVIEW EXERCISES

(4.1)   *Evaluate each expression.*

1. $4^2 + 4^3$
2. $5^0 + 7^0$
3. 2^{-4}

4. 6^{-1}
5. 7^{-2}
6. $\left(\dfrac{5}{8}\right)^{-2}$

7. $\left(\dfrac{1}{2}\right)^{-5}$
8. $2^{-1} + 4^{-1}$
9. $3^{-2} - 3^0$

Simplify. Write each answer in exponential form, using only positive exponents. Assume all variables are positive.

10. $5^4 \cdot 5^7$
11. $9^3 \cdot 9^{-5}$
12. $(-4)^5 \cdot (-4)^3$

13. $\dfrac{15^{17}}{15^{12}}$
14. $\dfrac{5^8}{5^{19}}$
15. $\dfrac{6^{-3}}{6^{-5}}$

16. $\dfrac{x^{-7}}{x^{-9}}$
17. $\dfrac{p^{-8}}{p^4}$
18. $\dfrac{r^{-2}}{r^{-6}}$

(4.2)

19. $(2^4)^2$
20. $(9^3)^{-2}$
21. $(5^{-2})^{-4}$

22. $(8^{-3})^4$
23. $\dfrac{(m^2)^3}{(m^4)^2}$
24. $\dfrac{y^4 \cdot y^{-2}}{y^{-5}}$

25. $\dfrac{r^9 \cdot r^{-5}}{r^{-2} \cdot r^{-7}}$
26. $(-5m^3)^2$
27. $(2y^{-4})^{-3}$

28. $(6r^{-2})^{-1}$
29. $(3p)^4(3p^{-7})$
30. $\left(\dfrac{6r^2s}{5}\right)^3$

31. $\dfrac{ab^{-3}}{a^4b^2}$
32. $\dfrac{(6r^{-1})^2 \cdot (2r^{-4})}{r^{-5}(r^2)^{-3}}$
33. $\dfrac{(2m^{-5}n^2)^3(3m^2)^{-1}}{m^{-2}n^{-4}(m^{-1})^2}$

Chapter 4   Review Exercises

(4.3) *Write each number in scientific notation.*

34. 64,000 **35.** 15,800,000 **36.** 26,954,000,000

37. .0004251 **38.** .0000976 **39.** .784

Write each number without exponents.

40. 1.2×10^4 **41.** 6.89×10^8 **42.** 4.253×10^{-4}

43. 8.77×10^{-1} **44.** $(6 \times 10^4) \times (1.5 \times 10^3)$ **45.** $(2 \times 10^{-3}) \times (4 \times 10^5)$

46. $\dfrac{9 \times 10^{-2}}{3 \times 10^2}$ **47.** $\dfrac{8 \times 10^4}{2 \times 10^{-2}}$

48. $\dfrac{12 \times 10^{-5} \times 5 \times 10^4}{4 \times 10^3 \times 6 \times 10^{-2}}$ **49.** $\dfrac{2.5 \times 10^5 \times 4.8 \times 10^{-4}}{7.5 \times 10^8 \times 1.6 \times 10^{-5}}$

(4.4) *Combine terms where possible in the following polynomials. Write answers in descending powers of the variable. Give the degree of the answer.*

	Simplified form	Degree

50. $9m^2 + 11m^2 + 2m^2$

51. $-4p + p^3 - p^2 + 8p + 2$

52. $2r^4 - r^3 + 8r^4 + r^3 - 6r^4$

53. $12a^5 - 9a^4 + 8a^3 + 2a^2 - a + 3$

54. $-7x^5 - 8x - x^5 + x + 9x^3$

55. $-5z^3 + 7 - 6z^2 + 8z$

218 Exponents and Polynomials

name   date   hour

Add or subtract as indicated.

56. Add.
$-2a^3 + 5a^2$
$\underline{\;3a^3 - \;\;a^2\;}$

57. Add.
$4r^3 - 8r^2 + 6r$
$\underline{-2r^3 + 5r^2 - 3r}$

58. Subtract.
$6y^2 - 8y + 2$
$\underline{5y^2 + 2y - 7}$

59. Subtract.
$-12k^4 - 8k^2 + 7k - 5$
$\underline{\;\;\;\;k^4 + 7k^2 - 11k + 1}$

60. $(2m^3 - 8m^2 + 4) + (3m^3 + 2m^2 - 7)$

61. $(-5y^2 + 3y - 11) + (4y^2 - 7y + 15)$

62. $(6p^2 - p - 8) - (-4p^2 + 2p - 3)$

63. $(12r^4 - 7r^3 + 2r^2) - (5r^4 - 3r^3 + 2r^2 - 1)$

64. $-8 - [6 - (3 + p)]$

65. $(5y^3 - 8y^2 + 7) - [(-3y^3 + y^2 + 2) + (y^3 - 8y^2 - 4)]$

(4.5) *Find each product.*

66. $5x(2x - 11)$

67. $-3p^3(2p^2 - 5p)$

68. $2y^2(-11y^2 + 2y + 9)$

69. $-m^5(8m^2 - 10m + 6)$

70. $(m - 9)(m + 2)$

71. $(3k - 6)(2k + 1)$

72. $(12a + 1)(12a - 1)$

73. $(-7 + 2k)^2$

74. $(a + 2)(a^2 - 4a + 1)$

75. $(3r - 2)(2r^2 + 4r - 3)$

76. $(5p + 3)(p + 2)(p - 1)$

77. $(r + 2)^3$

Chapter 4   Review Exercises   **219**

(4.6)

78. $(3k + 1)(2k - 3)$

79. $(2r + 5)(5r - 2)$

80. $(a + 3b)(2a - b)$

81. $(6k + 5q)(2k - 7q)$

82. $(a + 4)^2$

83. $(3p - 2)^2$

84. $(2r + 5s)^2$

85. $(8z - 3y)^2$

86. $(6m - 5)(6m + 5)$

87. $(2z + 7)(2z - 7)$

88. $(5a + 6b)(5a - 6b)$

89. $(9y + 8z)(9y - 8z)$

(4.7) Perform each division.

90. $\dfrac{-15y^4}{9y^2}$

91. $\dfrac{12x^3y^2}{6xy}$

92. $\dfrac{6y^4 - 12y^2 + 18y}{6y}$

93. $\dfrac{2p^3 - 6p^2 + 5p}{2p^2}$

94. $(-10m^4n^2 + 5m^3n^3 + 6m^2n^4) \div (5m^2n)$

95. $(25x^2y^3 - 8xy^2 + 15x^3y) \div (10x^2y^3)$

(4.8)

96. $\dfrac{2r^2 + 3r - 14}{r - 2}$

97. $\dfrac{12m^2 - 11m - 10}{3m - 5}$

98. $\dfrac{2y^3 + 17y^2 + 37y + 7}{2y + 7}$

99. $\dfrac{10a^3 + 9a^2 - 14a + 9}{5a - 3}$

100. $\dfrac{2k^4 + 3k^3 + 9k^2 - 8}{2k + 1}$

101. $\dfrac{2m^4 + 4m^3 - 4m^2 - 12m + 6}{2m^2 - 3}$

name _____ date _____ hour _____

CHAPTER 4 TEST

Evaluate each expression.

1. 3^{-4}

2. -17^0

Simplify. Write each answer using only positive exponents. Assume all variables are positive.

3. $5^{-3} \cdot 5^2$

4. $\dfrac{8^{-4}}{8^{-9}}$

5. $\dfrac{(p^{-2})^{-3}(p^4)^2}{(p^{-5})^2}$

6. $\left(\dfrac{a^2 b^3}{a^{-3} b}\right)^{-2}$

Write each number in scientific notation.

7. 245,000,000

8. .000379

Write each number without exponents.

9. 4.8×10^{-3}

10. $\dfrac{8 \times 10^{-4}}{2 \times 10^{-6}}$

For each polynomial, combine terms; then give the degree of the polynomial. Finally, select the most specific description from this list: trinomial, binomial, monomial, or none of these.

11. $3x^2 + 6x - 4x^2$

12. $11m^3 - m^2 + m^4 + m^4 - 7m^2$

13. $5x^3 - 4x^2 + 2x - 1$

1. _____
2. _____
3. _____
4. _____
5. _____
6. _____
7. _____
8. _____
9. _____
10. _____
11. _____
12. _____
13. _____

Chapter 4 Test   221

Perform the indicated operations.

14. $(2x^5 - 4x + 7) - (x^5 + x^2 - 2x - 5)$

15. $(y^2 - 5y - 3) + (3y^2 + 2y) - (y^2 - y - 1)$

16. $6m^2(m^3 + 2m^2 - 3m + 7)$

17. $(r - 5)(r + 2)$

18. $(3t + 4w)(2t - 3w)$

19. $(5r - 3s)^2$

20. $(6p - 8q)(6p + 8q)$

21. $(2x - 3)(x^2 + 2x - 5)$

22. $\dfrac{9y^3 - 15y^2 + 6y}{3y}$

23. $(10r^3 + 25r^2 - 15r + 8) \div (5r)$

24. $\dfrac{12y^2 - 15y - 11}{4y + 3}$

25. $\dfrac{3x^3 - 2x^2 - 6x - 4}{x - 2}$

Exponents and Polynomials

CUMULATIVE REVIEW EXERCISES

Simplify the following expressions by combining terms.

1. $4p - 6 + 3p - 8$

2. $3k - 5 + 9 - 5k$

3. $-4(k + 2) + 3(2k - 1)$

4. $-3(m + 1) + 5(3m - 4)$

Solve the following equations and check each solution.

5. $2r - 6 = 8$

6. $3k + 2 = 5$

7. $2(p - 1) = 3p + 2$

8. $3(a + 2) = 2a - 1$

9. $4 - 5(a + 2) = 3(a + 1) - 1$

10. $2 - 6(z + 1) = 4(z - 2) + 10$

11. $-(m - 1) = 3 - 2m$

12. $-(k + 2) = 4 - 2k$

13. $k + \dfrac{k + 2}{5} = 8$

14. $\dfrac{r - 3}{2} - 6 = 2r$

15. $\dfrac{y - 2}{3} = \dfrac{2y + 1}{5}$

16. $\dfrac{2x + 3}{5} = \dfrac{x - 4}{2}$

Solve for x.

17. $3y - x = 2$

18. $2x + y = 4$

19. $\dfrac{2x - a}{2} = 5$

20. $a(b + x) = 3$

Cumulative Review Exercises 223

Solve the following inequalities. Graph each solution.

21. $-5z \geq 4z - 18$

22. $-2k > 3k + 15$

23. $6(r - 1) + 2(3r - 5) \leq -4$

24. $4(m + 2) + 3(2m - 1) > 15$

25. $-3 < x + 4 < 9$

26. $4 \leq 4y < 20$

27. $-1 \leq 2x + 3 \leq 5$

28. $9 < 4x - 1 < 15$

Solve each word problem. Check the solution in the words of the original problem.

29. Three less than a number is multiplied by 5. The result is 45. Find the number.

30. The ratio of a number to the sum of the number and 4 is $\frac{2}{3}$. Find the number.

31. Ed Calvin bought textbooks at the college bookstore for $52.47, including 6% sales tax. What did the books cost?

32. Louise Palla received a bill from her credit card company for $104.93. The bill included interest at $1\frac{1}{2}\%$ per month for one month and a $5.00 late charge. How much did her purchases amount to?

224   Exponents and Polynomials

33. The perimeter of a rectangle is 98 centimeters. The width is 19 centimeters. Find the length.

34. The area of a triangle is 104 square inches. The base is 13 inches. Find the height.

Evaluate each expression.

35. $2^{-3} \cdot 2^2$

36. $(-3)^2(-3)$

37. $\left(\dfrac{3}{4}\right)^{-2}$

38. $(2^{-3} \cdot 3^2)^2$

39. $\dfrac{7^{-1}}{7}$

40. $\dfrac{6^5 \cdot 6^{-2}}{6^3}$

41. $\dfrac{9^3 \cdot 9^5}{9^4 \cdot 9^2}$

42. $\left(\dfrac{4^{-3} \cdot 4^4}{4^5}\right)^{-1}$

Simplify each expression. Write with only positive exponents.

43. $\dfrac{(4^{-2})^3}{4^6 \cdot 4^{-3}}$

44. $\dfrac{(2x^3)^{-1} \cdot x}{2^3 x^5}$

45. $\dfrac{(p^2)^3 p^{-4}}{(p^{-3})^{-1} p}$

46. $\dfrac{(m^{-2})^3 m}{m^5 m^{-4}}$

Perform the indicated operations.

47. $(x^3 + 3x^2 - 5x) + (4x^3 - x^2 + 7)$

48. $(4x^4 - 2x^2 + 7) - (6x^4 + 2x^2 + 8)$

Cumulative Review Exercises

49. $(5x^2 - 4x + 7) - (-2x^2 + 11x + 5)$

50. $(2k^2 + 3k) - (5k^2 - 2) - (k^2 + k - 1)$

51. $8x^2y^2(9x^4y^5)$

52. $3m^2(2m^5 - 5m^3 + m)$

53. $(2y + 1)(y - 4)$

54. $(3z - 2w)(4z + 2w)$

55. $(y^2 + 3y + 5)(3y - 1)$

56. $(r^3 + 2r^2 + 3r + 2)(r - 2)$

57. $(3p + 2)^2$

58. $(4a - b)^2$

59. $(5k - 4)(5k + 4)$

60. $(2p + 3q)(2p - 3q)$

61. $\dfrac{3a^3 - 9a^2 + 15a}{3a}$

62. $\dfrac{8x^4 + 12x^3 - 6x^2 + 20x}{2x}$

63. $\dfrac{12p^3 + 2p^2 - 12p + 4}{2p - 2}$

64. $\dfrac{15z^3 - 11z^2 + 22z + 8}{5z - 2}$

Factoring

5

5.1 FACTORS; THE GREATEST COMMON FACTOR

1 Prime factors of whole numbers were discussed in Section 1.1. Now the idea of factoring is extended to the set of integers. Recall that 6 and 2 are **factors** of 12 because the product of 6 and 2 is 12. The expression $6 \cdot 2$ is a **factored form** of 12. To **factor** means to write a quantity as a product. That is, factoring is the opposite of multiplication. For example,

Multiplication   *Factoring*

$6 \cdot 2 = 12$ and $12 = 6 \cdot 2$.
↑ ↑   ↑   ↑   ↑ ↑
Factors Product   Product Factors

Other factored forms of 12 are $(-6)(-2)$, $3 \cdot 4$, $(-3)(-4)$, $12 \cdot 1$, and $(-12)(-1)$. More than two factors may be used, so another factored form of 12 is $2 \cdot 2 \cdot 3$. The positive integer factors of 12 are

$$1, 2, 3, 4, 6, 12.$$

By definition,

> for integers a and b, the integer a is a **factor** of b if b can be divided by a with a remainder of zero.

EXAMPLE 1 (a) The positive integer factors of 36 are 1, 2, 3, 4, 6, 9, 12, 18, and 36.

(b) The positive integer factors of 11 are 1 and 11. ◀

Work Problem 1 at the side.

2 As shown in Example 1(b), the only positive integer factors of 11 are 11 and 1. Recall that a positive integer greater than 1 having only itself and 1 as factors is a **prime number.** The first few prime numbers are

$$2, 3, 5, 7, 11, 13, 17, 19, 23, 29, 31, 37, 41, 43$$

Objectives

▶ List the factors of an integer.
▶ Identify primes.
▶ Find the greatest common factor of a list of integers.
▶ Factor out the greatest common factor.
▶ Factor by grouping.

1. List the positive integer factors of each number.

   (a) 24

   (b) 40

   (c) 7

   (d) 19

ANSWERS
1. (a) 1, 2, 3, 4, 6, 8, 12, 24
   (b) 1, 2, 4, 5, 8, 10, 20, 40
   (c) 1, 7 (d) 1, 19

2. Tell whether each number is prime or composite.

   (a) 12

   (b) 13

   (c) 27

   (d) 59

   (e) 1,806,954

3. Write each number in prime factored form.

   (a) 70

   (b) 180

   (c) 400

   (d) 97

and so on. A positive integer (other than 1) that is not prime is called **composite.** The number 1 is neither composite nor prime.

EXAMPLE 2 Tell whether each number is prime or composite.

(a) 33

This number has factors of 3 and 11, as well as 1 and 33, so it is composite.

(b) 53

Try dividing 53 by various integers. It is divisible only by itself and 1, so it is prime.

(c) 14,976,083,922

This number is even and therefore divisible by 2. It is composite. ◄

Work Problem 2 at the side.

Each composite number may be expressed as a product of primes. For example,

$$30 = 2 \cdot 3 \cdot 5, \quad 55 = 5 \cdot 11, \quad 72 = 2^3 \cdot 3^2,$$

and so on. A number written as a product of prime factors is in **prime factored form.**

EXAMPLE 3 Write each number in prime factored form.

(a) 50

Divide 50 by the first prime, 2.

$$50 = 2 \cdot 25$$

We can't divide 25 by 2, or by the next prime, 3, but we can divide it by 5.

$$50 = 2 \cdot 5 \cdot 5$$
$$50 = 2 \cdot 5^2 \qquad \text{Prime factored form}$$

(b) $300 = 2 \cdot 150$
$ = 2 \cdot 2 \cdot 75$
$ = 2 \cdot 2 \cdot 3 \cdot 25$
$ = 2 \cdot 2 \cdot 3 \cdot 5 \cdot 5$
$ = 2^2 \cdot 3 \cdot 5^2 \qquad \text{Prime factored form}$

(c) 71

Since 71 is prime, its prime factored form is 71. ◄

Work Problem 3 at the side.

A table giving the prime factored form of all positive integers from 2 through 100 follows.

ANSWERS

2. 13 and 59 are prime; the others are composite.
3. (a) $2 \cdot 5 \cdot 7$ (b) $2^2 \cdot 3^2 \cdot 5$
   (c) $2^4 \cdot 5^2$ (d) 97

Prime Factors of the Numbers 2 Through 100

$2 = 2$	$26 = 2 \cdot 13$	$51 = 3 \cdot 17$	$76 = 2^2 \cdot 19$
$3 = 3$	$27 = 3^3$	$52 = 2^2 \cdot 13$	$77 = 7 \cdot 11$
$4 = 2^2$	$28 = 2^2 \cdot 7$	$53 = 53$	$78 = 2 \cdot 3 \cdot 13$
$5 = 5$	$29 = 29$	$54 = 2 \cdot 3^3$	$79 = 79$
$6 = 2 \cdot 3$	$30 = 2 \cdot 3 \cdot 5$	$55 = 5 \cdot 11$	$80 = 2^4 \cdot 5$
$7 = 7$	$31 = 31$	$56 = 2^3 \cdot 7$	$81 = 3^4$
$8 = 2^3$	$32 = 2^5$	$57 = 3 \cdot 19$	$82 = 2 \cdot 41$
$9 = 3^2$	$33 = 3 \cdot 11$	$58 = 2 \cdot 29$	$83 = 83$
$10 = 2 \cdot 5$	$34 = 2 \cdot 17$	$59 = 59$	$84 = 2^2 \cdot 3 \cdot 7$
$11 = 11$	$35 = 5 \cdot 7$	$60 = 2^2 \cdot 3 \cdot 5$	$85 = 5 \cdot 17$
$12 = 2^2 \cdot 3$	$36 = 2^2 \cdot 3^2$	$61 = 61$	$86 = 2 \cdot 43$
$13 = 13$	$37 = 37$	$62 = 2 \cdot 31$	$87 = 3 \cdot 29$
$14 = 2 \cdot 7$	$38 = 2 \cdot 19$	$63 = 3^2 \cdot 7$	$88 = 2^3 \cdot 11$
$15 = 3 \cdot 5$	$39 = 3 \cdot 13$	$64 = 2^6$	$89 = 89$
$16 = 2^4$	$40 = 2^3 \cdot 5$	$65 = 5 \cdot 13$	$90 = 2 \cdot 3^2 \cdot 5$
$17 = 17$	$41 = 41$	$66 = 2 \cdot 3 \cdot 11$	$91 = 7 \cdot 13$
$18 = 2 \cdot 3^2$	$42 = 2 \cdot 3 \cdot 7$	$67 = 67$	$92 = 2^2 \cdot 23$
$19 = 19$	$43 = 43$	$68 = 2^2 \cdot 17$	$93 = 3 \cdot 31$
$20 = 2^2 \cdot 5$	$44 = 2^2 \cdot 11$	$69 = 3 \cdot 23$	$94 = 2 \cdot 47$
$21 = 3 \cdot 7$	$45 = 3^2 \cdot 5$	$70 = 2 \cdot 5 \cdot 7$	$95 = 5 \cdot 19$
$22 = 2 \cdot 11$	$46 = 2 \cdot 23$	$71 = 71$	$96 = 2^5 \cdot 3$
$23 = 23$	$47 = 47$	$72 = 2^3 \cdot 3^2$	$97 = 97$
$24 = 2^3 \cdot 3$	$48 = 2^4 \cdot 3$	$73 = 73$	$98 = 2 \cdot 7^2$
$25 = 5^2$	$49 = 7^2$	$74 = 2 \cdot 37$	$99 = 3^2 \cdot 11$
	$50 = 2 \cdot 5^2$	$75 = 3 \cdot 5^2$	$100 = 2^2 \cdot 5^2$

3> An integer that is a factor of two or more integers is a **common factor** of those integers. For example, 6 is a common factor of 18 and 24 since 6 is a factor of both 18 and 24. Other common factors of 18 and 24 are 1, 2, and 3. The **greatest common factor** of a list of integers is the largest common factor of those integers. This means 6 is the greatest common factor of 18 and 24, since it is the largest of the common factors of these numbers.

Find the greatest common factor of a list of numbers as follows.

Finding the Greatest Common Factor

Step 1 Write each number in prime factored form.

Step 2 List each different prime factor that is in every prime factorization.

Step 3 Use as exponents on the prime factors the *smallest* exponent from the prime factored forms. (If a prime does not appear in one of the prime factored forms, it cannot appear in the greatest common factor.)

Step 4 Multiply together the primes from Step 3. If there are no primes left after Step 3, the greatest common factor is 1.

5.1 Factors; The Greatest Common Factor

4. Find the greatest common factor for each group of numbers.

   (a) 30, 20, 15

   (b) 42, 28, 35

   (c) 12, 18, 26, 32

   (d) 10, 15, 21

5. Find the greatest common factor for each group of terms.

   (a) $6m^4, 9m^2, 12m^5$

   (b) $25y^{11}, 30y^7$

   (c) $12p^5, 18q^4$

   (d) $11r^9, 10r^{15}, 8r^{12}$

   (e) y^4z^2, y^6z^8, z^9

   (f) $12p^{11}, 17q^5$

ANSWERS
4. (a) 5 (b) 7 (c) 2 (d) 1
5. (a) $3m^2$ (b) $5y^7$ (c) 6 (d) r^9
   (e) z^2 (f) 1

EXAMPLE 4 Find the greatest common factor for each group of numbers.

(a) 30, 45

First write each number in prime factored form.
$$30 = 2 \cdot 3 \cdot 5 \qquad 45 = 3^2 \cdot 5$$

Use each prime the *least* number of times it appears in all the factored forms. There is no 2 in the prime factored form of 45, so there will be no 2 in the greatest common factor. The least number of times 3 appears in all the factored forms is 1; the least number of times 5 appears is also 1. From this, the greatest common factor is

$$\boxed{3^1 \cdot 5^1} = 3 \cdot 5 = 15.$$

(b) 72, 120, 432

Find the prime factored form of each number.
$$72 = 2^3 \cdot 3^2 \qquad 120 = 2^3 \cdot 3 \cdot 5 \qquad 432 = 2^4 \cdot 3^3$$

The least number of times 2 appears in all the factored forms is 3, and the least number of times 3 appears is 1. There is no 5 in the prime factored form of either 72 or 432, so the greatest common factor is

$$\boxed{2^3 \cdot 3} = 24.$$

(c) 10, 11, 14

Write the prime factored form of each number.
$$10 = 2 \cdot 5; \qquad 11 = 11; \qquad 14 = 2 \cdot 7$$

There are no primes common to all three numbers, so the greatest common factor is 1. ◄

Work Problem 4 at the side.

The greatest common factor can also be found for a list of terms. For example, the terms x^4, x^5, x^6, and x^7 have x^4 as the greatest common factor, because 4 is the smallest exponent on x.

> The exponent on a variable in the greatest common factor is the *smallest* exponent that appears on that variable.

EXAMPLE 5 Find the greatest common factor for each list of terms.

(a) $21m^7, -18m^6, 45m^8, -24m^5$

First, 3 is the greatest common factor of the coefficients 21, -18, 45, and -24. The smallest exponent on m is 5, so the greatest common factor of the terms is $3m^5$.

(b) $x^2y^2, x^7y^5, x^3y^7, y^{15}$

There is no x in the last term, y^{15}, so x will not appear in the greatest common factor. There is a y in each term, however, and 2 is the smallest exponent on y. The greatest common factor is y^2. ◄

Work Problem 5 at the side.

▶ The idea of a greatest common factor can be used to write a polynomial in factored form. For example, the polynomial

$$3m + 12$$

consists of the two terms $3m$ and 12. The greatest common factor for these two terms is 3. Write $3m + 12$ so that each term is a product with 3 as one factor.

$$3m + 12 = 3 \cdot m + 3 \cdot 4$$

Now use the distributive property.

$$3m + 12 = 3 \cdot m + 3 \cdot 4 = 3(m + 4)$$

The factored form of $3m + 12$ is $3(m + 4)$. This process is called **factoring out the greatest common factor.**

EXAMPLE 6 Factor out the greatest common factor.

(a) $20m^5 + 10m^4 + 15m^3$

The greatest common factor for the terms of this polynomial is $5m^3$.

$$20m^5 + 10m^4 + 15m^3 = (5m^3)(4m^2) + (5m^3)(2m) + (5m^3)3$$
$$= 5m^3(4m^2 + 2m + 3)$$

Check this work by multiplying $5m^3$ and $4m^2 + 2m + 3$. You should get the original polynomial as your answer.

(b) $48y^{12} - 36y^{10} + 12y^7 = (12y^7)(4y^5) - (12y^7)(3y^3) + (12y^7)1$
$$= 12y^7(4y^5 - 3y^3 + 1)$$

Do not forget the 1 here; always be sure that the factored form can be multiplied out to yield the original polynomial.

(c) $x^5 + x^3 = (x^3)x^2 + (x^3)1 = x^3(x^2 + 1)$

(d) $20m^7p^2 - 36m^3p^4 = 4m^3p^2(5m^4 - 9p^2)$ ◀

Remember: Always look for a greatest common factor as the first step in factoring a polynomial.

Work Problem 6 at the side.

▶ Common factors are used in **factoring by grouping,** explained in the next example.

EXAMPLE 7 Factor by grouping.

(a) $2x + 6 + ax + 3a$

The first two terms have a common factor of 2, and the last two terms have a common factor of a.

$$(2x + 6) + (ax + 3a) = 2(x + 3) + a(x + 3)$$

Now $x + 3$ is a common factor.

$$2x + 6 + ax + 3a = 2(x + 3) + a(x + 3)$$
$$= (x + 3)(2 + a)$$

6. Factor out the greatest common factor.

(a) $32p^2 + 16p + 48$

(b) $10y^5 - 8y^4 + 6y^2$

(c) $27a^5 + 9a^4$

(d) $m^7 + m^9$

(e) $8p^5q^2 + 16p^6q^3 - 12p^4q^7$

ANSWERS
6. (a) $16(2p^2 + p + 3)$
   (b) $2y^2(5y^3 - 4y^2 + 3)$
   (c) $9a^4(3a + 1)$ (d) $m^7(1 + m^2)$
   (e) $4p^4q^2(2p + 4p^2q - 3q^5)$

5.1 Factors; The Greatest Common Factor

7. Factor by grouping.

   (a) $pq + 5q + 2p + 10$

   (b) $2mn - 8n + 3m - 12$

   (c) $6y^2 - 16y - 15y + 40$

(b) $m^2 + 6m + 2m + 12 = m(m + 6) + 2(m + 6)$
$= (m + 6)(m + 2)$

(c) $6y^2 - 21y - 8y + 28 = 3y(2y - 7) - 4(2y - 7)$
$= (2y - 7)(3y - 4)$

Since the quantities in parentheses in the second step must be the same, it was necessary here to factor out -4 rather than 4. ◀

Work Problem 7 at the side.

Use these steps when factoring by grouping.

> **Factoring by Grouping**
>
> *Step 1* Write the four terms so that the first two have a common factor and the last two have a common factor.
>
> *Step 2* Use the distributive property to factor each group of two terms.
>
> *Step 3* If possible, factor a common binomial factor from the results of Step 2.
>
> *Step 4* If Step 2 does not result in a common binomial factor, try grouping the terms of the original polynomial in a different way.

ANSWERS
7. (a) $(p + 5)(q + 2)$
   (b) $(m - 4)(2n + 3)$
   (c) $(3y - 8)(2y - 5)$

name   date   hour

5.1 EXERCISES

Find all positive integer factors of each number. See Example 1.

1. 14
2. 18
3. 27
4. 35

5. 45
6. 50
7. 60
8. 72

9. 100
10. 130
11. 29
12. 37

Find the prime factored form for the following. See Example 3.

13. 120
14. 150
15. 180
16. 225

17. 275
18. 350
19. 475
20. 650

Find the greatest common factor for each set of terms. See Examples 4 and 5.

21. $12y$, 24
22. $72m$, 12
23. $30p^2$, $20p^3$, $40p^5$

24. $14r^5$, $28r^2$, $56r^8$
25. $18m^2n^2$, $36m^4n^5$, $12m^3n$
26. $50p^5r^2$, $25p^4r^7$, $30p^7r^8$

Complete the factoring.

27. $12 = 6(\quad)$
28. $18 = 9(\quad)$
29. $3x^2 = 3x(\quad)$

30. $8x^3 = 8x(\quad)$
31. $9m^4 = 3m^2(\quad)$
32. $12p^5 = 6p^3(\quad)$

33. $-8z^9 = -4z^5(\quad)$
34. $-15k^{11} = -5k^8(\quad)$
35. $6m^4n^5 = 3m^3n(\quad)$

36. $27a^3b^2 = 9a^2b(\quad)$
37. $14x^4y^3 = 2xy(\quad)$
38. $-16m^3n^3 = 4mn^2(\quad)$

Factor out the greatest common factor. See Example 6.

39. $12x + 24$

40. $18m - 9$

41. $9a^2 - 18a$

42. $21m^5 - 14m^4$

43. $65y^9 - 35y^5$

44. $100a^4 + 16a^2$

45. $11z^2 - 100$

46. $12z^2 - 11y^4$

47. $19y^3p^2 + 38y^2p^3$

48. $4mn^2 - 12m^2n$

49. $13y^6 + 26y^5 - 39y^3$

50. $5x^4 + 25x^3 - 20x^2$

51. $45q^4p^5 - 36qp^6 + 81q^2p^3$

52. $a^5 + 2a^5b + 3a^5b^2 - 4a^5b^3$

53. $125z^5a^3 - 60z^4a^5 + 85z^3a^4$

54. $30a^2m^2 + 60a^3m + 180a^3m^2$

55. $33y^8 - 44y^{12} + 77y^3 + 11y^4$

56. $26g^6h^4 + 13g^3h^4 - 39g^4h^3$

Factor by grouping. See Example 7.

57. $p^2 + 4p + 3p + 12$

58. $m^2 + 2m + 5m + 10$

59. $a^2 - 2a + 5a - 10$

60. $y^2 - 6y + 4y - 24$

61. $7z^2 + 14z - az - 2a$

62. $8k^2 + 6kq + 12kq + 9q^2$

63. $5m^2 + 15mp - 2mp - 6p^2$

64. $18r^2 + 12ry - 3xr - 2xy$

65. $3a^3 + 3ab^2 + 2a^2b + 2b^3$

66. $16m^3 - 4m^2p^2 - 4mp + p^3$

67. $1 - a + ab - b$

68. $2pq^2 - 8q^2 + p - 4$

234 Factoring

5.2 FACTORING TRINOMIALS

Objectives

▸ Factor trinomials with a coefficient of 1 for the squared term.

▸ Factor such polynomials after factoring out the greatest common factor.

▸ The product of the polynomials $k - 3$ and $k + 1$ is
$$(k - 3)(k + 1) = k^2 - 2k - 3.$$

The polynomial $k^2 - 2k - 3$ can be rewritten as the product $(k - 3)(k + 1)$. The product is called the **factored form** of $k^2 - 2k - 3$, and the process of finding the factored form is called **factoring**. The discussion of factoring in this section is limited to trinomials like $x^2 - 2x - 24$ or $y^2 + 2y - 15$, where the coefficient of the squared term is 1.

When factoring polynomials with only integer coefficients, use only integers for numerical factors. For example, $x^2 + 5x + 6$ can be factored by finding integers a and b such that
$$x^2 + 5x + 6 = (x + \boxed{a})(x + \boxed{b}).$$

To find these integers a and b, first find the product of the two terms on the right-hand side:
$$(x + a)(x + b) = x^2 + ax + bx + ab.$$

Since $ax + bx = (a + b)x$ by the distributive property,
$$x^2 + ax + bx + ab = x^2 + (a + b)x + ab.$$

By this result, $x^2 + 5x + 6$ can be factored by finding integers a and b having a sum of 5 and a product of 6.

$$x^2 + \boxed{5}x + \boxed{6} = x^2 + (\boxed{a + b})x + \boxed{ab}$$

Product of a and b is 6
Sum of a and b is 5

Since many pairs of integers have a sum of 5, it is best to begin by listing those pairs of integers whose product is 6. Both 5 and 6 are positive, so only pairs in which both integers are positive need be considered.

1. (a) List all pairs of positive integers whose product is 6.

    (b) Find the pair from part (a) whose sum is 5.

Work Problem 1 at the side.

As in Problem 1 at the side, the numbers 1 and 6 and the numbers 2 and 3 both have a product of 6, but only the pair 2 and 3 has a sum of 5. So 2 and 3 are the needed integers, and
$$x^2 + 5x + 6 = (x + \boxed{2})(x + \boxed{3}).$$

Check by multiplying the binomials. Make sure that the sum of the outer and inner products produces the correct middle term.

$$(x + 2)(x + 3) = x^2 + \boxed{5x} + 6$$
$$2x$$
$$3x$$
$$\overline{5x}$$

This method of factoring can be used only for trinomials having the coefficient of the squared term equal to 1. Methods for factoring other trinomials will be given in the next section.

ANSWERS
1. (a) 1, 6; 2, 3 (b) 2, 3

2. Complete the given lists of numbers; then factor the given trinomial.

   (a) $m^2 + 11m + 30$
   $ab = 30$, $a + b = 11$
   30, 1   30 + 1 = 31
   15, 2   15 + 2 = ___
   10, 3   10 + 3 = ___
   6, 5   6 + 5 = ___

   (b) $y^2 + 12y + 20$
   $ab = 20$, $a + b = 12$
   20, 1   20 + 1 = 21
   10, ___   10 + ___ = ___
   5, ___   5 + ___ = ___

3. Factor each trinomial.

   (a) $p^2 + 7p + 6$

   (b) $y^2 + 4y + 3$

   (c) $a^2 - 9a - 22$

   (d) $r^2 - 6r - 16$

ANSWERS
2. (a) 17, 13; 11; $(m + 6)(m + 5)$
   (b) 2; 2; 12; 4; 4; 9; $(y + 10)(y + 2)$
3. (a) $(p + 6)(p + 1)$
   (b) $(y + 3)(y + 1)$
   (c) $(a - 11)(a + 2)$
   (d) $(r - 8)(r + 2)$

EXAMPLE 1 Factor $m^2 + 9m + 14$.

Look for two integers whose product is 14 and whose sum is 9. List the pairs of integers whose products are 14. Then examine the sums. Again, only positive integers are needed because all signs in $m^2 + 9m + 14$ are positive.

14, 1   14 + 1 = 15
7, 2   7 + 2 = **9**   Sum is 9

From the list, 7 and 2 are the required integers, since $7 \cdot 2 = 14$ and $7 + 2 = 9$. Thus

$$m^2 + 9m + 14 = (m + 2)(m + 7).$$

This answer also could have been written $(m + 7)(m + 2)$. Because of the commutative property of multiplication, the order of the factors does not matter. ◀

Work Problem 2 at the side.

EXAMPLE 2 Factor $p^2 - 2p - 15$.

Find two integers whose product is -15 and whose sum is -2. If these numbers do not come to mind right away, find them (if they exist) by listing all the pairs of integers whose product is -15. Because of the minus signs in $p^2 - 2p - 15$, negative integers as well as positive ones must be considered.

15, −1   15 + (−1) = 14
5, −3   5 + (−3) = 2
−15, 1   −15 + 1 = −14
−5, 3   −5 + 3 = **−2**   Sum is −2

The required integers are -5 and 3.

$$p^2 - 2p - 15 = (p - 5)(p + 3) \blacktriangleleft$$

Work Problem 3 at the side.

EXAMPLE 3 Factor each trinomial.

(a) $x^2 - 5x + 12$

First, list all pairs of integers whose product is 12. Then examine the sums.

12, 1   12 + 1 = 13
6, 2   6 + 2 = 8
3, 4   3 + 4 = 7
−12, −1   −12 + (−1) = −13
−6, −2   −6 + (−2) = −8
−3, −4   −3 + (−4) = −7

None of the pairs of integers has a sum of -5. Because of this, the trinomial $x^2 - 5x + 12$ *cannot be factored using only integer coefficients,* showing that it is a **prime polynomial.**

Factoring

(b) $k^2 - 8k + 11$

There is no pair of integers whose product is 11 and whose sum is -8, so $k^2 - 8k + 11$ is a prime polynomial. ◀

Work Problem 4 at the side.

EXAMPLE 4 Factor $z^2 - 2bz - 3b^2$.

Look for two expressions whose product is $-3b^2$ and whose sum is $-2b$. The expressions are $-3b$ and b, so
$$z^2 - 2bz - 3b^2 = (z - 3b)(z + b). ◀$$

Work Problem 5 at the side.

The trinomial in the next example does not fit the pattern used above. A preliminary step must be taken before using the pattern.

EXAMPLE 5 Factor $4x^5 - 28x^4 + 40x^3$.

First, factor out the greatest common factor, $4x^3$.
$$4x^5 - 28x^4 + 40x^3 = \mathbf{4x^3}(x^2 - 7x + 10)$$

Now factor $x^2 - 7x + 10$. The integers -5 and -2 have a product of 10 and a sum of -7. The complete factored form is
$$4x^5 - 28x^4 + 40x^3 = 4x^3(x - 5)(x - 2). ◀$$

> When factoring, always remember to look for a common factor first. Do not forget to include the common factor as part of the answer.

Work Problem 6 at the side.

4. Factor each trinomial, where possible.

   (a) $x^2 + x + 1$

   (b) $r^2 - 3r - 4$

   (c) $m^2 - 2m + 5$

   (d) $y^2 - 11y + 30$

5. Factor each trinomial.

   (a) $b^2 - 3ab - 4a^2$

   (b) $p^2 + 6pq + 5q^2$

   (c) $r^2 - 6rs + 8s^2$

6. Factor each trinomial as completely as possible.

   (a) $2p^3 + 6p^2 - 8p$

   (b) $3x^4 - 15x^3 + 18x^2$

   (c) $4m^7 + 28m^6 + 48m^5$

ANSWERS
4. (a) prime (b) $(r - 4)(r + 1)$
   (c) prime (d) $(y - 5)(y - 6)$
5. (a) $(b - 4a)(b + a)$
   (b) $(p + 5q)(p + q)$
   (c) $(r - 4s)(r - 2s)$
6. (a) $2p(p + 4)(p - 1)$
   (b) $3x^2(x - 3)(x - 2)$
   (c) $4m^5(m + 4)(m + 3)$

5.2 Factoring Trinomials

name   date   hour

5.2 EXERCISES

Complete the factoring.

1. $x^2 + 10x + 21 = (x + 7)(\quad)$
2. $p^2 + 11p + 30 = (p + 5)(\quad)$
3. $r^2 + 15r + 56 = (r + 7)(\quad)$
4. $x^2 + 15x + 44 = (x + 4)(\quad)$
5. $t^2 - 14t + 24 = (t - 2)(\quad)$
6. $x^2 - 9x + 8 = (x - 1)(\quad)$
7. $x^2 - 12x + 32 = (x - 4)(\quad)$
8. $y^2 - 2y - 15 = (y + 3)(\quad)$
9. $m^2 + 2m - 24 = (m - 4)(\quad)$
10. $x^2 + 9x - 22 = (x - 2)(\quad)$
11. $p^2 + 7p - 8 = (p + 8)(\quad)$
12. $y^2 - 7y - 18 = (y + 2)(\quad)$
13. $x^2 - 7xy + 10y^2 = (x - 2y)(\quad)$
14. $k^2 - 3kh - 28h^2 = (k - 7h)(\quad)$

Factor completely. If a polynomial cannot be factored, write prime. *See Examples 1–3.*

15. $y^2 + 9y + 8$
16. $a^2 + 9a + 20$
17. $b^2 + 8b + 15$
18. $x^2 - 6x - 7$
19. $m^2 + m - 20$
20. $p^2 + 4p + 5$
21. $n^2 + 4n - 12$
22. $y^2 - 6y + 8$
23. $r^2 - r - 30$
24. $s^2 + 2s - 35$
25. $h^2 + 11h + 12$
26. $n^2 - 12n - 35$
27. $a^2 - 2a - 99$
28. $b^2 - 11b + 24$
29. $x^2 - 9x + 20$
30. $k^2 - 10k + 25$
31. $z^2 - 14z + 49$
32. $y^2 - 12y - 45$
33. $r^2 + r - 42$
34. $z^2 - 3z - 40$
35. $p^2 + 5p - 66$

Factor completely. See Examples 4 and 5.

36. $x^2 + 4ax + 3a^2$

37. $x^2 - mx - 6m^2$

38. $y^2 - by - 30b^2$

39. $z^2 + 2zx - 15x^2$

40. $x^2 + xy - 30y^2$

41. $a^2 - ay - 56y^2$

42. $r^2 - 2rs + s^2$

43. $m^2 - 2mn - 3n^2$

44. $p^2 - 3pq - 10q^2$

45. $c^2 - 5cd + 4d^2$

46. $3m^3 + 12m^2 + 9m$

47. $3y^5 - 18y^4 + 15y^3$

48. $6a^2 - 48a - 120$

49. $h^7 - 5h^6 - 14h^5$

50. $3j^3 - 30j^2 + 72j$

51. $2x^6 - 8x^5 - 42x^4$

52. $3x^4 - 3x^3 - 90x^2$

53. $2y^3 - 8y^2 - 10y$

54. $a^5 + 3a^4b - 4a^3b^2$

55. $m^3n - 2m^2n^2 - 3mn^3$

56. $y^3z + y^2z^2 - 6yz^3$

57. $k^7 - 2k^6m - 15k^5m^2$

58. $z^{10} - 4z^9y - 21z^8y^2$

59. $x^9 + 5x^8w - 24x^7w^2$

60. Use the FOIL method from Section 4.6 to show that $(2x + 4)(x - 3) = 2x^2 - 2x - 12$. Why, then, is it incorrect to completely factor $2x^2 - 2x - 12$ as $(2x + 4)(x - 3)$?

61. Why is it incorrect to completely factor $3x^2 + 9x - 12$ as the product $(x - 1)(3x + 12)$?

62. What polynomial can be factored to give $(y - 7)(y + 6)$?

63. What polynomial can be factored to give $(a + 9)(a + 6)$?

5.3 MORE ON FACTORING TRINOMIALS

Objectives

Factor trinomials not having 1 as the coefficient of the squared term, using

▶ trial and error,

▶ factoring by grouping.

▶ Trinomials such as $2x^2 + 7x + 6$, in which the coefficient of the squared term is *not* 1, are factored with an extension of the method presented in the last section. Recall that a trinomial such as $m^2 + 3m + 2$ is factored by finding two numbers whose product is 2 and whose sum is 3.

To factor $2x^2 + 7x + 6$, find integers a, b, c, and d such that

$$2x^2 + 7x + 6 = (ax + b)(cx + d)$$

where, using FOIL, $ac = 2$, $bd = 6$, and $ad + bc = 7$. The possible factors of $2x^2$ are $2x$ and x, or $-2x$ and $-x$. Since the polynomial has only positive coefficients, use the factors with positive coefficients. Then the factored form of $2x^2 + 7x + 6$ can be set up as

$$2x^2 + 7x + 6 = (2x \quad)(x \quad).$$

The product 6 can be factored as $6 \cdot 1$, $1 \cdot 6$, $2 \cdot 3$, or $3 \cdot 2$. Try each pair to find the correct choices for b and d.

Work Problem 1 at the side.

Since $2x + 6 = 2(x + 3)$, the binomial $2x + 6$ has a common factor of 2, while $2x^2 + 7x + 6$ does not have a common factor other than 1. The product $(2x + 6)(x + 1)$ cannot be correct.

1. Is either factored form correct for $2x^2 + 7x + 6$?

(a) $(2x + 1)(x + 6)$

(b) $(2x + 6)(x + 1)$

> If the original polynomial has no common factor, then none of its binomial factors will either.

Now try 2 and 3 as factors of 6. Because of the common factor of 2 in $2x + 2$, $(2x + 2)(x + 3)$ will not work. Try $(2x + 3)(x + 2)$.

$$(2x + 3)(x + 2) = 2x^2 + \boxed{7x} + 6 \quad \text{Correct}$$
$$ \underset{\underline{7x}}{\overset{3x}{}\overset{4x}{}}$$

Finally, $2x^2 + 7x + 6$ factors as

$$2x^2 + 7x + 6 = (2x + 3)(x + 2).$$

Check by multiplying $2x + 3$ and $x + 2$.

EXAMPLE 1 Factor $8p^2 + 14p + 5$.

The number 8 has several possible pairs of factors, but 5 has only 1 and 5 or -1 and -5. For this reason, it is easier to begin by considering the factors of 5. Ignore the negative factors since all coefficients in the trinomial are positive. If $8p^2 + 14p + 5$ can be factored, it will be factored as

$$(\quad + 5)(\quad + 1).$$

ANSWERS

1. (a) incorrect (b) incorrect

2. Factor each trinomial.

   (a) $2p^2 + 9p + 9$

   (b) $8a^2 + 6a + 1$

   (c) $6p^2 + 19p + 10$

   (d) $2m^2 + 13m + 21$

3. Factor each trinomial.

   (a) $6x^2 + 5x - 4$

   (b) $3x^2 - 7x - 6$

   (c) $5p^2 + 13p - 6$

   (d) $6m^2 - 11m - 10$

   (e) $2p^2 - p - 21$

   (f) $15k^2 - k - 2$

The possible pairs of factors of $8p^2$ are $8p$ and p, or $4p$ and $2p$. Try various combinations.

$$(8p + 5)(p + 1) = 8p^2 + \boxed{13p} + 5 \quad \text{Incorrect}$$

$$(p + 5)(8p + 1) = 8p^2 + \boxed{41p} + 5 \quad \text{Incorrect}$$

$$(4p + 5)(2p + 1) = 8p^2 + \boxed{14p} + 5 \quad \text{Correct}$$

Finally, $8p^2 + 14p + 5$ factors as $(4p + 5)(2p + 1)$. ◀

Work Problem 2 at the side.

EXAMPLE 2 Factor $6x^2 - 11x + 3$.

There are several possible pairs of factors for 6, but 3 has only 1 and 3 or -1 and -3, so it is better to begin by factoring 3. The middle term of $6x^2 - 11x + 3$ has a negative coefficient, so negative factors must be considered. Try -3 and -1 as factors of 3:

$$(\quad - 3)(\quad - 1).$$

The factors of $6x^2$ are either $6x$ and x, or $2x$ and $3x$. Let us try $2x$ and $3x$.

$$(2x - 3)(3x - 1) = 6x^2 \boxed{- 11x} + 3 \quad \text{Correct}$$

Finally, $6x^2 - 11x + 3 = (2x - 3)(3x - 1)$. ◀

EXAMPLE 3 Factor $8x^2 + 6x - 9$.

The integer 8 has several possible pairs of factors, as does -9. Since the coefficient of the middle term is small, it is wise to avoid large factors such as 8 or 9. Let us begin by trying 4 and 2 as factors of 8, and 3 and -3 as factors of -9.

$$(4x + 3)(2x - 3) = 8x^2 \boxed{- 6x} - 9 \quad \text{Incorrect}$$

Try exchanging 3 and -3.

$$(4x - 3)(2x + 3) = 8x^2 \boxed{+ 6x} - 9 \quad \text{Correct} \blacktriangleleft$$

Work Problem 3 at the side.

ANSWERS
2. (a) $(2p + 3)(p + 3)$
   (b) $(4a + 1)(2a + 1)$
   (c) $(3p + 2)(2p + 5)$
   (d) $(m + 3)(2m + 7)$
3. (a) $(3x + 4)(2x - 1)$
   (b) $(3x + 2)(x - 3)$
   (c) $(5p - 2)(p + 3)$
   (d) $(2m - 5)(3m + 2)$
   (e) $(2p - 7)(p + 3)$
   (f) $(5k - 2)(3k + 1)$

Factoring

EXAMPLE 4 Factor $12a^2 - ab - 20b^2$.

There are several possible pairs of factors of $12a^2$, including $12a$ and a, $6a$ and $2a$, and $3a$ and $4a$, just as there are many possible pairs of factors of $-20b^2$, including $-20b$ and b, $10b$ and $-2b$, $-10b$ and $2b$, $4b$ and $-5b$, and $-4b$ and $5b$. Once again, since the desired middle term is small, it is better to avoid the larger factors. Let us try as factors $6a$ and $2a$ and $4b$ and $-5b$.

$$(6a + 4b)(2a - 5b)$$

This cannot be correct as mentioned before, since $6a + 4b$ has a common factor of 2 while the given trinomial has none. Let us try $3a$ and $4a$ with $4b$ and $-5b$.

$$(3a + 4b)(4a - 5b) = 12a^2 + ab - 20b^2 \quad \text{Incorrect}$$

Here the middle term has the wrong sign, so reverse the middle signs of the factors.

$$(3a - 4b)(4a + 5b) = 12a^2 - ab - 20b^2 \quad \text{Correct} \blacktriangleleft$$

Work Problem 4 at the side.

EXAMPLE 5 Factor $28x^5 - 58x^4 - 30x^3$.

First factor out the greatest common factor, $2x^3$.

$$28x^5 - 58x^4 - 30x^3 = 2x^3(14x^2 - 29x - 15)$$

Now try to factor $14x^2 - 29x - 15$. Let us try $7x$ and $2x$ as factors of $14x^2$ and -3 and 5 as factors of -15.

$$(7x - 3)(2x + 5) = 14x^2 + 29x - 15 \quad \text{Incorrect}$$

The middle term differs only in sign, so reverse the middle signs of the two factors.

$$(7x + 3)(2x - 5) = 14x^2 - 29x - 15 \quad \text{Correct}$$

Finally, the factored form of $28x^5 - 58x^4 - 30x^3$ is

$$28x^5 - 58x^4 - 30x^3 = 2x^3(7x + 3)(2x - 5).$$

Do not forget to include the common factor in the final result. ◀

Work Problem 5 at the side.

▶ The rest of this section shows an alternative method of factoring trinomials in which the coefficient of the squared term is not 1. This method uses factoring by grouping, introduced in Section 5.1. In the next example, we use the alternative method to factor $2x^2 + 7x + 6$, the same trinomial factored at the beginning of this section.

Recall that a trinomial such as $m^2 + 3m + 2$ is factored by finding two numbers whose product is 2 and whose sum is 3. To factor $2x^2 + 7x + 6$ by the alternate method, look for two integers whose product is 12 and whose sum is 7.

$$2x^2 + 7x + 6$$
Sum is 7
Product is $2 \cdot 6 = 12$

---

4. Factor each trinomial.

   (a) $2x^2 - 5xy - 3y^2$

   (b) $8a^2 + 2ab - 3b^2$

   (c) $3r^2 + 8rs + 5s^2$

   (d) $6m^2 + 11mn - 10n^2$

5. Factor each polynomial as completely as possible.

   (a) $4x^2 - 2x - 30$

   (b) $15y^3 + 55y^2 + 30y$

   (c) $18p^4 + 63p^3 + 27p^2$

   (d) $16m^5 - 8m^4 - 168m^3$

ANSWERS
4. (a) $(2x + y)(x - 3y)$
   (b) $(4a + 3b)(2a - b)$
   (c) $(3r + 5s)(r + s)$
   (d) $(3m - 2n)(2m + 5n)$
5. (a) $2(2x + 5)(x - 3)$
   (b) $5y(3y + 2)(y + 3)$
   (c) $9p^2(2p + 1)(p + 3)$
   (d) $8m^3(2m - 7)(m + 3)$

5.3 More on Factoring Trinomials

6. Find two integers whose product is 12 and whose sum is 7.

7. Factor each trinomial by grouping.

   (a) $2m^2 + 7m + 3$

   (b) $5p^2 - 2p - 3$

   (c) $3p^2 - 4p + 1$

   (d) $3r^2 + 11r - 4$

Work Problem 6 at the side.

As the problem in the margin shows, the necessary integers are 3 and 4. Use these integers to write the middle term, $7x$, as $7x = 3x + 4x$. With this, the trinomial $2x^2 + 7x + 6$ becomes

$$2x^2 + 7x + 6 = 2x^2 + \underbrace{3x + 4x}_{7x = 3x + 4x} + 6.$$

Factor the new polynomial by grouping:

$$2x^2 + 3x + 4x + 6 = x(2x + 3) + 2(2x + 3).$$

Factor out the common factor of $2x + 3$:

$$= (2x + 3)(x + 2),$$

so $2x^2 + 7x + 6 = (2x + 3)(x + 2).$

EXAMPLE 6 Factor each trinomial.

(a) $6r^2 + r - 1$

Find two integers whose product is $6(-1) = -6$ and whose sum is 1.

$$6r^2 + r - 1 = 6r^2 + 1r - 1$$

Sum is 1 / Product is -6

The integers are -2 and 3. Write the middle term, $+r$, as $-2r + 3r$, so that

$$6r^2 + r - 1 = 6r^2 - 2r + 3r - 1.$$

Factor by grouping on the right-hand side.

$$6r^2 + r - 1 = 6r^2 - 2r + 3r - 1$$
$$= 2r(3r - 1) + 1(3r - 1)$$
$$= (3r - 1)(2r + 1)$$

(b) $12z^2 - 5z - 2$

Look for two integers whose product is $12(-2) = -24$ and whose sum is -5. The required integers are 3 and -8, and

$$12z^2 - 5z - 2 = 12z^2 + 3z - 8z - 2$$
$$= 3z(4z + 1) - 2(4z + 1)$$
$$= (4z + 1)(3z - 2). \blacktriangleleft$$

Work Problem 7 at the side.

ANSWER
6. 3 and 4
7. (a) $(2m + 1)(m + 3)$
   (b) $(5p + 3)(p - 1)$
   (c) $(3p - 1)(p - 1)$
   (d) $(3r - 1)(r + 4)$

5.3 EXERCISES

Complete the factoring.

1. $2x^2 - x - 1 = (2x + 1)()$

2. $3a^2 + 5a + 2 = (3a + 2)()$

3. $5b^2 - 16b + 3 = (5b - 1)()$

4. $2x^2 + 11x + 12 = (2x + 3)()$

5. $4y^2 + 17y - 15 = (y + 5)()$

6. $7z^2 + 10z - 8 = (z + 2)()$

7. $15x^2 + 7x - 4 = (3x - 1)()$

8. $12c^2 - 7c - 12 = (4c + 3)()$

9. $2m^2 + 19m - 10 = (2m - 1)()$

10. $6x^2 + x - 12 = (2x + 3)()$

11. $6a^2 + 7ab - 20b^2 = (2a + 5b)()$

12. $9m^2 - 3mn - 2n^2 = (3m - 2n)()$

13. $4k^2 + 13km + 3m^2 = (4k + m)()$

14. $6x^2 - 13xy - 5y^2 = (3x + y)()$

15. $4x^3 - 10x^2 - 6x = 2x($ $) = 2x(2x + 1)($ $)$

16. $15r^3 - 39r^2 - 18r = 3r($ $) = 3r(5r + 2)($ $)$

17. $6m^6 + 7m^5 - 20m^4 = m^4($ $) = m^4(3m - 4)($ $)$

18. $16y^5 - 4y^4 - 6y^3 = 2y^3($ $) = 2y^3(4y - 3)($ $)$

Factor completely. Use either method.

19. $2x^2 + 7x + 3$ 20. $3y^2 + 13y + 4$ 21. $3a^2 + 10a + 7$

22. $7r^2 + 8r + 1$ 23. $4r^2 + r - 3$ 24. $3p^2 + 2p - 8$

25. $15m^2 + m - 2$ 26. $6x^2 + x - 1$ 27. $8m^2 - 10m - 3$

28. $2a^2 - 17a + 30$ 29. $5a^2 - 7a - 6$ 30. $12s^2 + 11s - 5$

31. $3r^2 + r - 10$

32. $20x^2 - 28x - 3$

33. $4y^2 + 69y + 17$

34. $21m^2 + 13m + 2$

35. $38x^2 + 23x + 2$

36. $20y^2 + 39y - 11$

37. $10x^2 + 11x - 6$

38. $6b^2 + 7b + 2$

39. $6w^2 + 19w + 10$

40. $20q^2 - 41q + 20$

41. $6q^2 + 23q + 21$

42. $8x^2 + 47x - 6$

43. $10m^2 - 23m + 12$

44. $4t^2 - 5t - 6$

45. $8k^2 + 2k - 15$

46. $15p^2 - p - 6$

47. $10m^2 - m - 24$

48. $16a^2 + 30a + 9$

49. $24x^2 - 42x + 9$

50. $48b^2 - 74b - 10$

51. $40m^2q + mq - 6q$

5.3 Exercises

52. $15a^2b + 22ab + 8b$

53. $2m^3 + 2m^2 - 40m$

54. $15n^4 - 39n^3 + 18n^2$

55. $24a^4 + 10a^3 - 4a^2$

56. $18x^5 + 15x^4 - 75x^3$

57. $32z^5 - 20z^4 - 12z^3$

58. $15x^2y^2 - 7xy^2 - 4y^2$

59. $12p^2 + 7pq - 12q^2$

60. $6m^2 - 5mn - 6n^2$

61. $25a^2 + 25ab + 6b^2$

62. $6x^2 - 5xy - y^2$

63. $6a^2 - 7ab - 5b^2$

64. $25g^2 - 5gh - 2h^2$

65. $6m^6n + 7m^5n^2 + 2m^4n^3$

66. $12k^3q^4 - 4k^2q^5 - kq^6$

67. $18z^3y - 3z^2y^2 - 105zy^3$

68. $5x^3m + 5x^2m^2 - 60xm^3$

5.4 SPECIAL FACTORIZATIONS

▶ Recall from the last chapter that
$$(a + b)(a - b) = a^2 - b^2.$$
Based on this product, a **difference of two squares** can be factored as
$$a^2 - b^2 = (a + b)(a - b).$$

For example,
$$m^2 - 16 = m^2 - 4^2 = (m + 4)(m - 4).$$

EXAMPLE 1 Factor each difference of two squares.

(a) $x^2 - 49 = (x + 7)(x - 7)$

(b) $z^2 - \dfrac{9}{16} = z^2 - \left(\dfrac{3}{4}\right)^2 = \left(z + \dfrac{3}{4}\right)\left(z - \dfrac{3}{4}\right)$

(c) $y^2 - m^2 = (y + m)(y - m)$

(d) $p^2 + 16$

Since $p^2 + 16$ is the *sum* of two squares, it is not equal to $(p + 4)(p - 4)$. Also, using FOIL,
$$(p - 4)(p - 4) = p^2 - 8p + 16 \neq p^2 + 16,$$
and
$$(p + 4)(p + 4) = p^2 + 8p + 16 \neq p^2 + 16,$$
so $p^2 + 16$ is a prime polynomial. ◀

As Example 1(d) suggests,

the sum of two squares usually cannot be factored.

EXAMPLE 2 Factor each difference of two squares.

(a) $25m^2 - 16$

This is the difference of two squares, since
$$25m^2 - 16 = (5m)^2 - 4^2.$$
Factor this as
$$(5m + 4)(5m - 4).$$

(b) $49z^2 - 64 = (7z)^2 - (8)^2 = (7z + 8)(7z - 8)$ ◀

Work Problem 1 at the side.

EXAMPLE 3 Factor completely.

(a) $9a^2 - 4b^2 = (3a)^2 - (2b)^2 = (3a + 2b)(3a - 2b)$

(b) $81y^2 - 36$

First factor out the common factor of 9.
$$81y^2 - 36 = 9(9y^2 - 4)$$
$$= 9(3y + 2)(3y - 2) \qquad \text{Difference of squares}$$

Objectives

Factor:

▶ the difference of two squares;

▶ a perfect square trinomial;

▶ the difference of two cubes;

▶ the sum of two cubes.

1. Factor.

(a) $p^2 - 100$

(b) $9m^2 - 49$

(c) $64a^2 - 25$

ANSWERS
1. (a) $(p + 10)(p - 10)$
   (b) $(3m + 7)(3m - 7)$
   (c) $(8a + 5)(8a - 5)$

2. Factor.

   (a) $50r^2 - 32$

   (b) $27y^2 - 75$

   (c) $k^4 - 49$

   (d) $9r^4 - 100$

   (e) $4z^2 - \dfrac{25}{49}$

3. Factor each trinomial that is a perfect square.

   (a) $p^2 + 14p + 49$

   (b) $m^2 + 8m + 16$

   (c) $k^2 + 20k + 100$

   (d) $z^2 + 10z + 16$

(c) $p^4 - 36 = (p^2)^2 - 6^2 = (p^2 + 6)(p^2 - 6)$

Neither $p^2 + 6$ nor $p^2 - 6$ can be factored further.

(d) $m^4 - 16 = (m^2)^2 - 4^2 = (m^2 + 4)(m^2 - 4)$

While $m^2 + 4$ cannot be factored, $m^2 - 4$ can be factored as $(m + 2)(m - 2)$.

$$m^4 - 16 = (m^2 + 4)(m + 2)(m - 2) \blacktriangleleft$$

Work Problem 2 at the side.

▶ The expressions 144, $4x^2$, and $81m^6$ are called perfect squares, since

$$144 = \boxed{12^2}, \quad 4x^2 = \boxed{(2x)^2}, \quad \text{and} \quad 81m^6 = \boxed{(9m^3)^2}.$$

A **perfect square trinomial** is a trinomial that is the square of a binomial. As an example, $x^2 + 8x + 16$ is a perfect square trinomial since it is the square of the binomial $x + 4$:

$$x^2 + 8x + 16 = \boxed{(x + 4)^2}.$$

For a trinomial to be a perfect square, two of its terms must be perfect squares. For this reason, $16x^2 + 4x + 15$ is not a perfect square trinomial since only the term $16x^2$ is a perfect square.

On the other hand, just because two of the terms are perfect squares, the trinomial may not be a perfect square trinomial. For example, $x^2 + 6x + 36$ has two perfect square terms, but it is not a perfect square trinomial. (Try to find a binomial that can be squared to give $x^2 + 6x + 36$.)

Multiply to see that the square of a binomial gives the following **perfect square trinomials.**

$$\boxed{a^2 + 2ab + b^2 = (a + b)^2}$$
$$\boxed{a^2 - 2ab + b^2 = (a - b)^2}$$

The middle term of a perfect square trinomial is always twice the product of the two terms in the squared binomial. (This was shown in Section 4.6.) Use this to check any attempt to factor a trinomial that appears to be a perfect square.

EXAMPLE 4 Factor $x^2 + 10x + 25$.

The term x^2 is a perfect square, and so is 25. Try to factor the trinomial as

$$x^2 + 10x + 25 = (x + 5)^2.$$

Check this by finding twice the product of the two terms in the squared binomial.

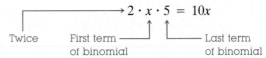

Since $10x$ is the middle term of the trinomial, the trinomial is a perfect square and can be factored as $(x + 5)^2$. ◀

Work Problem 3 at the side.

ANSWERS

2. (a) $2(5r + 4)(5r - 4)$
   (b) $3(3y + 5)(3y - 5)$
   (c) $(k^2 + 7)(k^2 - 7)$
   (d) $(3r^2 + 10)(3r^2 - 10)$
   (e) $\left(2z + \dfrac{5}{7}\right)\left(2z - \dfrac{5}{7}\right)$

3. (a) $(p + 7)^2$ (b) $(m + 4)^2$
   (c) $(k + 10)^2$ (d) not a perfect square trinomial

EXAMPLE 5 Factor each perfect square trinomial.

(a) $x^2 - 22x + 121$

The first and last terms are perfect squares ($121 = 11^2$). Check to see whether the middle term of $x^2 - 22x + 121$ is twice the product of the first and last terms of the binomial $(x - 11)$.

$$2 \cdot x \cdot 11 = 22x$$

Twice   First   Last
        term   term

Since twice the product of the first and last terms of the binomial is the middle term, $x^2 - 22x + 121$ is a perfect square trinomial and

$$x^2 - 22x + 121 = (x - 11)^2.$$

The middle sign in the binomial, a minus sign in this case, is always the same as the middle sign in the trinomial.

(b) $9m^2 - 24m + 16 = (3m)^2 - 2(3m)(4) + 4^2 = (3m - 4)^2$

Twice   First   Last
        term   term

(c) $25y^2 + 20y + 16$

The first and last terms are perfect squares.

$$25y^2 = (5y)^2 \quad \text{and} \quad 16 = 4^2$$

Twice the product of the first and last terms of the binomial $5y + 4$ is

$$2 \cdot 5y \cdot 4 = 40y,$$

which is not the middle term of $25y^2 + 20y + 16$. This polynomial is not a perfect square. In fact, the polynomial cannot be factored even with the methods of Section 5.3; it is a prime polynomial. ◀

Work Problem 4 at the side.

▶ The difference of two squares was factored above; it is also possible to factor the **difference of two cubes.** Use the pattern

$$a^3 - b^3 = (a - b)(a^2 + ab + b^2).$$

Check this rule by multiplying.

$$\begin{array}{r} a^2 + ab + b^2 \\ a - b \\ \hline -a^2b - ab^2 - b^3 \\ a^3 + a^2b + ab^2 \\ \hline a^3 \qquad\qquad\qquad - b^3 \end{array}$$

Memorize this pattern.

EXAMPLE 6 Factor.

(a) $m^3 - 125$

Let $a = m$ and $b = 5$ in the pattern for the difference of two cubes.

$$\begin{aligned} m^3 - 125 &= m^3 - 5^3 \\ &= (m - 5)(m^2 + 5m + 5^2) \\ &= (m - 5)(m^2 + 5m + 25) \end{aligned}$$

4. Factor each trinomial that is a perfect square.

(a) $p^2 - 8p + 16$

(b) $4m^2 + 20m + 25$

(c) $16a^2 + 56a + 49$

(d) $121p^2 + 110p + 100$

ANSWERS
4. (a) $(p - 4)^2$ (b) $(2m + 5)^2$
(c) $(4a + 7)^2$ (d) not a perfect square trinomial

5.4 Special Factorizations

5. Factor.

(a) $y^3 - 8$

(b) $27z^3 - 125$

(c) $64r^3 - 343z^3$
(*Hint*: $343 = 7^3$)

6. Factor.

(a) $q^3 + 125$

(b) $216m^3 + 1$

(c) $343a^3 + 27b^3$

(b) $8p^3 - 27$

Substitute into the rule using $2p$ for a and 3 for b.

$$\begin{aligned} 8p^3 - 27 &= (2p)^3 - 3^3 \\ &= (2p - 3)[(2p)^2 + (2p)3 + 3^2] \\ &= (2p - 3)(4p^2 + 6p + 9) \end{aligned}$$

(c) $\begin{aligned} 125t^3 - 216s^6 &= (5t)^3 - (6s^2)^3 \\ &= (5t - 6s^2)[(5t)^2 + (5t)(6s^2) + (6s^2)^2] \\ &= (5t - 6s^2)(25t^2 + 30ts^2 + 36s^4) \end{aligned}$ ◀

A common error in factoring the difference of two cubes, $a^3 - b^3 = (a - b)(a^2 + ab + b^2)$, is to try to factor $a^2 + ab + b^2$. It is easy to confuse this expression with the perfect square trinomial, $a^2 + 2ab + b^2$. Because of the lack of a 2 in the middle term of $a^2 + ab + b^2$, it is very unusual to be able to factor an expression of the form $a^2 + ab + b^2$.

Work Problem 5 at the side.

▶ A sum of two squares, such as $m^2 + 25$, cannot be factored, but the sum of two cubes can be factored by the following pattern, *which should be memorized.*

$$a^3 + b^3 = (a + b)(a^2 - ab + b^2)$$

EXAMPLE 7 Factor.

(a) $\begin{aligned} k^3 + 27 &= k^3 + 3^3 \\ &= (k + 3)(k^2 - 3k + 3^2) \\ &= (k + 3)(k^2 - 3k + 9) \end{aligned}$

(b) $\begin{aligned} 8m^3 + 125 &= (2m)^3 + 5^3 \\ &= (2m + 5)[(2m)^2 - (2m)(5) + 5^2] \\ &= (2m + 5)(4m^2 - 10m + 25) \end{aligned}$

(c) $\begin{aligned} 1000a^6 + 27b^3 &= (10a^2)^3 + (3b)^3 \\ &= (10a^2 + 3b)[(10a^2)^2 - (10a^2)(3b) + (3b)^2] \\ &= (10a^2 + 3b)(100a^4 - 30a^2b + 9b^2) \end{aligned}$ ◀

Work Problem 6 at the side.

The methods of factoring discussed in this section are summarized here. These rules should be memorized.

Special Factorizations

Difference of two squares	$a^2 - b^2 = (a + b)(a - b)$
Perfect square trinomials	$a^2 + 2ab + b^2 = (a + b)^2$
	$a^2 - 2ab + b^2 = (a - b)^2$
Difference of two cubes	$a^3 - b^3 = (a - b)(a^2 + ab + b^2)$
Sum of two cubes	$a^3 + b^3 = (a + b)(a^2 - ab + b^2)$

ANSWERS

5. (a) $(y - 2)(y^2 + 2y + 4)$
(b) $(3z - 5)(9z^2 + 15z + 25)$
(c) $(4r - 7z)(16r^2 + 28rz + 49z^2)$
6. (a) $(q + 5)(q^2 - 5q + 25)$
(b) $(6m + 1)(36m^2 - 6m + 1)$
(c) $(7a + 3b)(49a^2 - 21ab + 9b^2)$

name					date		hour

5.4 EXERCISES

Factor each binomial completely. The table of squares and square roots in the back of the book may be helpful. See Examples 1–3.

1. $x^2 - 16$
2. $m^2 - 25$
3. $p^2 - 4$
4. $r^2 - 9$

5. $9m^2 - 1$
6. $16y^2 - 9$
7. $25m^2 - \dfrac{16}{49}$
8. $25y^2 - \dfrac{9}{16}$

9. $36t^2 - 16$
10. $9 - 36a^2$
11. $25a^2 - 16r^2$
12. $100k^2 - 49m^2$

13. $81x^2 + 16$
14. $49m^2 + 100$
15. $p^4 - 36$

16. $r^4 - 9$
17. $a^4 - 1$
18. $x^4 - 16$

19. $m^4 - 81$
20. $p^4 - 256$
21. $16k^4 - 1$

Factor any expressions that are perfect square trinomials. See Examples 4 and 5.

22. $a^2 + 4a + 4$
23. $p^2 + 2p + 1$
24. $x^2 - 10x + 25$

25. $y^2 - 8y + 16$
26. $a^2 + 14a + 49$
27. $m^2 - 20m + 100$

28. $k^2 + k + \dfrac{1}{4}$
29. $r^2 + \dfrac{2}{3}r + \dfrac{1}{9}$
30. $b^2 - \dfrac{2}{5}b + \dfrac{1}{25}$

31. $9y^2 + 14y + 25$

32. $16m^2 + 42m + 49$

33. $16a^2 - 40ab + 25b^2$

34. $36y^2 - 60yp + 25p^2$

35. $100m^2 + 100m + 25$

36. $100a^2 - 140ab + 49b^2$

37. $49x^2 + 28xy + 4y^2$

38. $64y^2 - 48ya + 9a^2$

39. $4c^2 + 12cd + 9d^2$

40. $9t^2 - 24tr + 16r^2$

41. $25h^2 - 20hy + 4y^2$

42. $9x^2 + 24xy + 16y^2$

Factor each sum or difference of cubes. See Examples 6 and 7.

43. $y^3 + 1$

44. $m^3 - 1$

45. $8a^3 + 1$

46. $8a^3 - 1$

47. $27x^3 - 125$

48. $64p^3 + 27$

49. $8p^3 + q^3$

50. $y^3 - 8x^3$

51. $27a^3 - 64b^3$

52. $125t^3 + 8s^3$

53. $64x^3 + 125y^3$

54. $216z^3 - w^3$

55. $125m^3 - 8p^3$

56. $1000z^3 + 27x^3$

57. $64y^3 - 1331w^3$

58. $64y^6 + 1$

59. $m^6 - 8$

60. $8k^6 - 27q^3$

61. $125z^3 + 64r^6$

62. $1000a^3 - 343b^9$

63. $27r^9 + 125s^3$

SUPPLEMENTARY FACTORING EXERCISES

Follow the steps listed below to factor a polynomial.

> **Factoring a Polynomial**
> *Step 1* Is there a common factor?
> *Step 2* How many terms are in the polynomial?
>   *Two terms* Check to see whether it is either the difference of two squares or the sum or difference of two cubes.
>   *Three terms* Is it a perfect square trinomial? If the trinomial is not a perfect square, check to see whether the coefficient of the squared term is 1. If so, use the method of Section 5.2. If the coefficient of the squared term of the trinomial is not 1, use the general factoring methods of Section 5.3.
>   *Four terms* Can the polynomial be factored by grouping?
> *Step 3* Can any factors be factored further?

Factor as completely as possible.

1. $32m^9 + 16m^5 + 24m^3$
2. $2m^2 - 10m - 48$
3. $14k^3 + 7k^2 - 70k$

4. $9z^2 + 64$
5. $6z^2 + 31z + 5$
6. $m^2 - 3mn - 4n^2$

7. $49z^2 - 16y^2$
8. $100n^2r^2 + 30nr^3 - 50n^2r$
9. $8p^3 + 27$

10. $2m^2 + 5m - 3$
11. $10y^2 - 7yz - 6z^2$
12. $y^4 - 16$

13. $m^2 + 2m - 15$
14. $6y^2 - 5y - 4$
15. $32z^3 + 56z^2 - 16z$

16. $m^3 - n^3$
17. $z^2 - 12z + 36$
18. $9m^2 - 64$

19. $y^2 - 4yk - 12k^2$
20. $16z^2 - 8z + 1$
21. $6y^2 - 6y - 12$

22. $72y^3z^2 + 12y^2 - 24y^4z^2$ 23. $p^2 - 17p + 66$ 24. $a^2 + 17a + 72$

25. $k^2 + 9$ 26. $108m^2 - 36m + 3$ 27. $z^2 - 3za - 10a^2$

28. $45a^3b^5 - 60a^4b^2 + 75a^6b^4$ 29. $4k^2 - 12k + 9$ 30. $a^2 - 3ab - 28b^2$

31. $16r^2 + 24rm + 9m^2$ 32. $3k^2 + 4k - 4$ 33. $3k^3 - 12k^2 - 15k$

34. $a^4 - 625$ 35. $16k^2 - 48k + 36$ 36. $8k^2 - 10k - 3$

37. $36y^6 - 42y^5 - 120y^4$ 38. $8p^2 + 23p - 3$ 39. $5z^3 - 45z^2 + 70z$

40. $8k^2 - 2kh - 3h^2$ 41. $54m^2 - 24z^2$ 42. $4k^2 - 20kz + 25z^2$

43. $6a^2 + 10a - 4$ 44. $15h^2 + 11hg - 14g^2$ 45. $27r^3 - 8s^3$

46. $10z^2 - 7z - 6$ 47. $125m^4 - 400m^3n + 195m^2n^2$ 48. $9y^2 + 12y - 5$

49. $m^2 - 4m + 4$ 50. $27p^{10} - 45p^9 - 252p^8$ 51. $24k^4p + 60k^3p^2 + 150k^2p^3$

52. $10m^2 + 25m - 60$ 53. $12p^2 + pq - 6q^2$ 54. $125p^3 + 8q^3$

55. $64p^2 - 100m^2$ 56. $2m^2 + 7mn - 15n^2$ 57. $100a^2 - 81y^2$

58. $8a^2 + 23ab - 3b^2$ 59. $a^2 + 8a + 16$ 60. $4y^2 - 25$

5.5 SOLVING QUADRATIC EQUATIONS BY FACTORING

Objectives
1. Solve quadratic equations by factoring.
2. Solve other equations by factoring.

1 This section introduces **quadratic equations**, which are equations that contain a squared term and no terms of higher degree.

> **Quadratic Equations**
> Quadratic equations can be written in the form
> $$ax^2 + bx + c = 0,$$
> where a, b, and c are real numbers, with $a \neq 0$.

For example,

$$x^2 + 5x + 6 = 0, \quad 2a^2 - 5a = 3, \quad \text{and} \quad y^2 = 4$$

are all quadratic equations.

Some quadratic equations can be solved by factoring. A more general method for solving those equations that cannot be solved by factoring is given in Chapter 10.

Use the **zero-factor property** to solve a quadratic equation by factoring.

> **Zero-Factor Property**
> If a and b represent real numbers and if $ab = 0$, then $a = 0$ or $b = 0$.

In other words, if the product of two numbers is zero, then at least one of the numbers must be zero.

EXAMPLE 1 Solve the equation $(x + 3)(2x - 1) = 0$.

The product $(x + 3)(2x - 1)$ is equal to zero. By the zero-factor property, the only way that the product of these two factors can be zero is if at least one of the factors is zero. Therefore, either $x + 3 = 0$ or $2x - 1 = 0$. Solve each of these two linear equations as in Chapter 3.

$$x + 3 = 0 \qquad\qquad 2x - 1 = 0$$
$$x = -3 \qquad\qquad 2x = 1$$
$$x = \frac{1}{2}$$

The given equation $(x + 3)(2x - 1) = 0$ has two solutions, $x = -3$ and $x = \frac{1}{2}$. Check these answers by substituting -3 for x in the original equation, $(x + 3)(2x - 1) = 0$. Then start over and substitute $\frac{1}{2}$ for x.

If $x = -3$, then
$$(-3 + 3)[2(-3) - 1] = 0$$
$$0(-7) = 0. \quad \text{True}$$

If $x = \frac{1}{2}$, then
$$\left(\frac{1}{2} + 3\right)\left(2 \cdot \frac{1}{2} - 1\right) = 0$$
$$\frac{7}{2}(1 - 1) = 0$$
$$\frac{7}{2} \cdot 0 = 0. \quad \text{True}$$

1. Solve each equation. Check your answers.

   (a) $(x - 5)(x + 2) = 0$

   (b) $(3x - 2)(x + 6) = 0$

   (c) $(5x + 7)(2x + 3) = 0$

   (d) $(2x + 9)(3x - 5) = 0$

2. Solve each equation.

   (a) $m^2 - 3m - 10 = 0$

   (b) $a^2 + 6a + 8 = 0$

   (c) $y^2 - y = 6$

   (d) $r^2 = 8 - 2r$

ANSWERS

1. (a) $5, -2$ (b) $\frac{2}{3}, -6$
   (c) $-\frac{7}{5}, -\frac{3}{2}$ (d) $-\frac{9}{2}, \frac{5}{3}$
2. (a) $5, -2$ (b) $-4, -2$
   (c) $3, -2$ (d) $-4, 2$

Both -3 and $\frac{1}{2}$ result in true equations, so they are solutions to the original equation. ◀

Work Problem 1 at the side.

In Example 1 the equation to be solved was presented with the polynomial in factored form. If the polynomial in an equation is not already factored, first make sure that all terms are on one side of the equals sign, with 0 alone on the other side. Then factor.

EXAMPLE 2 Solve each equation.

(a) $x^2 - 5x = -6$

First, rewrite the equation with all terms on one side by adding 6 to both sides.

$$x^2 - 5x + 6 = -6 + 6 \quad \text{Add 6}$$
$$x^2 - 5x + 6 = 0$$

Now factor $x^2 - 5x + 6$. Find two numbers whose product is 6 and whose sum is -5. These two numbers are -2 and -3, so the equation becomes

$$(x - 2)(x - 3) = 0.$$

Proceed as in Example 1. Set each factor equal to 0.

$$x - 2 = 0 \quad \text{or} \quad x - 3 = 0$$

Solve the equation on the left by adding 2 to both sides. In the equation on the right, add 3 to both sides. Doing this gives

$$x = 2 \quad \text{or} \quad x = 3.$$

Check both solutions by substituting first 2 and then 3 for x in the original equation.

(b) $y^2 = y + 20$

To get 0 alone on one side of the equation, subtract y and 20 from both sides.

$$y^2 = y + 20$$
$$y^2 - y - 20 = 0 \quad \text{Subtract } y \text{ and 20}$$
$$(y - 5)(y + 4) = 0 \quad \text{Factor}$$
$$y - 5 = 0 \quad \text{or} \quad y + 4 = 0 \quad \text{Zero-factor property}$$

Solve each of these two equations to get the solutions.

$$y = 5 \quad \text{or} \quad y = -4$$

Check these solutions by substituting in the original equation. ◀

Work Problem 2 at the side.

EXAMPLE 3 Solve the equation $2p^2 - 13p + 20 = 0$.

Factor $2p^2 - 13p + 20$ as $(2p - 5)(p - 4)$, giving

$$(2p - 5)(p - 4) = 0.$$

Set each of these two factors equal to 0.

$$2p - 5 = 0 \quad \text{or} \quad p - 4 = 0$$

Factoring

Solve the equation on the left by adding 5 to both sides of the equation. Then divide both sides by 2. Solve the equation on the right by adding 4 to both sides.

$$2p - 5 = 0 \quad \text{or} \quad p - 4 = 0$$
$$2p = 5$$
$$p = \frac{5}{2} \quad \text{or} \quad p = 4$$

The solutions of $2p^2 - 13p + 20 = 0$ are $\frac{5}{2}$ and 4; check them by substituting in the original equation. ◀

Work Problem 3 at the side.

EXAMPLE 4 Solve each equation.

(a) $16m^2 - 25 = 0$

Factor the left-hand side of the equation as the difference of two squares.

$$(4m + 5)(4m - 5) = 0$$

Set each factor equal to 0.

$$4m + 5 = 0 \quad \text{or} \quad 4m - 5 = 0$$

Solve each equation.

$$4m = -5 \quad \text{or} \quad 4m = 5$$
$$m = -\frac{5}{4} \quad \text{or} \quad m = \frac{5}{4}$$

The two solutions, $-\frac{5}{4}$ and $\frac{5}{4}$, can be checked in the original equation.

(b) $k(2k + 5) = 3$

Multiply on the left-hand side and then get all terms on one side.

$$k(2k + 5) = 3$$
$$2k^2 + 5k = 3$$
$$2k^2 + 5k - 3 = 0 \quad \text{Subtract 3}$$

Now factor.

$$(2k - 1)(k + 3) = 0$$

Place each factor equal to 0 and solve the equations.

$$2k - 1 = 0 \quad \text{or} \quad k + 3 = 0$$
$$2k = 1$$
$$k = \frac{1}{2} \quad \text{or} \quad k = -3$$

The two solutions are $\frac{1}{2}$ and -3. ◀

Work Problem 4 at the side.

In Example 4(b) the zero-factor property could not be used to solve the original equation because of the 3 on the right.

The zero-factor property applies only to a product that equals 0.

3. Solve each equation.

(a) $2a^2 - a - 3 = 0$

(b) $3r^2 + 10r - 8 = 0$

(c) $3x^2 = 11x + 4$

(d) $12m^2 - 17m = 5$

4. Solve each equation.

(a) $p^2 - 36 = 0$

(b) $49m^2 - 9 = 0$

(c) $36y^2 - 25 = 0$

(d) $p(4p + 7) = 2$

(e) $k(6k + 11) = 10$

ANSWERS

3. (a) $\frac{3}{2}, -1$ (b) $\frac{2}{3}, -4$
   (c) $-\frac{1}{3}, 4$ (d) $-\frac{1}{4}, \frac{5}{3}$

4. (a) $6, -6$ (b) $\frac{3}{7}, -\frac{3}{7}$
   (c) $\frac{5}{6}, -\frac{5}{6}$ (d) $\frac{1}{4}, -2$ (e) $\frac{2}{3}, -\frac{5}{2}$

5.5 Solving Quadratic Equations by Factoring

5. Solve each equation. Check each solution.

   (a) $2a(a - 1)(a + 3) = 0$

   (b) $r^3 - 16r = 0$

6. Solve each equation. Check each solution.

   (a) $(m + 3)(m^2 - 11m + 10) = 0$

   (b) $(2x + 1)(2x^2 + 7x - 15) = 0$

ANSWERS
5. (a) 0, 1, −3 (b) 0, 4, −4
6. (a) −3, 1, 10 (b) $-\frac{1}{2}, \frac{3}{2}, -5$

In summary, go through the following steps to solve quadratic equations by factoring.

Solving a Quadratic Equation by Factoring

Step 1 Get all terms on one side of the equals sign, with 0 on the other side.

Step 2 Factor completely.

Step 3 Set each factor containing a variable equal to 0, and solve the resulting equations.

Step 4 Check each solution in the original equation.

Remember: Not all quadratic equations can be solved by factoring.

▶ The zero-factor property can also be used to solve equations that result in more than two factors, as shown in Example 5. (These equations are not quadratic equations. Why not?)

EXAMPLE 5 Solve the equation $6z^3 - 6z = 0$.

First, factor out the greatest common factor in $6z^3 - 6z$.

$$6z^3 - 6z = 0$$
$$6z(z^2 - 1) = 0$$

Now factor $z^2 - 1$ as $(z + 1)(z - 1)$ to get

$$6z(z + 1)(z - 1) = 0.$$

By an extension of the zero-factor property, this product can equal 0 only if at least one of the factors is 0. Write three equations, one for each factor with a variable.

$$6z = 0 \quad \text{or} \quad z + 1 = 0 \quad \text{or} \quad z - 1 = 0$$

Solving these three equations gives the three solutions

$$z = 0 \quad \text{or} \quad z = -1 \quad \text{or} \quad z = 1.$$

Check by substituting, in turn, 0, −1, and 1 in the original equation. ◀

Work Problem 5 at the side.

EXAMPLE 6 Solve the equation $(2x - 1)(x^2 - 9x + 20) = 0$.

Factor $x^2 - 9x + 20$ as $(x - 5)(x - 4)$. Then rewrite the original equation as

$$(2x - 1)(x - 5)(x - 4) = 0.$$

Set each of these three factors equal to 0.

$$2x - 1 = 0 \quad \text{or} \quad x - 5 = 0 \quad \text{or} \quad x - 4 = 0$$

Solving these three equations gives

$$x = \frac{1}{2} \quad \text{or} \quad x = 5 \quad \text{or} \quad x = 4$$

as the solutions of the original equation. Check each solution. ◀

Work Problem 6 at the side.

name date hour

5.5 EXERCISES

Solve each equation and check your answers. See Example 1.

1. $(x - 2)(x + 4) = 0$
2. $(y - 3)(y + 5) = 0$
3. $(3x + 5)(2x - 1) = 0$

4. $(2a + 3)(a - 2) = 0$
5. $(5p + 1)(2p - 1) = 0$
6. $(3k - 8)(k + 7) = 0$

Solve each equation and check your answers. See Examples 2–4.

7. $x^2 + 5x + 6 = 0$
8. $y^2 - 3y + 2 = 0$
9. $m^2 + 3m - 28 = 0$

10. $p^2 - p - 6 = 0$
11. $a^2 - 24 - 5a$
12. $r^2 = 2r + 15$

13. $x^2 = 3 + 2x$
14. $m^2 = 3m + 4$
15. $z^2 = -2 - 3z$

16. $p^2 = 2p + 3$
17. $m^2 + 8m + 16 = 0$
18. $b^2 - 6b + 9 = 0$

19. $3a^2 + 5a - 2 = 0$
20. $6r^2 - r - 2 = 0$
21. $2k^2 - k - 10 = 0$

22. $6x^2 - 7x - 5 = 0$
23. $6p^2 = 4 - 5p$
24. $6x^2 - 5x = 4$

25. $6a^2 = 5 - 13a$
26. $9s^2 + 12s = -4$
27. $2z^2 + 3z = 20$

28. $25p^2 + 20p + 4 = 0$
29. $3a^2 + 7a = 20$
30. $6z^2 + 11z + 3 = 0$

31. $15r^2 = r + 2$ **32.** $3m^2 = 5m + 28$ **33.** $16r^2 - 25 = 0$ **34.** $4k^2 - 9 = 0$

In each of the following exercises, first simplify and then solve. Check each solution. See Example 4(b).

35. $m(m - 7) = -10$ **36.** $z(2z + 7) = 4$ **37.** $b(2b + 3) = 9$

38. $5b^2 = 4(2b + 1)$ **39.** $2(x^2 - 66) = -13x$ **40.** $3(m^2 + 4) = 20m$

41. $3r(r + 1) = (2r + 3)(r + 1)$ **42.** $(3k + 1)(k + 1) = 2k(k + 3)$

43. $12k(k - 4) = 3(k - 4)$ **44.** $y^2 = 4(y - 1)$

Solve each equation and check your answers. See Example 5.

45. $(2r - 5)(3r^2 - 16r + 5) = 0$ **46.** $(3m - 4)(6m^2 + m - 2) = 0$

47. $(2x + 7)(x^2 - 2x - 3) = 0$ **48.** $(x - 1)(6x^2 + x - 12) = 0$

49. $9y^3 - 49y = 0$ **50.** $16r^3 - 9r = 0$ **51.** $r^3 - 2r^2 - 8r = 0$

52. $x^3 - x^2 - 6x = 0$ **53.** $a^3 + a^2 - 20a = 0$ **54.** $y^3 - 6y^2 + 8y = 0$

5.6 APPLICATIONS OF QUADRATIC EQUATIONS

Objectives

Solve word problems about
1. area;
2. perimeter;
3. consecutive integers;
4. the Pythagorean formula.

▶ Quadratic equations can be applied in solving different kinds of word problems by using the techniques discussed in this section.

EXAMPLE 1 The width of a rectangle is 4 centimeters less than the length. The area is 96 square centimeters. Find the length and width of the rectangle.

Let x represent the length of the rectangle. Then, according to the statement of the problem, the width is 4 less than the length, so $x - 4$ represents the width. See Figure 1. The area of a rectangle is given by the formula

$$\text{area} = LW = \text{length} \times \text{width}.$$

Figure 1

Substitute 96 for the area, x for the length, and $x - 4$ for the width into the formula

$$A = LW.$$
$$96 = x(x - 4) \quad \text{Let } A = 96, L = x, W = x - 4$$

Multiply on the right.

$$96 = x^2 - 4x$$
$$0 = x^2 - 4x - 96 \quad \text{Subtract 96}$$
$$0 = (x - 12)(x + 8) \quad \text{Factor}$$

Set each factor equal to 0.

$$x - 12 = 0 \quad \text{or} \quad x + 8 = 0$$

Solve each equation.

$$x = 12 \quad \text{or} \quad x = -8$$

The solutions of the equations are $x = 12$ or $x = -8$. Always be careful, however, to check solutions against physical facts. Since a rectangle cannot have a negative length, discard the solution -8. Then 12 centimeters is the length of the rectangle, and $12 - 4 = 8$ centimeters is the width. As a check, the width is 4 less than the length, and the area is $8 \cdot 12 = 96$ square centimeters. ◀

Work Problem 1 at the side.

1. Solve each word problem.

    (a) The length of a rectangle is 2 meters more than the width. The area is 48 square meters. Find the length and width of the rectangle.

    (b) The length of a rectangle is five times the width. The area is 45 square meters. Find the length and width of the rectangle.

▶ The next word problem involves **perimeter**, the distance around a figure.

ANSWERS

1. (a) 8 meters, 6 meters   (b) 15 meters, 3 meters

2. Solve each problem.

(a) The length of a rectangle is 5 more than the width. The area is numerically 32 more than the perimeter. Find the length and width of the rectangle.

(b) The width of a rectangle is 4 less than the length. The area is numerically 92 more than the perimeter. Find the length and width of the rectangle.

EXAMPLE 2 The length of a rectangle is 4 feet more than the width. The area of the rectangle is numerically 1 more than the perimeter. (See Figure 2.) Find the length and width of the rectangle.

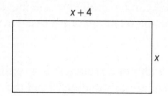

Figure 2

Let x be the width of the rectangle. Then the length is $x + 4$. The area is the product of the length and width, or

$$A = LW.$$

Substituting $x + 4$ for the length and x for the width gives

$$A = (x + 4)x.$$

Now substitute into the formula for perimeter,

$$P = 2L + 2W$$
$$P = 2(x + 4) + 2x.$$

According to the information given in the problem, the area is numerically 1 more than the perimeter.

The area	is	1	more than	the perimeter.
↓	↓	↓	↓	↓
$x(x + 4)$	=	1	+	$2(x + 4) + 2x$

Simplify and solve this equation.

$$x^2 + 4x = 1 + 2x + 8 + 2x$$
$$x^2 + 4x = 9 + 4x$$
$$x^2 = 9 \qquad \text{Subtract } 4x \text{ from both sides}$$
$$x^2 - 9 = 0 \qquad \text{Subtract 9 from both sides}$$
$$(x + 3)(x - 3) = 0 \qquad \text{Factor}$$
$$x + 3 = 0 \qquad \text{or} \qquad x - 3 = 0$$
$$x = -3 \qquad \text{or} \qquad x = 3$$

A rectangle cannot have a negative width, so ignore -3. The only valid solution is 3, so the width is 3 feet and the length is $3 + 4 = 7$ feet. (Check to see that the area is numerically 1 more than the perimeter.) The rectangle is 3 by 7. ◀

Work Problem 2 at the side.

③ As mentioned in Section 3.4, **consecutive integers** are integers that are next to each other on a number line, such as 5 and 6, or -11 and -10. **Consecutive odd integers** are odd integers that are next to each other, such as 21 and 23, or -17 and -15. The next example shows how quadratic equations can occur in work with consecutive integers.

The following list, which was given in Section 3.4, may be helpful in working with consecutive integers. Here x represents the first of the integers.

ANSWERS
2. (a) 11, 6 (b) 14, 10

Consecutive Integers	
Two consecutive integers	$x, x+1$
Three consecutive integers	$x, x+1, x+2$
Two consecutive odd integers	$x, x+2$
Two consecutive even integers	$x, x+2$

EXAMPLE 3 The product of two consecutive odd integers is 1 less than five times their sum. Find the integers.

Let s represent the smaller of the two integers. Since the problem mentions consecutive *odd* integers, use $s + 2$ for the larger of the two integers. According to the problem, the product is 1 less than five times the sum.

$$\underbrace{s(s+2)}_{\text{The product}} \underbrace{=}_{\text{is}} \underbrace{5(s+s+2)}_{\text{five times the sum}} \underbrace{-1}_{\text{less 1.}}$$

Simplify this equation and solve it.

$$s^2 + 2s = 5s + 5s + 10 - 1$$
$$s^2 + 2s = 10s + 9$$
$$s^2 - 8s - 9 = 0 \qquad \text{Subtract } 10s \text{ and } 9$$
$$(s-9)(s+1) = 0 \qquad \text{Factor}$$
$$s - 9 = 0 \quad \text{or} \quad s + 1 = 0$$
$$s = 9 \qquad \qquad s = -1$$

We need to find two consecutive odd integers.

If $s = 9$ is the first, then $s + 2 = 11$ is the second.
If $s = -1$ is the first, then $s + 2 = 1$ is the second.

Check that two pairs of integers satisfy the problem: 9 and 11 or -1 and 1. ◀

Work Problem 3 at the side.

The next example requires the **Pythagorean formula** from geometry.

Pythagorean Formula

If a right triangle (a triangle with a 90° angle) has longest side of length c and two other sides of length a and b, then

$$a^2 + b^2 = c^2.$$

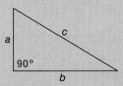

The longest side is called the **hypotenuse** and the two shorter sides are the **legs** of the triangle.

3. (a) The product of two consecutive even integers is 4 more than two times their sum. Find the integers.

(b) Find three consecutive integers such that the product of the first two is 2 more than six times the third.

ANSWERS
3. (a) 4, 6 or $-2, 0$
   (b) 7, 8, 9 or $-2, -1, 0$

5.6 Applications of Quadratic Equations

4. The hypotenuse of a right triangle is 3 inches longer than the longer leg. The shorter leg is 3 inches shorter than the longer leg. Find the lengths of the sides of the triangle.

EXAMPLE 4 The hypotenuse of a right triangle is 2 feet more than the shorter leg. The longer leg is 1 foot more than the shorter leg. Find the lengths of the sides of the triangle.

Let x be the length of the shorter leg. Then

$$x = \text{shorter leg}$$
$$x + 1 = \text{longer leg}$$
$$x + 2 = \text{hypotenuse.}$$

Place these on a right triangle, as in Figure 3.

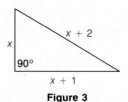

Figure 3

Substitute into the Pythagorean formula.

$$a^2 + b^2 = c^2$$
$$x^2 + (x + 1)^2 = (x + 2)^2$$

Since $(x + 1)^2 = x^2 + 2x + 1$, and since $(x + 2)^2 = x^2 + 4x + 4$, the equation becomes

$$x^2 + x^2 + 2x + 1 = x^2 + 4x + 4.$$

Get 0 on one side of the equation.

$$x^2 - 2x - 3 = 0$$

Factor.

$$(x - 3)(x + 1) = 0$$

Place each factor equal to 0.

$$x - 3 = 0 \quad \text{or} \quad x + 1 = 0$$
$$x = 3 \quad \text{or} \quad x = -1$$

Since -1 cannot be the length of a side of a triangle, 3 is the only possible answer. The triangle has a shorter leg of length 3 feet, a longer leg of length $3 + 1 = 4$ feet, and a hypotenuse of length $3 + 2 = 5$ feet. Check that $3^2 + 4^2 = 5^2$. ◀

Work Problem 4 at the side.

ANSWER
4. 9 inches, 12 inches, 15 inches

5.6 EXERCISES

Solve each problem. Check your answers. See Examples 1 and 2.

1. The length of a rectangle is 5 centimeters more than the width. The area is 66 square centimeters. Find the length and width of the rectangle.

2. The length of a rectangle is 1 foot more than the width. The area is 56 square feet. Find the length and width of the rectangle.

3. The length of a rectangle is 3 feet more than the width. The area is numerically 4 less than the perimeter. Find the dimensions of the rectangle.

4. The width of a rectangle is 5 meters less than the length. The area is numerically 10 more than the perimeter. Find the dimensions of the rectangle.

5. The length of a rectangle is twice its width. If the width were increased by 2 inches while the length remained the same, the resulting rectangle would have an area of 48 square inches. Find the dimensions of the original rectangle.

6. The length of a rectangle is three times its width. If the length were decreased by 1 while the width stayed the same, the area of the new figure would be 44 square centimeters. Find the length and width of the original rectangle.

Problems 7 and 8 require the formula for the area of a triangle,

$$\text{area} = \frac{1}{2}bh.$$

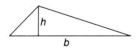

7. The area of a triangle is 25 square centimeters. The base is twice the height. Find the length of the base and the height of the triangle.

8. The height of a triangle is 3 inches more than the base. The area of the triangle is 27 square inches. Find the length of the base and the height of the triangle.

Problems 9 and 10 require the formula for the volume of a pyramid,

$V = \frac{1}{3}Bh$, where B is the area of the base.

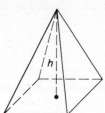

9. The volume of a pyramid is 32 cubic meters. Suppose the numerical value of the height is 10 meters less than the numerical value of the area of the base. Find the height and the area of the base.

10. Suppose a pyramid has a rectangular base whose width is 3 centimeters less than the length. If the height is 8 centimeters and the volume is 144 cubic centimeters, find the dimensions of the base.

Work the following problems. Check your answers.

11. One square has sides 1 foot less than the length of the sides of a second square. If the difference of the areas of the two squares is 37 square feet, find the lengths of the sides of the two squares.

12. The sides of one square have a length 2 meters more than the sides of another square. If the area of the larger square is subtracted from three times the area of the smaller square, the answer is 12 square meters. Find the lengths of the sides of each square.

13. John wishes to build a box to hold his tools. The box is to be 4 feet high, and the width of the box is to be 1 foot less than the length. The volume of the box will be 120 cubic feet. Find the length and width of the box. (*Hint:* The formula for the volume of a box is $V = LWH$.)

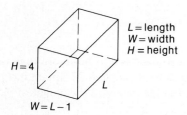

14. The volume of a box must be 315 cubic meters. The length of the box is to be 7 meters, and the height is to be 4 meters more than the width. Find the width and height of the box.

name	date	hour

Work the following problems. Check your answers. See Example 3.

15. The product of two consecutive integers is 2 more than twice their sum. Find the integers.

16. The product of two consecutive even integers is 60 more than twice the larger. Find the integers.

17. Find three consecutive even integers such that four times the sum of all three equals the product of the smaller two.

18. One number is 4 more than another. The square of the smaller increased by three times the larger is 66. Find the numbers.

19. If the square of the sum of two consecutive integers is reduced by three times their product, the result is 31. Find the integers.

20. If the square of the larger of two numbers is reduced by six times the smaller, the result is five times the larger. The larger is twice the smaller. Find the numbers.

21. The sum of three times the square of an integer and twice the integer is 8. Find the integer.

22. When four times an integer is subtracted from twice the square of the integer, the result is 16. Find the integer.

The following exercises require the Pythagorean formula. See Example 4.

23. The hypotenuse of a right triangle is 1 centimeter longer than the longer leg. The shorter leg is 7 centimeters shorter than the longer leg. Find the length of the longer leg of the triangle.

24. The longer leg of a right triangle is 1 meter longer than the shorter leg. The hypotenuse is 1 meter shorter than twice the shorter leg. Find the length of the shorter leg of the triangle.

5.6 Exercises

25. The hypotenuse of a right triangle is 1 foot longer than twice the shorter leg. The longer leg is 1 foot shorter than twice the shorter leg. Find the length of the shorter leg of the triangle.

26. If the shorter leg of a right triangle is tripled, with 4 inches added to the result, the result is the length of the hypotenuse. The longer leg is 10 inches longer than twice the shorter leg. Find the length of the shorter leg of the triangle.

Work the following problems involving formulas.

If an object is dropped, the distance d it falls in t seconds (disregarding air resistance) is given by

$$d = \frac{1}{2}gt^2,$$

where g is approximately 32 feet per second per second. Find the distance an object would fall in the following periods of time.

27. 4 seconds

28. 8 seconds

How long would it take an object to fall the following distances?

29. 1600 feet

30. 2304 feet

If an object is projected straight up with an initial velocity of v_0 feet per second, its height h after t seconds is given by

$$h = v_0 t - 16t^2.$$

Suppose an object is thrown upward with an initial velocity of 64 feet per second. Find its height after the following periods of time.

31. 1 second

32. 2 seconds

33. 3 seconds

34. When will the object hit the ground? (*Hint:* It will hit the ground when $h = 0$.)

name   date   hour

CHAPTER 5   REVIEW EXERCISES

(5.1)   *Find the prime factored form for each number.*

1. 12   **2.** 48   **3.** 110

4. 180   **5.** 600   **6.** 29

Factor out the greatest common factor.

7. $6p + 12$ **8.** $40r^2 + 20$ **9.** $25m^4 - 50m^3$

10. $32z + 48z^2$ **11.** $6 - 18r^5 + 12r^3$ **12.** $100y^6 - 50y^3 + 300y^4$

13. $15m^3n^4 - 20m^2n^5 + 50m^3n^6$ **14.** $32y^4r^3 - 48y^5r^2 + 24y^7r^5$

15. $6p^2 + 9p + 4p + 6$ **16.** $4y^2 + 3y + 8y + 6$

17. $15m^2 + 20mp - 12mp - 16p^2$ **18.** $12r^2 + 18rq - 10rq - 15q^2$

19. $12x^2yz^3 + 12xy^2z - 30x^3y^2z^4$ **20.** $24ab^3c^2 - 56a^2bc^3 + 72a^2b^2c$

(5.2)   *Factor completely.*

21. $m^2 + 5m + 6$ **22.** $y^2 + 13y + 40$ **23.** $r^2 - 6r - 27$

24. $p^2 + p - 30$
25. $z^2 - 7z - 44$
26. $k^2 + 2k - 63$

27. $z^2 - 11zx + 10x^2$
28. $r^2 - 4rs - 96s^2$
29. $p^2 + 2pq - 120q^2$

30. $4p^3 - 12p^2 - 40p$
31. $3z^4 - 30z^3 + 48z^2$
32. $m^2 - 3mn - 18n^2$

33. $y^2 - 8yz + 15z^2$
34. $p^7 - p^6q - 2p^5q^2$
35. $3r^5 - 6r^4s - 45r^3s^2$

36. $2a^5 - 8a^4 - 24a^3$
37. $5p^6 - 45p^5 + 70p^4$
38. $6m^9 - 84m^8 + 270m^7$

(5.3)

39. $2k^2 - 5k + 2$
40. $3z^2 + 11z - 4$
41. $6r^2 - 5r - 6$

42. $10p^2 - 3p - 1$
43. $8y^2 + 17y - 21$
44. $3k^2 + 11k + 10$

45. $6m^3 - 21m^2 - 45m$
46. $24k^5 - 20k^4 + 4k^3$
47. $7m^2 + 19mn - 6n^2$

48. $10r^3s + 17r^2s^2 + 6rs^3$
49. $2z^3 + 9z^2 - 5z$
50. $3p^4 - 2p^3 - 8p^2$

(5.4)

51. $m^2 - 25$
52. $25p^2 - 121$
53. $100a^2 - 9$

name date hour

54. $49y^2 - 25z^2$

55. $144p^2 - 36q^2$

56. $y^4 - 625$

57. $m^2 + 9$

58. $z^2 - 8z + 16$

59. $9r^2 - 42r + 49$

60. $16m^2 + 40mn + 25n^2$

61. $25a^2 + 15ab + 9b^2$

62. $54x^3 - 72x^2 + 24x$

63. $p^3 - 27$

64. $y^3 + 8$

65. $8m^3 + 27p^3$

66. $125r^3 - 216s^3$

67. $(x + y)^3 + 64$

68. $(a + b)^2 - (a - b)^2$

(5.5) *Solve each equation. Check each solution.*

69. $(3k - 1)(k + 2) = 0$

70. $(5m - 2)(m + 4) = 0$

71. $(2a + 5)(3a - 7) = 0$

72. $z^2 + 4z + 3 = 0$

73. $r^2 + 3r = 10$

74. $m^2 - 5m + 4 = 0$

75. $k^2 = 8k - 15$

76. $y^2 = 11y - 30$

77. $8p^2 = 10p + 3$

78. $3k^2 - 11k - 20 = 0$

79. $25m^2 - 20m + 4 = 0$

80. $100b^2 - 49 = 0$

81. $m(m - 5) = 6$

82. $p^2 = 12(p - 3)$

83. $(3z - 1)(z^2 + 3z + 2) = 0$

Chapter 5 Review Exercises **273**

84. $(2p + 3)(p^2 - 4p + 3) = 0$ **85.** $x^3 - 9x = 0$ **86.** $y^3 - 8 = 0$

(5.6) *Solve each word problem. Check your answers.*

87. The length of a rectangle is 6 meters more than the width. The area is 40 square meters. Find the length and width of the rectangle.

88. The length of a rectangle is three times the width. If the width were increased by 3 meters while the length remained the same, the new rectangle would have an area of 30 square meters. Find the length and width of the original rectangle.

89. The length of a rectangle is 2 centimeters more than the width. The area is numerically 44 more than the perimeter. Find the length and width of the rectangle.

90. The area of a triangle is 12 square meters. The base is 2 meters longer than the height. Find the base and height of the triangle.

91. A pyramid has a rectangular base with a length that is 2 meters more than the width. The height of the pyramid is 6 meters, and its volume is 48 cubic meters. Find the length and width of the base.

92. The volume of a box is to be 120 cubic meters. The width of the box is to be 4 meters, and the height 1 meter less than the length. Find the length and height of the box.

93. The product of two consecutive even integers is 4 more than twice their sum. Find the integers.

94. One number is 2 more than another. If the square of the smaller is added to the square of the larger, the result is 100. Find the two numbers.

95. The length of a rectangle is 4 meters more than the width. The area is numerically 1 more than the perimeter. Find the dimensions of the rectangle.

96. The width of a rectangle is 5 meters less than the length. The area is numerically 10 more than the perimeter. Find the dimensions of the rectangle.

97. The sum of twice the square of an integer and three times the integer is 5. Find the integer.

98. When five times an integer is subtracted from the square of the integer, the result is -6. Find the integer.

99. If 3 times the square of a number is subtracted from the square of the sum of the number and 2, the result is -2. Find the number.

100. The product of the smaller two of three consecutive integers equals the largest plus 23. Find the integers.

101. The difference of the squares of two consecutive even integers is 28 less than the square of the smaller integer. Find the two integers.

102. When 10 times the larger of two consecutive even integers is subtracted from the square of the smaller, the result is 76. Find the integers.

103. A lot is shaped like a right triangle. The hypotenuse is 3 meters longer than the longer leg. The longer leg is 6 meters more than twice the length of the shorter leg. Find the lengths of the sides of the lot.

104. A bicyclist heading east and a motorist traveling south left an intersection at the same time. When the motorcyclist had gone 17 miles further than the bicyclist, the distance between them was 1 mile more than the distance traveled by the motorist. How far apart were they then? (*Hint*: Draw a sketch.)

Chapter 5 Review Exercises

105. Two cars left an intersection at the same time. One traveled north. The other traveled 14 miles further, but to the west. How far apart were they then if the distance between them was 4 miles more than the distance traveled west?

106. A ladder is leaning against a building. The distance from the bottom of the ladder to the building is 4 feet less than the length of the ladder. How high up the side of the building is the top of the ladder if that distance is 2 feet less than the length of the ladder?

If an object is thrown straight up with an initial velocity of 128 feet per second, its height h after t seconds is

$$h = 128t - 16t^2.$$

Find the height of the object after the following periods of time.

107. 1 second   **108.** 2 seconds   **109.** 4 seconds

110. When does the object hit the ground?

111. A 9-inch by 12-inch picture is to be placed on a cardboard mat so that there is an equal border around the picture. The area of the finished mat and picture is to be 208 square inches. How wide will the border be?

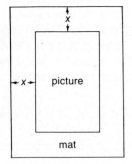

112. A box is made from a 12-centimeter by 10-centimeter piece of cardboard by cutting a square from each corner and folding up the sides. The area of the bottom of the box is to be 48 square centimeters. Find the length of a side of the cutout squares.

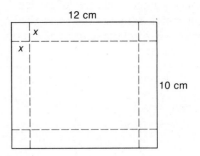

name _____ date _____ hour _____

CHAPTER 5   TEST

Factor as completely as possible.

1. $16m^2 - 24m$

2. $6xy + 12y^2$

3. $28pq + 14p + 56p^2$

4. $3m^2n + 9mn + 6mn^2$

5. $12p + 11r$

6. $x^2 + 11x + 30$

7. $p^2 + 6p - 7$

8. $2y^2 - 7y - 15$

9. $4m^2 + 4m - 3$

10. $3x^2 + 13x - 10$

11. $10z^2 + 7z + 1$

12. $10a^2 - 23a - 5$

13. $12r^2 + 19r + 5$

14. $m^2 + 11m + 14$

15. $a^2 + 3ab - 10b^2$

16. $6r^2 - rs - 2s^2$

1. _____
2. _____
3. _____
4. _____
5. _____
6. _____
7. _____
8. _____
9. _____
10. _____
11. _____
12. _____
13. _____
14. _____
15. _____
16. _____

17. $x^2 - 25$

18. $25m^2 - 49$

19. $4p^2 + 12p + 9$

20. $25z^2 - 10z + 1$

21. $4p^3 + 16p^2 + 16p$

22. $10m^4 + 55m^3 + 25m^2$

Solve each equation. Check each solution.

23. $y^2 + 3y + 2 = 0$

24. $3x^2 + 5x = 2$

25. $2p^2 + 3p = 20$

26. $x^2 - 4 = 0$

27. $z^3 = 16z$

Write an equation for each problem and solve it. Check your answers.

28. The length of a certain rectangle is 1 inch less than twice the width. The area is 15 square inches. Find the length and width of the rectangle.

29. Find two consecutive integers such that the square of the sum of the two integers is the negative of the smaller integer.

30. The length of the hypotenuse of a right triangle is twice the length of the shorter leg plus 3 meters. The longer leg is 7 meters longer than the shorter leg. Find the lengths of the sides.

Rational Expressions   6

6.1 THE FUNDAMENTAL PROPERTY OF RATIONAL EXPRESSIONS

Objectives

▶ Find the values for which a rational expression is meaningless.

▶ Find the numerical value of a rational expression.

▶ Write rational expressions in lowest terms.

▶ The quotient of two integers (with denominator not zero) is called a rational number. In the same way, the quotient of two polynomials with denominator not equal to zero is called a *rational expression*.

> A **rational expression** is an expression of the form
> $$\frac{P}{Q},$$
> where P and Q are polynomials, with $Q \neq 0$.

Examples of rational expressions include

$$\frac{-6x}{x^3 + 8}, \quad \frac{9x}{y + 3}, \quad \text{and} \quad \frac{2m^3}{8}.$$

A number with a zero denominator is *not* a rational expression, since division by zero is not possible. For that reason, be careful when substituting a number in the denominator of a rational expression. For example, in

$$\frac{8x^2}{x - 3}$$

the variable x can take on any value except 3. When $x = 3$ the denominator becomes $3 - 3 = 0$, making the expression meaningless.

EXAMPLE 1 Find any values for which the following are meaningless.

(a) $\dfrac{p + 5}{3p + 2}$

This rational expression is meaningless for any value of p that makes the denominator equal to zero. Find these values by solving the equation

$$3p + 2 = 0$$
$$3p = -2$$
$$p = -\frac{2}{3}.$$

Since $p = -\frac{2}{3}$ will make the denominator zero, the given expression is meaningless for $-\frac{2}{3}$.

1. Find all values for which the following rational expressions are meaningless.

   (a) $\dfrac{x+2}{x-5}$

   (b) $\dfrac{3r}{r^2+6r+8}$

   (c) $\dfrac{-5m}{m^2+4}$

2. Find the value of each rational expression when $x = 3$.

   (a) $\dfrac{x}{2x+1}$

   (b) $\dfrac{2x+6}{x-3}$

(b) $\dfrac{9m^2}{m^2-5m+6}$

Find the numbers that make the denominator zero by solving the equation
$$m^2 - 5m + 6 = 0.$$
Factor the polynomial and set each factor equal to zero.
$$(m-2)(m-3) = 0$$
$$m - 2 = 0 \quad \text{or} \quad m - 3 = 0$$
$$m = 2 \quad \text{or} \quad m = 3$$

The original expression is meaningless for 2 and for 3.

(c) $\dfrac{2r}{r^2+1}$

This denominator is never equal to zero, so there are no values for which the rational expression is meaningless. ◄

Work Problem 1 at the side.

▶ The next example shows how to find the numerical value of a rational expression for a given value of the variable.

EXAMPLE 2 Find the numerical value of $\dfrac{3x+6}{2x-4}$ for each of the following values of x.

(a) $x = 1$

Find the value of the rational expression by substituting 1 for x.
$$\dfrac{3x+6}{2x-4} = \dfrac{3(1)+6}{2(1)-4} \quad \text{Let } x = 1$$
$$= \dfrac{9}{-2}$$
$$= -\dfrac{9}{2}$$

(b) $x = 2$

Substituting 2 for x makes the denominator zero, so there is no value for the rational expression when $x = 2$. ◄

Work Problem 2 at the side.

▶ A rational expression represents a number for each value of the variable that does not make the denominator zero. For this reason, the properties of rational numbers also apply to rational expressions. For example, the **fundamental property of rational expressions** permits rational expressions to be written in lowest terms.

Fundamental Property of Rational Expressions

If $\dfrac{P}{Q}$ is a rational expression and if K represents any factor, where $K \neq 0$, then
$$\dfrac{PK}{QK} = \dfrac{P}{Q}.$$

ANSWERS

1. (a) 5 (b) $-4, -2$ (c) never meaningless
2. (a) $\dfrac{3}{7}$ (b) meaningless

This property is based on the identity property of multiplication:

$$\frac{PK}{QK} = \frac{P}{Q} \cdot \frac{K}{K} = \frac{P}{Q} \cdot 1 = \frac{P}{Q}.$$

The next example shows how to write both a rational number and a rational expression in lowest terms. Notice the similarity in the procedures.

EXAMPLE 3 Write in lowest terms.

(a) $\dfrac{30}{72}$ (b) $\dfrac{14k^2}{2k^3}$

Begin by factoring.

$$\frac{30}{72} = \frac{2 \cdot 3 \cdot 5}{2 \cdot 2 \cdot 2 \cdot 3 \cdot 3}$$

Write k^2 as $k \cdot k$.

$$\frac{14k^2}{2k^3} = \frac{2 \cdot 7 \cdot k \cdot k}{2 \cdot k \cdot k \cdot k}$$

Group any factors common to the numerator and denominator.

$$\frac{30}{72} = \frac{5 \cdot (2 \cdot 3)}{2 \cdot 2 \cdot 3 \cdot (2 \cdot 3)} \qquad \frac{14k^2}{2k^3} = \frac{7(2 \cdot k \cdot k)}{k(2 \cdot k \cdot k)}$$

Use the fundamental property.

$$\frac{30}{72} = \frac{5}{2 \cdot 2 \cdot 3} = \frac{5}{12} \qquad \frac{14k^2}{2k^3} = \frac{7}{k} \blacktriangleleft$$

Work Problem 3 at the side.

EXAMPLE 4 Write $\dfrac{3x - 12}{5x - 20}$ in lowest terms.

Begin by factoring both numerator and denominator. Then use the fundamental property.

$$\frac{3x - 12}{5x - 20} = \frac{3(x - 4)}{5(x - 4)}$$

$$= \frac{3}{5} \blacktriangleleft$$

Work Problem 4 at the side.

EXAMPLE 5 Write $\dfrac{m^2 + 2m - 8}{2m^2 - m - 6}$ in lowest terms.

Always begin by factoring both numerator and denominator, if possible. Then use the fundamental property.

$$\frac{m^2 + 2m - 8}{2m^2 - m - 6} = \frac{(m + 4)(m - 2)}{(2m + 3)(m - 2)}$$

$$= \frac{m + 4}{2m + 3} \blacktriangleleft$$

Work Problem 5 at the side.

---

3. Use the fundamental property to write the following rational expressions in lowest terms.

   (a) $\dfrac{5x^4}{15x^2}$

   (b) $\dfrac{6p^3}{2p^2}$

4. Write each rational expression in lowest terms.

   (a) $\dfrac{8p + 8q}{5p + 5q}$

   (b) $\dfrac{4y + 2}{6y + 3}$

5. Write each rational expression in lowest terms.

   (a) $\dfrac{a^2 - b^2}{a^2 + 2ab + b^2}$

   (b) $\dfrac{x^2 + 4x + 4}{4x + 8}$

ANSWERS

3. (a) $\dfrac{x^2}{3}$ (b) $3p$

4. (a) $\dfrac{8}{5}$ (b) $\dfrac{2}{3}$

5. (a) $\dfrac{a - b}{a + b}$ (b) $\dfrac{x + 2}{4}$

6.1 The Fundamental Property of Rational Expressions

6. Write each rational expression in lowest terms.

(a) $\dfrac{5-y}{y-5}$

(b) $\dfrac{m-n}{n-m}$

(c) $\dfrac{x-3}{3-x}$

(d) $\dfrac{9-k}{9+k}$

(e) $\dfrac{z^2-5}{5-z^2}$

EXAMPLE 6 Write $\dfrac{x-y}{y-x}$ in lowest terms.

At first glance, there does not seem to be any way in which $x-y$ and $y-x$ can be factored to get a common factor. However,
$$y - x = -1(-y + x)$$
$$= -1(x - y).$$
With these factors, use the fundamental property to simplify the rational expression.
$$\dfrac{x-y}{y-x} = \dfrac{1(x-y)}{-1(x-y)}$$
$$= \dfrac{1}{-1}$$
$$= -1 \blacktriangleleft$$

As suggested by Example 6,

the quotient of two nonzero expressions that differ only in sign is -1.

EXAMPLE 7 Write each rational expression in lowest terms.

(a) $\dfrac{2-m}{m-2}$

Since $2-m$ and $m-2$ (or $-2+m$) differ only in sign,
$$\dfrac{2-m}{m-2} = -1.$$

(b) $\dfrac{3+r}{3-r}$

The quantities $3+r$ and $3-r$ do not differ only in sign. This rational expression cannot be written in simpler form. $\blacktriangleleft$

Work Problem 6 at the side.

ANSWERS

6. (a) -1 (b) -1 (c) -1
   (d) cannot be written in simpler form (e) -1

6.1 EXERCISES

Find all values for which the following are meaningless. See Example 1.

1. $\dfrac{3}{4x}$
2. $\dfrac{5}{2x}$
3. $\dfrac{8}{x-4}$
4. $\dfrac{6}{x+3}$

5. $\dfrac{x^2}{x+5}$
6. $\dfrac{3x^2}{2x-1}$
7. $\dfrac{a+4}{a^2-8a+15}$
8. $\dfrac{p+6}{p^2-p-12}$

9. $\dfrac{8r+2}{2r^2-r-3}$
10. $\dfrac{7k+2}{3k^2-k-10}$
11. $\dfrac{9y}{y^2+16}$
12. $\dfrac{12z}{z^2+100}$

Find the numerical value of each rational expression when (a) $x = 2$ *and* (b) $x = -3$. *See Example 2.*

13. $\dfrac{4x-2}{3x}$
    (a)   (b)

14. $\dfrac{-5x+1}{2x}$
    (a)   (b)

15. $\dfrac{4x^2-2x}{3x}$
    (a)   (b)

16. $\dfrac{x^2-1}{x}$
    (a)   (b)

17. $\dfrac{(-8x)^2}{3x+9}$
    (a)   (b)

18. $\dfrac{2x^2+5}{3+x}$
    (a)   (b)

19. $\dfrac{x+8}{x^2-4x+2}$
    (a)   (b)

20. $\dfrac{2x-1}{x^2-7x+3}$
    (a)   (b)

21. $\dfrac{5x^2}{6-3x-x^2}$
    (a)   (b)

22. $\dfrac{-2x^2}{8+x-x^2}$
    (a)   (b)

23. $\dfrac{2x + 5}{x^2 + 3x - 10}$
    (a)   (b)

24. $\dfrac{3x - 7}{2x^2 - 3x - 2}$
    (a)   (b)

Write each rational expression in lowest terms. See Examples 3–5.

25. $\dfrac{12k^2}{6k}$

26. $\dfrac{9m^3}{3m}$

27. $\dfrac{-8y^6}{6y^3}$

28. $\dfrac{16x^4}{-8x^2}$

29. $\dfrac{12m^2p}{9mp^2}$

30. $\dfrac{6a^2b^3}{24a^3b^2}$

31. $\dfrac{8r - 12}{4}$

32. $\dfrac{9z + 6}{3}$

33. $\dfrac{12m^2 - 9}{3}$

34. $\dfrac{15p^2 - 10}{5}$

35. $\dfrac{32y + 20}{24}$

36. $\dfrac{40q - 25}{20}$

37. $\dfrac{6y + 12}{8y + 16}$

38. $\dfrac{9m + 18}{5m + 10}$

39. $\dfrac{x^2 - 1}{(x + 1)^2}$

40. $\dfrac{3t + 15}{t^2 + 4t - 5}$

41. $\dfrac{5m^2 - 5m}{10m - 10}$

42. $\dfrac{3y^2 - 3y}{2(y - 1)}$

43. $\dfrac{16r^2 - 4s^2}{4r - 2s}$

44. $\dfrac{11s^2 - 22s^3}{6 - 12s}$

45. $\dfrac{m^2 - 4m + 4}{m^2 + m - 6}$

46. $\dfrac{a^2 - a - 6}{a^2 + a - 12}$

47. $\dfrac{x^2 + 3x - 4}{x^2 - 1}$

48. $\dfrac{8m^2 + 6m - 9}{16m^2 - 9}$

Write each rational expression in lowest terms. See Examples 6 and 7.

49. $\dfrac{m - 5}{5 - m}$

50. $\dfrac{3 - p}{p - 3}$

51. $\dfrac{x^2 - 1}{1 - x}$

52. $\dfrac{p^2 - q^2}{q - p}$

53. $\dfrac{m^2 - 4m}{4m - m^2}$

54. $\dfrac{s^2 - r^2}{r^2 - s^2}$

284   Rational Expressions

6.2 MULTIPLICATION AND DIVISION OF RATIONAL EXPRESSIONS

▶ The product of two fractions is found by multiplying the numerators and multiplying the denominators. Rational expressions are multiplied in the same way.

Objectives

▶ Multiply rational expressions.
▶ Divide rational expressions.

> **Multiplying Rational Expressions**
> The product of the rational expressions $\frac{P}{Q}$ and $\frac{R}{S}$ is
> $$\frac{P}{Q} \cdot \frac{R}{S} = \frac{PR}{QS}.$$

The next example shows the multiplication of both two rational numbers and two rational expressions. This parallel discussion lets you compare the steps.

EXAMPLE 1 Multiply. Write answers in lowest terms.

(a) $\dfrac{3}{10} \cdot \dfrac{5}{9}$ (b) $\dfrac{6}{x} \cdot \dfrac{x^2}{12}$

Find the product of the numerators and the product of the denominators.

$$\frac{3}{10} \cdot \frac{5}{9} = \frac{3 \cdot 5}{10 \cdot 9} \qquad \frac{6}{x} \cdot \frac{x^2}{12} = \frac{6 \cdot x^2}{x \cdot 12}$$

Use the fundamental property to write each product in lowest terms.

$$\frac{3}{10} \cdot \frac{5}{9} = \frac{3 \cdot 5}{2 \cdot 5 \cdot 3 \cdot 3} = \frac{1}{6} \qquad \frac{6}{x} \cdot \frac{x^2}{12} = \frac{6 \cdot x \cdot x}{2 \cdot 6 \cdot x} = \frac{x}{2}$$

Notice in the second step above that the products were left in factored form, since common factors are needed to write the product in lowest terms. ◀

Work Problem 1 at the side.

EXAMPLE 2 Find the product of $\dfrac{x+y}{2x}$ and $\dfrac{x^2}{(x+y)^2}$.

Use the definition of multiplication.

$$\frac{x+y}{2x} \cdot \frac{x^2}{(x+y)^2} = \frac{(x+y)x^2}{2x(x+y)^2}$$
$$= \frac{(x+y)x \cdot x}{2x(x+y)(x+y)}$$
$$= \frac{x}{2(x+y)} \cdot \frac{x(x+y)}{x(x+y)}$$
$$= \frac{x}{2(x+y)}$$

Write the product in lowest terms by factoring and using the fundamental property of rational expressions. ◀

Work Problem 2 at the side.

1. Multiply.

(a) $\dfrac{3m^2}{2} \cdot \dfrac{10}{m}$

(b) $\dfrac{8p^2q}{3} \cdot \dfrac{9}{pq^2}$

2. Multiply.

(a) $\dfrac{a+b}{5} \cdot \dfrac{30}{2(a+b)}$

(b) $\dfrac{3(p-q)}{p} \cdot \dfrac{q}{2(p-q)}$

ANSWERS

1. (a) $15m$ (b) $\dfrac{24p}{q}$

2. (a) 3 (b) $\dfrac{3q}{2p}$

3. Multiply.

(a) $\dfrac{x^2 + 7x + 10}{3x + 6} \cdot \dfrac{6x - 6}{x^2 + 2x - 15}$

(b) $\dfrac{m^2 + 4m - 5}{m + 5} \cdot \dfrac{m^2 + 8m + 15}{m - 1}$

4. Divide.

(a) $\dfrac{r}{r - 1} \div \dfrac{3r}{r + 4}$

(b) $\dfrac{9p^2}{3p + 4} \div \dfrac{6p^3}{3p + 4}$

(c) $\dfrac{6x - 4}{3} \div \dfrac{15x - 10}{9}$

ANSWERS

3. (a) $\dfrac{2(x - 1)}{x - 3}$ (b) $(m + 5)(m + 3)$

4. (a) $\dfrac{r + 4}{3(r - 1)}$ (b) $\dfrac{3}{2p}$ (c) $\dfrac{6}{5}$

286

EXAMPLE 3 Find the product of $\dfrac{x^2 + 3x}{x^2 - 3x - 4}$ and $\dfrac{x^2 - 5x + 4}{x^2 + 2x - 3}$.

Before multiplying, factor the numerators and denominators whenever possible. Then use the fundamental property to write the product in lowest terms.

$$\dfrac{x^2 + 3x}{x^2 - 3x - 4} \cdot \dfrac{x^2 - 5x + 4}{x^2 + 2x - 3} = \dfrac{x(x + 3)}{(x - 4)(x + 1)} \cdot \dfrac{(x - 4)(x - 1)}{(x + 3)(x - 1)}$$

$$= \dfrac{x(x + 3)(x - 4)(x - 1)}{(x - 4)(x + 1)(x + 3)(x - 1)}$$

$$= \dfrac{x}{x + 1} \blacktriangleleft$$

Work Problem 3 at the side.

▶ The fraction $\dfrac{a}{b}$ is divided by the nonzero fraction $\dfrac{c}{d}$ by multiplying $\dfrac{a}{b}$ and the reciprocal of $\dfrac{c}{d}$, which is $\dfrac{d}{c}$. Division of rational expressions is defined in the same way.

Dividing Rational Expressions

If $\dfrac{P}{Q}$ and $\dfrac{R}{S}$ are any two rational expressions, with $\dfrac{R}{S} \neq 0$, then

$$\dfrac{P}{Q} \div \dfrac{R}{S} = \dfrac{P}{Q} \cdot \dfrac{S}{R} = \dfrac{PS}{QR}.$$

The next example shows the division of two rational numbers and the division of two rational expressions.

EXAMPLE 4 Divide. Write answers in lowest terms.

(a) $\dfrac{5}{8} \div \dfrac{7}{16}$ (b) $\dfrac{y}{y + 3} \div \dfrac{4y}{y + 5}$

Multiply the first expression and the reciprocal of the second.

$\dfrac{5}{8} \div \dfrac{7}{16} = \dfrac{5}{8} \cdot \boxed{\dfrac{16}{7}}$

Reciprocal of $\dfrac{7}{16}$

$= \dfrac{5 \cdot 16}{8 \cdot 7}$

$= \dfrac{5 \cdot 8 \cdot 2}{8 \cdot 7}$

$= \dfrac{5 \cdot 2}{7}$

$= \dfrac{10}{7}$

$\dfrac{y}{y + 3} \div \dfrac{4y}{y + 5}$

$= \dfrac{y}{y + 3} \cdot \boxed{\dfrac{y + 5}{4y}}$

Reciprocal of $\dfrac{4y}{y + 5}$

$= \dfrac{y(y + 5)}{(y + 3)(4y)}$

$= \dfrac{y + 5}{4(y + 3)} \blacktriangleleft$

Work Problem 4 at the side.

Rational Expressions

EXAMPLE 5 Divide: $\dfrac{(3m)^2}{(2p)^3} \div \dfrac{6m^3}{16p^2}$.

Use the properties of exponents as necessary.

$$\dfrac{(3m)^2}{(2p)^3} \div \dfrac{6m^3}{16p^2} = \dfrac{9m^2}{8p^3} \div \dfrac{6m^3}{16p^2}$$

$$= \dfrac{9m^2}{8p^3} \cdot \dfrac{16p^2}{6m^3} \quad \text{Multiply by reciprocal}$$

$$= \dfrac{9 \cdot 16m^2 p^2}{8 \cdot 6p^3 m^3} \quad \text{Multiply the fractions}$$

$$= \dfrac{3}{mp} \quad \text{Fundamental property} \blacktriangleleft$$

Work Problem 5 at the side.

EXAMPLE 6 Divide: $\dfrac{x^2 - 4}{(x+3)(x-2)} \div \dfrac{(x+2)(x+3)}{2x}$.

First, use the definition of division.

$$\dfrac{x^2 - 4}{(x+3)(x-2)} \div \dfrac{(x+2)(x+3)}{2x} = \dfrac{x^2 - 4}{(x+3)(x-2)} \cdot \boxed{\dfrac{2x}{(x+2)(x+3)}}$$

Reciprocal of second expression

Next, be sure all numerators and all denominators are factored.

$$\dfrac{x^2 - 4}{(x+3)(x-2)} \div \dfrac{(x+2)(x+3)}{2x} = \dfrac{(x+2)(x-2)}{(x+3)(x-2)} \cdot \dfrac{2x}{(x+2)(x+3)}$$

Now multiply numerators and denominators and simplify.

$$\dfrac{(x+2)(x-2)}{(x+3)(x-2)} \cdot \dfrac{2x}{(x+2)(x+3)} = \dfrac{(x+2)(x-2)(2x)}{(x+3)(x-2)(x+2)(x+3)}$$

$$= \dfrac{2x}{(x+3)^2} \blacktriangleleft$$

Work Problem 6 at the side.

EXAMPLE 7 Divide: $\dfrac{m^2 - 4}{m^2 - 1} \div \dfrac{2m^2 + 4m}{1 - m}$.

Use the definition of division. Then factor.

$$\dfrac{m^2 - 4}{m^2 - 1} \div \dfrac{2m^2 + 4m}{1 - m} = \dfrac{m^2 - 4}{m^2 - 1} \cdot \dfrac{1 - m}{2m^2 + 4m}$$

$$= \dfrac{(m+2)(m-2)}{(m+1)(m-1)} \cdot \dfrac{1 - m}{2m(m+2)}$$

---

5. Divide.

(a) $\dfrac{5a^2 b}{2} \div \dfrac{10ab^2}{8}$

(b) $\dfrac{(3t)^2}{w} \div \dfrac{3t^2}{5w^4}$

6. Divide.

(a) $\dfrac{y^2 + 4y + 3}{y + 3} \div \dfrac{y^2 - 4y - 5}{y - 3}$

(b) $\dfrac{4x(x+3)}{2x + 1} \div \dfrac{x^2(x+3)}{4x^2 - 1}$

ANSWERS

5. (a) $\dfrac{2a}{b}$ (b) $15w^3$

6. (a) $\dfrac{y - 3}{y - 5}$ (b) $\dfrac{4(2x - 1)}{x}$

6.2 Multiplication and Division of Rational Expressions

7. Divide.

(a) $\dfrac{x^2 - y^2}{x^2 - 1} \div \dfrac{x^2 + 2xy + y^2}{x^2 + x}$

(b) $\dfrac{ab - a^2}{a^2 - 1} \div \dfrac{b - a}{a^2 + 2a + 1}$

As shown in Section 6.1, $\dfrac{1 - m}{m - 1} = -1$, so

$$\dfrac{(m + 2)(m - 2)}{(m + 1)(m - 1)} \cdot \dfrac{1 - m}{2m(m + 2)} = \dfrac{-1(m - 2)}{2m(m + 1)}$$
$$= \dfrac{2 - m}{2m(m + 1)}. \blacktriangleleft$$

Work Problem 7 at the side.

ANSWERS

7. (a) $\dfrac{x(x - y)}{(x - 1)(x + y)}$

(b) $\dfrac{a(a + 1)}{a - 1}$

name   date   hour

6.2 EXERCISES

Multiply or divide. Write each answer in lowest terms. See Examples 1 and 5.

1. $\dfrac{9m^2}{16} \cdot \dfrac{4}{3m}$

2. $\dfrac{21z^4}{8} \cdot \dfrac{12}{7z^3}$

3. $\dfrac{4p^2}{8p} \cdot \dfrac{3p^3}{16p^4}$

4. $\dfrac{6x^3}{9x} \cdot \dfrac{12x}{x^2}$

5. $\dfrac{8a^4}{12a^3} \cdot \dfrac{9a^5}{3a^2}$

6. $\dfrac{14p^5}{2p^2} \cdot \dfrac{8p^6}{28p^3}$

7. $\dfrac{3r^2}{9r^3} \div \dfrac{8r^4}{6r^5}$

8. $\dfrac{25m^{10}}{9m^5} \div \dfrac{15m^6}{10m^4}$

9. $\dfrac{3m^2}{(4m)^3} \div \dfrac{9m^3}{32m^4}$

10. $\dfrac{5x^3}{(4x)^2} \div \dfrac{15x^2}{8x^4}$

11. $\dfrac{-6r^4}{3r^5} \div \dfrac{(2r^2)^2}{-4}$

12. $\dfrac{-10a^6}{3a^2} \div \dfrac{(3a)^3}{81a}$

Multiply or divide. Write each answer in lowest terms. See Examples 2, 3, 4, 6, and 7.

13. $\dfrac{a+b}{2} \cdot \dfrac{12}{(a+b)^2}$

14. $\dfrac{3(x-1)}{y} \cdot \dfrac{2y}{5(x-1)}$

15. $\dfrac{a-3}{16} \div \dfrac{a-3}{32}$

16. $\dfrac{9}{8-2y} \div \dfrac{3}{4-y}$

17. $\dfrac{2k+8}{6} \div \dfrac{3k+12}{2}$

18. $\dfrac{5m+25}{10} \cdot \dfrac{12}{6m+30}$

19. $\dfrac{9y-18}{6y+12} \cdot \dfrac{3y+6}{15y-30}$

20. $\dfrac{12p+24}{36p-36} \div \dfrac{6p+12}{8p-8}$

21. $\dfrac{3r+12}{8} \cdot \dfrac{16r}{9r+36}$

22. $\dfrac{2r+2p}{8z} \div \dfrac{r^2+rp}{72}$

23. $\dfrac{y^2-16}{y+3} \div \dfrac{y-4}{y^2-9}$

24. $\dfrac{9(y-4)^2}{8(z+3)^2} \cdot \dfrac{16(z+3)}{3(y-4)}$

25. $\dfrac{6(m+2)}{3(m-1)^2} \div \dfrac{(m+2)^2}{9(m-1)}$

26. $\dfrac{4y+12}{2y-10} \div \dfrac{y^2-9}{y^2-y-20}$

27. $\dfrac{2-y}{8} \cdot \dfrac{7}{y-2}$

28. $\dfrac{9-2z}{3} \cdot \dfrac{9}{2z-9}$

29. $\dfrac{8-r}{8+r} \div \dfrac{r-8}{r+8}$

30. $\dfrac{6r-18}{3r^2+2r-8} \cdot \dfrac{12r-16}{4r-12}$

31. $\dfrac{a^2+2a-15}{a^2+3a-10} \div \dfrac{a-3}{a+1}$

32. $\dfrac{y^2+y-2}{y^2+3y-4} \div \dfrac{y+2}{y+3}$

33. $\dfrac{k^2-k-6}{k^2+k-12} \div \dfrac{k^2+2k-3}{k^2+3k-4}$

34. $\dfrac{m^2+3m+2}{m^2+5m+4} \div \dfrac{m^2+10m+24}{m^2+5m+6}$

35. $\dfrac{z^2-z-6}{z^2-2z-8} \cdot \dfrac{z^2+7z+12}{z^2-9}$

36. $\dfrac{p^2-4p+3}{p^2-3p+2} \div \dfrac{p-3}{p+5}$

37. $\dfrac{2k^2+3k-2}{6k^2-7k+2} \cdot \dfrac{4k^2-5k+1}{k^2+k-2}$

38. $\dfrac{6n^2-5n-6}{6n^2+5n-6} \cdot \dfrac{12n^2-17n+6}{12n^2-n-6}$

39. $\dfrac{2m^2-5m-12}{m^2-10m+24} \div \dfrac{4m^2-9}{m^2-9m+18}$

40. $\dfrac{2m^2+7m+3}{m^2-9} \cdot \dfrac{m^2-3m}{2m^2+11m+5}$

41. $\dfrac{m^2+2mp-3p^2}{m^2-3mp+2p^2} \div \dfrac{m^2+4mp+3p^2}{m^2+2mp-8p^2}$

42. $\dfrac{r^2+rs-12s^2}{r^2-rs-20s^2} \div \dfrac{r^2-2rs-3s^2}{r^2+rs-30s^2}$

43. $\left(\dfrac{x^2+10x+25}{x^2+10x} \cdot \dfrac{10x}{x^2+15x+50}\right) \div \dfrac{x+5}{x+10}$

44. $\left(\dfrac{m^2-12m+32}{8m} \cdot \dfrac{m^2-8m}{m^2-8m+16}\right) \div \dfrac{m-8}{m-4}$

6.3 LEAST COMMON DENOMINATORS

▶ Just as with rational numbers, adding or subtracting rational expressions (to be discussed in the next section) often requires a **least common denominator**, the least expression that all denominators divide into without a remainder. For example, the least common denominator for $\frac{2}{9}$ and $\frac{5}{12}$ is 36, since 36 is the smallest number that both 9 and 12 divide into. Least common denominators are found by a procedure similar to that used in Chapter 5 for finding the greatest common factor, except that

> least common denominators are found by multiplying together each different factor the *greatest* number of times it appears in any denominator.

In Example 1, the least common denominator is found both for numerical denominators and algebraic denominators.

EXAMPLE 1 Find the least common denominator for each pair of fractions.

(a) $\frac{1}{24}, \frac{7}{15}$ (b) $\frac{1}{8x}, \frac{3}{10x}$

Write each denominator in factored form, with numerical coefficients in prime factored form.

$24 = 2 \cdot 2 \cdot 2 \cdot 3$ $8x = 2 \cdot 2 \cdot 2 \cdot x$
$15 = 3 \cdot 5$ $10x = 2 \cdot 5 \cdot x$

The least common denominator is found by taking each different factor the greatest number of times it appears as a factor in any of the denominators.

The factor 2 appears three times in one product and not at all in the other, so the greatest number of times 2 appears is three. The greatest number of times both 3 and 5 appear is one.

least common denominator
$= 2 \cdot 2 \cdot 2 \cdot 3 \cdot 5 = 120$

Here 2 appears three times in one product and once in the other, so the greatest number of times the 2 appears is three. The greatest number of times the 5 appears is one, and the greatest number of times x appears in either product is one.

least common denominator
$= 2 \cdot 2 \cdot 2 \cdot 5 \cdot x = 40x$ ◀

Work Problem 1 at the side.

EXAMPLE 2 Find the least common denominator.

(a) $\frac{6}{5m}, \frac{4}{m^2 - 3m}$

Factor each denominator.

$5m = 5 \cdot m \qquad m^2 - 3m = m(m - 3)$

Take each different factor the greatest number of times it appears as a factor.

least common denominator $= 5 \cdot m \cdot (m - 3) = 5m(m - 3)$

Objectives

▶ Find least common denominators.

▶ Rewrite rational expressions with the least common denominator.

1. Find the least common denominator.

(a) $\frac{7}{20p}, \frac{11}{30p}$

(b) $\frac{13}{12m}, \frac{17}{30m}$

(c) $\frac{19}{60r}, \frac{5}{72r}$

(d) $\frac{9}{8m^4}, \frac{11}{12m^6}$

(e) $\frac{8}{15p^8}, \frac{7}{20p^2}$

ANSWERS
1. (a) $60p$ (b) $60m$ (c) $360r$
   (d) $24m^6$ (e) $60p^8$

6.3 Least Common Denominators

2. Find the least common denominator.

(a) $\dfrac{1}{12a}, \dfrac{5}{a^2 - 4a}$

(b) $\dfrac{6}{x^2 - 4x}, \dfrac{3x - 1}{x^2 - 16}$

(c) $\dfrac{2m}{m^2 - 3m + 2}, \dfrac{5m - 3}{m^2 + 3m - 10}$

3. Rewrite each rational expression with the indicated denominator.

(a) $\dfrac{7k}{5} = \dfrac{}{30}$

(b) $\dfrac{9}{2a + 5} = \dfrac{}{6a + 15}$

(c) $\dfrac{5k + 1}{k^2 + 2k} = \dfrac{}{k(k + 2)(k - 1)}$

ANSWERS
2. (a) $12a(a - 4)$
   (b) $x(x - 4)(x + 4)$
   (c) $(m - 1)(m - 2)(m + 5)$
3. (a) $\dfrac{42k}{30}$ (b) $\dfrac{27}{6a + 15}$
   (c) $\dfrac{(5k + 1)(k - 1)}{k(k + 2)(k - 1)}$

Since m is *not* a factor of $m - 3$, both factors, m and $m - 3$, must appear in the least common denominator.

(b) $\dfrac{1}{r^2 - 4r - 5}, \dfrac{3}{r^2 - r - 20}$

Factor each denominator.
$$r^2 - 4r - 5 = (r - 5)(r + 1)$$
$$r^2 - r - 20 = (r - 5)(r + 4)$$

The least common denominator is
$$(r - 5)(r + 1)(r + 4).$$

(c) $\dfrac{1}{q - 5}, \dfrac{3}{5 - q}$

The expression $5 - q$ can be written as $-1(q - 5)$, since
$$-1(q - 5) = -q + 5 = 5 - q.$$

Because of this, either $q - 5$ or $5 - q$ can be used as the least common denominator. ◀

Work Problem 2 at the side.

The steps in finding a least common denominator are listed below.

Finding the Least Common Denominator

Step 1 Factor all denominators completely.

Step 2 Take each different factor the *greatest* number of times that it appears as a factor in any of the denominators.

Step 3 The least common denominator is the product of all factors to the greatest power found in Step 2.

▶ Once the least common denominator has been found, use the fundamental property to rewrite rational expressions with the least common denominator.

EXAMPLE 3 Rewrite each rational expression with the indicated denominator.

(a) $\dfrac{9k}{25} = \dfrac{}{50k}$

The given rational expression has a denominator of 25. To get a denominator of $50k$, multiply both numerator and denominator by $2k$.
$$\dfrac{9k}{25} = \dfrac{9k\,(2k)}{25\,(2k)} = \dfrac{18k^2}{50k}$$

(b) $\dfrac{12p}{p^2 + 8p} = \dfrac{}{p(p + 8)(p - 4)}$

Factor $p^2 + 8p$ as $p(p + 8)$; then multiply by $p - 4$.
$$\dfrac{12p}{p^2 + 8p} = \dfrac{12p}{p(p + 8)} \cdot \dfrac{p - 4}{p - 4} = \dfrac{12p(p - 4)}{p(p + 8)(p - 4)} \blacktriangleleft$$

Work Problem 3 at the side.

6.3 EXERCISES

Find the least common denominator for each list of rational expressions. See Examples 1 and 2.

1. $\dfrac{5}{12}, \dfrac{7}{10}$

2. $\dfrac{1}{4}, \dfrac{5}{6}$

3. $\dfrac{7}{15}, \dfrac{11}{20}, \dfrac{5}{24}$

4. $\dfrac{9}{10}, \dfrac{12}{25}, \dfrac{11}{35}$

5. $\dfrac{17}{100}, \dfrac{13}{120}, \dfrac{29}{180}$

6. $\dfrac{17}{250}, \dfrac{1}{300}, \dfrac{127}{360}$

7. $\dfrac{9}{x^2}, \dfrac{8}{x^5}$

8. $\dfrac{2}{m^7}, \dfrac{3}{m^8}$

9. $\dfrac{2}{5p}, \dfrac{5}{6p}$

10. $\dfrac{4}{15k}, \dfrac{3}{4k}$

11. $\dfrac{2}{25m}, \dfrac{17}{30m}$

12. $\dfrac{9}{40a}, \dfrac{11}{60a}$

13. $\dfrac{7}{15y^2}, \dfrac{5}{36y^4}$

14. $\dfrac{4}{25m^3}, \dfrac{7}{10m^4}$

15. $\dfrac{3}{50r^5}, \dfrac{7}{60r^3}$

16. $\dfrac{9}{25a^5}, \dfrac{17}{80a^6}$

17. $\dfrac{1}{5a^2b^3}, \dfrac{2}{15a^5b}$

18. $\dfrac{7}{3r^4s^5}, \dfrac{2}{9r^6s^8}$

19. $\dfrac{7}{6p}, \dfrac{3}{4p-8}$

20. $\dfrac{7}{8k}, \dfrac{13}{12k-24}$

21. $\dfrac{5}{32r^2}, \dfrac{9}{16r-32}$

22. $\dfrac{13}{18m^3}, \dfrac{8}{9m-36}$

23. $\dfrac{7}{6r-12}, \dfrac{4}{9r-18}$

24. $\dfrac{4}{5p-30}, \dfrac{5}{6p-36}$

25. $\dfrac{5}{12p+60}, \dfrac{1}{p^2+5p}$

26. $\dfrac{1}{r^2+7r}, \dfrac{3}{5r+35}$

27. $\dfrac{9}{p-q}, \dfrac{3}{q-p}$

28. $\dfrac{4}{z-x}, \dfrac{8}{x-z}$

29. $\dfrac{6}{a^2 + 6a}$, $\dfrac{4}{a^2 + 3a - 18}$

30. $\dfrac{3}{y^2 - 5y}$, $\dfrac{2}{y^2 - 2y - 15}$

31. $\dfrac{5}{k^2 + 2k - 35}$, $\dfrac{3}{k^2 + 3k - 40}$

32. $\dfrac{9}{z^2 + 4z - 12}$, $\dfrac{1}{z^2 + z - 30}$

33. $\dfrac{3}{2y^2 + 7y - 4}$, $\dfrac{7}{2y^2 - 7y + 3}$

34. $\dfrac{2}{5a^2 + 13a - 6}$, $\dfrac{6}{5a^2 - 22a + 8}$

35. $\dfrac{1}{6r^2 - r - 15}$, $\dfrac{4}{3r^2 - 8r + 5}$

36. $\dfrac{8}{2m^2 - 11m + 14}$, $\dfrac{2}{2m^2 - m - 21}$

Rewrite each rational expression with the given denominator. See Example 3.

37. $\dfrac{7}{11} = \dfrac{}{66}$

38. $\dfrac{5}{8} = \dfrac{}{56}$

39. $\dfrac{9}{r} = \dfrac{}{6r}$

40. $\dfrac{7}{k} = \dfrac{}{9k}$

41. $\dfrac{-11}{m} = \dfrac{}{8m}$

42. $\dfrac{-5}{z} = \dfrac{}{6z}$

43. $\dfrac{12}{35y} = \dfrac{}{70y^3}$

44. $\dfrac{17}{9r} = \dfrac{}{36r^2}$

45. $\dfrac{15m^2}{8k} = \dfrac{}{32k^4}$

46. $\dfrac{5t^2}{3y} = \dfrac{}{9y^2}$

47. $\dfrac{19z}{2z - 6} = \dfrac{}{6z - 18}$

48. $\dfrac{2r}{5r - 5} = \dfrac{}{15r - 15}$

49. $\dfrac{-2a}{9a - 18} = \dfrac{}{18a - 36}$

50. $\dfrac{-5y}{6y + 18} = \dfrac{}{24y + 72}$

51. $\dfrac{6}{k^2 - 4k} = \dfrac{}{k(k - 4)(k + 1)}$

52. $\dfrac{15}{m^2 - 9m} = \dfrac{}{m(m - 9)(m + 8)}$

Rational Expressions

6.4 ADDITION AND SUBTRACTION OF RATIONAL EXPRESSIONS

Objectives
1. Add rational expressions having the same denominator.
2. Add rational expressions having different denominators.
3. Subtract rational expressions.

▶ The sum of two rational expressions is found with a procedure similar to that for adding two fractions.

Adding Rational Expressions

If $\dfrac{P}{Q}$ and $\dfrac{R}{Q}$ are rational expressions, then

$$\dfrac{P}{Q} + \dfrac{R}{Q} = \dfrac{P+R}{Q}.$$

Again, the first example shows how the addition of rational expressions compares with that of rational numbers.

EXAMPLE 1 Add.

(a) $\dfrac{4}{7} + \dfrac{2}{7}$

(b) $\dfrac{3x}{x+1} + \dfrac{2x}{x+1}$

The denominators are the same, so the sum is found by adding the two numerators and keeping the same (common) denominator.

$$\dfrac{4}{7} + \dfrac{2}{7} = \dfrac{4+2}{7} = \dfrac{6}{7}$$

$$\dfrac{3x}{x+1} + \dfrac{2x}{x+1} = \dfrac{3x+2x}{x+1} = \dfrac{5x}{x+1} \blacktriangleleft$$

Work Problem 1 at the side.

1. Find each sum.

(a) $\dfrac{3}{y+4} + \dfrac{2}{y+4}$

(b) $\dfrac{x}{x+y} + \dfrac{1}{x+y}$

▶ Use the steps given below to add two rational expressions with different denominators. These are the same steps that are used to add fractions with different denominators.

Adding Rational Expressions with Different Denominators

Step 1 Find the least common denominator.

Step 2 Rewrite each rational expression as a fraction with the least common denominator as the denominator.

Step 3 Add the numerators to get the numerator of the sum. The least common denominator is the denominator of the sum.

Step 4 Write the answer in lowest terms.

EXAMPLE 2 Add.

(a) $\dfrac{1}{12} + \dfrac{7}{15}$

(b) $\dfrac{2}{3y} + \dfrac{1}{4y}$

First find the least common denominator, using the methods of the last section.

least common denominator
$= 2^2 \cdot 3 \cdot 5 = 60$

least common denominator
$= 2^2 \cdot 3 \cdot y = 12y$

ANSWERS

1. (a) $\dfrac{5}{y+4}$ (b) $\dfrac{x+1}{x+y}$

2. Find each sum.

   (a) $\dfrac{6}{5x} + \dfrac{9}{2x}$

   (b) $\dfrac{m}{3n} + \dfrac{2}{7n}$

3. Find the sums.

   (a) $\dfrac{2p}{3p+3} + \dfrac{5p}{2p+2}$

   (b) $\dfrac{4}{y^2-1} + \dfrac{6}{y+1}$

   (c) $\dfrac{-9}{p+1} + \dfrac{p}{p^2-1}$

ANSWERS

2. (a) $\dfrac{57}{10x}$ (b) $\dfrac{7m+6}{21n}$

3. (a) $\dfrac{19p}{6(p+1)}$

   (b) $\dfrac{6y-2}{(y+1)(y-1)}$

   (c) $\dfrac{-8p+9}{(p+1)(p-1)}$

Now rewrite each rational expression as a fraction with the least common denominator, either 60 or 12y, as the denominator.

$$\dfrac{1}{12} + \dfrac{7}{15} = \dfrac{1(5)}{12(5)} + \dfrac{7(4)}{15(4)} \qquad \dfrac{2}{3y} + \dfrac{1}{4y} = \dfrac{2(4)}{3y(4)} + \dfrac{1(3)}{4y(3)}$$

$$= \dfrac{5}{60} + \dfrac{28}{60} \qquad\qquad\qquad = \dfrac{8}{12y} + \dfrac{3}{12y}$$

Since the fractions now have common denominators, add the numerators. (Write in lowest terms.)

$$\dfrac{5}{60} + \dfrac{28}{60} = \dfrac{5+28}{60} \qquad\qquad \dfrac{8}{12y} + \dfrac{3}{12y} = \dfrac{8+3}{12y}$$

$$= \dfrac{33}{60} = \dfrac{11}{20} \qquad\qquad\qquad = \dfrac{11}{12y} \blacktriangleleft$$

Work Problem 2 at the side.

EXAMPLE 3 Add $\dfrac{x}{x^2-1}$ and $\dfrac{x}{x+1}$.

Find the least common denominator by factoring both denominators.

$$x^2 - 1 = (x+1)(x-1); \qquad x+1 \text{ cannot be factored}$$

Write the sum with denominators in factored form as

$$\dfrac{x}{(x+1)(x-1)} + \dfrac{x}{x+1}.$$

The least common denominator is $(x+1)(x-1)$. Here only the second fraction must be changed. Multiply the numerator and denominator of the second fraction by $x-1$.

$$\dfrac{x}{(x+1)(x-1)} + \dfrac{x(x-1)}{(x+1)(x-1)} \qquad \text{Multiply by } x-1$$

With both denominators now the same, add the numerators.

$$\dfrac{x + x(x-1)}{(x+1)(x-1)} = \dfrac{x + x^2 - x}{(x+1)(x-1)} \qquad \text{Add numerators}$$

$$= \dfrac{x^2}{(x+1)(x-1)}$$

It is usually most convenient to leave the denominator in factored form. ◀

Work Problem 3 at the side.

EXAMPLE 4 Add $\dfrac{2x}{x^2+5x+6}$ and $\dfrac{x+1}{x^2+2x-3}$.

Begin by factoring the denominators.

$$\dfrac{2x}{(x+2)(x+3)} + \dfrac{x+1}{(x+3)(x-1)}$$

Rational Expressions

The least common denominator is $(x + 2)(x + 3)(x - 1)$. By the fundamental property,

$$\frac{2x}{(x + 2)(x + 3)} + \frac{x + 1}{(x + 3)(x - 1)}$$

$$= \frac{2x(x - 1)}{(x + 2)(x + 3)(x - 1)} + \frac{(x + 1)(x + 2)}{(x + 3)(x - 1)(x + 2)}$$

Since the two rational expressions above have the same denominator, add their numerators.

$$\frac{2x(x - 1)}{(x + 2)(x + 3)(x - 1)} + \frac{(x + 1)(x + 2)}{(x + 3)(x - 1)(x + 2)}$$

$$= \frac{2x(x - 1) + (x + 1)(x + 2)}{(x + 2)(x + 3)(x - 1)}$$

$$= \frac{2x^2 - 2x + x^2 + 3x + 2}{(x + 2)(x + 3)(x - 1)}$$

$$= \frac{3x^2 + x + 2}{(x + 2)(x + 3)(x - 1)} \blacktriangleleft$$

Work Problem 4 at the side.

3 Subtract rational expressions as follows.

> **Subtracting Rational Expressions**
> If $\frac{P}{Q}$ and $\frac{R}{Q}$ are rational expressions, then
> $$\frac{P}{Q} - \frac{R}{Q} = \frac{P - R}{Q}.$$

EXAMPLE 5 Subtract: $\frac{12}{m^2} - \frac{8}{m^2}$.

By the definition of subtraction,

$$\frac{12}{m^2} - \frac{8}{m^2} = \frac{12 - 8}{m^2} = \frac{4}{m^2}. \blacktriangleleft$$

Work Problem 5 at the side.

EXAMPLE 6 Subtract: $\frac{9}{x - 2} - \frac{3}{x}$.

The least common denominator is $x(x - 2)$.

$$\frac{9}{x - 2} - \frac{3}{x} = \frac{9x}{x(x - 2)} - \frac{3(x - 2)}{x(x - 2)} \quad \text{Get least common denominator}$$

$$= \frac{9x - 3(x - 2)}{x(x - 2)} \quad \text{Subtract}$$

$$= \frac{9x - 3x + 6}{x(x - 2)} \quad \text{Distributive property}$$

$$= \frac{6x + 6}{x(x - 2)}$$

$$= \frac{6(x + 1)}{x(x - 2)} \quad \text{Factor} \blacktriangleleft$$

4. Add.

(a) $\dfrac{2k}{k^2 - 5k + 4} + \dfrac{3}{k^2 - 1}$

(b) $\dfrac{4m}{m^2 + 3m + 2} + \dfrac{2m - 1}{m^2 + 6m + 5}$

5. Find each difference. Write answers in lowest terms.

(a) $\dfrac{5}{2y} - \dfrac{9}{2y}$

(b) $\dfrac{3}{m^2} - \dfrac{2}{m^2}$

ANSWERS

4. (a) $\dfrac{2k^2 + 5k - 12}{(k - 4)(k - 1)(k + 1)}$

(b) $\dfrac{6m^2 + 23m - 2}{(m + 2)(m + 1)(m + 5)}$

5. (a) $\dfrac{-2}{y}$ (b) $\dfrac{1}{m^2}$

6.4 Addition and Subtraction of Rational Expressions

6. Subtract.

   (a) $\dfrac{9}{5p} - \dfrac{2}{3p}$

   (b) $\dfrac{4}{m^2} - \dfrac{3}{m}$

   (c) $\dfrac{1}{k+4} - \dfrac{2}{k}$

   (d) $\dfrac{6}{a+2} - \dfrac{1}{a-3}$

7. Subtract.

   (a) $\dfrac{4y}{y^2-1} - \dfrac{5}{y^2+2y+1}$

   (b) $\dfrac{3r}{r^2-5r} - \dfrac{4}{r^2-10r+25}$

8. Subtract.

   (a) $\dfrac{2}{p^2-5p+4} - \dfrac{3}{p^2-1}$

   (b) $\dfrac{q}{2q^2+5q-3} - \dfrac{3q+4}{3q^2+10q+3}$

ANSWERS

6. (a) $\dfrac{17}{15p}$ (b) $\dfrac{4-3m}{m^2}$
   (c) $\dfrac{-k-8}{k(k+4)}$ (d) $\dfrac{5a-20}{(a+2)(a-3)}$

7. (a) $\dfrac{4y^2-y+5}{(y+1)^2(y-1)}$
   (b) $\dfrac{3r-19}{(r-5)^2}$

8. (a) $\dfrac{14-p}{(p-4)(p-1)(p+1)}$
   (b) $\dfrac{-3q^2-4q+4}{(2q-1)(q+3)(3q+1)}$

298

Work Problem 6 at the side.

EXAMPLE 7 Find $\dfrac{6x}{x^2-2x+1} - \dfrac{1}{x^2-1}$.

Begin by factoring.

$$\dfrac{6x}{x^2-2x+1} - \dfrac{1}{x^2-1} = \dfrac{6x}{(x-1)(x-1)} - \dfrac{1}{(x-1)(x+1)}$$

From the factored denominators, identify the common denominator, $(x-1)(x-1)(x+1)$. Use the factor $x-1$ twice, since it appears twice in the first denominator.

$$\dfrac{6x}{(x-1)(x-1)} - \dfrac{1}{(x-1)(x+1)}$$

$$= \dfrac{6x(x+1)}{(x-1)(x-1)(x+1)} - \dfrac{1(x-1)}{(x-1)(x-1)(x+1)} \quad \text{Fundamental property}$$

$$= \dfrac{6x(x+1) - 1(x-1)}{(x-1)(x-1)(x+1)} \quad \text{Subtract}$$

$$= \dfrac{6x^2+6x-x+1}{(x-1)(x-1)(x+1)} \quad \text{Distributive property}$$

$$= \dfrac{6x^2+5x+1}{(x-1)(x-1)(x+1)}$$

If desired, write the difference as

$$\dfrac{6x^2+5x+1}{(x-1)^2(x+1)}. \blacktriangleleft$$

Work Problem 7 at the side.

EXAMPLE 8 Find $\dfrac{q}{q^2-4q-5} - \dfrac{3}{2q^2-13q+15}$.

Start by factoring each denominator.

$$\dfrac{q}{q^2-4q-5} - \dfrac{3}{2q^2-13q+15}$$

$$= \dfrac{q}{(q+1)(q-5)} - \dfrac{3}{(q-5)(2q-3)}$$

Now rewrite each of the two rational expressions with the least common denominator, $(q+1)(q-5)(2q-3)$. Then subtract numerators.

$$\dfrac{q}{(q+1)(q-5)} - \dfrac{3}{(q-5)(2q-3)}$$

$$= \dfrac{q(2q-3)}{(q+1)(q-5)(2q-3)} - \dfrac{3(q+1)}{(q+1)(q-5)(2q-3)}$$

$$= \dfrac{q(2q-3) - 3(q+1)}{(q+1)(q-5)(2q-3)} \quad \text{Subtract}$$

$$= \dfrac{2q^2-3q-3q-3}{(q+1)(q-5)(2q-3)} \quad \text{Distributive property}$$

$$= \dfrac{2q^2-6q-3}{(q+1)(q-5)(2q-3)} \blacktriangleleft$$

Work Problem 8 at the side.

Rational Expressions

6.4 EXERCISES

Find the sums or differences. Write each answer in lowest terms. See Examples 1 and 5.

1. $\dfrac{2}{p} + \dfrac{5}{p}$

2. $\dfrac{3}{r} + \dfrac{6}{r}$

3. $\dfrac{9}{k} - \dfrac{12}{k}$

4. $\dfrac{15}{z} - \dfrac{25}{z}$

5. $\dfrac{y}{y+1} + \dfrac{1}{y+1}$

6. $\dfrac{3m}{m-4} - \dfrac{12}{m-4}$

7. $\dfrac{p-q}{3} - \dfrac{p+q}{3}$

8. $\dfrac{a+b}{2} - \dfrac{a-b}{2}$

9. $\dfrac{m^2}{m+6} + \dfrac{6m}{m+6}$

10. $\dfrac{y^2}{y-1} + \dfrac{-y}{y-1}$

11. $\dfrac{q^2 - 4q}{q-2} + \dfrac{4}{q-2}$

12. $\dfrac{z^2 - 10z}{z-5} + \dfrac{25}{z-5}$

Find the sums or differences. Write each answer in lowest terms. See Examples 2, 3, 4, 6, 7, and 8.

13. $\dfrac{m}{3} + \dfrac{1}{2}$

14. $\dfrac{p}{6} - \dfrac{2}{3}$

15. $\dfrac{7}{5} - \dfrac{z}{4}$

16. $\dfrac{9}{10} - \dfrac{r}{2}$

17. $\dfrac{4}{3} - \dfrac{1}{y}$

18. $\dfrac{8}{5} - \dfrac{2}{a}$

19. $\dfrac{5m}{6} - \dfrac{2m}{3}$

20. $\dfrac{3}{x} + \dfrac{5}{2x}$

21. $\dfrac{4 + 2k}{5} + \dfrac{2 + k}{10}$

22. $\dfrac{5 - 4r}{8} - \dfrac{2 - 3r}{6}$

23. $\dfrac{3m + 5}{3} - \dfrac{m + 2}{6}$

24. $\dfrac{5r + 4}{3} - \dfrac{2r - 3}{9}$

25. $\dfrac{m + 2}{m} + \dfrac{m + 3}{4m}$

26. $\dfrac{2x - 5}{x} + \dfrac{x - 1}{2x}$

27. $\dfrac{6}{y^2} - \dfrac{2}{y}$

28. $\dfrac{3}{p} + \dfrac{5}{p^2}$

29. $\dfrac{9}{2p} - \dfrac{4}{p^2}$

30. $\dfrac{15}{4k^2} - \dfrac{3}{k}$

31. $\dfrac{-1}{x^2} + \dfrac{3}{xy}$

32. $\dfrac{9}{p^2} + \dfrac{4}{px}$

33. $\dfrac{8}{x-2} - \dfrac{4}{x+2}$

34. $\dfrac{6}{m-5} - \dfrac{2}{m+5}$

35. $\dfrac{2x}{x+y} - \dfrac{3x}{2x+2y}$

36. $\dfrac{1}{a-b} - \dfrac{a}{4a-4b}$

37. $\dfrac{1}{m^2-9} + \dfrac{1}{3m+9}$

38. $\dfrac{-6}{y^2-4} - \dfrac{3}{2y+4}$

39. $\dfrac{1}{m^2-1} - \dfrac{1}{m^2+3m+2}$

40. $\dfrac{1}{y^2-4} + \dfrac{3}{y^2+5y+6}$

41. $\dfrac{8}{m-2} + \dfrac{3}{5m} + \dfrac{7}{5m(m-2)}$

42. $\dfrac{-1}{7z} + \dfrac{3}{z+2} + \dfrac{4}{7z(z+2)}$

43. $\dfrac{4}{r^2-r} + \dfrac{6}{r^2+2r} - \dfrac{1}{r^2+r-2}$

44. $\dfrac{6}{k^2+3k} - \dfrac{1}{k^2-k} + \dfrac{2}{k^2+2k-3}$

45. $\dfrac{k-1}{k^2+2k-8} + \dfrac{3k-2}{k^2+3k-4}$

46. $\dfrac{r+1}{r^2-3r-10} + \dfrac{r-1}{r^2+r-30}$

47. $\dfrac{4y-1}{2y^2+5y-3} - \dfrac{y+3}{6y^2+y-2}$

48. $\dfrac{2q+1}{3q^2+10q-8} - \dfrac{3q+5}{2q^2+5q-12}$

6.5 COMPLEX FRACTIONS

A rational expression with fractions in the numerator, denominator, or both, is called a **complex fraction.** Examples of complex fractions include

$$\frac{3+\frac{4}{x}}{5}, \quad \frac{\frac{3x^2-5x}{6x^2}}{2x-\frac{1}{x}}, \quad \text{and} \quad \frac{3+x}{5-\frac{2}{x}}.$$

The parts of a complex fraction are named as follows.

$$\frac{\frac{2}{p}-\frac{1}{q}}{\frac{3}{p}+\frac{5}{q}} \quad \begin{array}{l}\leftarrow \text{Numerator of complex fraction} \\ \leftarrow \text{Main fraction bar} \\ \leftarrow \text{Denominator of complex fraction}\end{array}$$

▶ Since a fraction represents a quotient, one method of simplifying complex fractions is to rewrite both the numerator and denominator as single fractions, and then to perform the indicated division.

EXAMPLE 1 Simplify each complex fraction.

(a) $\dfrac{\frac{2}{3}+\frac{5}{9}}{\frac{1}{4}+\frac{1}{12}}$ 　　(b) $\dfrac{6+\frac{3}{x}}{\frac{x}{4}+\frac{1}{8}}$

First, write each numerator as a single fraction.

$$\frac{2}{3}+\frac{5}{9}=\frac{2(3)}{3(3)}+\frac{5}{9} \qquad 6+\frac{3}{x}=\frac{6}{1}+\frac{3}{x}$$

$$=\frac{6}{9}+\frac{5}{9}=\boxed{\frac{11}{9}} \qquad =\frac{6x}{x}+\frac{3}{x}=\boxed{\frac{6x+3}{x}}$$

Do the same thing with each denominator.

$$\frac{1}{4}+\frac{1}{12}=\frac{1(3)}{4(3)}+\frac{1}{12} \qquad \frac{x}{4}+\frac{1}{8}=\frac{x(2)}{4(2)}+\frac{1}{8}$$

$$=\frac{3}{12}+\frac{1}{12}=\boxed{\frac{4}{12}} \qquad =\frac{2x}{8}+\frac{1}{8}=\boxed{\frac{2x+1}{8}}$$

The original complex fraction can now be written as follows.

$$\frac{\frac{11}{9}}{\frac{4}{12}} \qquad \frac{\frac{6x+3}{x}}{\frac{2x+1}{8}}$$

Now simplify by using the definition of division and the fundamental property.

$$\frac{11}{9} \div \frac{4}{12} = \frac{11}{9} \cdot \frac{12}{4} \qquad \frac{6x+3}{x} \div \frac{2x+1}{8}$$

$$= \frac{11 \cdot 2 \cdot 2 \cdot 3}{3 \cdot 3 \cdot 2 \cdot 2} \qquad = \frac{6x+3}{x} \cdot \frac{8}{2x+1}$$

$$= \frac{11}{3} \qquad = \frac{3(2x+1)}{x} \cdot \frac{8}{2x+1} = \frac{24}{x} \blacktriangleleft$$

Objectives

Simplify complex fractions by

▶ simplifying numerator and denominator;

▶ multiplying by the least common denominator.

6.5 Complex Fractions　　301

1. Simplify the complex fractions.

   (a) $\dfrac{6 + \dfrac{1}{x}}{5 - \dfrac{2}{x}}$

   (b) $\dfrac{9 - \dfrac{4}{p}}{\dfrac{2}{p} + 1}$

   (c) $\dfrac{m + \dfrac{1}{2}}{\dfrac{6m + 3}{4m}}$

2. Simplify the complex fractions.

   (a) $\dfrac{\dfrac{rs^2}{t}}{\dfrac{r^2 s}{t^2}}$

   (b) $\dfrac{\dfrac{m^2 n^3}{p}}{\dfrac{m^4 n}{p^2}}$

ANSWERS

1. (a) $\dfrac{6x+1}{5x-2}$ (b) $\dfrac{9p-4}{2+p}$ (c) $\dfrac{2m}{3}$

2. (a) $\dfrac{st}{r}$ (b) $\dfrac{n^2 p}{m^2}$

Work Problem 1 at the side.

EXAMPLE 2 Simplify the complex fraction $\dfrac{\dfrac{xp}{q^3}}{\dfrac{p^2}{qx^2}}$.

Here the numerator and denominator are already single fractions, so use the definition of division and then the fundamental property.

$$\frac{xp}{q^3} \div \frac{p^2}{qx^2} = \frac{xp}{q^3} \cdot \frac{qx^2}{p^2} = \frac{x^3}{q^2 p} \blacktriangleleft$$

Work Problem 2 at the side.

▶ As an alternative method, complex fractions may be simplified by multiplying both numerator and denominator by the least common denominator of all the denominators appearing in the complex fraction.

EXAMPLE 3 Simplify each complex fraction. (These are the same ones simplified in Example 1 by the other method.)

(a) $\dfrac{\dfrac{2}{3} + \dfrac{5}{9}}{\dfrac{1}{4} + \dfrac{1}{12}}$ (b) $\dfrac{6 + \dfrac{3}{x}}{\dfrac{x}{4} + \dfrac{1}{8}}$

Find the least common denominator for all the denominators in the complex fraction.

The least common denominator for 3, 9, 4, and 12 is 36.

The least common denominator for x, 4, and 8 is $8x$.

Multiply numerator and denominator of the complex fraction by the least common denominator.

$$\dfrac{\dfrac{2}{3} + \dfrac{5}{9}}{\dfrac{1}{4} + \dfrac{1}{12}}$$

$$= \dfrac{36\left(\dfrac{2}{3} + \dfrac{5}{9}\right)}{36\left(\dfrac{1}{4} + \dfrac{1}{12}\right)}$$

$$= \dfrac{36\left(\dfrac{2}{3}\right) + 36\left(\dfrac{5}{9}\right)}{36\left(\dfrac{1}{4}\right) + 36\left(\dfrac{1}{12}\right)}$$

$$= \dfrac{24 + 20}{9 + 3}$$

$$= \dfrac{44}{12} = \dfrac{11}{3}$$

$$\dfrac{6 + \dfrac{3}{x}}{\dfrac{x}{4} + \dfrac{1}{8}}$$

$$= \dfrac{8x\left(6 + \dfrac{3}{x}\right)}{8x\left(\dfrac{x}{4} + \dfrac{1}{8}\right)}$$

$$= \dfrac{8x(6) + 8x\left(\dfrac{3}{x}\right)}{8x\left(\dfrac{x}{4}\right) + 8x\left(\dfrac{1}{8}\right)}$$

$$= \dfrac{48x + 24}{2x^2 + x}$$

$$= \dfrac{24(2x+1)}{x(2x+1)} = \dfrac{24}{x} \blacktriangleleft$$

Work Problem 3 at the side.

The two methods for simplifying a complex fraction are summarized below.

Simplifying Complex Fractions

Method 1 Simplify the numerator and denominator of the complex fraction separately. Then divide the simplified numerator by the simplified denominator.

Method 2 Multiply numerator and denominator of the complex fraction by the least common denominator of all the denominators appearing in the complex fraction.

3. Simplify by the second method.

(a) $\dfrac{2 - \dfrac{6}{a}}{3 + \dfrac{4}{a}}$

(b) $\dfrac{\dfrac{5}{p} - 6}{\dfrac{2p + 1}{p}}$

(c) $\dfrac{3 + x}{5 - \dfrac{2}{x}}$

ANSWERS

3. (a) $\dfrac{2a - 6}{3a + 4}$ (b) $\dfrac{5 - 6p}{2p + 1}$
(c) $\dfrac{x(3 + x)}{5x - 2}$

6.5 Complex Fractions

6.5 EXERCISES

Simplify each complex fraction. Use either method. See Examples 1–3.

1. $\dfrac{\dfrac{5}{8} + \dfrac{2}{3}}{\dfrac{7}{3} - \dfrac{1}{4}}$

2. $\dfrac{\dfrac{6}{5} - \dfrac{1}{9}}{\dfrac{2}{5} + \dfrac{5}{3}}$

3. $\dfrac{1 - \dfrac{3}{8}}{2 + \dfrac{1}{4}}$

4. $\dfrac{3 + \dfrac{5}{4}}{1 - \dfrac{7}{8}}$

5. $\dfrac{\dfrac{m^3 p^4}{5m}}{\dfrac{8mp^5}{p^2}}$

6. $\dfrac{\dfrac{9z^5 x^3}{2x}}{\dfrac{8z^2 x^5}{3z}}$

7. $\dfrac{\dfrac{x+1}{4}}{\dfrac{x-3}{x}}$

8. $\dfrac{\dfrac{m+6}{m}}{\dfrac{m-6}{2}}$

9. $\dfrac{\dfrac{3}{y} + 1}{\dfrac{3+y}{2}}$

10. $\dfrac{6 + \dfrac{2}{r}}{\dfrac{r+2}{3}}$

11. $\dfrac{\dfrac{1}{x} + x}{\dfrac{x^2 + 1}{8}}$

12. $\dfrac{\dfrac{3}{m} - m}{\dfrac{3 - m^2}{4}}$

13. $\dfrac{x + \dfrac{1}{x}}{\dfrac{4}{x} + y}$

14. $\dfrac{y - \dfrac{6}{y}}{y + \dfrac{2}{y}}$

15. $\dfrac{\dfrac{p+3}{p}}{\dfrac{1}{p} + \dfrac{1}{5}}$

16. $\dfrac{r + \dfrac{1}{r}}{\dfrac{1}{r} - r}$

17. $\dfrac{\dfrac{2}{p^2} - \dfrac{3}{5p}}{\dfrac{4}{p} + \dfrac{1}{4p}}$

18. $\dfrac{-\dfrac{2}{m^2} - \dfrac{3}{m}}{\dfrac{2}{5m^2} + \dfrac{1}{3m}}$

19. $\dfrac{\dfrac{5}{x^2 y} - \dfrac{2}{xy^2}}{\dfrac{3}{x^2 y^2} + \dfrac{4}{xy}}$

20. $\dfrac{\dfrac{1}{m^3p} + \dfrac{2}{mp^2}}{\dfrac{4}{mp} + \dfrac{1}{m^2p}}$

21. $\dfrac{\dfrac{1}{4} - \dfrac{1}{a^2}}{\dfrac{1}{2} + \dfrac{1}{a}}$

22. $\dfrac{\dfrac{1}{9} - \dfrac{1}{m^2}}{\dfrac{1}{3} + \dfrac{1}{m}}$

23. $\dfrac{\dfrac{1}{z+5}}{\dfrac{4}{z^2-25}}$

24. $\dfrac{\dfrac{a}{a+1}}{\dfrac{2}{a^2-1}}$

25. $\dfrac{\dfrac{1}{m+1} - 1}{\dfrac{1}{m+1} + 1}$

26. $\dfrac{\dfrac{2}{x-1} + 2}{\dfrac{2}{x-1} - 2}$

27. $\dfrac{\dfrac{1}{m-1} + \dfrac{2}{m+2}}{\dfrac{2}{m+2} - \dfrac{1}{m-3}}$

28. $\dfrac{\dfrac{5}{r+3} - \dfrac{1}{r-1}}{\dfrac{2}{r+2} + \dfrac{3}{r+3}}$

The following exercises are real "head-scratchers." (Hint: Begin at the lower right and work upward.)

29. $1 - \dfrac{1}{1 + \dfrac{1}{1+1}}$

30. $3 - \dfrac{2}{4 + \dfrac{2}{4-2}}$

31. $r + \dfrac{r}{4 - \dfrac{2}{6+2}}$

32. $\dfrac{2q}{7} - \dfrac{q}{6 + \dfrac{8}{4+4}}$

6.6 EQUATIONS INVOLVING RATIONAL EXPRESSIONS

Objectives

▶ Solve equations involving rational expressions.
▶ Solve for a specified variable.

▶ To solve an equation with fractions, begin by simplifying the equation using the multiplication property of equality. The goal is to obtain a new equation that does not have fractions. Choose as multiplier the least common denominator of all denominators in the fractions of the equation.

EXAMPLE 1 Solve $\dfrac{x}{3} + \dfrac{x}{4} = 10 + x$.

Since the least common denominator of the two fractions is 12, begin by multiplying both sides of the equation by 12.

$$12\left(\dfrac{x}{3} + \dfrac{x}{4}\right) = 12(10 + x)$$

$$12\left(\dfrac{x}{3}\right) + 12\left(\dfrac{x}{4}\right) = 12(10) + 12x \quad \text{Distributive property}$$

$$\dfrac{12x}{3} + \dfrac{12x}{4} = 120 + 12x$$

$$4x + 3x = 120 + 12x$$

This equation has no fractions. Solve it using the methods of solving linear equations that were given earlier.

$$7x = 120 + 12x$$
$$-5x = 120 \quad \text{Subtract } 12x$$
$$x = -24 \quad \text{Divide by } -5$$

Check this solution by substituting -24 for x in the original equation. ◀

Work Problems 1 and 2 at the side.

1. Check the solution to Example 1 by substituting -24 for x in the original equation,

$$\dfrac{x}{3} + \dfrac{x}{4} = 10 + x.$$

Is -24 the solution?

2. Solve each equation and check your answer.

(a) $\dfrac{x}{5} + 3 = \dfrac{3}{5}$

(b) $\dfrac{x}{2} - \dfrac{x}{3} = \dfrac{5}{6}$

(c) $\dfrac{k}{7} - \dfrac{k}{2} = -5$

EXAMPLE 2 Solve $\dfrac{p}{2} - \dfrac{p-1}{3} = 1$.

Multiply both sides by the common denominator, 6.

$$6\left(\dfrac{p}{2} - \dfrac{p-1}{3}\right) = 6 \cdot 1$$

$$6\left(\dfrac{p}{2}\right) - 6\left(\dfrac{p-1}{3}\right) = 6 \quad \text{Distributive property}$$

$$3p - 2(p - 1) = 6$$

Be very careful to put parentheses around $p - 1$; otherwise an incorrect solution may be found. Continue simplifying and solve.

$$3p - 2p + 2 = 6 \quad \text{Distributive property}$$
$$p + 2 = 6$$
$$p = 4 \quad \text{Subtract 2}$$

Check to see that 4 is correct by replacing p with 4 in the original equation. ◀

Work Problem 3 at the side.

3. Solve each equation and check your answer.

(a) $\dfrac{k}{6} - \dfrac{k+1}{4} = -\dfrac{1}{2}$

(b) $\dfrac{2m-3}{5} - \dfrac{m}{3} = -\dfrac{6}{5}$

(c) $\dfrac{r-2}{3} + \dfrac{r+1}{5} = \dfrac{3}{5}$

ANSWERS
1. yes
2. (a) -12 (b) 5 (c) 14
3. (a) 3 (b) -9 (c) 2

4. Solve $1 - \dfrac{2}{x+1} = \dfrac{2x}{x+1}$ and check your answer.

When solving equations that have a variable in the denominator, remember that the number 0 cannot be used as a denominator. Therefore, the solution cannot be a number that will make the denominator equal 0.

EXAMPLE 3 Solve $\dfrac{x}{x-2} = \dfrac{2}{x-2} + 2$.

The common denominator is $x - 2$. Multiply both sides of the equation by $x - 2$.

$$(x-2)\left(\dfrac{x}{x-2}\right) = (x-2)\left(\dfrac{2}{x-2}\right) + (x-2)(2)$$

$$x = 2 + 2x - 4$$
$$x = -2 + 2x$$
$$-x = -2 \qquad \text{Subtract } 2x$$
$$x = 2 \qquad \text{Divide by } -1$$

The proposed solution is 2. However, 2 cannot be a solution because 2 makes a denominator equal 0. The equation has no solution. (Equations with no solutions are one of the main reasons that it is important to always check proposed solutions.) ◀

Work Problem 4 at the side.

In summary, solve equations with rational expressions as follows.

Solving Equations with Rational Expressions

Step 1 Multiply both sides of the equation by the least common denominator. (This clears the equation of fractions.)

Step 2 Solve the resulting equation.

Step 3 Check each solution by substituting it in the original equation.

EXAMPLE 4 Solve $\dfrac{2m}{m^2-4} + \dfrac{1}{m-2} = \dfrac{2}{m+2}$.

Since $m^2 - 4 = (m+2)(m-2)$, use $(m+2)(m-2)$ as the common denominator. Multiply both sides by $(m+2)(m-2)$.

$$(m+2)(m-2)\left(\dfrac{2m}{m^2-4} + \dfrac{1}{m-2}\right) = (m+2)(m-2)\dfrac{2}{m+2}$$

$$(m+2)(m-2)\dfrac{2m}{m^2-4} + (m+2)(m-2)\dfrac{1}{m-2}$$
$$= (m+2)(m-2)\dfrac{2}{m+2}$$

$$2m + m + 2 = 2(m-2)$$
$$3m + 2 = 2m - 4 \qquad \text{Distributive property}$$
$$m + 2 = -4 \qquad \text{Subtract } 2m$$
$$m = -6 \qquad \text{Subtract } 2$$

Check to see that -6 is indeed a valid solution for the given equation. ◀

ANSWERS

4. When the equation is solved, -1 is found. However, since $x = -1$ leads to a 0 denominator in the original equation, there is no solution.

Work Problems 5 and 6 at the side.

EXAMPLE 5 Solve $\dfrac{2}{x^2 - x} = \dfrac{1}{x^2 - 1}$.

Begin by finding a least common denominator. Since $x^2 - x$ can be factored as $x(x - 1)$, and $x^2 - 1$ can be factored as $(x + 1)(x - 1)$, the least common denominator is $x(x + 1)(x - 1)$. Multiply both sides of the equation by $x(x + 1)(x - 1)$.

$$x(x + 1)(x - 1)\dfrac{2}{x(x - 1)} = x(x + 1)(x - 1)\dfrac{1}{(x + 1)(x - 1)}$$

$$2(x + 1) = x$$
$$2x + 2 = x \qquad \text{Distributive property}$$
$$x + 2 = 0 \qquad \text{Subtract } x$$
$$x = -2 \qquad \text{Subtract 2}$$

To be sure that $x = -2$ is a solution, substitute -2 for x in the original equation. Since -2 satisfies the equation, the solution is -2. ◀

Work Problem 7 at the side.

EXAMPLE 6 Solve $\dfrac{1}{x - 1} + \dfrac{1}{2} = \dfrac{2}{x^2 - 1}$.

The least common denominator is $2(x + 1)(x - 1)$. Multiply both sides of the equation by this common denominator.

$$2(x + 1)(x - 1)\left(\dfrac{1}{x - 1} + \dfrac{1}{2}\right) = 2(x + 1)(x - 1)\dfrac{2}{(x + 1)(x - 1)}$$

$$2(x + 1)(x - 1)\dfrac{1}{x - 1} + 2(x + 1)(x - 1)\dfrac{1}{2}$$

$$= 2(x + 1)(x - 1)\dfrac{2}{(x + 1)(x - 1)}$$

$$2(x + 1) + (x + 1)(x - 1) = 4$$
$$2x + 2 + x^2 - 1 = 4 \qquad \text{Distributive property}$$
$$x^2 + 2x + 1 = 4$$
$$x^2 + 2x - 3 = 0 \qquad \text{Subtract 4}$$

Factoring gives

$$(x + 3)(x - 1) = 0.$$

Solving this equation suggests that $x = -3$ or $x = 1$. But 1 makes a denominator of the original equation equal 0, so 1 is not a solution. However, -3 is a solution, as can be shown by substituting -3 for x in the original equation.

$$\dfrac{1}{x - 1} + \dfrac{1}{2} = \dfrac{2}{x^2 - 1}$$

$$\dfrac{1}{-3 - 1} + \dfrac{1}{2} = \dfrac{2}{(-3)^2 - 1} \qquad \text{Let } x = -3$$

$$\dfrac{1}{-4} + \dfrac{1}{2} = \dfrac{2}{9 - 1}$$

$$\dfrac{1}{4} = \dfrac{1}{4} \qquad \text{True}$$

5. Check -6 as a solution to Example 4.

Is the solution correct?

6. Solve each equation and check your answer.

   (a) $\dfrac{2p}{p^2 - 1} = \dfrac{2}{p + 1} - \dfrac{1}{p - 1}$

   (b) $\dfrac{8r}{4r^2 - 1} = \dfrac{3}{2r + 1} + \dfrac{3}{2r - 1}$

7. Solve each equation and check your answer.

   (a) $\dfrac{4}{3m + 3} = \dfrac{m + 1}{m^2 + m}$

   (b) $\dfrac{2}{p^2 - 2p} = \dfrac{3}{p^2 - p}$

ANSWERS
5. yes
6. (a) -3 (b) 0
7. (a) 3   (b) 4

6.6 Equations Involving Rational Expressions

8. Solve each equation and check your answer.

   (a) $\dfrac{1}{x-2} + \dfrac{1}{5} = \dfrac{2}{5(x^2-4)}$

   (b) $\dfrac{2}{k-1} - 1 = \dfrac{3}{k^2-1}$

9. Solve each equation and check your answer.

   (a) $\dfrac{2}{m^2-3m+2} + \dfrac{4}{3m-6} = \dfrac{5}{2m-2}$

   (b) $\dfrac{6}{5a+10} - \dfrac{1}{a-5} = \dfrac{4}{a^2-3a-10}$

10. Solve each equation for the specified variable.

    (a) $\dfrac{2}{m} = \dfrac{1}{p} + \dfrac{1}{q}$ for q

    (b) $z = \dfrac{x}{x+y}$ for y

    (c) $\dfrac{1}{R} = \dfrac{1}{r_1} + \dfrac{1}{r_2}$ for R

ANSWERS

8. (a) $-4, -1$ (b) $0, 2$
9. (a) $\dfrac{34}{7}$ (b) 60
10. (a) $q = \dfrac{mp}{2p-m}$
    (b) $y = \dfrac{x-zx}{z}$
    (c) $R = \dfrac{r_1 r_2}{r_1 + r_2}$

310

The check shows that -3 is a solution. ◀

Work Problem 8 at the side.

EXAMPLE 7 Solve $\dfrac{1}{k^2+4k+3} + \dfrac{1}{2k+2} = \dfrac{3}{4k+12}$.

Factor the three denominators to get the common denominator, $4(k+1)(k+3)$. Multiply both sides by this product.

$$4(k+1)(k+3)\left(\dfrac{1}{(k+1)(k+3)} + \dfrac{1}{2(k+1)}\right)$$
$$= 4(k+1)(k+3)\dfrac{3}{4(k+3)}$$

$$4(k+1)(k+3)\dfrac{1}{(k+1)(k+3)} + 4(k+1)(k+3)\dfrac{1}{2(k+1)}$$
$$= 4(k+1)(k+3)\dfrac{3}{4(k+3)}$$

$$4 + 2(k+3) = 3(k+1)$$
$$4 + 2k + 6 = 3k + 3$$
$$2k + 10 = 3k + 3$$
$$7 = k$$

Check to see that 7 actually is a solution for the given equation. ◀

Work Problem 9 at the side.

▶ Solving a formula for a specified variable was discussed in Chapter 3. In the next example this process is applied to formulas with fractions.

EXAMPLE 8 Solve the formula $\dfrac{1}{R} = \dfrac{1}{r_1} + \dfrac{1}{r_2}$ for r_2.

The variables r_1 (read "r-sub one") and r_2 are different. The least common denominator for R, r_1, and r_2 is Rr_1r_2, so multiply both sides by Rr_1r_2.

$$Rr_1r_2\left(\dfrac{1}{R}\right) = Rr_1r_2\left(\dfrac{1}{r_1} + \dfrac{1}{r_2}\right)$$

$$Rr_1r_2\left(\dfrac{1}{R}\right) = Rr_1r_2\left(\dfrac{1}{r_1}\right) + Rr_1r_2\left(\dfrac{1}{r_2}\right)$$

$$r_1r_2 = Rr_2 + Rr_1$$

Get the terms containing r_2 (the desired variable) alone on one side of the equals sign. Do this here by subtracting Rr_2 from both sides.

$$r_1r_2 - Rr_2 = Rr_1$$

Factor out the common factor of r_2 on the left.

$$r_2(r_1 - R) = Rr_1$$

Finally, divide both sides of the equation by $r_1 - R$.

$$r_2 = \dfrac{Rr_1}{r_1 - R}$$ ◀

Work Problem 10 at the side.

Rational Expressions

6.6 EXERCISES

Solve each equation and check your answers. See Examples 1 and 2.

1. $\dfrac{6}{x} - \dfrac{4}{x} = 5$

2. $\dfrac{3}{x} + \dfrac{2}{x} = 5$

3. $\dfrac{x}{2} - \dfrac{x}{4} = 6$

4. $\dfrac{4}{y} + \dfrac{2}{3} = 1$

5. $\dfrac{9}{m} = 5 - \dfrac{1}{m}$

6. $\dfrac{3x}{5} + 2 = \dfrac{1}{4}$

7. $\dfrac{2t}{7} - 5 = t$

8. $\dfrac{1}{2} + \dfrac{2}{m} = 1$

9. $\dfrac{x+1}{2} = \dfrac{x+2}{3}$

10. $\dfrac{t-4}{3} = \dfrac{t+2}{4}$

11. $\dfrac{3m}{2} + m = 5$

12. $\dfrac{2z}{5} + z = \dfrac{7}{5}$

13. $\dfrac{2p+8}{9} = \dfrac{10p+4}{27}$

14. $\dfrac{2k+3}{k} = \dfrac{3}{2}$

15. $\dfrac{5-y}{y} + \dfrac{3}{4} = \dfrac{7}{y}$

16. $\dfrac{x}{x-4} = \dfrac{2}{x-4} + 5$

17. $\dfrac{a-4}{4} = \dfrac{a}{16} + \dfrac{1}{2}$

18. $\dfrac{m-2}{5} = \dfrac{m}{10} + \dfrac{4}{5}$

19. $\dfrac{2y-1}{y} + 2 = \dfrac{1}{2}$

20. $\dfrac{a}{2} - \dfrac{17+a}{5} = 2a$

21. $\dfrac{m-2}{4} + \dfrac{m+1}{3} = \dfrac{10}{3}$

22. $\dfrac{y+2}{5} + \dfrac{y-5}{3} = \dfrac{7}{5}$

23. $\dfrac{a+7}{8} - \dfrac{a-2}{3} = \dfrac{4}{3}$

24. $\dfrac{m+2}{5} - \dfrac{m-6}{7} = 2$

25. $\dfrac{p}{2} - \dfrac{p-1}{4} = \dfrac{5}{4}$

26. $\dfrac{r}{6} - \dfrac{r-2}{3} = \dfrac{-4}{3}$

27. $\dfrac{5y}{3} - \dfrac{2y-1}{4} = \dfrac{1}{4}$

28. $\dfrac{8k}{5} - \dfrac{3k-4}{2} = \dfrac{5}{2}$

29. $\dfrac{y-1}{2} - \dfrac{y-3}{4} = 1$

30. $\dfrac{r+5}{3} - \dfrac{r-1}{4} = \dfrac{7}{4}$

Solve each equation and check your answers. See Examples 3–7.

31. $\dfrac{4}{k-2} + \dfrac{3}{5k-10} = \dfrac{23}{5}$

32. $\dfrac{1}{3p+15} + \dfrac{5}{4p+20} = \dfrac{19}{24}$

33. $\dfrac{m}{2m+2} = \dfrac{2m-3}{m+1} - \dfrac{m}{2m+2}$

34. $\dfrac{5p+1}{3p+3} = \dfrac{5p-5}{5p+5} + \dfrac{3p-1}{p+1}$

312 Rational Expressions

35. $\dfrac{x+1}{x-3} = \dfrac{4}{x-3} + 6$

36. $\dfrac{p}{p-2} + 4 = \dfrac{2}{p-2}$

37. $\dfrac{2}{y} = \dfrac{y}{5y-12}$

38. $\dfrac{8x+3}{x} = 3x$

39. $\dfrac{2}{k-3} - \dfrac{3}{k+3} = \dfrac{12}{k^2-9}$

40. $\dfrac{1}{r+5} - \dfrac{3}{r-5} = \dfrac{-10}{r^2-25}$

41. $\dfrac{3y}{y^2+5y+6} = \dfrac{5y}{y^2+2y-3} - \dfrac{2}{y^2+y-2}$

42. $\dfrac{x+4}{x^2-3x+2} - \dfrac{5}{x^2-4x+3} = \dfrac{x-4}{x^2-5x+6}$

43. $\dfrac{3}{r^2+r-2} - \dfrac{1}{r^2-1} = \dfrac{7}{2(r^2+3r+2)}$

6.6 Exercises

44. $\dfrac{4}{p} - \dfrac{2}{p+1} = 3$

45. $\dfrac{6}{r} + \dfrac{1}{r-2} = 3$

46. $\dfrac{2}{m-1} + \dfrac{1}{m+1} = \dfrac{5}{4}$

47. $\dfrac{5}{z-2} + \dfrac{10}{z+2} = 7$

48. $\dfrac{1}{x} = \dfrac{1}{y} - \dfrac{1}{z}$ for y

49. $\dfrac{3}{k} = \dfrac{1}{p} + \dfrac{1}{q}$ for q

50. $9x + \dfrac{3}{z} = \dfrac{5}{y}$ for z

51. $2a - \dfrac{5}{b} + \dfrac{1}{c} = 0$ for b

52. $I = \dfrac{E}{R+r}$ for R

53. $I = \dfrac{E}{R+r}$ for r

54. $S = \dfrac{a}{1-r}$ for r

55. $h = \dfrac{2A}{B+b}$ for b

56. $\dfrac{1}{R} = \dfrac{1}{r_1} + \dfrac{1}{r_2}$ for r_1

SUPPLEMENTARY EXERCISES ON RATIONAL EXPRESSIONS

A common error when working with rational expressions is to confuse *operations* on rational expressions with the *solution of equations* with rational expressions. For example, the four possible operations on the rational expressions

$$\frac{1}{x} \quad \text{and} \quad \frac{1}{x-2}$$

can be performed as follows.

Add: $\dfrac{1}{x} + \dfrac{1}{x-2} = \dfrac{x-2}{x(x-2)} + \dfrac{x}{x(x-2)} = \dfrac{x-2+x}{x(x-2)} = \dfrac{2x-2}{x(x-2)}.$

Subtract: $\dfrac{1}{x} - \dfrac{1}{x-2} = \dfrac{x-2}{x(x-2)} - \dfrac{x}{x(x-2)} = \dfrac{x-2-x}{x(x-2)} = \dfrac{-2}{x(x-2)}.$

Multiply: $\dfrac{1}{x} \cdot \dfrac{1}{x-2} = \dfrac{1}{x(x-2)}.$

Divide: $\dfrac{1}{x} \div \dfrac{1}{x-2} = \dfrac{1}{x} \cdot \dfrac{x-2}{1} = \dfrac{x-2}{x}.$

On the other hand, the *equation*

$$\frac{1}{x} + \frac{1}{x-2} = \frac{3}{4}$$

is solved by multiplying both sides by the least common denominator, $4x(x-2)$, giving an equation with no denominators.

$$4x(x-2)\frac{1}{x} + 4x(x-2)\frac{1}{x-2} = 4x(x-2)\frac{3}{4}$$
$$4x - 8 + 4x = 3x^2 - 6x$$
$$0 = 3x^2 - 14x + 8$$
$$0 = (3x-2)(x-4)$$
$$x = \frac{2}{3} \quad \text{or} \quad x = 4$$

In each of the following exercises, first decide whether the given rational expressions should be added, subtracted, multiplied, or divided; then perform the operation, or solve the given equation.

1. $\dfrac{6}{m} + \dfrac{2}{m}$

2. $\dfrac{b^2 c^3}{b^5 c^4} \cdot \dfrac{c^5}{b^7}$

3. $\dfrac{2}{x^2 + 2x - 3} \div \dfrac{8x^2}{3x - 3}$

4. $\dfrac{4}{m-2} = 1$

Supplementary Exercises on Rational Expressions

5. $\dfrac{2r^2 - 3r - 9}{2r^2 - r - 6} \cdot \dfrac{r^2 + 2r - 8}{r^2 - 2r - 3}$

6. $\dfrac{1}{m^2 - 3m} + \dfrac{4}{m^2 - 9}$

7. $\dfrac{p + 3}{8} = \dfrac{p - 2}{9}$

8. $\dfrac{4t^2 - t}{6t^2 + 10t} \div \dfrac{8t^2 + 2t - 1}{3t^2 + 11t + 10}$

9. $\dfrac{5}{y - 1} + \dfrac{2}{3y - 3}$

10. $\dfrac{1}{z} + \dfrac{1}{z + 2} = \dfrac{8}{15}$

11. $\dfrac{2}{r - 1} + \dfrac{1}{r} = \dfrac{5}{2}$

12. $\dfrac{2}{y} - \dfrac{7}{5y}$

13. $\dfrac{4}{9z} - \dfrac{3}{2z}$

14. $\dfrac{r - 3}{2} = \dfrac{2r - 5}{5}$

15. $\dfrac{1}{m^2 + 5m + 6} + \dfrac{2}{m^2 + 4m + 3}$

16. $\dfrac{2k^2 - 3k}{20k^2 - 5k} \div \dfrac{2k^2 - 5k + 3}{4k^2 + 11k - 3}$

6.7 APPLICATIONS OF RATIONAL EXPRESSIONS

Objectives

Solve word problems with rational expressions about

▶ numbers;
▶ distance;
▶ work;
▶ variation.

▶ Rational expressions are often useful in solving word problems. Example 1, involving numbers, is included in this section to show the basic steps in solving a word problem with fractions.

EXAMPLE 1 If the same number is added to both the numerator and denominator of the fraction $\frac{3}{4}$, the result is $\frac{5}{6}$. Find the number.

If x represents the number added to numerator and denominator, then

$$\frac{3 + x}{4 + x} \quad \leftarrow \text{Same number added to numerator and denominator}$$

represents the result of adding the same number to both the numerator and denominator. Since this result is $\frac{5}{6}$,

$$\frac{3 + x}{4 + x} = \frac{5}{6}.$$

Solve this equation by multiplying both sides by the common denominator, $6(4 + x)$.

$$6(4 + x)\frac{3 + x}{4 + x} = 6(4 + x)\frac{5}{6}$$
$$6(3 + x) = 5(4 + x)$$
$$18 + 6x = 20 + 5x \quad \text{Distributive property}$$
$$x = 2$$

Check this solution in the words of the original problem: if 2 is added to both the numerator and denominator of $\frac{3}{4}$, the result is $\frac{5}{6}$, as needed. ◀

Work Problem 1 at the side.

1. Solve each word problem.

   (a) A certain number is added to the numerator and subtracted from the denominator of $\frac{5}{8}$. The new number equals the reciprocal of $\frac{5}{8}$. Find the number.

   (b) The denominator of a fraction is 1 more than the numerator. If 6 is added to the numerator and subtracted from the denominator, the result is $\frac{15}{4}$. Find the original fraction.

▶ The next example shows how to solve word problems involving distance.

EXAMPLE 2 The Big Muddy River has a current of 3 miles per hour. A motorboat takes as long to go 12 miles downstream as to go 8 miles upstream. What is the speed of the boat in still water?

This problem requires the distance formula,

$$d = rt \,(\text{distance} = \text{rate} \cdot \text{time}).$$

Let x represent the speed of the boat in still water. Since the current pushes the boat when the boat is going downstream, the speed of the boat downstream will be the sum of the speed of the boat and the speed of the current, or $x + 3$ miles per hour. Also, the boat's speed going upstream is given by $x - 3$ miles per hour.

The information given in the problem is summarized in this chart.

	d	r	t
Downstream	12	$x + 3$	
Upstream	8	$x - 3$	

ANSWERS

1. (a) 3 (b) $\frac{9}{10}$

6.7 Applications of Rational Expressions   317

Fill in the last column, representing time, by solving the formula $d = rt$ for t.

$$d = rt$$
$$\frac{d}{r} = t \quad \text{Divide by } r$$

Then the time upstream is the distance divided by the rate, or

$$\frac{8}{x-3},$$

and the time downstream is also the distance divided by the rate, or

$$\frac{12}{x+3}.$$

Now complete the chart.

	d	r	t
Downstream	12	$x + 3$	$\frac{12}{x+3}$
Upstream	8	$x - 3$	$\frac{8}{x-3}$

Times are equal

According to the original problem, the time upstream equals the time downstream. The two times from the chart must therefore be equal, giving the equation

$$\frac{12}{x+3} = \frac{8}{x-3}.$$

Solve this equation by multiplying both sides by $(x + 3)(x - 3)$.

$$(x+3)(x-3)\frac{12}{x+3} = (x+3)(x-3)\frac{8}{x-3}$$
$$12(x - 3) = 8(x + 3)$$
$$12x - 36 = 8x + 24 \quad \text{Distributive property}$$
$$4x = 60 \quad \text{Subtract } 8x; \text{ add } 36$$
$$x = 15 \quad \text{Divide by 4}$$

The speed of the boat in still water is 15 miles per hour.

Check the solution by first finding the speed of the boat downstream, which is $15 + 3 = 18$ miles per hour. Traveling 12 miles would take

$$d = rt$$
$$12 = 18t$$
$$t = \frac{2}{3} \text{ hour.}$$

On the other hand, the speed of the boat upstream is $15 - 3 = 12$ miles per hour, and traveling 8 miles would take

$$d = rt$$
$$8 = 12t$$
$$t = \frac{2}{3} \text{ hour.}$$

The time upstream equals the time downstream, as required. ◀

Work Problem 2 at the side.

▶ The third example shows a word problem about the length of time needed to do a job. This type of problem is often called a **work problem.**

EXAMPLE 3 With a riding lawn mower, John, the grounds keeper in a large park, can cut the lawn in 8 hours. With a small mower, his assistant Walt needs 14 hours to cut the same lawn. If both John and Walt work on the lawn, how long will it take to cut it?

Let x be the number of hours that it takes John and Walt to cut the lawn, working together. Certainly x will be less than 8, since John alone can cut the lawn in 8 hours. In one hour, John can do $\frac{1}{8}$ of the lawn, and in one hour Walt can do $\frac{1}{14}$ of the lawn. Since it takes them x hours to cut the lawn when working together, in one hour together they can do $\frac{1}{x}$ of the lawn. The amount of the lawn cut by John in one hour plus the amount cut by Walt in one hour must equal the amount they can do together in one hour, or

Amount by John → $\frac{1}{8} + \frac{1}{14} = \frac{1}{x}$ ← Amount together
                              ↑
                    Amount by Walt

Since $56x$ is the least common denominator for 8, 14, and x, multiply both sides of the equation by $56x$.

$$56x\left(\frac{1}{8} + \frac{1}{14}\right) = 56x \cdot \frac{1}{x}$$

$$56x \cdot \frac{1}{8} + 56x \cdot \frac{1}{14} = 56x \cdot \frac{1}{x}$$

$$7x + 4x = 56$$

$$11x = 56$$

$$x = \frac{56}{11}$$

Working together, John and Walt can cut the lawn in $\frac{56}{11}$ hours, or $5\frac{1}{11}$ hours, about 5 hours and 5 minutes. ◀

Work Problem 3 at the side.

▶ Equations with fractions often result when discussing **variation.** Two variables **vary directly** if one is a constant multiple of the other, as stated below.

> **Direct Variation**
>
> **y varies directly as x** if there exists a constant k such that
>
> $$y = kx.$$

2. **(a)** A boat can go 20 miles against a wind in the same time it can go 60 miles with the wind. The wind is blowing at 4 miles per hour. Find the speed of the boat with no wind.

   (b) An airplane, maintaining a constant airspeed, takes as long to go 450 miles with the wind as it does to go 375 miles against the wind. If the wind is blowing at 15 miles per hour, what is the speed of the plane?

3. **(a)** Jerry and Louise operate a small roofing company. Louise can roof an average house alone in 9 hours. Jerry can roof a house alone in 8 hours. How long will it take them to do the job if they work together?

   (b) Walt can do a job in 7 hours, but Tammie needs only 6 hours. How long would it take them working together?

ANSWERS

2. **(a)** 8 miles per hour **(b)** 165 miles per hour

3. **(a)** $\frac{72}{17}$ hours **(b)** $\frac{42}{13}$ hours

6.7 Applications of Rational Expressions

4. (a) If z varies directly as t, and $z = 11$ when $t = 4$, find z when $t = 32$.

(b) Suppose q varies directly as m, and $q = \frac{1}{2}$ when $m = 2$. Find q when $m = 5$.

5. (a) Suppose z varies inversely as t, and $z = 8$ when $t = 2$. Find z when $t = 32$.

(b) If p varies inversely as q, and $p = 5$ when $q = 1$, find p when $q = 4$.

EXAMPLE 4 Suppose y varies directly as x, and $y = 20$ when $x = 4$. Find y when $x = 9$.

Since y varies directly as x, there is a constant k such that $y = kx$. We know that $y = 20$ when $x = 4$. Substituting these values into $y = kx$ gives

$$y = kx$$
$$20 = k \cdot 4,$$

from which $k = 5.$

Since $y = kx$ and $k = 5$,

$$y = 5x. \quad \text{Let } k = 5$$

When $x = 9$,

$$y = 5x = 5 \cdot 9 = 45. \quad \text{Let } x = 9$$

Thus, $y = 45$ when $x = 9$. ◀

Work Problem 4 at the side.

In another common type of variation, the value of one variable increases while the value of another decreases.

Inverse Variation

y varies inversely as x if there exists a constant k such that

$$y = \frac{k}{x}.$$

EXAMPLE 5 Suppose z varies inversely as t, and $z = 8$ when $t = 5$. Find z when $t = 20$.

Since z varies inversely as t, there is a constant k such that

$$z = \frac{k}{t}.$$

Find k by replacing z with 8 and t with 5.

$$8 = \frac{k}{5} \quad \text{Let } z = 8 \text{ and } t = 5$$

Multiply both sides by 5 to get

$$k = 40,$$

so that

$$z = \frac{40}{t}.$$

When $t = 20$,

$$z = \frac{40}{20} = 2. \quad ◀$$

Work Problem 5 at the side.

ANSWERS

4. (a) 88 (b) $\frac{5}{4}$

5. (a) $\frac{1}{2}$ (b) $\frac{5}{4}$

6.7 EXERCISES

Solve each problem. See Example 1.

1. One-half of a number is 3 more than one-sixth of the same number. What is the number?

2. The numerator of the fraction $\frac{4}{7}$ is increased by an amount such that the value of the resulting fraction is $\frac{27}{21}$. By what amount was the numerator increased?

3. In a certain fraction, the denominator is 5 larger than the numerator. If 3 is added to both the numerator and the denominator, the result equals $\frac{3}{4}$. Find the original fraction.

4. The denominator of a certain fraction is three times the numerator. If 1 is added to the numerator and subtracted from the denominator, the result equals $\frac{1}{2}$. Find the original fraction.

5. One number is 3 more than another. If the smaller is added to two-thirds the larger, the result is four-fifths the sum of the original numbers. Find the numbers.

6. The sum of a number and its reciprocal is $\frac{5}{2}$. Find the number.

7. If twice the reciprocal of a number is subtracted from the number, the result is $-\frac{7}{3}$. Find the number.

8. The sum of the reciprocals of two consecutive integers is $\frac{5}{6}$. Find the integers.

9. If three times a number is added to twice its reciprocal, the answer is 5. Find the number.

10. If twice a number is subtracted from three times its reciprocal, the result is 1. Find the number.

6.7 Exercises   **321**

11. A man and his son worked four days at a job. The son's daily wage was $\frac{2}{5}$ that of the father. If together they earned $672, what were their daily wages?

12. The profits from a student show are to be given to two scholarships so that one scholarship receives $\frac{3}{2}$ as much money as the other. If the total amount given to the two scholarships is $390, find the amount that goes to the scholarship that receives the lesser amount.

13. A new instructor is paid $\frac{3}{4}$ the salary of an experienced professor. In a certain college, the total salary paid an instructor and a professor was $56,000. Find the salary paid the professor.

14. A child takes $\frac{5}{8}$ the number of pills that an adult takes for the same illness. Together the child and the adult use 26 pills. Find the number used by the adult.

Solve each problem. See Example 2.

15. Sam can row 4 miles per hour in still water. It takes as long to row 8 miles upstream as 24 miles downstream. How fast is the current?

	d	r	t
Upstream	8	$4 - x$	
Downstream	24	$4 + x$	

16. Mary flew from Philadelphia to Des Moines at 180 miles per hour and from Des Moines to Philadelphia at 150 miles per hour. The trip at the slower speed took 1 hour longer than the trip at the higher speed. Find the distance between the two cities. (Assume there was no wind in either direction.)

	d	r	t
P to D	x	180	
D to P	x	150	

17. On a business trip, Arlene traveled to her destination at an average speed of 60 miles per hour. Coming home, her average speed was 50 miles per hour and the trip took $\frac{1}{2}$ hour longer. How far did she travel each way?

18. Rae flew her airplane 500 miles against the wind in the same time it took her to fly it 600 miles with the wind. If the speed of the wind was 10 miles per hour, what was the average speed of her plane?

19. The distance from Seattle, Washington, to Victoria, British Columbia, is about 148 miles by ferry. It takes about 4 hours less to travel by ferry from Victoria to Vancouver, British Columbia, a distance of about 74 miles. What is the average speed of the ferry?

20. A boat goes 210 miles downriver in the same time it can go 140 miles upriver. The speed of the current is 5 miles per hour. Find the speed of the boat in still water.

Solve each problem. See Example 3.

21. Paul can tune up his Toyota in 2 hours. His friend Marco can do the job in 3 hours. How long would it take them if they worked together?

22. George can paint a room, working alone, in 8 hours. Jenny can paint the same room, working alone, in 6 hours. How long will it take them if they work together?

23. Machine A can do a certain job in 7 hours, and machine B takes 12 hours. How long will the job take the two machines working together?

24. One pipe can fill a swimming pool in 6 hours, and another pipe can do it in 9 hours. How long will it take the two pipes working together to fill the pool $\frac{3}{4}$ full?

25. Dennis can do a job in 4 days. When Dennis and Sue work together, the job takes $2\frac{1}{3}$ days. How long would the job take Sue if she worked alone?

26. An inlet pipe can fill a swimming pool in 9 hours, and an outlet pipe can empty the pool in 12 hours. Through an error, both pipes are left open. How long will it take to fill the pool?

The next four exercises are more challenging.

27. A cold water faucet can fill a sink in 12 minutes, and a hot water faucet can fill it in 15. The drain can empty the sink in 25 minutes. If both faucets are on and the drain is open, how long will it take to fill the sink?

28. Refer to Exercise 26. Assume the error was discovered after both pipes had been running for 3 hours, and the outlet pipe was then closed. How much more time would then be required to fill the pool? (*Hint:* How much of the job had been done when the error was discovered?)

29. An experienced employee can enter tax data into a computer twice as fast as a new employee. Working together, it takes the employees 2 hours. How long would it take the experienced employee working alone?

30. One painter can paint a house three times faster than another. Working together they can paint a house in 4 days. How long would it take the faster painter working alone?

Solve the following problems about variation. See Examples 4 and 5.

31. If x varies directly as y, and $x = 9$ when $y = 2$, find x when y is 7.

32. If z varies directly as x, and $z = 15$ when $x = 4$, find z when x is 11.

33. If m varies directly as p, and $m = 20$ when $p = 2$, find m when p is 5.

34. If a varies directly as b, and $a = 48$ when $b = 4$, find a when $b = 7$.

35. If y varies inversely as x, and $y = 10$ when $x = 3$, find y when $x = 12$.

36. If r varies inversely as s, and $r = 7$ when $s = 8$, find r when $s = 12$.

37. The circumference of a circle varies directly as the radius. A circle with a radius of 7 centimeters has a circumference of 43.96 centimeters. Find the circumference if the radius changes to 11 centimeters.

38. The pressure exerted by a certain liquid at a given point varies directly as the depth of the point beneath the surface of the liquid. The pressure at 10 feet is 50 pounds per square inch. What is the pressure at 20 feet?

39. The current in a simple electrical circuit varies inversely as the resistance. If the current is 50 amp (an *ampere* is a unit for measuring current) when the resistance is 10 ohm (an *ohm* is a unit for measuring resistance), find the current if the resistance is 5 ohm.

40. The force required to compress a spring varies directly as the change in the length of the spring. If a force of 12 pounds is required to compress a certain spring 3 inches, how much force is required to compress the spring 5 inches?

CHAPTER 6   REVIEW EXERCISES

(6.1) Find any values of the variables for which the following rational expressions are meaningless.

1. $\dfrac{2}{7x}$

2. $\dfrac{3}{m-5}$

3. $\dfrac{r-3}{r^2-2r-8}$

4. $\dfrac{3z+5}{2z^2+5z-3}$

Find the numerical value of each rational expression when **(a)** $x = 3$ and **(b)** $x = -1$.

5. $\dfrac{x^2}{x+2}$
   (a)   (b)

6. $\dfrac{5x+3}{2x-1}$
   (a)   (b)

7. $\dfrac{8x}{x^2-2}$
   (a)   (b)

8. $\dfrac{x-5}{x-3}$
   (a)   (b)

Write each rational expression in lowest terms.

9. $\dfrac{6y^2z^3}{9y^4z^2}$

10. $\dfrac{9x^2-16}{6x+8}$

11. $\dfrac{m-5}{5-m}$

12. $\dfrac{6k^2+11ky-10y^2}{12k^2-11ky+2y^2}$

(6.2) Find each product or quotient. Write each answer in lowest terms.

13. $\dfrac{12y^2}{8} \cdot \dfrac{24}{y}$

14. $\dfrac{6z^2}{9z^4} \cdot \dfrac{8z^3}{3z}$

15. $\dfrac{10p^5}{5} \div \dfrac{3p^7}{20}$

16. $\dfrac{8z^2}{(4z)^3} \div \dfrac{4z^5}{32z}$

17. $\dfrac{m-1}{2} \cdot \dfrac{5}{2m-2}$

18. $\dfrac{p+4}{6} \div \dfrac{3p+12}{2}$

19. $\dfrac{7y+14}{8y-5} \div \dfrac{4y+8}{16y-10}$

20. $\dfrac{3k+5}{k+2} \cdot \dfrac{k^2-4}{18k^2-50}$

21. $\dfrac{2a^2 - a - 6}{a^2 + 4a - 5} \div \dfrac{2a + 3}{a + 5}$

22. $\dfrac{z^2 + z - 2}{z^2 + 7z + 10} \div \dfrac{z - 3}{z + 5}$

23. $\dfrac{2p^2 + 3p - 2}{p^2 + 5p + 6} \div \dfrac{p^2 - 2p - 15}{2p^2 - 7p - 15}$

24. $\dfrac{8r^2 + 23r - 3}{64r^2 - 1} \div \dfrac{r^2 - 4r - 21}{64r^2 + 16r + 1}$

(6.3) *Find the least common denominator for each list of fractions.*

25. $\dfrac{1}{15}, \dfrac{7}{30}, \dfrac{4}{45}$

26. $\dfrac{3}{8y}, \dfrac{7}{12y^2}, \dfrac{1}{16y^3}$

27. $\dfrac{1}{y^2 + 2y}, \dfrac{4}{y^2 + 6y + 8}$

28. $\dfrac{3}{z^2 + z - 6}, \dfrac{2}{z^2 + 4z + 3}$

Rewrite each rational expression with the given denominator.

29. $\dfrac{4}{9}, \dfrac{}{45}$

30. $\dfrac{12}{m}, \dfrac{}{5m}$

31. $\dfrac{3}{8m^2}, \dfrac{}{24m^3}$

32. $\dfrac{12}{y - 4}, \dfrac{}{8y - 32}$

33. $\dfrac{-2k}{3k + 15}, \dfrac{}{15k + 75}$

34. $\dfrac{12y}{y^2 - y - 2}, \dfrac{}{(y - 2)(y + 1)(y - 4)}$

(6.4) *Add or subtract as indicated. Write each answer in lowest terms.*

35. $\dfrac{5}{m} + \dfrac{8}{m}$

36. $\dfrac{11}{3r} - \dfrac{8}{3r}$

37. $\dfrac{4}{p} + \dfrac{1}{3}$

38. $\dfrac{7}{k} - \dfrac{2}{5}$

39. $\dfrac{3 + 5m}{2} - \dfrac{m}{4}$

40. $\dfrac{8}{r^2} - \dfrac{3}{2r}$

41. $\dfrac{4}{2p - q} + \dfrac{3}{2p + q}$

42. $\dfrac{2}{y + 1} - \dfrac{3}{y - 1}$

43. $\dfrac{2}{p^2 - 4} - \dfrac{3}{5p + 10}$

44. $\dfrac{4}{z^2 - 2z + 1} - \dfrac{3}{z^2 - 1}$

326 Rational Expressions

45. $\dfrac{6}{m^2 + 5m} - \dfrac{1}{m^2 + 3m - 10}$
46. $\dfrac{10}{p^2 - 2p} - \dfrac{2}{p^2 - 5p + 6}$

(6.5) *Simplify each complex fraction.*

47. $\dfrac{\dfrac{r^2}{q^4}}{\dfrac{r^5}{q}}$
48. $\dfrac{\dfrac{m-2}{m}}{\dfrac{m+2}{3m}}$
49. $\dfrac{\dfrac{5k-1}{k}}{\dfrac{4k+3}{8k}}$
50. $\dfrac{\dfrac{6}{r} - 1}{\dfrac{6-r}{4r}}$

51. $\dfrac{\dfrac{1}{m} - \dfrac{1}{n}}{\dfrac{1}{n-m}}$
52. $\dfrac{z + \dfrac{1}{x}}{z - \dfrac{1}{x}}$
53. $\dfrac{\dfrac{1}{a+b} - 1}{\dfrac{1}{a+b} + 1}$
54. $\dfrac{\dfrac{2}{p-q} + 3}{5 - \dfrac{4}{p+q}}$

(6.6) *Solve each equation. Check your answer.*

55. $\dfrac{p}{4} - \dfrac{1}{2} = \dfrac{1}{3}$
56. $\dfrac{5+m}{m} + \dfrac{3}{4} = \dfrac{-2}{m}$

57. $\dfrac{y}{3} - \dfrac{y-2}{8} = -1$
58. $\dfrac{2}{z} - \dfrac{z}{z+3} = \dfrac{1}{z+3}$

59. $\dfrac{3y-1}{y-2} = \dfrac{5}{y-2} + 1$
60. $\dfrac{3}{m-2} + \dfrac{1}{m-1} = \dfrac{7}{m^2 - 3m + 2}$

Solve for the specified variable.

61. $\dfrac{1}{k} + \dfrac{3}{r} = \dfrac{5}{z}$ for r
62. $\dfrac{2}{a} + \dfrac{4}{b} = \dfrac{1}{c}$ for b

63. $x = \dfrac{3y}{2y+z}$ for y
64. $p^2 = \dfrac{4m}{3m-q}$ for m

(6.7) *Solve each word problem.*

65. When half a number is subtracted from two-thirds of the number, the result is 2. Find the number.

66. A certain fraction has a numerator that is 4 more than the denominator. If 6 is subtracted from the denominator, the result is 3. Find the original fraction.

67. The sum of a number and its reciprocal is 2. Find the number.

68. Five times a number is added to three times the reciprocal of the number, giving $\frac{17}{2}$. Find the number.

69. The commission received by a salesperson for selling a small car is $\frac{2}{3}$ that received for selling a large car. On a recent day, Linda sold one of each, earning a commission of $300. Find the commission for each type of car.

70. A boat goes 7 miles per hour in still water. It takes as long to go 20 miles upstream as 50 miles downstream. Find the speed of the current.

71. Kerrie flew her plane 400 kilometers with the wind in the same time it took her to go 200 kilometers against the wind. The speed of the wind is 50 kilometers per hour. Find the speed of the plane in still air.

72. A man can plant his garden in 5 hours, working alone. His daughter can do the same job in 8 hours. How long would it take them if they worked together?

73. When Mary and Sue work together on a job, they can do it in $3\frac{3}{7}$ days. Mary can do the job working alone in 8 days. How long would it take Sue working alone?

74. One painter can paint a house twice as fast as another. Working together, they can paint the house in $1\frac{1}{3}$ days. How long would it take the faster painter working alone?

75. If x varies directly as y, and $x = 12$ when $y = 5$, find x when $y = 3$.

76. If m varies directly as q, and $m = 8$ when $q = 4$, find m when $q = 6$.

77. If r varies inversely as s, and $r = 9$ when $s = 5$, find r when $s = 2$.

78. If z varies inversely as y, and $z = 5$ when $y = 2$, find z when $y = \frac{3}{4}$.

CHAPTER 6 TEST

1. Find any values for which $\dfrac{8k + 1}{k^2 - 4k + 3}$ is meaningless.

Write each rational expression in lowest terms.

2. $\dfrac{5s^3 - 5s}{2s + 2}$

3. $\dfrac{4p^2 + pq - 3q^2}{p^2 + 2pq + q^2}$

Multiply or divide. Write all answers in lowest terms.

4. $\dfrac{x^6 y}{x^3} \cdot \dfrac{y^2}{x^2 y^3}$

5. $\dfrac{8y - 16}{9} \div \dfrac{3y - 6}{5}$

6. $\dfrac{6m^2 - m - 2}{8m^2 + 10m + 3} \cdot \dfrac{4m^2 + 7m + 3}{3m^2 + 5m + 2}$

7. $\dfrac{5a^2 + 7a - 6}{2a^2 + 3a - 2} \div \dfrac{5a^2 + 17a - 12}{2a^2 + 5a - 3}$

Rewrite each rational expression with the given denominator.

8. $\dfrac{11}{7r},\ \dfrac{}{49r^2}$

9. $\dfrac{5}{8m - 16},\ \dfrac{}{24m - 48}$

Chapter 6 Test   329

Add or subtract as indicated. Write each answer in lowest terms.

10. $\dfrac{5}{x} - \dfrac{6}{x}$

11. $\dfrac{-3}{a+1} + \dfrac{5}{6a+6}$

12. $\dfrac{3}{2k^2 + 3k - 2} + \dfrac{1}{k^2 + 3k + 2}$

13. $\dfrac{5}{2m^2 - m - 10} - \dfrac{3}{2m^2 + 5m + 2}$

Simplify each complex fraction.

14. $\dfrac{\dfrac{1}{k} - 2}{\dfrac{1}{3} + k}$

15. $\dfrac{\dfrac{1}{p+4} - 2}{\dfrac{1}{p+4} + 2}$

Solve each equation.

16. $\dfrac{3}{2p} + \dfrac{12}{5p} = \dfrac{13}{20}$

17. $\dfrac{2}{p^2 - 2p - 3} = \dfrac{3}{p - 3} + \dfrac{2}{p + 1}$

For each problem, write an equation and solve it.

18. If four times a number is added to the reciprocal of twice the number, the result is 3. Find the number.

19. A man can paint a room in his house, working alone, in 5 hours. His wife can do the job in 4 hours. How long would it take them working together?

20. If x varies directly as y, and $x = 8$ when $y = 12$, find x when $y = 28$.

CUMULATIVE REVIEW EXERCISES

Factor completely.

1. $8x^2 - 24x$
2. $55q^3 - 88q^4$
3. $9rq^5 + 18q^6p$

4. $11t^6w^2 - 12r^5s$
5. $8r^2 - 9rs + 12s^3$
6. $54y^4z^3 + 48y^5z^2 - 6y^3z^4$

7. $x^2 + 6x + 5$
8. $m^2 + 9m + 14$
9. $p^2 - 4p - 5$

10. $r^2 - 3r - 40$
11. $p^2 - 7pq - 18q^2$
12. $b^2 - 9bd - 22d^2$

13. $2a^2 + 7a - 4$
14. $4m^2 + m - 14$
15. $10m^2 + 19m + 6$

16. $15x^2 - xy - 6y^2$
17. $8t^2 + 10tv + 3v^2$
18. $18r^2 - 25rs - 3s^2$

19. $9x^2 + 6x + 1$
20. $4p^2 - 12p + 9$
21. $25m^2 - 30mp + 9p^2$

22. $16t^2 + 56tz + 49z^2$
23. $25r^2 - 81t^2$
24. $100p^2 - 49q^2$

25. $100y^2 - 169p^4$
26. $64z^2 + 121y^2$
27. $144m^2 - 64p^2$

28. $100p^2 - 625q^2$
29. $6a^2m + am - 2m$
30. $2pq + 6p^3q + 8p^2q$

31. $216z^3 + 125$
32. $64r^3 - 27$
33. $8p^3 - 125q^3$

34. $27m^3 + 64p^3$
35. $1000a^3 - 125m^3$
36. $125y^3 + 216z^3$

Solve each of the following equations.

37. $(2p - 3)(p + 2) = 0$
38. $y^2 + y - 6 = 0$
39. $r^2 = 2r + 15$

40. $q^2 = 4q + 77$
41. $6m^2 + m - 2 = 0$
42. $4z^2 + 23z + 15 = 0$

43. $8m^2 = 64m$
44. $9p^2 = 36$
45. $(y - 2)(y + 1)(y - 3) = 0$

46. $(r - 5)(2r + 1)(3r - 5) = 0$
47. $x^3 = 16x$
48. $25a = a^3$

Solve each word problem.

49. One number is 4 more than another. The product of the numbers is 2 less than the smaller number. Find the smaller number.

50. One number is 8 less than another. The product of the numbers is 10 more than the larger number. Find the larger number.

51. The sum of two numbers is 8. The product of the numbers is 10 more than the larger number. Find the numbers. (*Hint:* Write the numbers as x and $8 - x$.)

52. The length of a rectangle is 2 meters less than twice the width. The area is 60 square meters. Find the width of the rectangle.

53. The length of a rectangle is 3 centimeters more than twice the width. The area of the rectangle is 44 square centimeters. Find the width of the rectangle.

54. The length of a rectangle is 1 centimeter less than twice the width. The area is 15 square centimeters. Find the width of the rectangle.

Find any values for which the following are meaningless.

55. $\dfrac{4}{3k-5}$

56. $\dfrac{7m}{2m+5}$

57. $\dfrac{1}{a^2+3a-10}$

58. $\dfrac{3m-7}{2m^2+3m-20}$

59. $\dfrac{4q-15}{9}$

60. $\dfrac{3r^2-5r+2}{8}$

Write each expression in lowest terms.

61. $\dfrac{5m}{20}$

62. $\dfrac{2p^3}{8p^5}$

63. $\dfrac{y^2-4}{6y-12}$

64. $\dfrac{5r-40}{r^2-7r-8}$

Perform each operation.

65. $\dfrac{x^6 y^2}{y^5 x^9} \cdot \dfrac{x^2}{y^3}$

66. $\dfrac{r^6 s^4}{r^3 s^5} \div \dfrac{r^{14}}{s^{10}}$

67. $\dfrac{2}{m} + \dfrac{5}{m}$

68. $\dfrac{5}{q} - \dfrac{1}{q}$

69. $\dfrac{4}{3} - \dfrac{1}{z}$

70. $\dfrac{3}{7} + \dfrac{4}{r}$

71. $\dfrac{5}{m} - \dfrac{3}{2m}$

72. $\dfrac{6}{y} + \dfrac{3}{8y}$

73. $\dfrac{2}{m+1} + \dfrac{m}{m+1}$

74. $\dfrac{p}{p-5} - \dfrac{7}{p-5}$

75. $\dfrac{3}{a-2} + \dfrac{2}{3a-6}$

76. $\dfrac{5}{2m+6} + \dfrac{2}{m+3}$

77. $\dfrac{4}{3y+9} + \dfrac{2}{5y+15}$

78. $\dfrac{4}{5q-20} - \dfrac{1}{3q-12}$

79. $\dfrac{1}{a^2-2a} + \dfrac{2}{a^2+a}$

80. $\dfrac{2}{k^2+k} - \dfrac{3}{k^2-k}$

Cumulative Review Exercises

81. $\dfrac{2m-2}{3m^2+3m-6} \cdot \dfrac{5m+10}{10m-10}$

82. $\dfrac{7z^2+49z+70}{16z^2+72z-40} \div \dfrac{3z+6}{4z^2-1}$

83. $\dfrac{3x^2+5x-2}{6x^2+x-1} \cdot \dfrac{2x^2+7x+3}{x^2+3x+2}$

84. $\dfrac{5a^2-2a}{12a^2-20a} \div \dfrac{5a^2+4a-1}{3a^2-2a-5}$

Simplify each complex fraction.

85. $\dfrac{\dfrac{5}{x}}{1+\dfrac{1}{x}}$

86. $\dfrac{\dfrac{5z^3y}{y^2z}}{\dfrac{15y^4}{z^2}}$

87. $\dfrac{2-\dfrac{5}{x}}{3+\dfrac{4}{x}}$

88. $\dfrac{\dfrac{4}{a}+\dfrac{5}{2a}}{\dfrac{7}{6a}-\dfrac{1}{5a}}$

Solve each of the following equations.

89. $\dfrac{r+2}{5} = \dfrac{r-3}{3}$

90. $\dfrac{m-4}{6} = \dfrac{m+2}{2}$

91. $\dfrac{1}{x} = \dfrac{1}{x+1} + \dfrac{1}{2}$

92. $\dfrac{1}{p} + \dfrac{1}{p-1} = \dfrac{5}{6}$

Solve each word problem.

93. Janet can weed the yard in 3 hours. Russ can weed the yard in 2 hours. How long would it take them if they worked together?

94. Tom can paint a room in 9 hours. Allen can paint the same room in 6 hours. How long would it take them if they worked together on the room?

95. When Mary and Joann work together, a job takes 2 days. Joann can do the job alone in 6 days. How long would the job take Mary if she worked alone?

96. When working together, Fred and Alan can do a job in 2 hours. Alan needs 4 hours when working alone. How long would it take Fred to do the job if he worked alone?

97. If z varies inversely as y, and $z=5$ when $y=4$, find z when y is 8.

98. Suppose y varies directly as x, and y is 20 when x is 2. Find y when x is 6.

Graphing Linear Equations 7

7.1 LINEAR EQUATIONS IN TWO VARIABLES

Most of the equations discussed so far, such as

$$3x + 5 = 12 \quad \text{or} \quad 2x^2 + x + 5 = 0,$$

have had only one variable. Equations in two variables, such as

$$y = 4x + 5 \quad \text{or} \quad 2x + 3y = 6,$$

are discussed in this chapter. Both of these equations are examples of *linear equations*.

> A **linear equation** in two variables is an equation that can be put in the form
> $$ax + by = c,$$
> where a, b, and c are real numbers and a and b are not both 0.

▶ A solution of a linear equation in two variables requires two numbers, one for each variable. For example, the equation $y = 4x + 5$ is satisfied by replacing x with 2 and y with 13, since

$$13 = 4(2) + 5. \quad \text{Let } x = 2, y = 13$$

The pair of numbers $x = 2$ and $y = 13$ gives a solution of the equation $y = 4x + 5$. The phrase "$x = 2$ and $y = 13$" is abbreviated

$$(2, 13),$$

with the x-value, 2, and the y-value, 13, given as a pair of numbers written inside parentheses. The x-value is always given first. A pair of numbers such as (2, 13) is called an **ordered pair.**

Of course, letters other than x and y may be used in the equation, with the numbers in the ordered pair usually placed in alphabetical order. For example, one solution to the equation $3p - q = -11$ is $p = -2$ and $q = 5$. This solution is written as the ordered pair $(-2, 5)$.

Work Problem 1 at the side.

▶ The next example shows how to decide whether an ordered pair is a solution of an equation.

EXAMPLE 1 Decide whether the given ordered pair is a solution of the given equation.

(a) (3, 2); $2x + 3y = 12$

Objectives

▶ Write a solution as an ordered pair.

▶ Decide whether a given ordered pair is a solution of a given equation.

▶ Complete ordered pairs for a given equation.

1. Write each solution as an ordered pair.

   (a) $x = 5$ and $y = 7$

   (b) $y = 6$ and $x = -1$

   (c) $q = 4$ and $p = -3$

   (d) $r = 3$ and $s = 12$

ANSWERS
1. (a) (5, 7) (b) (-1, 6)
   (c) (-3, 4) (d) (3, 12)

2. Decide whether or not the ordered pairs are solutions of the equation $5x + 2y = 20$.

   (a) $(0, 10)$

   (b) $(2, -5)$

   (c) $(3, 2)$

   (d) $(-4, 20)$

   (e) $(8, 10)$

3. Complete the given ordered pairs for the equation $y = 2x - 9$.

   (a) $(5, \ )$

   (b) $(2, \ )$

   (c) $(\ , 7)$

   (d) $(\ , -13)$

To see whether $(3, 2)$ is a solution of the equation $2x + 3y = 12$, substitute 3 for x and 2 for y in the given equation.

$$2x + 3y = 12$$
$$2(3) + 3(2) = 12 \quad \text{Let } x = 3; \text{ let } y = 2$$
$$6 + 6 = 12$$
$$12 = 12 \quad \text{True}$$

This result is true, so $(3, 2)$ is a solution of $2x + 3y = 12$.

(b) $(-2, -7); \quad 2m + 3n = 12$

$$2(-2) + 3(-7) = 12 \quad \text{Let } m = -2; \text{ let } n = -7$$
$$-4 + (-21) = 12$$
$$-25 = 12 \quad \text{False}$$

This result is false, so $(-2, -7)$ is *not* a solution of $2m + 3n = 12$. ◀

Work Problem 2 at the side.

▶ Choosing a number for one variable in a linear equation makes it possible to find the value of the other variable, as shown in the next example.

EXAMPLE 2 Complete the given ordered pairs for the equation $y = 4x + 5$.

(a) $(7, \ )$

In this ordered pair, $x = 7$. (Remember that x always comes first.) Find the corresponding value of y by replacing x with 7 in the equation $y = 4x + 5$.

$$y = 4(7) + 5 = 28 + 5 = 33$$

The ordered pair is $(7, 33)$.

(b) $(\ , -3)$

In this ordered pair $y = -3$. Find the value of x by replacing y with -3 in the equation; then solve for x.

$$y = 4x + 5$$
$$-3 = 4x + 5 \quad \text{Let } y = -3$$
$$-8 = 4x \quad \text{Subtract 5 from both sides}$$
$$-2 = x \quad \text{Divide both sides by 4}$$

The ordered pair is $(-2, -3)$. ◀

Work Problem 3 at the side.

EXAMPLE 3 Complete the given ordered pairs for the equation $5x - y = 24$.

Equation	Ordered pairs
$5x - y = 24$	$(5, \ ) \ (-3, \ ) \ (0, \ )$

ANSWERS

2. (a) yes (b) no (c) no (d) yes
   (e) no
3. (a) $(5, 1)$ (b) $(2, -5)$ (c) $(8, 7)$
   (d) $(-2, -13)$

Find the y-value for the ordered pair (5,) by replacing x with 5 in the given equation and solving for y.

$$5x - y = 24$$
$$5(5) - y = 24 \quad \text{Let } x = 5$$
$$25 - y = 24$$
$$-y = -1 \quad \text{Subtract 25 from both sides}$$
$$y = 1$$

The ordered pair is (5, 1).

Complete the ordered pair $(-3, \ )$ by letting $x = -3$ in the given equation. Also, complete $(0, \ )$ by letting $x = 0$.

If $x = -3$,
then $5x - y = 24$
becomes $5(-3) - y = 24$
$-15 - y = 24$
$-y = 39$
$y = -39$.

If $x = 0$,
then $5x - y = 24$
becomes $5(0) - y = 24$
$0 - y = 24$
$-y = 24$
$y = -24$.

The completed ordered pairs are as follows.

Equation	Ordered pairs
$5x - y = 24$	(5, 1) (−3, −39) (0, −24) ◀

Work Problem 4 at the side.

EXAMPLE 4 Complete the given ordered pairs for the equation $x - 2y = 8$.

Equation	Ordered pairs
$x - 2y = 8$	(2,) (, 0)
	(10,) (, −2)

Complete the two ordered pairs on the left by letting $x = 2$ and $x = 10$, respectively.

If $x = 2$,
then $x - 2y = 8$
becomes $2 - 2y = 8$
$-2y = 6$
$y = -3$.

If $x = 10$,
then $x - 2y = 8$
becomes $10 - 2y = 8$
$-2y = -2$
$y = 1$.

Now complete the two ordered pairs on the right by letting $y = 0$ and $y = -2$, respectively.

If $y = 0$,
then $x - 2y = 8$
becomes $x - 2(0) = 8$
$x - 0 = 8$
$x = 8$.

If $y = -2$,
then $x - 2y = 8$
becomes $x - 2(-2) = 8$
$x + 4 = 8$
$x = 4$.

The completed ordered pairs are as follows.

Equation	Ordered pairs
$x - 2y = 8$	(2, −3) (8, 0)
	(10, 1) (4, −2) ◀

4. Complete the given ordered pairs for the equation $x + 2y = 12$.

(a) (0, ___)

(b) (___ , 0)

(c) (4, ___)

(d) (−6, ___)

ANSWERS
4. (a) (0, 6) (b) (12, 0)
   (c) (4, 4) (d) (−6, 9)

7.1 Linear Equations in Two Variables

5. Complete the given ordered pairs for the equation $2x - 3y = 12$.

   (a) (0,)

   (b) (, 0)

   (c) (3,)

   (d) (−6,)

   (e) (, −3)

   (f) (4,)

6. Complete the ordered pairs for each equation.

   (a) $x = 3$
   (, 2) (, −4) (, 0)

   (b) $y = −4$
   (2,) (6,) (−5,)

Work Problem 5 at the side.

EXAMPLE 5 Complete the given ordered pairs for the equation $x = 5$.

Equation	Ordered pairs
$x = 5$	(, −2) (, 6) (, 3)

The given equation is $x = 5$. No matter which value of y might be chosen, the value of x is always the same, 5. Each ordered pair can be completed by placing 5 in the first position.

Equation	Ordered pairs
$x = 5$	(5, −2) (5, 6) (5, 3) ◀

When an equation such as $x = 5$ is discussed along with equations in two variables, think of $x = 5$ as an equation in two variables by rewriting $x = 5$ as $x + 0y = 5$. This form of the equation shows that for any value of y, the value of x is 5.

Each of the equations in this section has many ordered pairs as solutions. Each choice of a real number for one variable will lead to a particular real number for the other variable. This is true of linear equations in general: linear equations in two variables have an infinite number of solutions.

Work Problem 6 at the side.

ANSWERS
5. (a) (0, −4) (b) (6, 0)
   (c) (3, −2) (d) (−6, −8)
   (e) $\left(\frac{3}{2}, -3\right)$ (f) $\left(4, -\frac{4}{3}\right)$
6. (a) (3, 2), (3, −4), (3, 0)
   (b) (2, −4), (6, −4), (−5, −4)

7.1 EXERCISES

Decide whether the given ordered pair is a solution of the given equation. See Example 1.

1. $x + y = 9$; $(2, 7)$
2. $3x + y = 8$; $(0, 8)$
3. $2p - q = 6$; $(2, -2)$

4. $2v + w = 5$; $(2, 1)$
5. $4x - 3y = 6$; $(1, 2)$
6. $5x - 3y = 1$; $(0, 1)$

7. $y = 3x$; $(1, 3)$
8. $x = -4y$; $(8, -2)$
9. $x = -6$; $(-6, 8)$

10. $y = 2$; $(9, 2)$
11. $x + 4 = 0$; $(-5, 1)$
12. $x - 6 = 0$; $(5, -1)$

Complete the given ordered pairs for the equation $y = 3x + 5$. See Example 2.

13. $(2, \underline{\ \ })$
14. $(5, \underline{\ \ })$
15. $(8, \underline{\ \ })$

16. $(0, \underline{\ \ })$
17. $(-3, \underline{\ \ })$
18. $(-4, \underline{\ \ })$

19. $(\underline{\ \ }, 14)$
20. $(\underline{\ \ }, -10)$
21. $(\underline{\ \ }, 8)$

Complete the given ordered pairs for the equation $y = -4x + 8$. See Example 2.

22. $(0, \underline{\ \ })$
23. $(2, \underline{\ \ })$
24. $(\underline{\ \ }, 16)$

25. $(\underline{\ \ }, 24)$
26. $(\underline{\ \ }, -4)$
27. $(\underline{\ \ }, -8)$

Complete the given ordered pairs for the given equations. Round to the nearest thousandth in Exercises 38–41. See Examples 2–4.

28. $y = 2x + 1$ $(3, \underline{\ \ })$ $(0, \underline{\ \ })$ $(-1, \underline{\ \ })$

29. $y = 3x - 5$ $(2, \underline{\ \ })$ $(0, \underline{\ \ })$ $(-3, \underline{\ \ })$

30. $y = 8 - 3x$ $(2, \underline{\ \ })$ $(0, \underline{\ \ })$ $(-3, \underline{\ \ })$

31. $y = -2 - 5x$ $(4, \underline{\ \ })$ $(0, \underline{\ \ })$ $(-4, \underline{\ \ })$

32. $2x + y = 9$ $(0, \underline{\ \ })$ $(3, \underline{\ \ })$ $(12, \underline{\ \ })$

33. $-3m + n = 4$ (1, ___) (0, ___) (−2, ___)

34. $2p + 3q = 6$ (0, ___) (___, 0) (___, 4)

35. $4t + 3w = 12$ (0, ___) (___, 0) (___, 8)

36. $3u - 5v = 15$ (0, ___) (___, 0) (___, −6)

37. $4x - 9y = 36$ (___, 0) (0, ___) (___, 4)

38. $4x + 5y = 10$ (1.1, ___) (___, 2.3) (___, −4.2)

39. $2x - 3y = 4$ (___, 3.51) (−2.48, ___) (___, 1.77)

40. $6x - 4y = 5$ (3.82, ___) (___, −4.71) (.853, ___)

41. $4x - 3y = 7$ (2.9, ___) (___, 9.3) (___, −1.5)

Complete the given ordered pairs using the given equations. See Example 5.

42. $x = -4$ (___, 6) (___, 2) (___, −3)

43. $x = 2$ (___, 3) (___, 8) (___, 0)

44. $y = 3$ (8, ___) (4, ___) (−2, ___)

45. $y = -8$ (4, ___) (0, ___) (−4, ___)

46. $x + 9 = 0$ (___, 8) (___, 3) (___, 0)

47. $y + 4 = 0$ (9, ___) (2, ___) (0, ___)

48. $y - 5 = 0$ (6, ___) (4, ___) (−8, ___)

7.2 GRAPHING ORDERED PAIRS

As shown earlier, a number line is used to graph the solutions of an equation in one variable. Techniques for graphing the solutions of an equation in *two* variables will be shown in this section. The solutions of such an equation are *ordered pairs* of numbers of the form (x, y), so *two* number lines are needed, one for x and one for y. These two number lines can be drawn as shown in Figure 1. The horizontal line is called the *x-axis* and the vertical line is called the *y-axis*. Together, the x-axis and y-axis form a **coordinate system.**

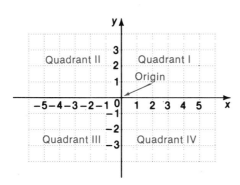

Figure 1

▶ The coordinate system is divided into four regions, called **quadrants.** These quadrants are numbered counterclockwise, as shown in Figure 1. Points on the axes themselves are not in any quadrant. The point at which the x-axis and y-axis meet is called the **origin.**

Work Problem 1 at the side.

▶ By referring to the two axes, every point in the plane can be associated with an ordered pair. The numbers in the ordered pair are called the **coordinates** of the point. For example, locate the point associated with the ordered pair (2, 3) by starting at the origin. Since the x-coordinate is 2, go 2 units to the right along the x-axis. Then, since the y-coordinate is 3, turn and go up 3 units on a line parallel to the y-axis. The point (2, 3) is **plotted** in Figure 2. From now on we will refer to the point with x-coordinate 2 and y-coordinate 3 as the point (2, 3).

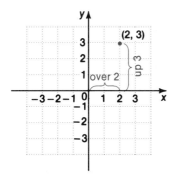

Figure 2

Objectives

▶ Decide in which quadrant a point is located.

▶ Graph ordered pairs.

▶ Given an equation, find and graph ordered pairs.

1. Name the quadrant in which each point in the figure is located.

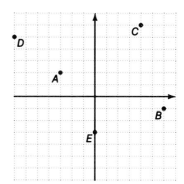

ANSWERS

1. *A*, II; *B*, IV; *C*, I; *D*, II; *E*, no quadrant

7.2 Graphing Ordered Pairs 341

2. Plot the given ordered pairs on a coordinate system.

   (a) (3, 5)

   (b) (−2, 6)

   (c) (−4, 0)

   (d) (−5, −2)

   (e) (5, −2)

   (f) (0, −6)

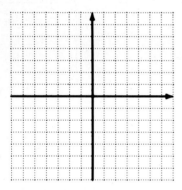

EXAMPLE 1 Plot the given ordered pairs on a coordinate system.

(a) (1, 5)

(b) (−2, 3)

(c) (−1, −4)

(d) (7, −2)

(e) $\left(\frac{3}{2}, 2\right)$

Locate the point (−1, −4), for example, by first going 1 unit to the left along the x-axis. Then turn and go 4 units down, parallel to the y-axis. Plot the point $(\frac{3}{2}, 2)$ by first going $\frac{3}{2}$ (or $1\frac{1}{2}$) units to the right along the x-axis. Then turn and go 2 units up, parallel to the y-axis. Figure 3 shows the graphs of the points in this example. ◂

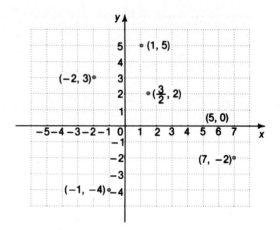

Figure 3

Work Problem 2 at the side.

3▸ The next section shows how to graph a linear equation. The first step in graphing these equations is to find and graph some ordered pairs that satisfy the equation, as shown in the next example.

EXAMPLE 2 Complete the given ordered pairs. Then plot the ordered pairs on a coordinate system.

Equation	Ordered pairs
$x + 2y = 7$	(1,) (−3,) (3,) (7,)

Complete the ordered pairs by substituting the given x-values into the equation $x + 2y = 7$.

When $x =$ **1**,
$x + 2y = 7$
$1 + 2y = 7$
$2y = 6$
$y =$ **3**.

When $x =$ **−3**,
$x + 2y = 7$
$-3 + 2y = 7$
$2y = 10$
$y =$ **5**.

The ordered pairs are (1, 3) and (−3, 5).

ANSWERS

2.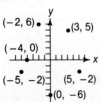

In the same way, if $x = 3$, then $y = 2$, giving $(3, 2)$. Finally, if $x = 7$, then $y = 0$, giving $(7, 0)$. The completed ordered pairs are as follows.

Equation	Ordered pairs
$x + 2y = 7$	$(1, 3)$ $(-3, 5)$ $(3, 2)$ $(7, 0)$

The graph of these ordered pairs is shown in Figure 4. ◀

3. Complete the given ordered pairs for the equation $2x + y = 6$ and graph the results.

(a) $(0, \ )$

(b) $(2, \ )$

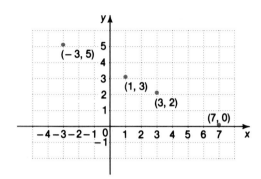

(c) $(4, \ )$

Figure 4

(d) $(\ , 1)$

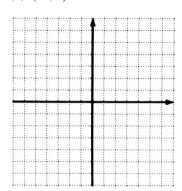

Work Problem 3 at the side.

EXAMPLE 3 In a certain city, the cost of a taxi ride of x miles is given by

$$y = 25x + 250,$$

where y represents the cost in cents. Complete the given ordered pairs for this equation.

$(1, \ )$ $(2, \ )$ $(3, \ )$

Complete the ordered pair $(1, \ )$ by letting $x = 1$.

$y = 25x + 250$
$y = 25(1) + 250$ Let $x = 1$
$y = 25 + 250$
$y = 275$

This gives the ordered pair $(1, 275)$, which says that a taxi ride of 1 mile costs 275 cents, or $2.75.

Complete $(2, \ )$ and $(3, \ )$ as follows.

$y = 25x + 250$ $y = 25x + 250$
$y = 25(2) + 250$ Let $x = 2$ $y = 25(3) + 250$ Let $x = 3$
$y = 50 + 250$ $y = 75 + 250$
$y = 300$ $y = 325$

These ordered pairs $(2, 300)$ and $(3, 325)$, along with the ones given in Problem 4 on the next page, are graphed in Figure 5. In this graph, different scales are used on the two axes, since the y-values in the ordered pairs are much larger than the x-values. Here, each square represents 100 units in the vertical direction. ◀

ANSWERS
3. (a) $(0, 6)$ (b) $(2, 2)$ (c) $(4, -2)$
(d) $\left(\dfrac{5}{2}, 1\right)$

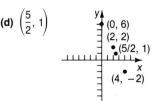

7.2 Graphing Ordered Pairs   343

4. Complete the given ordered pairs using the equation $y = 25x + 250$ in Example 3.

(a) (4,)

(b) (5,)

(c) (6,)

(d) (7,)

(e) (8,)

(f) (9,)

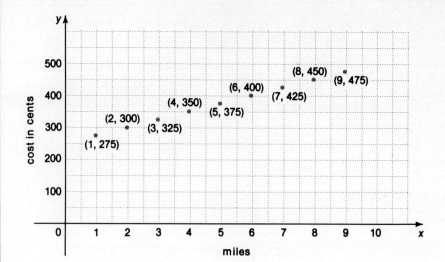

Figure 5

Work Problem 4 at the side.

ANSWERS

4. (a) (4, 350) (b) (5, 375)
   (c) (6, 400) (d) (7, 425)
   (e) (8, 450) (f) (9, 475)

7.2 EXERCISES

Write the x- and y-coordinates of the points labeled A through F in the figure.

1. A
2. B
3. C
4. D
5. E
6. F

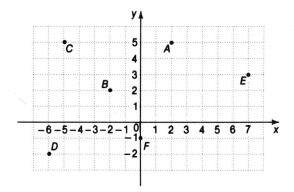

Plot the ordered pairs on the coordinate system provided. See Example 1.

7. (6, 1)
8. (4, −2)
9. (3, 5)

10. (−4, −5)
11. (−2, 4)
12. (−5, −1)

13. (−3, 5)
14. (3, −5)
15. (4, 0)

16. (1, 0)
17. (−2, 0)
18. (0, 3)

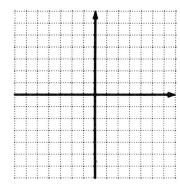

Without plotting the given point, name the quadrant in which each point lies.

19. (2, 3)
20. (2, −3)
21. (−2, 3)
22. (−2, −3)
23. (−1, −1)

24. (4, 7)
25. (−3, 6)
26. (−2, 0)
27. (5, −4)
28. (0, 4)

Complete the given ordered pairs using the given equation; then plot the ordered pairs. See Example 2.

29. $x + 2y = 6$
 (0,) (, 0)
 (4,) (−1,)

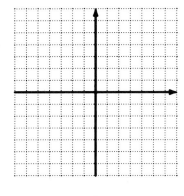

30. $3x − y = 6$
 (0,) (, 3)
 (, 0) (, −3)

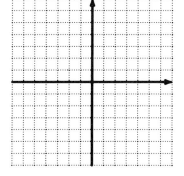

31. $4x - 3y = 12$
(0,) (, 0)
(6,) ($1\frac{1}{2}$,)

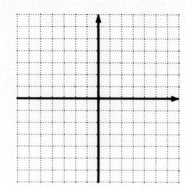

32. $3x + 4y = 12$
(0,) (, 0)
(2,) (, $-1\frac{1}{2}$)

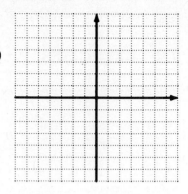

33. $y + 2 = 0$
(5,) (0,)
(-3,) (-2,)

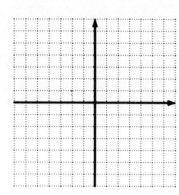

34. $x - 4 = 0$
(, 6) (, 0)
(, -4) (, 4)

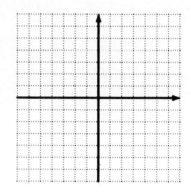

35. In statistics we often want to know whether two quantities (such as the height and weight of an individual) are related in such a way that we can predict one given the other. To find out, ordered pairs that give these quantities for a number of individuals are plotted on a graph called a *scatter diagram*. If the points lie approximately in a line, the variables have a linear relationship.

(a) Make a scatter diagram by plotting on the given axes the following pairs of heights and weights for six men: (70, 162), (72, 160), (75, 180), (69, 147), (73, 175), (73, 184). (The break on the horizontal and vertical axes shows that numbers have been skipped.)

(b) Does there seem to be a linear relationship between height and weight?

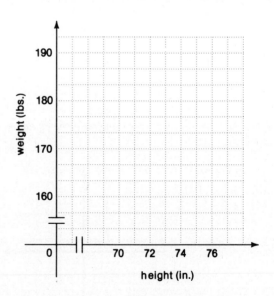

346 Graphing Linear Equations

7.3 GRAPHING LINEAR EQUATIONS

Objectives

1. Graph linear equations by completing ordered pairs.
2. Find intercepts.
3. Graph linear equations with just one intercept.
4. Graph linear equations of the form $y =$ a number or $x =$ a number.

▶ There are an infinite number of ordered pairs that satisfy an equation in two variables. For example, find ordered pairs that are solutions of the equation $x + 2y = 7$ by choosing as many values of x (or y) as desired and then completing each ordered pair.

For example, if we choose $x = 1$, then $y = 3$, so that the ordered pair $(1, 3)$ is a solution of the equation $x + 2y = 7$.

$$1 + 2(3) = 1 + 6 = 7$$

Work Problem 1 at the side.

Figure 6 shows a graph of all the ordered pairs found for $x + 2y = 7$ in Problem 1 at the side.

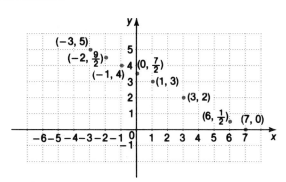

Figure 6

1. Complete the given ordered pairs for the equation $x + 2y = 7$.

 (a) $(-3, \ )$

 (b) $(3, \ )$

 (c) $(0, \ )$

 (d) $(-2, \ )$

 (e) $(-1, \ )$

 (f) $(6, \ )$

 (g) $(7, \ )$

Notice that the points plotted in this figure all appear to lie on a straight line. The line that goes through these points is shown in Figure 7. In fact, all ordered pairs satisfying the equation $x + 2y = 7$ correspond to points that lie on this same straight line. This graph gives a "picture" of all the solutions of the equation $x + 2y = 7$. Only a portion of the line is shown here, but it extends indefinitely in both directions, as suggested by the arrowhead on each end of the line.

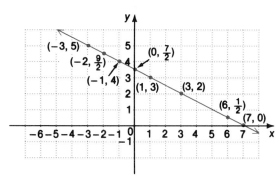

Figure 7

It can be shown that

the graph of any linear equation in two variables is a straight line.

(Remember that the word *line* appears in the name "*linear* equation.")

ANSWERS

1. (a) $(-3, 5)$ (b) $(3, 2)$ (c) $\left(0, \dfrac{7}{2}\right)$
 (d) $\left(-2, \dfrac{9}{2}\right)$ (e) $(-1, 4)$
 (f) $\left(6, \dfrac{1}{2}\right)$ (g) $(7, 0)$

7.3 Graphing Linear Equations

2. Complete the given ordered pairs and then graph the line.

$x + y = 6$

(0,   )
(, 0)
(2,   )

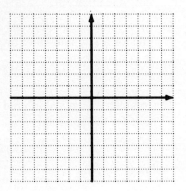

Since two distinct points determine a line, a straight line can be graphed by finding any two different points on the line. However, it is a good idea to plot a third point as a check.

EXAMPLE 1 Graph the linear equation $3x + 2y = 6$.

For most linear equations, two different points on the graph can be found by first letting $x = 0$ and then letting $y = 0$.

If $x = 0$,

$3x + 2y = 6$
$3(0) + 2y = 6$ Let $x = 0$
$0 + 2y = 6$
$2y = 6$
$y = 3$.

If $y = 0$,

$3x + 2y = 6$
$3x + 2(0) = 6$ Let $y = 0$
$3x + 0 = 6$
$3x = 6$
$x = 2$.

This gives the ordered pairs $(0, 3)$ and $(2, 0)$. Get a third point (as a check) by letting x or y equal some other number. For example, let $x = -2$. (Just about any other number could have been used instead.) Replace x with -2 in the given equation.

$3x + 2y = 6$
$3(-2) + 2y = 6$ Let $x = -2$
$-6 + 2y = 6$
$2y = 12$
$y = 6$

Now plot these three ordered pairs $(0, 3)$, $(2, 0)$, and $(-2, 6)$, which should lie on a line, and draw a line through them. This line, shown in Figure 8, is the graph of $3x + 2y = 6$. ◀

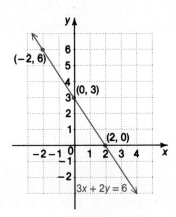

Figure 8

Work Problem 2 at the side.

EXAMPLE 2 Graph the linear equation $4x - 5y = 20$.

As above, at least two different points are needed to draw the graph. First let $x = 0$ and then let $y = 0$ to complete two ordered pairs.

ANSWERS

2. (0, 6), (6, 0), (2, 4)

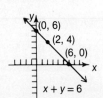

Graphing Linear Equations

$$4x - 5y = 20 \qquad\qquad 4x - 5y = 20$$
$$4(0) - 5y = 20 \quad \text{Let } x = 0 \qquad 4x - 5(0) = 20 \quad \text{Let } y = 0$$
$$-5y = 20 \qquad\qquad\qquad 4x = 20$$
$$y = -4 \qquad\qquad\qquad\quad x = 5$$

The ordered pairs are $(0, -4)$ and $(5, 0)$. Get a third ordered pair (as a check) by choosing some number other than 0 for x or y. This time, let us choose $y = 2$. Replacing y with 2 in the equation $4x - 5y = 20$ leads to the ordered pair $(\frac{15}{2}, 2)$, or $(7\frac{1}{2}, 2)$.

Plot the three ordered pairs we have found, $(0, -4)$, $(5, 0)$, and $(7\frac{1}{2}, 2)$, and draw a line through them. This line, shown in Figure 9, is the graph of $4x - 5y = 20$. ◀

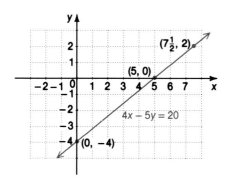

Figure 9

Work Problem 3 at the side.

▶ In Figure 9 the graph crosses the y-axis at $(0, -4)$ and crosses the x-axis at $(5, 0)$. For this reason, $(0, -4)$ is called the **y-intercept,** and $(5, 0)$ is called the **x-intercept** of the graph. The intercepts are particularly useful for graphing linear equations, as in Examples 1 and 2. The intercepts are found by replacing, in turn, each variable with 0 in the equation and solving for the value of the other variable.

> Find the x-intercept by letting $y = 0$ in the given equation and solving for x.
> Find the y-intercept by letting $x = 0$ in the given equation and solving for y.

EXAMPLE 3 Find the intercepts for the graph of $2x + y = 4$. Draw the graph.

Find the y-intercept by letting $x = 0$; find the x-intercept by letting $y = 0$.

$$2x + y = 4 \qquad\qquad 2x + y = 4$$
$$2(0) + y = 4 \quad \text{Let } x = 0 \qquad 2x + 0 = 4 \quad \text{Let } y = 0$$
$$0 + y = 4 \qquad\qquad\qquad 2x = 4$$
$$y = 4 \qquad\qquad\qquad\quad x = 2$$

3. Complete three ordered pairs and graph each line.

(a) $2x - 3y = 6$

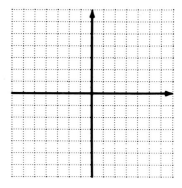

(b) $3y - 2x = 6$

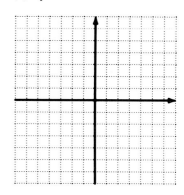

ANSWERS

3. (a) 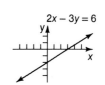   (b)

7.3 Graphing Linear Equations

349

4. Find the intercepts for
   $5x + 2y = 10$.

The y-intercept is (0, 4). The x-intercept is (2, 0). The graph with the two intercepts shown in color is given in Figure 10. Get a third point as a check. For example, choosing $x = 1$ gives $y = 2$. Plot (0, 4), (2, 0), and (1, 2) and draw a line through them. This line, shown in Figure 10, is the graph of $2x + y = 4$. ◀

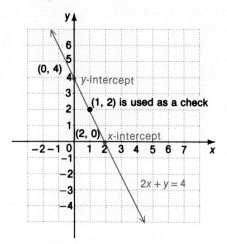

Figure 10

Work Problem 4 at the side.

▶ In the examples above, the x- and y-intercepts were used to help draw the graphs. This is not always possible, as the following examples show. Example 4 shows what to do when the x- and y-intercepts are the same point.

EXAMPLE 4 Graph the linear equation $x = 3y$.

If we let $x = 0$, then $y = 0$, giving the ordered pair (0, 0). Letting $y = 0$ also gives (0, 0). This is the same ordered pair, so choose two additional values for x or y. Choosing 2 for y gives $x = 3 \cdot 2 = 6$, giving the ordered pair (6, 2). For a check point, choose -6 for x and get -2 for y. This ordered pair, $(-6, -2)$, along with (0, 0) and (6, 2), was used to get the graph shown in Figure 11. ◀

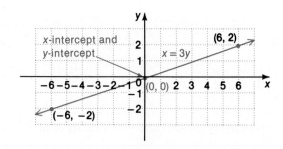

Figure 11

Generalizing from Example 4,

> If a is a real number, the graph of a linear equation of the form
>
> $x = ay$ or $y = ax$
>
> goes through the origin (0, 0).

ANSWER
4. x-intercept (2, 0); y-intercept (0, 5)

Work Problem 5 at the side.

▶ The equation $y = -4$ is a linear equation with the coefficient of x equal to 0. (To see this, write $y = -4$ as $y = 0x - 4$.) Also, $x = 3$ is a linear equation with the coefficient of y equal to 0. These equations lead to horizontal or vertical straight lines, as the next examples show.

EXAMPLE 5 Graph the linear equation $y = -4$.

As the equation states, for any value of x that might be chosen, y is always equal to -4. Get ordered pairs that are solutions of this equation by choosing different numbers for x but always using -4 for y. Three ordered pairs that can be used are $(-2, -4)$, $(0, -4)$, and $(3, -4)$. Drawing a line through these points gives the horizontal line shown in Figure 12. ◀

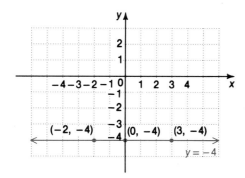

Figure 12

Horizontal Line

The graph of the linear equation $y = k$, where k is a real number, is the horizontal line going through the point $(0, k)$.

EXAMPLE 6 Graph the linear equation $x - 3 = 0$.

First add 3 to both sides of $x - 3 = 0$ to get $x = 3$. All the ordered pairs that are solutions of this equation have an x-coordinate of 3. Any number can be used for y. Three ordered pairs that work are $(3, 3)$, $(3, 0)$ and $(3, -2)$. Drawing a line through these points gives the vertical line shown in Figure 13. ◀

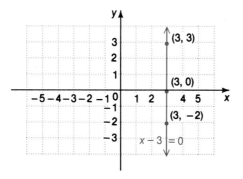

Figure 13

5. Graph each equation.

(a) $2x = y$

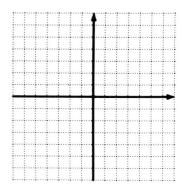

(b) $x = -4y$

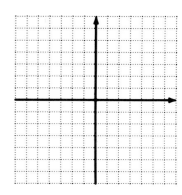

ANSWERS

5. (a) 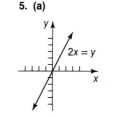 (b)

7.3 Graphing Linear Equations

6. Graph each equation.

(a) $y = -5$

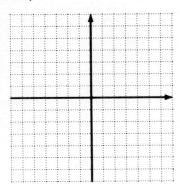

(b) $x = 2$

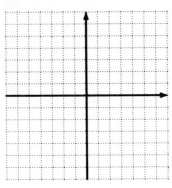

Vertical Line

The graph of the linear equation $x = k$, where k is a real number, is the vertical line going through the point $(k, 0)$.

Work Problem 6 at the side.

The different forms of straight-line equations and the methods of graphing them are summarized below.

Equation	Graphing methods	Example
$y = k$	Draw a horizontal line, through $(0, k)$.	$y = -2$
$x = k$	Draw a vertical line, through $(k, 0)$	$x = 4$
$y = ax$ or $x = ay$	Graph goes through $(0, 0)$. To get additional points that lie on the graph, choose any value of x, or y, except 0.	$y = -3x$ $x = 2y$
$ax + by = c$ but not of the types above	Find any two points the line goes through. A good choice is to find the intercepts: let $x = 0$, and find the corresponding value of y; then let $y = 0$, and find x. As a check, get a third point by choosing a value of x or y that has not yet been used.	$3x - 2y = 6$

ANSWERS

6. (a) 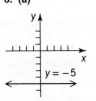 (b)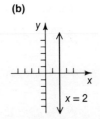

7.3 EXERCISES

Complete the given ordered pairs using the given equation. Then graph the equation by plotting the points and drawing a line through them. See Examples 1 and 2.

1. $x + y = 5$
   (0,) (, 0) (2,)

2. $y = x - 3$
   (0,) (, 0) (5,)

3. $y = x + 4$
   (0,) (, 0) (−2,)

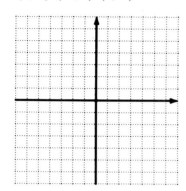
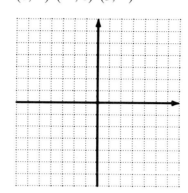
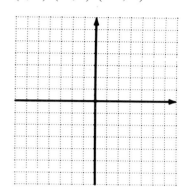

4. $y + 5 = x$
   (0,) (, 0) (6,)

5. $y = 3x - 6$
   (0,) (, 0) (3,)

6. $x = 2y + 1$
   (, 2) (, 0) (, −3)

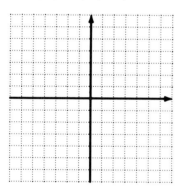
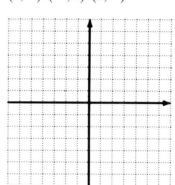
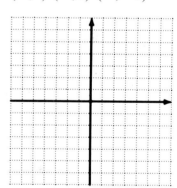

Find the intercepts for each equation. See Example 3.

7. $2x + 3y = 6$
   x-intercept:
   y-intercept:

8. $7x + 2y = 14$
   x-intercept:
   y-intercept:

9. $3x - 5y = 9$
   x-intercept:
   y-intercept:

10. $6x - 5y = 12$
    x-intercept:
    y-intercept:

11. $2y = 5x$
    x-intercept:
    y-intercept:

12. $6y = 12$
    x-intercept:
    y-intercept:

Graph each linear equation. See Examples 1, 2, and 4–6.

13. $x - y = 2$

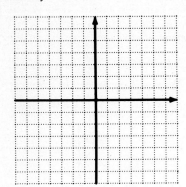

14. $x + y = 6$

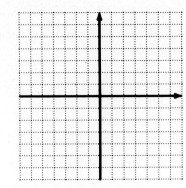

15. $y = x + 4$

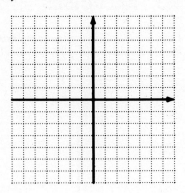

16. $y = x - 5$

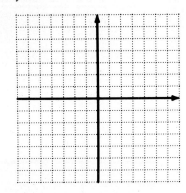

17. $x + 2y = 6$

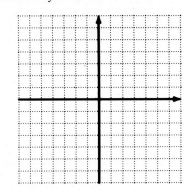

18. $3x - y = 6$

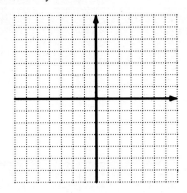

19. $4x = 3y - 12$

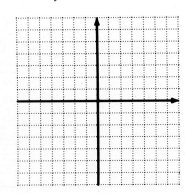

20. $5x = 2y - 10$

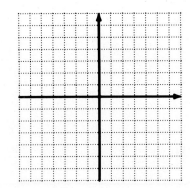

21. $3x = 6 - 2y$

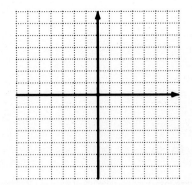

name   date   hour

22. $2x + 3y = 12$ **23.** $2x - 7y = 14$ **24.** $3x + 5y = 15$

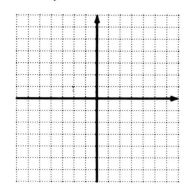 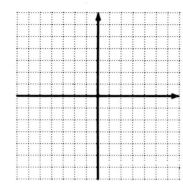 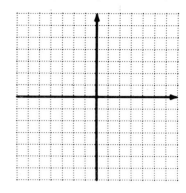

25. $3x + 7y = 14$ **26.** $6x - 5y = 18$ **27.** $y = 2x$

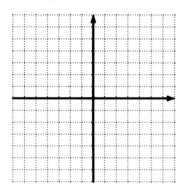 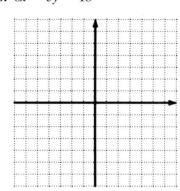 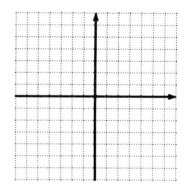

28. $y = -3x$ **29.** $y + 6x = 0$ **30.** $y - 4x = 0$

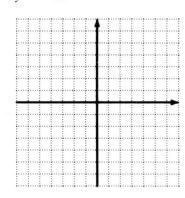 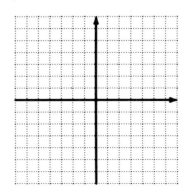 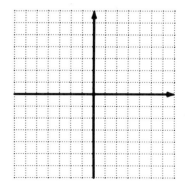

7.3 Exercises

31. $2x + 3y = 0$

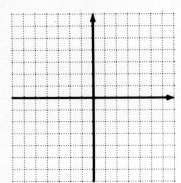

32. $3x - 4y = 0$

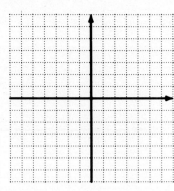

33. $x + 2 = 0$

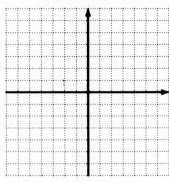

34. $y - 3 = 0$

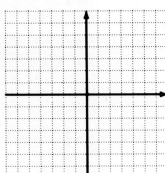

35. $y = 6$

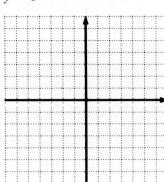

36. $y = 0$

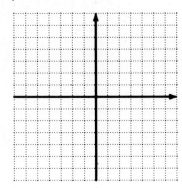

37. The graph* shows the average cost of attending college for the school years ending as shown.

(a) Use the graph to estimate the cost of attending a private school in 1986.

(b) Use the graph to estimate the cost of attending a public school in 1987.

(c) In 1989, how much more will a private school cost than a public school?

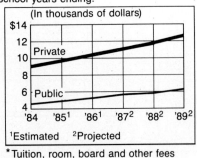

The average costs* of attending college for school years ending:

¹Estimated ²Projected

*Tuition, room, board and other fees

38. The demand for an item is closely related to its price. As price goes up, demand goes down. On the other hand, when price goes down, demand goes up. Suppose the demand for a certain small calculator is 500 when its price is $30 and 4000 when it costs $15.

(a) Let x be the price and y be the demand (in thousands) for the calculator. Graph the two given pairs of prices and demands.

(b) Assume the relationship is linear. Draw a line through the two points from part (a). From your graph estimate the demand if the price drops to $10.

(c) Use the graph to estimate the price if the demand is 5000.

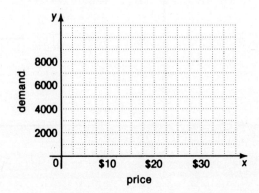

---

*Reprinted by permission of *The Wall Street Journal*, March 4, 1986. © Dow Jones & Company, Inc. 1986. All Rights Reserved.

7.4 THE SLOPE OF A LINE

A straight line can be graphed if at least two different points on the line are known. A line can also be graphed by using one point that the line goes through, along with the "steepness" of the line.

▶ One way to get a measure of the "steepness" of a line is to compare the vertical change in the line (the rise) to the horizontal change (the run) while moving along the line from one fixed point to another. This measure of steepness is called the **slope** of the line. Slope is defined as follows. (See Figure 14.)

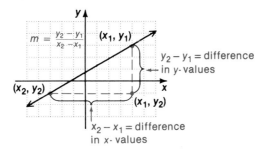

Figure 14

> The **slope** of a line is the quotient of the difference of the *y*-values and the difference of the *x*-values, or if (x_1, y_1) (read "*x* sub one, *y* sub one") and (x_2, y_2) are any two *different* points on a line, then the slope, *m*, of the line is
>
> $$m = \frac{y_2 - y_1}{x_2 - x_1} \quad \text{if } x_2 \neq x_1.$$

Work Problem 1 at the side.

EXAMPLE 1 Find the slope of each of the following lines.

(a) The line through $(-4, 7)$ and $(1, -2)$.

Use the definition given above. Let $(-4, 7) = (x_2, y_2)$ and $(1, -2) = (x_1, y_1)$. Then

$$\text{slope} = m = \frac{y_2 - y_1}{x_2 - x_1} = \frac{7 - (-2)}{-4 - 1} = \frac{9}{-5} = -\frac{9}{5}.$$

See Figure 15.

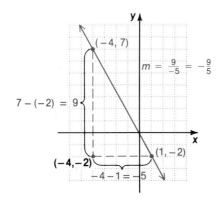

Figure 15

Objectives

▶ Given two points, find the slope of the line on which they lie.

▶ Given the equation of a line, find the slope.

▶ Use slopes to determine whether two lines are parallel, perpendicular, or neither.

1. Evaluate $\dfrac{y_2 - y_1}{x_2 - x_1}$ for the following values.

(a) $y_2 = 4,\ y_1 = -1$
$x_2 = 3,\ x_1 = 4$

(b) $x_1 = 3,\ x_2 = -5,$
$y_1 = 7,\ y_2 = -9$

(c) $x_1 = 2,\ x_2 = 7,$
$y_1 = 4,\ y_2 = 9$

ANSWERS
1. (a) -5 (b) 2 (c) 1

2. Find the slope of each of the following lines.

   (a) Through $(6, -2)$ and $(5, 4)$

   (b) Through $(-3, 5)$ and $(-4, 7)$

   (c) Through $(6, -8)$ and $(-2, 4)$

(b) The line through $(12, -5)$ and $(-9, -2)$

$$m = \frac{y_2 - y_1}{x_2 - x_1} = \frac{-5 - (-2)}{12 - (-9)} = \frac{-3}{21} = -\frac{1}{7}$$

The same slope is obtained by subtracting in reverse order.

$$\frac{-2 - (-5)}{-9 - 12} = \frac{3}{-21} = -\frac{1}{7} \blacktriangleleft$$

It makes no difference which point is (x_1, y_1) or (x_2, y_2); however,

it is important to be consistent: start with the *x*- and *y*-values of *one* point (either one) and subtract the corresponding values of the *other* point.

Work Problem 2 at the side.

The next examples show how to look for slopes of horizontal and vertical lines.

EXAMPLE 2 Find the slope of the line through $(-8, 4)$ and $(2, 4)$.

Use the definition of slope.

$$m = \frac{y_2 - y_1}{x_2 - x_1} = \frac{4 - 4}{-8 - 2} = \frac{0}{-10} = 0$$

As shown in Figure 16 by a sketch of the graph through these two points, the line through the given points is horizontal. *All horizontal lines have a slope of 0,* since the difference in *y*-values will always be 0. ◀

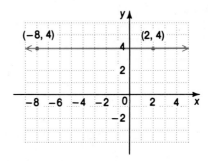

Figure 16

EXAMPLE 3 Find the slope of the line through $(6, 2)$ and $(6, -9)$.

$$m = \frac{y_2 - y_1}{x_2 - x_1} = \frac{2 - (-9)}{6 - 6} = \frac{11}{0} \quad \text{Undefined}$$

Since division by zero is meaningless, this line has an undefined slope. (This is why the formula for slope at the beginning of this section had the restriction $x_1 \neq x_2$.) The graph in Figure 17 shows that this line is vertical. Since all points on a vertical line have the same *x*-value, *vertical lines have undefined slope.* ◀

ANSWERS

2. (a) -6 (b) -2 (c) $-\dfrac{3}{2}$

Graphing Linear Equations

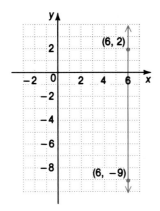

Figure 17

3. Find the slope of each line.

   (a) Through (2, 5) and (−1, 5)

   (b) Through (3, 1) and (3, −4)

   (c) With equation $y = -1$

   (d) With equation $x - 4 = 0$

Horizontal lines, which have equations of the form $y = k$, have a slope of 0.

Vertical lines, which have equations of the form $x = k$, have an undefined slope.

Work Problem 3 at the side.

▶ The slope of a line can be found directly from its equation. For example, let us find the slope of the line

$$y = -3x + 5.$$

The definition of slope requires two different points on the line. Get these two points by first choosing two different values of x and then finding the corresponding values of y. Let us choose $x = -2$ and $x = 4$.

$y = -3(-2) + 5$ Let $x = -2$ $y = -3(4) + 5$ Let $x = 4$
$y = 6 + 5$ $y = -12 + 5$
$y = 11$ $y = -7$

The ordered pairs are $(-2, 11)$ and $(4, -7)$. Now find the slope.

$$\text{Slope} = \frac{11 - (-7)}{-2 - 4} = \frac{18}{-6} = -3$$

The slope, -3, is the same number as the coefficient of x in the equation $y = -3x + 5$. It can be shown that this always happens, as long as the equation is solved for y. This fact is used to find the slope of a line from its equation.

Finding the Slope of a Line From Its Equation

Step 1 Solve the equation for y.

Step 2 The slope is given by the coefficient of x.

EXAMPLE 4 Find the slope of each of the following lines.

(a) $2x - 5y = 4$

ANSWERS

3. (a) 0   (b) undefined slope
   (c) 0   (d) undefined slope

7.4 The Slope of a Line

4. Find the slope of each line.

   (a) $y = 2x - 7$

   (b) $y = -\frac{7}{2}x + 1$

   (c) $3x + 2y = 9$

   (d) $4x - 5y = 12$

   (e) $y + 4 = 0$

Solve the equation for y.

$$2x - 5y = 4$$
$$-5y = -2x + 4 \quad \text{Subtract } 2x \text{ from both sides}$$
$$y = \frac{2}{5}x - \frac{4}{5} \quad \text{Divide both sides by } -5$$

The slope is given by the coefficient of x, so

$$\text{slope} = \frac{2}{5}.$$

(b) $8x + 4y = 1$

Solve for y; you should get

$$y = -2x + \frac{1}{4}.$$

The slope of this line is given by the coefficient of x, which is -2. ◀

Work Problem 4 at the side.

3 Slopes can be used to tell whether two lines are parallel. For example, Figure 18 shows the graph of $x + 2y = 4$ and the graph of $x + 2y = -6$. These lines appear to be parallel. Solve for y to find that both $x + 2y = 4$ and $x + 2y = -6$ have a slope of $-\frac{1}{2}$. It turns out that parallel lines always have equal slopes.

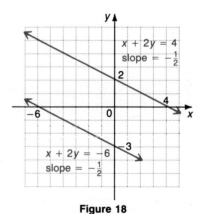

Figure 18

Also, Figure 19 shows the graph of $x + 2y = 4$ and the graph of $2x - y = 6$. These lines appear to be perpendicular (meet at a 90° angle). Solving for y shows that the slope of $x + 2y = 4$ is $-\frac{1}{2}$, while the slope of $2x - y = 6$ is 2. The product of $-\frac{1}{2}$ and 2 is

$$\left(-\frac{1}{2}\right)(2) = -1.$$

It turns out that the product of the slopes of two perpendicular lines is always -1. In summary,

two lines with the same slope are parallel; two lines whose slopes have a product of -1 are perpendicular.

ANSWERS

4. (a) 2 (b) $-\frac{7}{2}$ (c) $-\frac{3}{2}$ (d) $\frac{4}{5}$
   (e) 0

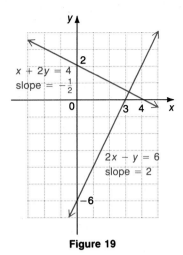

Figure 19

EXAMPLE 5 Decide whether each pair of lines is *parallel, perpendicular,* or *neither.*

(a) $\quad x + 2y = 7$
$\quad -2x + y = 3$

Find the slope of each line by first solving each equation for y.

$$x + 2y = 7$$
$$2y = -x + 7$$
$$y = -\frac{1}{2}x + \frac{7}{2}$$

Slope is $-\frac{1}{2}$.

$$-2x + y = 3$$
$$y = 2x + 3$$

Slope is 2.

Since the slopes are not equal, the lines are not parallel. Check the product of the slopes: $(-\frac{1}{2})(2) = -1$. The two lines are perpendicular because the products of their slopes is -1. See Figure 20.

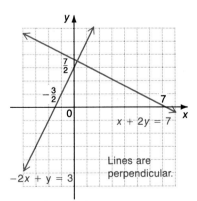

Lines are perpendicular.

Figure 20

7.4 The Slope of a Line

5. Write *parallel*, *perpendicular*, or *neither* for each pair of lines.

(a) $x + y = 6$
$x + y = 1$

(b) $3x - y = 4$
$x + 3y = 9$

(c) $2x - y = 5$
$2x + y = 3$

(b) $3x - y = 4$
$6x - 2y = -12$

Find the slopes. Both lines have a slope of 3, so the lines are parallel. The graphs of these lines are shown in Figure 21.

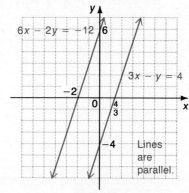

Figure 21

(c) $4x + 3y = 6$
$2x - y = 5$

Here the slopes are $-\frac{4}{3}$ and 2. These lines are neither parallel nor perpendicular. ◀

Work Problem 5 at the side.

ANSWERS
5. (a) parallel **(b)** perpendicular
  (c) neither

7.4 EXERCISES

Find the slope of each of the following lines. See Example 1.

1.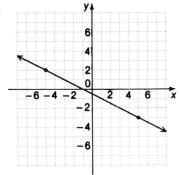

2.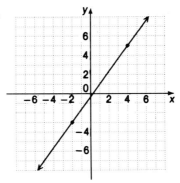

3.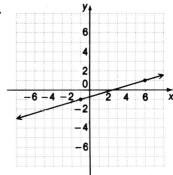

4.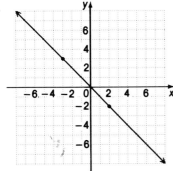

5.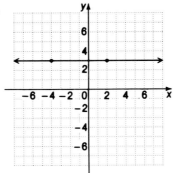

6.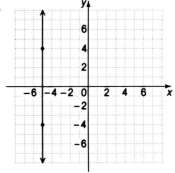

Find the slope of the line going through each pair of points. Round to the nearest thousandth in Exercises 20 and 21. See Examples 1–3.

7. $(-4, 1)$ $(2, 8)$

8. $(3, 7)$ $(5, 2)$

9. $(-1, 2)$ $(-3, -7)$

10. $(5, -4)$ $(-5, -9)$

11. $(8, 0)$ $(0, 5)$

12. $(0, -3)$ $(2, 0)$

13. $(-1, 6)$ $(4, 6)$

14. $\left(2, \frac{1}{4}\right) \left(3, -\frac{1}{4}\right)$

15. $\left(5, \frac{2}{3}\right) \left(-1, -\frac{4}{3}\right)$

16. $\left(\frac{7}{5}, -\frac{3}{10}\right) \left(-\frac{1}{5}, \frac{1}{2}\right)$

17. $\left(\frac{3}{7}, -\frac{1}{4}\right) \left(\frac{5}{7}, -\frac{5}{4}\right)$

18. $(-9, 1)$ $(-9, 0)$

7.4 Exercises 363

19. (4, −11) (3, −11) **20.** (1.23, 4.80) (2.56, −3.75) **21.** (0.03, 1.57) (3.54, −2.01)

Find the slope of each of the following lines. See Example 4.

22. $y = 2x - 1$ **23.** $y = 5x + 2$ **24.** $y = -x + 4$

25. $y = x + 1$ **26.** $y = 6 - 5x$ **27.** $y = 3 + 9x$

28. $-6x + 4y = 1$ **29.** $3x - 2y = 5$ **30.** $5x - 7y = 8$

31. $6x - 9y = 5$ **32.** $2x + 5y = 4$

33. $9x + 7y = 5$ **34.** $y + 4 = 0$

Give the slope of each of the following lines that has a slope and then decide whether each pair of lines is parallel, perpendicular, or neither. See Example 5.

35. $y = 4x - 3$
$y = 4x + 1$

36. $y = 5 - 5x$
$y = 4 + \frac{1}{5}x$

37. $y - x = 3$
$y - x = 5$

38. $x + y = 5$
$x - y = 1$

39. $y - x = 4$
$y + x = 3$

40. $2x - 5y = 4$
$4x - 10y = 1$

41. $3x - 2y = 4$
$2x + 3y = 1$

42. $3x - 5y = 2$
$5x + 3y = -1$

43. $4x - 3y = 4$
$8x - 6y = 0$

44. $x - 4y = 2$
$2x + 4y = 1$

45. $8x - 9y = 2$
$8x + 6y = 1$

46. $5x - 3y = 8$
$3x - 5y = 10$

47. $6x + y = 12$
$x - 6y = 12$

48. $2x - 5y = 11$
$4x + 5y = 2$

49. $y = 2$
$y - 4 = 8$

50. $x + 5 = 3$
$x = -1$

7.5 EQUATIONS OF A LINE

The last section showed how to find the slope of a line from its equation. For example, the slope of the line having the equation $y = 2x + 3$ is 2, the coefficient of x. What does the number 3 represent? If we let $x = 0$, the equation becomes $y = 2(0) + 3 = 0 + 3 = 3$. Since we found $y = 3$ by letting $x = 0$, $(0, 3)$ is the y-intercept of the graph of $y = 2x + 3$.

▶ An equation solved for y is said to be in **slope-intercept form** because both the slope and the y-intercept of the line can be found from the equation.

> **Slope-Intercept Form**
>
> When the equation of a line is in slope-intercept form,
>
> $$y = mx + b,$$
>
> where m and b are any real numbers, the slope of the line is m and the y-intercept is $(0, b)$.

EXAMPLE 1 Find an equation for each of the following lines.

(a) With slope 5 and y-intercept $(0, 3)$

Use the slope-intercept form. Let $m = 5$ and $b = 3$.

$$y = \boxed{m}x + \boxed{b}$$
$$y = \boxed{5}x + \boxed{3}$$

(b) With slope $\frac{2}{3}$ and y-intercept $(0, -1)$

Here $m = \frac{2}{3}$ and $b = -1$.

$$y = \frac{2}{3}x - 1 \blacktriangleleft$$

Work Problem 1 at the side.

▶ The slope and y-intercept can be used to graph a line. For example, graph $y = \frac{2}{3}x - 1$ by first locating the y-intercept, $(0, -1)$, on the y-axis. From the definition of slope and the fact that the slope of this line is $\frac{2}{3}$,

$$\frac{\text{difference in } y\text{-values}}{\text{difference in } x\text{-values}} = \frac{2}{3},$$

so another point on the graph can be found by counting from the y-intercept 2 units up and then counting 3 units to the right. The line is then drawn through this point and the y-intercept, as shown in Figure 22. This method can be extended to graph a line given its slope and a point on the line.

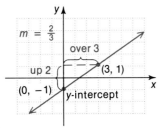

Figure 22

Objectives

▶ Write an equation of a line, given its slope and y-intercept.

▶ Graph a line, given its slope and a point on the line.

▶ Write an equation of a line, given its slope and a point on the line.

▶ Write an equation of a line, given two points on the line.

1. Find the equation of the line with the given slope and value of b.

(a) $m = \frac{1}{2}$; $b = -4$

(b) $m = -1$; $b = 8$

(c) $m = 3$; $b = 0$

(d) $m = 0$, $b = 2$

ANSWERS

1. (a) $y = \frac{1}{2}x - 4$ (b) $y = -x + 8$
   (c) $y = 3x$ (d) $y = 2$

2. Graph each line.

(a) Through $(-1, 4)$, with slope $\frac{5}{2}$

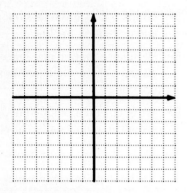

(b) Through $(2, -3)$, with slope $-\frac{1}{3}$

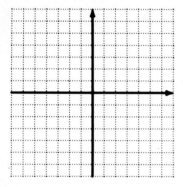

ANSWERS

2. (a) (b)

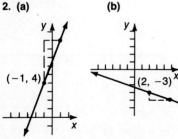

EXAMPLE 2 Graph the line through $(-2, 3)$ with slope $-\frac{5}{4}$.

First, locate the point $(-2, 3)$. Write the slope as

$$\frac{\text{difference in } y\text{-values}}{\text{difference in } x\text{-values}} = \frac{-5}{4}.$$

(We could have used $\frac{5}{-4}$ instead.) Another point on the line is located by counting 5 units *down* (because of the negative sign) and then 4 units to the right. Finally, draw the line through this new point and the given point $(-2, 3)$. See Figure 23. ◀

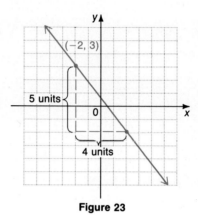

Figure 23

Work Problem 2 at the side.

③ An equation for a line can also be found from a point on the line and the slope of the line. To do this, let m represent the slope of the line and let (x_1, y_1) represent the given point on the line. Let (x, y) represent any other point on the line. Then by the definition of slope,

$$\frac{y - y_1}{x - x_1} = m$$

or $\quad y - y_1 = m(x - x_1).$

This result is the **point-slope form** of the equation of a line.

> **Point-Slope Form**
>
> An equation of the line with slope m going through (x_1, y_1) is given by the point-slope form of the equation of a line,
>
> $$y - y_1 = m(x - x_1).$$

This very important result should be memorized.

EXAMPLE 3 Find an equation of each of the following lines. Write the equation in the form $ax + by = c$.

(a) Through $(-2, 4)$, with slope -3

The given point is $(-2, 4)$ so $x_1 = -2$ and $y_1 = 4$. Also, $m = -3$. Substitute these values into the point-slope form.

$$y - y_1 = m(x - x_1)$$
$$y - 4 = -3[x - (-2)]$$
$$y - 4 = -3(x + 2)$$
$$y - 4 = -3x - 6 \quad \text{Distributive property}$$
$$y = -3x - 2 \quad \text{Add 4}$$
$$3x + y = -2 \quad \text{Add } 3x$$

(b) Through $(4, 2)$ with slope $\frac{3}{5}$

Use $x_1 = 4$, $y_1 = 2$, and $m = \frac{3}{5}$ in the point-slope form.

$$y - y_1 = m(x - x_1)$$
$$y - 2 = \frac{3}{5}(x - 4)$$

Multiply both sides by 5 to clear the fractions.

$$5(y - 2) = 5 \cdot \frac{3}{5}(x - 4)$$
$$5(y - 2) = 3(x - 4)$$
$$5y - 10 = 3x - 12 \quad \text{Distributive property}$$
$$5y = 3x - 2 \quad \text{Add 10}$$
$$-3x + 5y = -2 \quad \text{Subtract } 3x$$

Work Problem 3 at the side.

▶ The point-slope form also can be used to find an equation of a line when two different points on the line are known.

EXAMPLE 4 Find an equation for the line through the points $(-2, 5)$ and $(3, 4)$. Write the equation in the form $ax + by = c$.

First find the slope of the line, using the definition of slope.

$$\text{Slope} = \frac{5 - 4}{-2 - 3} = \frac{1}{-5} = -\frac{1}{5}$$

Now use either $(-2, 5)$ or $(3, 4)$ and the point-slope form. Using $(3, 4)$ gives

$$y - y_1 = m(x - x_1)$$
$$y - 4 = -\frac{1}{5}(x - 3)$$
$$5(y - 4) = -1(x - 3) \quad \text{Multiply by 5}$$
$$5y - 20 = -x + 3 \quad \text{Distributive property}$$
$$5y = -x + 23 \quad \text{Add 20 on each side}$$
$$x + 5y = 23. \quad \text{Add } x \text{ on each side}$$

The same result would be found by using $(-2, 5)$ for (x_1, y_1). ◀

3. Find an equation for each line. Write answers in the form $ax + by = c$.

(a) Through $(-1, 3)$, with slope -2

(b) Through $(5, 2)$, with slope $-\frac{1}{3}$

ANSWERS

3. (a) $2x + y = 1$ **(b)** $x + 3y = 11$

7.5 Equations of a Line

4. Write an equation for the line through the following points. Write answers in the form $ax + by = c$.

(a) $(-3, 1)$ and $(2, 4)$

(b) $(2, 5)$ and $(-1, 6)$

Work Problem 4 at the side.

A summary of the types of *linear equations* is given here.

Linear Equations

$ax + by = c$ — **General form** (neither a nor b is 0)
Slope is $-\dfrac{a}{b}$.
x-intercept is $\left(\dfrac{c}{a}, 0\right)$.
y-intercept is $\left(0, \dfrac{c}{b}\right)$.

$x = k$ — **Vertical line**
Slope is undefined.
x-intercept is $(k, 0)$.

$y = k$ — **Horizontal line**
Slope is 0.
y-intercept is $(0, k)$.

$y = mx + b$ — **Slope-intercept form**
Slope is m.
y-intercept is $(0, b)$.

$y - y_1 = m(x - x_1)$ — **Point-slope form**
Slope is m.
Line goes through (x_1, y_1).

ANSWERS
4. (a) $3x - 5y = -14$
(b) $x + 3y = 17$

7.5 EXERCISES

Write an equation for each line given its slope and y-intercept. See Example 1.

	Slope	y-intercept	Equation		Slope	y-intercept	Equation
1.	3	(0, 5)		**2.**	-2	(0, 4)	
3.	-1	(0, -6)		**4.**	2	(0, -3)	
5.	$\frac{5}{3}$	$\left(0, \frac{1}{2}\right)$		**6.**	$\frac{2}{5}$	$\left(0, -\frac{1}{4}\right)$	
7.	8	(0, 0)		**8.**	0	(0, -5)	

Graph the line going through the given point and having the given slope. (In Exercises 20–23, recall the type of lines having 0 slope and undefined slope.) See Example 2.

9. (2, 5), $m = \frac{1}{2}$

10. (-4, -3), $m = -\frac{2}{5}$

11. (-1, -1), $m = -\frac{3}{8}$

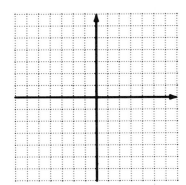
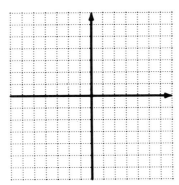

12. $(0, 2)$, $m = \dfrac{3}{4}$

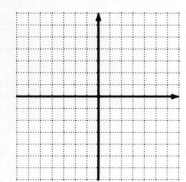

13. $(-3, 0)$, $m = -\dfrac{5}{4}$

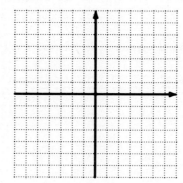

14. $(6, 4)$, $m = -\dfrac{2}{3}$

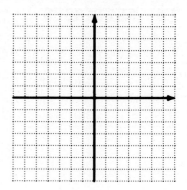

15. $(-2, -1)$, $m = 1$

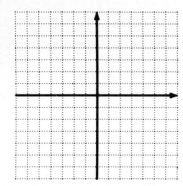

16. $(-4, 7)$, $m = -3$

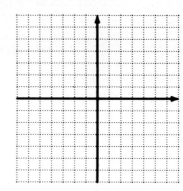

17. $(2, -3)$, $m = -4$

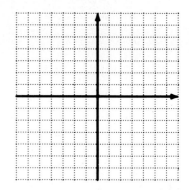

18. $(4, -1)$, $m = 2$

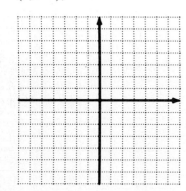

19. $(3, -5)$, $m = 1$

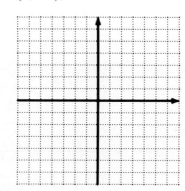

20. $(1, 2)$, $m = 0$

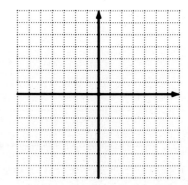

21. $(-4, 3)$, $m = 0$

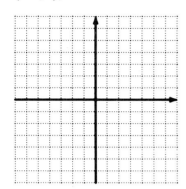

22. $(3, 5)$, undefined slope

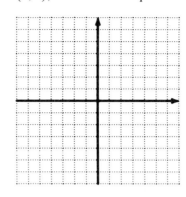

23. $(2, 3)$, undefined slope

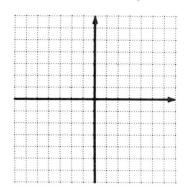

Write an equation for the line passing through the given point and having the given slope. Write the equation in the form $ax + by = c$. See Example 3.

24. $(5, 3)$, $m = 2$

25. $(1, 4)$, $m = 3$

26. $(2, -8)$, $m = -2$

27. $(-1, 7)$, $m = -4$

28. $(3, 5)$, $m = \dfrac{2}{3}$

29. $(2, -4)$, $m = \dfrac{4}{5}$

30. $(-3, -2)$, $m = -\dfrac{3}{4}$

31. $(-8, -2)$, $m = -\dfrac{5}{9}$

32. $(6, 0)$, $m = -\dfrac{8}{11}$

Write an equation for the line passing through each of the given pairs of points. Write the equations in the form $ax + by = c$. See Example 4.

33. $(7, 4), (8, 5)$

34. $(-2, 1), (3, 4)$

35. $(-8, -2), (-1, -7)$

36. $(3, -4), (-2, -1)$

37. $(-7, -5), (-9, -2)$

38. $(0, 2), (3, 0)$

39. $(4, 0), (0, -2)$

40. $(2, -5), (-4, 7)$

41. $(3, -7), (-5, 0)$

7.5 Exercises

42. $\left(\dfrac{1}{2}, \dfrac{3}{2}\right), \left(-\dfrac{1}{4}, \dfrac{5}{4}\right)$　　43. $\left(-\dfrac{2}{3}, \dfrac{8}{3}\right), \left(\dfrac{1}{3}, \dfrac{7}{3}\right)$　　44. $\left(-1, \dfrac{5}{8}\right), \left(\dfrac{1}{8}, 2\right)$

The cost to produce x items is often expressed as $y = mx + b$, where x is the number of items produced. The number b gives the fixed cost (the cost that is the same no matter how many items are produced), and the number m is the variable cost (the cost to produce an additional item). Write the cost equation for each of the following.

45. Fixed cost 50, variable cost 9

46. Fixed cost 100, variable cost 12

47. Fixed cost 70.5, variable cost 3.5

48. Refer to Exercise 46 and find the total cost to make
    (a) 50 items   and   (b) 125 items.

The sales of a company for a given year can be written as an ordered pair in which the first number gives the year (perhaps since the company started business) and the second number gives the sales for that year. If the sales increase at a steady rate, a linear equation for sales can be found. Sales for two years are given for each of two companies below.

Company A		Company B	
Year x	Sales y	Year x	Sales y
1	24	1	18
5	48	4	27

49. (a) Write two ordered pairs of the form (year, sales) for Company A.

　　(b) Write an equation of the line through the two pairs in part (a). This is a sales equation for Company A.

50. (a) Write two ordered pairs of the form (year, sales) for Company B.

　　(b) Write an equation of the line through the two pairs in part (a). This is a sales equation for Company B.

7.6 GRAPHING LINEAR INEQUALITIES IN TWO VARIABLES

Objectives

1. Graph $\leq$ or $\geq$ linear inequalities.
2. Graph $>$ or $<$ linear inequalities.
3. Graph inequalities with a boundary through the origin.

▶ The discussion in Section 7.3 covered methods for graphing linear equations, such as $2x + 3y = 6$. Now this discussion is extended to **linear inequalities in two variables,** such as

$$2x + 3y \leq 6.$$

(Recall that $\leq$ is read "is less than or equal to.")

The inequality $2x + 3y \leq 6$ means that

$$2x + 3y < 6 \quad \text{or} \quad 2x + 3y = 6.$$

As found at the beginning of this chapter, the graph of $2x + 3y = 6$ is a line. This line divides the plane into two regions. The graph of the solutions of the inequality $2x + 3y < 6$ will include only *one* of these regions. Find the required region by solving the given inequality for y.

$$2x + 3y \leq 6$$
$$3y \leq -2x + 6 \quad \text{Subtract } 2x$$
$$y \leq -\frac{2}{3}x + 2 \quad \text{Divide by 3}$$

By this last statement, ordered pairs in which y is *less than or equal to* $-\frac{2}{3}x + 2$ will be solutions to the inequality. The ordered pairs in which y is equal to $-\frac{2}{3}x + 2$ are on a line, so the ordered pairs in which y is less than $-\frac{2}{3}x + 2$ will be *below* that line. Indicate the solution by shading the region below the line, as in Figure 24. The shaded region, along with the original line, is the desired graph.

1. Shade the appropriate region for each linear inequality.

(a) $x + 2y \geq 6$

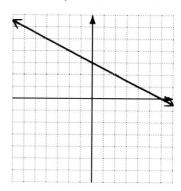

(b) $3x + 4y \leq 12$

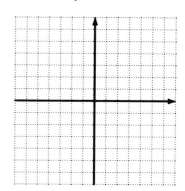

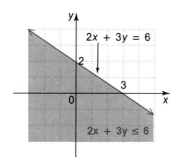

Figure 24

Work Problem 1 at the side.

A test point gives a quick way to find the correct region to shade. Choose any point *not* on the line. Because $(0, 0)$ is easy to substitute into an inequality, it is often a good choice, and we will use it here.

Substitute 0 for x and 0 for y in the given inequality to see whether the resulting statement is true or false. In the example above,

$$2x + 3y \leq 6$$
$$2(0) + 3(0) \leq 6 \quad \text{Let } x = 0 \text{ and } y = 0$$
$$0 + 0 \leq 6$$
$$0 \leq 6. \quad \text{True}$$

Since the last statement is true, shade the region that includes the test point $(0, 0)$. This agrees with the result shown in Figure 24.

ANSWERS

1. (a)

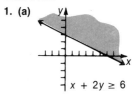

$x + 2y \geq 6$

(b)

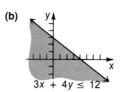

$3x + 4y \leq 12$

7.6 Graphing Linear Inequalities in Two Variables

2. Use (0, 0) as a test point to shade the proper region for each inequality.

(a) $4x - 5y \leq 20$

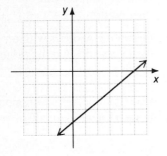

(b) $3x + 5y \geq 15$

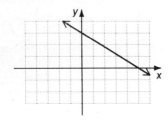

Work Problem 2 at the side.

▶ Inequalities that do not include the equals sign are graphed in a similar way.

EXAMPLE 1 Graph the inequality $x - y > 5$.

This inequality does not include the equals sign. Therefore, the points on the line

$$x - y = 5$$

do not belong to the graph. However, the line still serves as a boundary for two regions, one of which satisfies the inequality. To graph the inequality, first graph the equation $x - y = 5$. Use a dashed line to show that the points on the line are *not* solutions of the inequality $x - y > 5$. Choose a test point to see which side of the line satisfies the inequality. Let us choose (0, 0) again.

$$\boxed{x} - \boxed{y} > 5 \quad \text{Original inequality}$$
$$\boxed{0} - \boxed{0} > 5 \quad \text{Let } x = 0 \text{ and } y = 0$$
$$0 > 5 \quad \text{False}$$

Since $0 > 5$ is false, the graph of the inequality includes the region that does not contain (0, 0). Shade this region, as shown in Figure 25. This shaded region is the desired graph. Check that the proper region is shaded by selecting points in the shaded region and substituting for x and y in the inequality $x - y > 5$. For example, use $(4, -3)$ from the shaded region, as follows.

$$x - y > 5$$
$$4 - (-3) > 5 \quad \text{Let } x = 4 \text{ and } y = -3$$
$$7 > 5 \quad \text{True} \blacktriangleleft$$

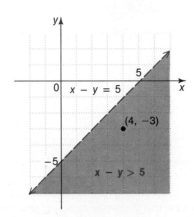

Figure 25

EXAMPLE 2 Graph the inequality $2x - 5y \geq 10$.

Start by graphing the equation

$$2x - 5y = 10.$$

ANSWERS
2. (a) (b)

 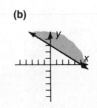

Graphing Linear Equations

Use a solid line to show that the points on the line are solutions of the inequality $2x - 5y \geq 10$. Choose any test point not on the line. Again, choose $(0, 0)$.

$$2x - 5y \geq 10 \quad \text{Original inequality}$$
$$2(0) - 5(0) \geq 10 \quad \text{Let } x = 0 \text{ and } y = 0$$
$$0 - 0 \geq 10$$
$$0 \geq 10 \quad \text{False}$$

Since $0 \geq 10$ is false, shade the region *not* containing $(0, 0)$. (See Figure 26.) ◀

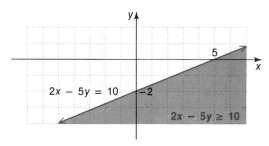

Figure 26

Work Problem 3 at the side.

EXAMPLE 3 Graph the inequality $x \leq 3$.

First graph $x = 3$, a vertical line going through the point $(3, 0)$. Use a solid line. (Why?) Choose $(0, 0)$ as a test point.

$$x \leq 3 \quad \text{Original inequality}$$
$$0 \leq 3 \quad \text{Let } x = 0$$
$$0 \leq 3 \quad \text{True}$$

Since $0 \leq 3$ is true, shade the region containing $(0, 0)$, as in Figure 27. ◀

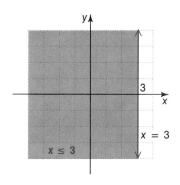

Figure 27

Work Problem 4 at the side.

▶ The next example shows how to graph an inequality having a boundary line going through the origin, an inequality in which $(0, 0)$ cannot be used as a test point.

3. Graph $2x - y \geq -4$.

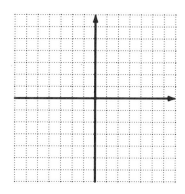

4. Graph $y \leq 4$.

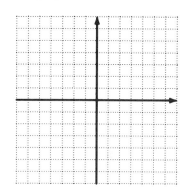

ANSWERS

3.

4.

7.6 Graphing Linear Inequalities in Two Variables

5. Graph each linear inequality.

(a) $x \geq -3y$

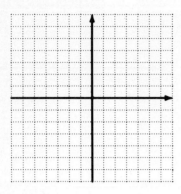

(b) $3x < y$

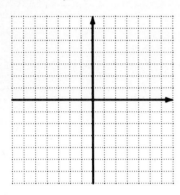

EXAMPLE 4 Graph the inequality $x \leq 2y$.

Begin by graphing $x = 2y$. Some ordered pairs that can be used to graph this line are $(0, 0)$, $(6, 3)$, and $(4, 2)$. Use a solid line. We cannot use $(0, 0)$ as a test point since $(0, 0)$ is on the line $x = 2y$. Instead, choose a test point off the line $x = 2y$. Let us choose $(1, 3)$, which is not on the line.

$$x \leq 2y \quad \text{Original inequality}$$
$$1 \leq 2(3) \quad \text{Let } x = 1 \text{ and } y = 3$$
$$1 \leq 6 \quad \text{True}$$

Since $1 \leq 6$ is true, shade the side of the graph containing the test point $(1, 3)$. (See Figure 28.) ◀

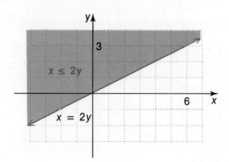

Figure 28

Work Problem 5 at the side.

A summary of the steps used to graph a linear inequality is given below.

Graphing a Linear Inequality in Two Variables

Step 1 Graph the line that is the boundary of the region. Use the methods of Section 7.3. Make the line solid if the inequality has $\leq$ or $\geq$; make the line dashed if the inequality has $<$ or $>$.

Step 2a If the boundary line *does not* go through the origin, use $(0, 0)$ as a test point. Replace x with 0 and y with 0 in the original inequality. If a true statement results, shade the side containing $(0, 0)$. If a false statement results, shade the other side.

Step 2b If the boundary *does* go through the origin, use any point not on the line as a test point.

ANSWERS

5. (a) 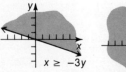 (b)

7.6 EXERCISES

In Exercises 1–12, the straight line for each inequality has been drawn. Complete each graph by shading the correct region. See Examples 1–4.

1. $x + y \leq 4$

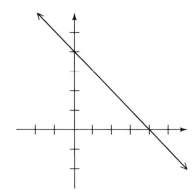

2. $x + y \geq 2$

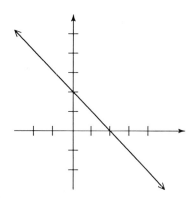

3. $x + 2y \leq 7$

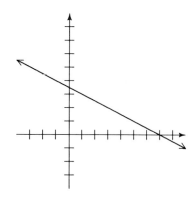

4. $2x + y \leq 5$

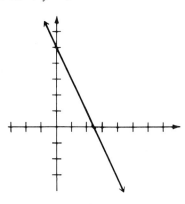

5. $-3x + 4y < 12$

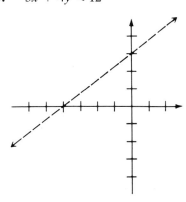

6. $4x - 5y > 20$

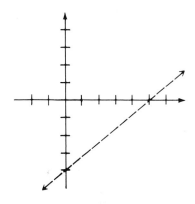

7. $5x + 3y > 15$

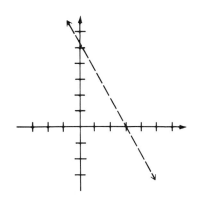

8. $6x - 5y < 30$

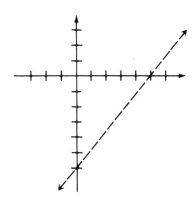

9. $x < 4$

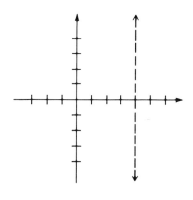

7.6 Exercises 377

10. $y > -1$

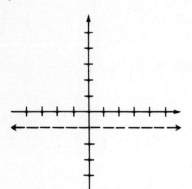

11. $x \leq 4y$

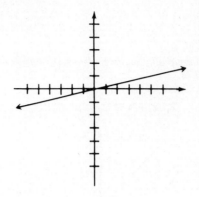

12. $-2x > y$

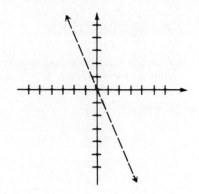

Graph each linear inequality. See Examples 1–4.

13. $x + y \leq 8$

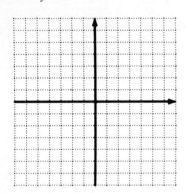

14. $x + y \geq 2$

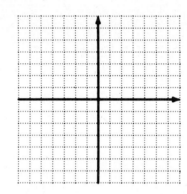

15. $x - y \leq -2$

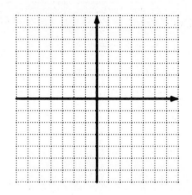

16. $x - y \leq 3$

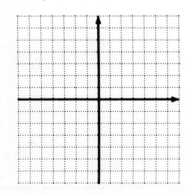

17. $x + 2y \geq 4$

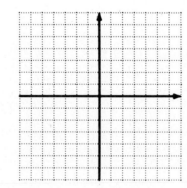

18. $x + 3y \leq 6$

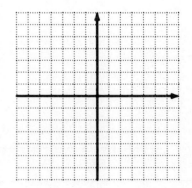

name   date   hour

19. $2x + 3y > 6$ **20.** $3x + 4y > 12$ **21.** $3x - 4y < 12$

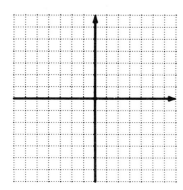 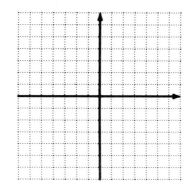 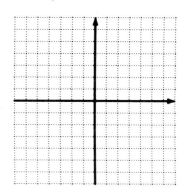

22. $2x - 3y < -6$ **23.** $2x + 5y \geq 10$ **24.** $x < 4$

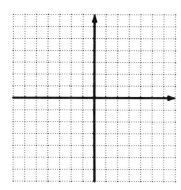 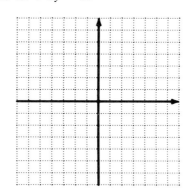 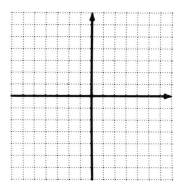

25. $x < -2$ **26.** $y \leq 2$ **27.** $y \leq -3$

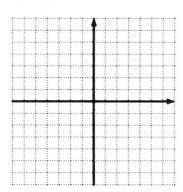 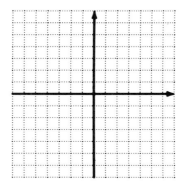 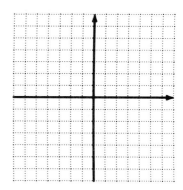

7.6 Exercises

28. $x \leq 3y$

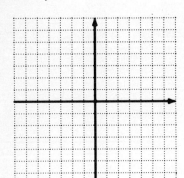

29. $x \leq 5y$

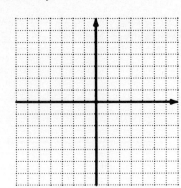

30. $x \geq -2y$

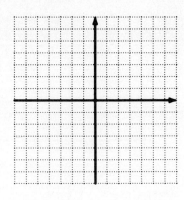

31. A company will ship x units of merchandise to outlet I and y units of merchandise to outlet II. The company must ship a total of at least 500 units to these two outlets. Express this by writing

$$x + y \geq 500.$$

(a) Graph this inequality. (Here $x \geq 0$ and $y \geq 0$.)

(b) Give some sample values of x and y that satisfy the inequality.

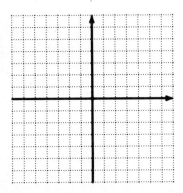

32. Ms. Branson takes x vitamin C tablets each day at a cost of 10¢ each and y multivitamins each day at a cost of 15¢ each. She wants the total cost to be no more than 50¢ a day. Express this as

$$10x + 15y \leq 50.$$

(a) Graph this inequality. (Here $x \geq 0$ and $y \geq 0$.)

(b) Give some sample values of x and y that satisfy the inequality.

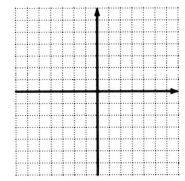

CHAPTER 7 REVIEW EXERCISES

(7.1) Complete the given ordered pairs for each equation.

1. $y = 3x - 2$ $(-1, \underline{})$ $(0, \underline{})$ $(\underline{}, 5)$

2. $4x + 3y = 9$ $(0, \underline{})$ $(\underline{}, 0)$ $(-2, \underline{})$

3. $x = 2y$ $(0, \underline{})$ $(8, \underline{})$ $(\underline{}, -3)$

4. $x + 4 = 0$ $(\underline{}, -3)$ $(\underline{}, 0)$ $(\underline{}, 5)$

Decide whether the given ordered pair is a solution of the given equation.

5. $x + y = 7$; $(3, 4)$

6. $2x + y = 5$; $(1, 4)$

7. $3x - y = 4$; $(1, -1)$

8. $5x - 3y = 16$; $(2, -2)$

(7.2) Plot the ordered pairs on the given coordinate system.

9. $(4, 2)$
10. $(-5, 1)$
11. $(2, 0)$

12. $(-4, 3)$
13. $(0, 3)$
14. $(-2, -5)$

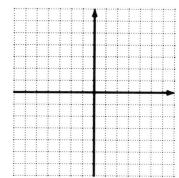

Without plotting the given point, name the quadrant in which each point lies.

15. $(-2, 3)$
16. $(-1, -4)$
17. $(0, 4)$
18. $(0, 0)$

(7.3) Find the intercepts for each equation.

19. $y = 2x - 5$
    x-intercept:
    y-intercept:

20. $2x + y = 7$
    x-intercept:
    y-intercept:

21. $3x - y = 8$
    x-intercept:
    y-intercept:

Graph each linear equation.

22. $y = 2x + 3$

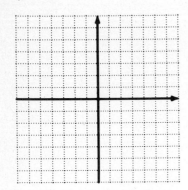

23. $x + y = 5$

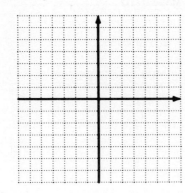

24. $2x - y = 5$

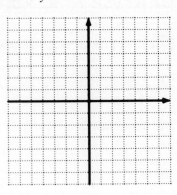

25. $x + 2y = 0$

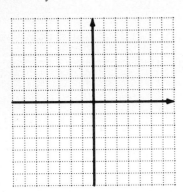

26. $y + 3 = 0$

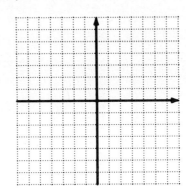

27. $x - y = 0$

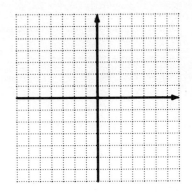

(7.4) *Find the slope of each line.*

28. Through $(2, 3)$ and $(-1, 1)$

29. Through $(4, -1)$ and $(-2, -3)$

30. Through $(0, 0)$ and $(-1, -2)$

31. Through $(2, 5)$ and $(2, -2)$

32. $y = 3x - 1$

33. $y = \frac{2}{3}x + 5$

34. Through $(0, -1)$ and $(9, -5)$

35. Through $(0, 6)$ and $(5, 6)$

36. $y = 8$

37. $x = 2$

38. $6x + 7y = 9$

39. $11x - 3y = 4$

40. $8x = 6 - 3y$

Decide whether the lines of each pair are parallel, perpendicular, *or* neither.

41. $3x + 2y = 5$
$6x + 4y = 12$

42. $x - 3y = 8$
$3x + y = 6$

43. $4x + 3y = 10$
$3x - 4y = 12$

44. $x - 2y = 3$
$x + 2y = 3$

(7.5) *Write an equation for each line. Write equations in the form* $ax + by = c$.

45. $m = 3;\ b = -2$

46. $m = -1;\ b = \dfrac{3}{4}$

47. $m = \dfrac{2}{3};\ b = 5$

48. $m = \dfrac{-1}{4};\ b = \dfrac{-5}{4}$

49. Through $(8, 6);\ m = -3$

50. Through $(5, -2);\ m = 1$

51. Through $(-1, 4);\ m = \dfrac{2}{3}$

52. Through $(1, -1);\ m = \dfrac{-3}{4}$

53. Through $(2, 1)$ and $(-2, 2)$

54. Through $(3, -5)$ and $(-4, -1)$

55. Through $(5, 0)$ and $(5, -1)$

56. Through $(-2, 6)$ with slope 0

57. Through $\left(\dfrac{1}{2}, -\dfrac{3}{4}\right)$ with undefined slope

(7.6) *Complete the graph of each linear inequality by shading the correct region.*

58. $x - y \leq 3$

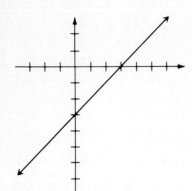

59. $3x - y \geq 5$

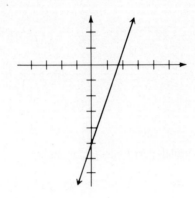

60. $x + 2y > 6$

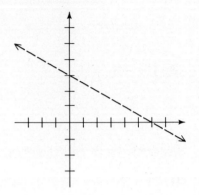

Graph each linear inequality.

61. $x + 2y \leq 4$

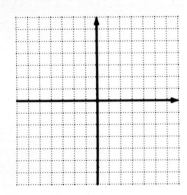

62. $3x + 5y \geq 9$

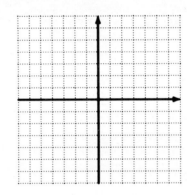

63. $2x - 3y \geq -6$

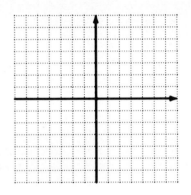

64. $y \leq -2$

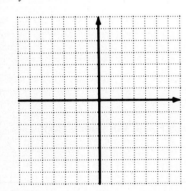

65. $y > 3x$

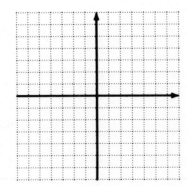

66. $x - 2y > 0$

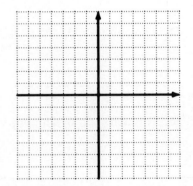

name

CHAPTER 7 TEST

Complete the ordered pairs using the given equations.

      Equation   Ordered pairs

1. $y = 5x - 6$ $(0, \ )\ (-2, \ )\ (\ , 14)$

2. $3x - 5y = 30$ $(0, \ )\ (\ , 0)\ (5, \ )$

3. $x = 3y$ $(0, \ )\ (\ , 2)\ (8, \ )\ (-12, \ )$

4. $y - 2 = 0$ $(5, \ )\ (4, \ )\ (0, \ )\ (-3, \ )$

Graph each linear equation. Give the x- and y-intercepts.

5. $x + y = 4$

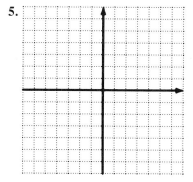

6. $2x + y = 6$

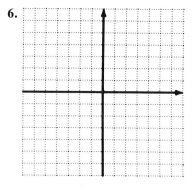

date   hour

1. _____

2. _____

3. _____

4. _____

7.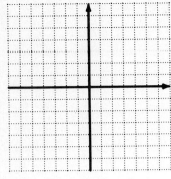

7. $3x - 4y = 18$

8.

8. $2x + y = 0$

9.

9. $x + 5 = 0$

10.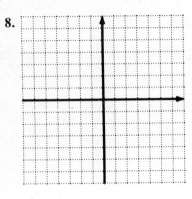

10. $y = 2$

Graphing Linear Equations

name _____   date _____ hour _____

Find the slope of each line.

11. Through $(-2, 4)$ and $(5, 1)$

11. _____

12. $y = \dfrac{-3}{4}x + 6$

12. _____

Write an equation for each line. Write the equations in the form $ax + by = c$.

13. Through $(1, -3)$; $m = -4$

13. _____

14. $m = 3$; $b = -1$

14. _____

15. $m = \dfrac{-3}{4}$; $b = 2$

15. _____

16. Through $(-2, -6)$ and $(-1, 3)$

16. _____

Graph each linear inequality.

17. $x + y \leq 6$

17.

Chapter 7 Test   387

18.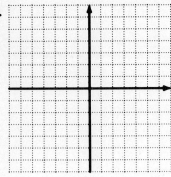

18. $3x - 4y > 12$

19.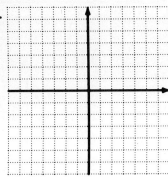

19. $x < 3y$

20.

20. $y \leq -3$

Linear Systems

8

8.1 SOLVING SYSTEMS OF LINEAR EQUATIONS BY GRAPHING

A **system of linear equations** consists of two or more linear equations with the same variables. Examples of systems of two linear equations with two variables are shown below.

$$2x + 3y = 4 \qquad x + 3y = 1 \qquad x - y = 1$$
$$3x - y = -5 \qquad -y = 4 - 2x \qquad y = 3$$

In the system on the right, think of $y = 3$ as an equation in two variables by writing it as $0x + y = 3$.

▶ The **solution of a system** of linear equations includes all the ordered pairs that make all the equations of the system true at the same time.

EXAMPLE 1 Is $(4, -3)$ a solution of the following systems?

(a) $x + 4y = -8$
$3x + 2y = 6$

Decide whether or not $(4, -3)$ is a solution of the system by substituting 4 for x and -3 for y in each equation.

$$\begin{array}{rl} x + 4y &= -8 \\ 4 + 4(-3) &= -8 \\ 4 + (-12) &= -8 \\ -8 &= -8 \quad \text{True} \end{array} \qquad \begin{array}{rl} 3x + 2y &= 6 \\ 3(4) + 2(-3) &= 6 \\ 12 + (-6) &= 6 \\ 6 &= 6 \quad \text{True} \end{array}$$

Since $(4, -3)$ satisfies both equations, it is a solution of the system.

(b) $2x + 5y = -7$
$3x + 4y = 2$

Again, substitute 4 for x and -3 for y in both equations.

$$\begin{array}{rl} 2x + 5y &= -7 \\ 2(4) + 5(-3) &= -7 \\ 8 + (-15) &= -7 \\ -7 &= -7 \quad \text{True} \end{array} \qquad \begin{array}{rl} 3x + 4y &= 2 \\ 3(4) + 4(-3) &= 2 \\ 12 + (-12) &= 2 \\ 0 &= 2 \quad \text{False} \end{array}$$

The ordered pair $(4, -3)$ is not a solution, because it does not satisfy the second equation. ◀

Work Problem 1 at the side.

Applications often require solving a system of equations. In this chapter methods of solving a system of two linear equations in two variables are discussed.

Objectives

▶ Decide whether a given ordered pair is a solution of a system.

▶ Solve linear systems by graphing.

▶ Identify systems with no solutions or with an infinite number of solutions.

1. Decide whether the given ordered pair is a solution of the system.

(a) $(2, 5)$
$3x - 2y = -4$
$5x + y = 15$

(b) $(1, -2)$
$x - 3y = 7$
$4x + y = 5$

(c) $(-3, 3)$
$4x + 6y = 5$
$10x + 3y = -11$

(d) $(4, -1)$
$5x + 6y = 14$
$2x + 5y = 3$

ANSWERS
1. (a) yes (b) no (c) no (d) yes

8.1 Solving Systems of Linear Equations by Graphing   389

▶ One way to find the solution of a system of two linear equations is to graph both equations on the same axes. The graph of each line shows points whose coordinates satisfy the equation of that line. The coordinates of any point where the lines intersect give a solution of the system. Since two different straight lines can intersect at no more than one point, there can never be more than one solution for such a system.

EXAMPLE 2 Solve each system of equations by graphing both equations on the same axes.

(a) $2x + 3y = 4$
$3x - y = -5$

Use methods from the previous chapter to graph both $2x + 3y = 4$ and $3x - y = -5$ on the same axes. (It is important to draw the graphs carefully, since the solution will be read from the graph.) The lines in Figure 1 suggest that the graphs intersect at the point $(-1, 2)$. Check this by substituting -1 for x and 2 for y in both equations. Since the ordered pair $(-1, 2)$ satisfies both equations, the solution of this system is $(-1, 2)$.

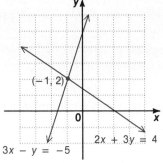

Figure 1

(b) $2x + y = 0$
$4x - 3y = 10$

Find the solution of the system by graphing the two lines on the same axes. As suggested by Figure 2, the solution is $(1, -2)$, the point at which the graphs of the two lines intersect. Check by substituting 1 for x and -2 for y in both equations of the system. ◀

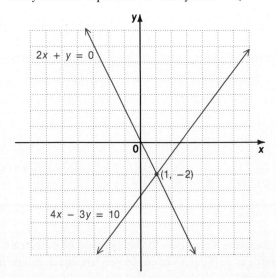

Figure 2

Linear Systems

Work Problem 2 at the side.

▶ Sometimes the graphs of the two equations in a system either do not intersect at all or are the same line, as in the systems in Example 3.

EXAMPLE 3 Solve each system by graphing.

(a) $2x + y = 2$
$2x + y = 8$

The graphs of these lines are shown in Figure 3. The two lines are parallel and have no points in common. For a system whose equations lead to graphs with no points in common, we will write "no solution."

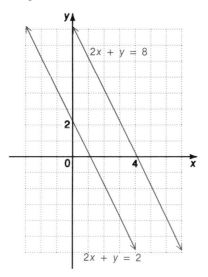

Figure 3

(b) $2x + 5y = 1$
$6x + 15y = 3$

The graphs of these two equations are the same line. See Figure 4. Here the second equation can be obtained by multiplying both sides of the first equation by 3. In this case, every point on the line is a solution of the system, and the solution is an infinite number of ordered pairs. We will write "infinite number of solutions" or "same line" to indicate this situation. ◀

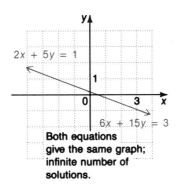

Figure 4

2. Solve each system of equations by graphing both equations on the same axes. Check your answers.

(a) $5x - 3y = 9$
$x + 2y = 7$

(One of the lines is already graphed.)

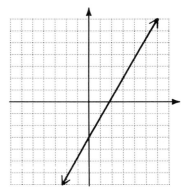

(b) $x + y = 4$
$2x - y = -1$

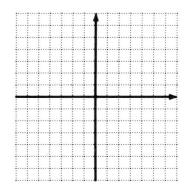

ANSWERS
2. (a) (3, 2) (b) (1, 3)

8.1 Solving Systems of Linear Equations by Graphing   391

3. Solve each system of equations by graphing both equations on the same axes.

(a) $3x - y = 4$
$6x - 2y = 12$

(One of the lines is already graphed.)

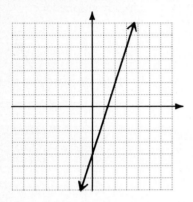

(b) $-x + 3y = 2$
$2x - 6y = -4$

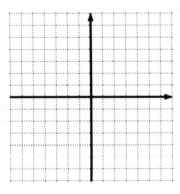

A system of equations with no solutions, such as the one in Example 3(a), is called an **inconsistent system.** A system with a solution is a **consistent system.** Finally, the equations of the system in Example 3(b) have the same graph; these equations are called **dependent equations.**

Work Problem 3 at the side.

Examples 2 and 3 show the three cases that may occur in a system of two equations with two unknowns.

1. The graphs intersect at exactly one point, which gives the (single) solution of the system. The system is consistent and the equations are independent.

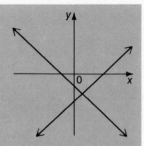

2. The graphs are parallel lines, so there is no solution. The system is inconsistent.

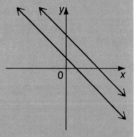

3. The graphs are the same line. The solution is an infinite set of ordered pairs. The equations are dependent.

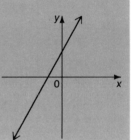

ANSWERS

3. (a) no solution (b) same line

8.1 EXERCISES

Decide whether the given ordered pair is the solution of the given system. See Example 1.

1. $(2, -5)$
   $3x + y = 1$
   $2x + 3y = -11$

2. $(-1, 6)$
   $2x + y = 4$
   $3x + 2y = 9$

3. $(4, -2)$
   $x + y = 2$
   $2x + 5y = 2$

4. $(-6, 3)$
   $x + 2y = 0$
   $3x + 5y = 3$

5. $(2, 0)$
   $3x + 5y = 6$
   $4x + 2y = 5$

6. $(0, -4)$
   $2x - 5y = 20$
   $3x + 6y = -20$

7. $(5, 2)$
   $4x + 3y = 26$
   $3x + 7y = 29$

8. $(9, 1)$
   $2x + 5y = 23$
   $3x + 2y = 29$

9. $(6, -8)$
   $x + 2y + 10 = 0$
   $2x - 3y + 30 = 0$

10. $(-5, 2)$
    $3x - 5y + 20 = 0$
    $2x + 3y + 4 = 0$

11. $(5, -2)$
    $x - 5 = 0$
    $y + 2 = 0$

12. $(-8, 3)$
    $x - 8 = 0$
    $y - 3 = 0$

Solve each system of equations by graphing both equations on the same axes. See Example 2.

13. $x + y = 6$
    $x - y = 2$

14. $x + y = -1$
    $x - y = 3$

15. $x + y = 4$
    $y - x = 4$

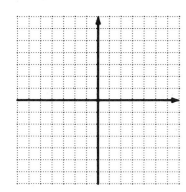
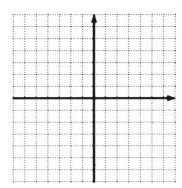
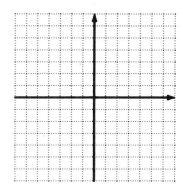

16. $y - x = -5$
$x + y = 1$

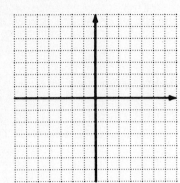

17. $2x + y = 6$
$x - 3y = -4$

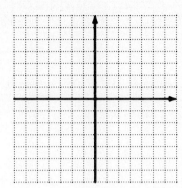

18. $4x + y = 2$
$2x - y = 4$

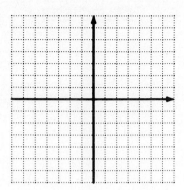

19. $3x - 2y = -3$
$3x + y = 6$

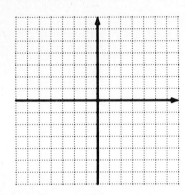

20. $2x + 3y = 12$
$2x - y = 4$

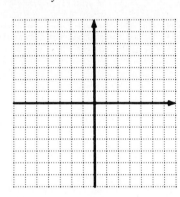

21. $5x + 4y = 7$
$2x - 3y = 12$

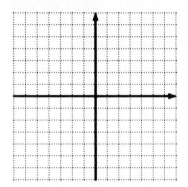

22. $2x + 5y = 17$
$3x - 4y = -9$

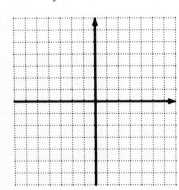

23. $4x + 5y = 3$
$2x - 5y = 9$

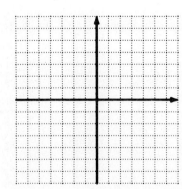

24. $2x + y = 5$
$3x - 4y = 24$

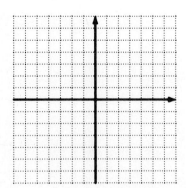

25. $3x + 2y = -10$
$x - 2y = -6$

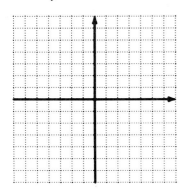

26. $4x + y = 5$
$3x - 2y = 12$

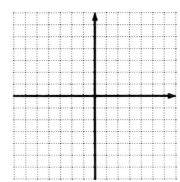

27. $3x - 4y = -24$
$5x + 2y = -14$

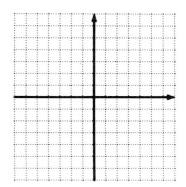

28. $\frac{1}{2}x - y = -\frac{5}{2}$
$4x + 3y = -9$

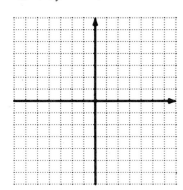

29. $x - 2y = 5$
$x + \frac{3}{4}y = 5$

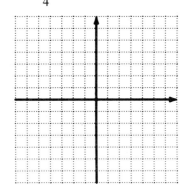

30. $-4x + 3y = 16$
$\frac{2}{3}x - y = \frac{-8}{3}$

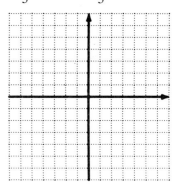

31. $3x - 4y = 8$
$4x + 5y = -10$

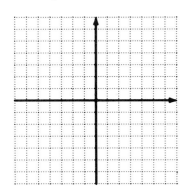

32. $3x + 2y = 10$
$4x - 3y = -15$

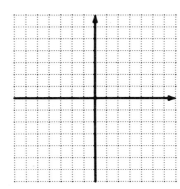

33. $2x + 5y = 12$
$x - 2y = -3$

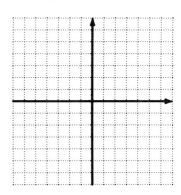

8.1 Exercises

Solve each system by graphing. If the two equations produce parallel lines, write no solution. *If the two equations produce the same line, write* same line. *See Example 3.*

34. $2x + 3y = 5$
$4x + 6y = 9$

35. $5x - 4y = 5$
$10x - 8y = 23$

36. $3x = y + 5$
$6x - 2y = 5$

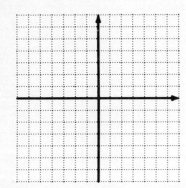
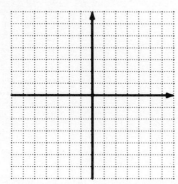
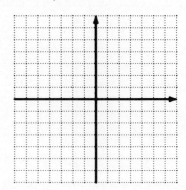

37. $4y + 1 = x$
$2x - 3 = 8y$

38. $2x - y = 4$
$4x = 2y + 8$

39. $3x = 5 - y$
$6x + 2y = 10$

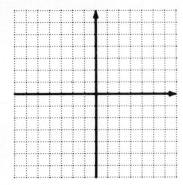
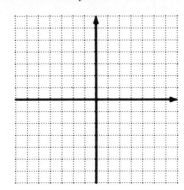
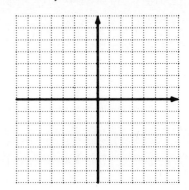

40. $x + 2y = 0$
$4y = -2x$

41. $y = 3x$
$y + 3 = 3x$

42. $x + y = 4$
$2x = 8 - 2y$

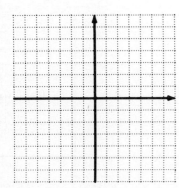
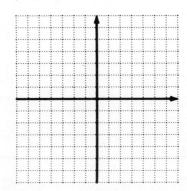
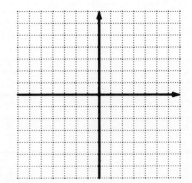

8.2 SOLVING SYSTEMS OF LINEAR EQUATIONS BY ADDITION

Graphing to solve a system of equations has a serious drawback: It is difficult to accurately find a solution such as $(\frac{1}{3}, -\frac{5}{6})$ from a graph.

▶ An algebraic method that depends on the addition property of equality can be used to solve systems. As mentioned earlier, adding the same quantity to each side of an equation results in equal sums.

$$\text{If } A = B, \text{ then } A + C = B + C.$$

This addition can be taken a step further. Adding *equal* quantities, rather than the *same* quantity, to both sides of an equation also results in equal sums.

$$\text{If } A = B \text{ and } C = D, \text{ then } A + C = B + D.$$

The use of the addition property to solve systems is called the **addition method** for solving systems of equations. For most systems, this method is more efficient than graphing.

EXAMPLE 1 Use the addition method to solve the system

$$x + y = 5$$
$$x - y = 3.$$

Each equation in this system is a statement of equality, so, as discussed above, the sum of the right-hand sides equals the sum of the left-hand sides. Adding in this way gives

$$(x + y) + (x - y) = 5 + 3.$$

Combine terms and simplify to get

$$2x = 8 \quad \text{Divide by 2}$$
$$x = 4.$$

The result, $x = 4$, gives the x-value of the solution of the given system. Find the y-value of the solution by substituting 4 for x in either of the two equations in the system.

Work Problem 1 at the side.

The solution found at the side, (4, 1), can be checked by substituting 4 for x and 1 for y into both equations in the given system.

$$\begin{array}{ll} x + y = 5 & x - y = 3 \\ \boxed{4} + \boxed{1} = 5 & \boxed{4} - \boxed{1} = 3 \\ 5 = 5 \quad \text{True} & 3 = 3 \quad \text{True} \end{array}$$

Since both results are true, the solution of the given system is (4, 1). ◀

Work Problem 2 at the side.

EXAMPLE 2 Solve the system

$$-2x + y = -11$$
$$5x - y = 26.$$

Objectives

▶ 1 Solve linear systems by addition.

▶ 2 Multiply one or both equations of a system so the addition method can be used.

▶ 3 Write equations in the proper form to use the addition method.

1. (a) Choose the equation $x + y = 5$ and let $x = 4$ to find the value of y.

   (b) Give the solution of the system.

2. Solve each system by the addition method. Check each solution.

   (a) $x + y = 8$
       $x - y = 2$

   (b) $3x - y = 7$
       $2x + y = 3$

   (c) $x - 8y = -8$
       $-x + 2y = 2$

ANSWERS
1. (a) $y = 1$ (b) (4, 1)
2. (a) (5, 3) (b) (2, −1) (c) (0, 1)

8.2 Solving Systems of Linear Equations by Addition

3. Solve each system by the addition method. Check each solution.

   (a) $2x - y = 2$
   $4x + y = 10$

   (b) $x + 3y = -2$
   $-x - 2y = 0$

   (c) $8x - 5y = 32$
   $4x + 5y = 4$

4. Solve each system by the addition method. Check each solution.

   (a) $x - 3y = -7$
   $3x + 2y = 23$

   (b) $8x + 2y = 2$
   $3x - y = 6$

As above, add left-hand sides and add right-hand sides. This may be done most easily by drawing a line under the second equation and adding vertically. (Like terms must be placed in columns.)

$$-2x + y = -11$$
$$5x - y = 26$$
$$3x = 15 \quad \text{Add in columns}$$
$$x = 5 \quad \text{Divide by 3}$$

Substitute 5 for x in either of the original equations. We choose the first.

$$-2x + y = -11$$
$$-2(5) + y = -11 \quad \text{Let } x = 5$$
$$-10 + y = -11$$
$$y = -1 \quad \text{Add 10}$$

The solution is $(5, -1)$. Check this solution by substituting 5 for x and -1 for y in both of the original equations. ◀

Work Problem 3 at the side.

▶ In both examples above, the addition step caused a variable to be eliminated. Sometimes one or both equations in a system must be multiplied by some number before the addition step will eliminate a variable.

EXAMPLE 3 Solve the system

$$x + 3y = 7 \quad (1)$$
$$2x + 5y = 12. \quad (2)$$

Adding the two equations gives $3x + 8y = 19$, which does not help to solve the system. However, if both sides of equation (1) are first multiplied by -2, the terms with the variable x will drop out after adding.

$$-2(x + 3y) = -2(7)$$
$$-2x - 6y = -14 \quad (3)$$

Now add equations (3) and (2).

$$-2x - 6y = -14 \quad (3)$$
$$2x + 5y = 12 \quad (2)$$
$$-y = -2$$
$$y = 2$$

Substituting into equation (1) gives

$$x + 3y = 7$$
$$x + 3(2) = 7 \quad \text{Let } y = 2$$
$$x + 6 = 7$$
$$x = 1.$$

The solution of this system is $(1, 2)$. Check that this ordered pair satisfies both of the original equations. ◀

Work Problem 4 at the side.

ANSWERS

3. (a) $(2, 2)$ (b) $(4, -2)$
   (c) $\left(3, -\dfrac{8}{5}\right)$

4. (a) $(5, 4)$ (b) $(1, -3)$

EXAMPLE 4 Solve the system

$$2x + 3y = -15 \quad (1)$$
$$5x + 2y = 1. \quad (2)$$

Here the multiplication property of equality is used with both equations instead of just one. Multiply by numbers that will cause the coefficients of x (or of y) in the two equations to be additive inverses of each other. For example, multiply both sides of equation (1) by 5, and both sides of equation (2) by -2.

$$\begin{array}{ll} 10x + 15y = -75 & \text{Multiply by 5} \\ \underline{-10x - 4y = -2} & \text{Multiply by } -2 \\ 11y = -77 & \text{Add} \\ y = -7 & \text{Divide by 11} \end{array}$$

Substituting -7 for y in either equation (1) or (2) gives $x = 3$. The solution of the system is $(3, -7)$. Check this solution.

The same result would have been obtained by multiplying both sides of equation (1) by 2 and both sides of equation (2) by -3. This process would eliminate the y terms so that the value of x would have been found first. ◀

Work Problem 5 at the side.

▶ Before a system can be solved by the addition method, the two equations of the system must have like terms in the same positions. When this is not the case, the terms should first be rearranged, as the next example shows. This example also shows an alternative way to get the second number when finding the solution of a system.

EXAMPLE 5 Solve the system

$$4x = 9 - 3y \quad (1)$$
$$5x - 2y = 8. \quad (2)$$

Rearrange the terms in equation (1) so that the like terms can be aligned in columns. Add $3y$ to both sides to get the following system.

$$4x + 3y = 9 \quad (3)$$
$$5x - 2y = 8 \quad (2)$$

Let us eliminate y by multiplying both sides of equation (3) by 2 and both sides of equation (2) by 3, and then adding.

$$\begin{array}{ll} 8x + 6y = 18 & \text{Multiply by 2} \\ \underline{15x - 6y = 24} & \text{Multiply by 3} \\ 23x = 42 & \text{Add} \\ x = \dfrac{42}{23} & \text{Divide by 23} \end{array}$$

Substituting $\frac{42}{23}$ for x in one of the given equations would give y, but the arithmetic involved would be messy. Instead, solve for y by starting again with the original equations and eliminating x. Do this by multi-

---

5. Solve each system of equations. Check each solution.

   (a) $4x - 5y = -18$
       $3x + 2y = -2$

   (b) $6x + 7y = 4$
       $5x + 8y = -1$

ANSWERS

5. (a) $(-2, 2)$ (b) $(3, -2)$

6. Solve each system of equations.

   (a) $5x = 7 + 2y$
   $5y = 5 - 3x$

   (b) $3y = 8 + 4x$
   $6x = 9 - 2y$

plying both sides of equation (3) by 5 and both sides of equation (2) by -4, and then adding.

$$\begin{array}{rl} 20x + 15y = & 45 \quad \text{Multiply by 5} \\ -20x + 8y = & -32 \quad \text{Multiply by } -4 \\ \hline 23y = & 13 \quad \text{Add} \\ y = & \dfrac{13}{23} \quad \text{Divide by 23} \end{array}$$

The solution is $\left(\dfrac{42}{23}, \dfrac{13}{23}\right)$. ◀

When the value of the first variable is a complicated fraction, the method used in Example 5 avoids errors that often occur when working with fractions. (Of course, this method could be used in solving any system of equations.)

Work Problem 6 at the side.

The solution of a linear system of equations having exactly one solution can be found by the addition method. A summary of the steps is given below.

Solving a Linear System

Step 1 Write both equations of the system in the form $ax + by = c$.

Step 2 Multiply one or both equations by appropriate numbers so that the coefficients of x (or y) are negatives of each other.

Step 3 Add the two equations to get an equation with only one variable.

Step 4 Solve the equation from Step 3.

Step 5 Substitute the solution from Step 4 into either of the original equations.

Step 6 Solve the resulting equation from Step 5 for the remaining variable.

Step 7 Check the answer.

ANSWERS

6. (a) $\left(\dfrac{45}{31}, \dfrac{4}{31}\right)$ (b) $\left(\dfrac{11}{26}, \dfrac{42}{13}\right)$

8.2 EXERCISES

Solve each system by the addition method. Check your answers. See Examples 1 and 2.

1. $x - y = 3$
   $x + y = -1$

2. $x + y = 7$
   $x - y = -3$

3. $x + y = 2$
   $2x - y = 4$

4. $3x - y = 8$
   $x + y = 4$

5. $2x + y = 14$
   $x - y = 4$

6. $2x + y = 2$
   $-x - y = 1$

7. $3x + 2y = 6$
   $-3x - y = 0$

8. $5x - y = 9$
   $-5x + 2y = -8$

9. $6x - y = 1$
   $-6x + 5y = 7$

10. $6x + y = -2$
    $-6x + 3y = -14$

11. $2x - y = 5$
    $4x + y = 4$

12. $x - 4y = 13$
    $-x + 6y = -18$

Solve each system by the addition method. Check your answers. See Example 3.

13. $2x - y = 7$
$3x + 2y = 0$

14. $x + y = 7$
$-3x + 2y = -11$

15. $x + 3y = 16$
$2x - y = 4$

16. $4x - 3y = 8$
$2x + y = 14$

17. $x + 4y = -18$
$3x + 5y = -19$

18. $2x + y = 3$
$5x - 2y = -15$

19. $5x - 3y = 15$
$-3x + 6y = -9$

20. $4x + 3y = -9$
$5x - 6y = 18$

21. $3x - 2y = -6$
$-5x + 4y = 16$

22. $-4x + 3y = 0$
$5x - 6y = 9$

23. $2x - y = -8$
$5x + 2y = -20$

24. $5x + 3y = -9$
$7x + y = -3$

402 Linear Systems

25. $2x + y = 5$
$5x + 3y = 11$

26. $2x + 7y = -53$
$4x + 3y = -7$

Solve each system by the addition method. Check your answers. See Examples 4 and 5.

27. $5x - 4y = -1$
$-7x + 5y = 8$

28. $3x + 2y = 12$
$5x - 3y = 1$

29. $3x + 5y = 33$
$4x - 3y = 15$

30. $2x + 5y = 3$
$5x - 3y = 23$

31. $3x + 5y = -7$
$5x + 4y = 10$

32. $2x + 3y = -11$
$5x + 2y = 22$

33. $2x + 3y = -12$
$5x - 7y = -30$

34. $2x + 9y = 16$
$5x - 6y = 40$

35. $4x - 3y = -7$
    $6x + 5y = 18$

36. $8x + 3y = -4$
    $12x + 7y = 4$

37. $24x + 12y = 19$
    $16x - 18y = -9$

38. $9x + 4y = -4$
    $6x + 6y = -11$

39. $3x - 2y = 3$
    $\frac{4}{3}x + y = \frac{1}{3}$

40. $3x - 2y = 27$
    $x - \frac{7}{2}y = -25$

41. $5x - 7y = 6$
    $3x - 6y = 2$

42. $3x + 4y = 2$
    $4x + 3y = 6$

43. $6.5x + 2.3y = 6.1$
    $5.4x - 4.6y = 24.6$

44. $-2.2x + 7.1y = -1.7$
    $3.8x + 14.2y = 29.4$

45. $.13x - .52y = -.39$
    $.39x + .08y = -2.81$

46. $8.12x + 4.03y = 12.21$
    $2.03x - 1.52y = 5.58$

8.3 TWO SPECIAL CASES

▶ In Section 8.1 some of the systems had equations with graphs that were parallel lines (from an inconsistent system), while the equations of other systems had graphs that were the same line (from dependent equations). This section shows how to solve these systems with the addition method discussed in the previous section. The first example shows the solution of an inconsistent system, where the graphs of the equations are parallel lines.

EXAMPLE 1 Solve by the addition method.
$$2x + 4y = 5$$
$$4x + 8y = -9$$

Multiply both sides of $2x + 4y = 5$ by -2 and then add.
$$-4x - 8y = -10$$
$$\underline{4x + 8y = -9}$$
$$0 = -19 \quad \text{False}$$

The false statement $0 = -19$ shows that the given system is self-contradictory. *It has no solution.* This means that the graphs of the equations of this system are parallel lines, as shown in Figure 5. Since this system has no solution, it is inconsistent. ◀

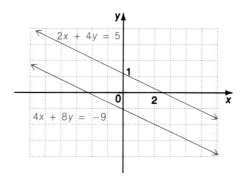

Figure 5

Work Problem 1 at the side.

▶ The next example shows the result of using the addition method when the equations of the system are dependent, with the graphs of the equations in the system the same line.

EXAMPLE 2 Solve by the addition method.
$$3x - y = 4$$
$$-9x + 3y = -12$$

Multiply both sides of the first equation by 3 and then add the two equations to get
$$9x - 3y = 12$$
$$\underline{-9x + 3y = -12}$$
$$0 = 0. \quad \text{True}$$

It can be shown that this means that every solution of one equation is also a solution of the other, so the system has an infinite number of solutions: all the ordered pairs corresponding to points that lie on the

Objectives

▶ Solve linear systems having parallel lines as their graphs.
▶ Solve linear systems having the same line as their graphs.

1. Solve each system by the addition method.

   (a) $4x + 3y = 10$
   $2x + \dfrac{3}{2}y = 12$

   (b) $-2x - 4y = -1$
   $5x + 10y = 15$

ANSWERS
1. (a) no solution (b) no solution

2. Solve each system by the addition method.

   (a) $\quad 6x + 3y = 9$
   $\quad\quad -8x - 4y = -12$

   (b) $\quad\quad 4x - 6y = 10$
   $\quad\quad -10x + 15y = -25$

common graph. As mentioned in Section 8.1, the equations of this system are dependent. In the answers at the back of this book, a solution of such a system of dependent equations is indicated by the words *same line*. A graph of the equations of this system is shown in Figure 6. ◀

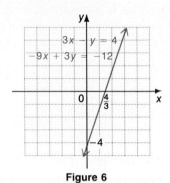

Figure 6

Work Problem 2 at the side.

One of three situations may occur when the addition method is used to solve a linear system of equations.

1. The result of the addition step is a statement such as $x = 2$ or $y = -3$. The solution will be exactly one ordered pair. The graphs of the equations of the system will intersect at exactly one point.

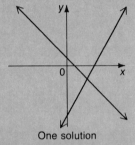

One solution

2. The result of the addition step is a false statement, such as $0 - 4$. In this case, the graphs are parallel lines and there is no solution for the system.

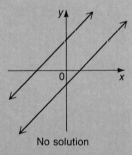

No solution

3. The result of the addition step is a true statement, such as $0 = 0$. The graphs of the equations of the system are the same line, and an infinite number of ordered pairs are solutions.

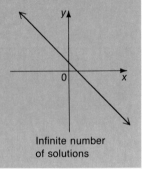

Infinite number of solutions

ANSWERS

2. **(a)** infinite number of solutions, same line
   (b) infinite number of solutions, same line

8.3 EXERCISES

Use the addition method to solve each system. See Examples 1 and 2.

1. $x + y = 4$
   $x + y = -2$

2. $2x - y = 1$
   $2x - y = 4$

3. $5x - 2y = 6$
   $10x - 4y = 10$

4. $3x - 5y = 2$
   $6x - 10y = 8$

5. $x + 3y = 5$
   $2x + 6y = 10$

6. $6x - 2y = 12$
   $-3x + y = -6$

7. $2x + 3y = 8$
   $4x + 6y = 12$

8. $4x + y = 6$
   $-8x - 2y = 21$

9. $5x = y + 4$
   $5x = y - 4$

10. $4y = 3x - 2$
    $4y = 3x + 5$

11. $6x + 3y = 0$
    $-12x - 6y = 0$

12. $3x - 5y = 0$
    $6x - 10y = 0$

13. $2x - 3y = 0$
    $4x + 5y = 0$

14. $3x - 5y = 0$
    $6x + 10y = 0$

15. $3x + 5y = 19$
    $4x - 3y = 6$

16. $2x + 5y = 17$
    $4x + 3y = -1$

17. $4x - 2y = 1$
    $8x - 4y = 1$

18. $-2x + 3y = 5$
    $4x - 6y = 5$

19. $3x - 2y = 8$
    $-3x + 2y = -8$

20. $4x + y = 4$
    $-8x - 2y = -8$

21. $4x - y = 3$
    $-2x + \frac{1}{2}y = -\frac{3}{2}$

22. $5x - 2y = 8$
    $-\frac{5}{2}x + y = -4$

8.4 SOLVING SYSTEMS OF LINEAR EQUATIONS BY SUBSTITUTION

Objectives

▷ Solve linear systems by substitution.

▷ Solve linear systems with fractions as coefficients.

▷ As discussed in the preceding sections, the graphing method and the addition method can be used for solving systems of linear equations. A third method, the **substitution method,** is particularly useful for solving systems where one equation can be solved quickly for one of the variables.

EXAMPLE 1 Solve the system

$$3x + 5y = 26$$
$$y = 2x.$$

The second of these two equations is already solved for y: this equation says that $y = 2x$. Substituting $2x$ for y in the first equation gives

$$3x + 5y = 26$$
$$3x + 5(2x) = 26 \quad \text{Let } y = 2x$$
$$3x + 10x = 26$$
$$13x = 26$$
$$x = 2. \quad \text{Divide by 13}$$

Find y by using $y = 2x$ and $x = 2$ to get $y = 2(2) = 4$. Check that the solution of the given system is $(2, 4)$. ◀

Work Problem 1 at the side.

1. Solve by the substitution method.

   (a) $3x + 5y = 69$
   $y = 4x$

   (b) $-x + 4y = 26$
   $y = -3x$

   (c) $2x + 7y = -12$
   $x = -2y$

EXAMPLE 2 Use substitution to solve the system

$$2x + 5y = 7$$
$$x = -1 - y.$$

The second equation gives x in terms of y. Substitute $-1 - y$ for x in the first equation.

$$2x + 5y = 7$$
$$2(-1 - y) + 5y = 7 \quad \text{Let } x = -1 - y$$
$$-2 - 2y + 5y = 7 \quad \text{Distributive property}$$
$$-2 + 3y = 7$$
$$3y = 9$$
$$y = 3 \quad \text{Divide by 3}$$

Use $x = -1 - y$ and $y = 3$ to get $x = -1 - 3 = -4$. Check that the solution of the given system is $(-4, 3)$. ◀

EXAMPLE 3 Use substitution to solve the system

$$x = 5 - 2y$$
$$2x + 4y = 6.$$

Substitute $5 - 2y$ for x in the second equation.

$$2x + 4y = 6$$
$$2(5 - 2y) + 4y = 6 \quad \text{Let } x = 5 - 2y$$
$$10 - 4y + 4y = 6 \quad \text{Distributive property}$$
$$10 = 6 \quad \text{False}$$

ANSWERS
1. (a) (3, 12) (b) (−2, 6)
   (c) (8, −4)

2. Solve each system by substitution. Check each solution.

(a) $3x - 4y = -11$
$x = y - 2$

(b) $5x + 2y = -2$
$y = 1 - 2x$

(c) $8x - y = 4$
$y = 8x + 4$

3. Solve each system by substitution. Check each solution.

(a) $x + 4y = -1$
$2x - 5y = 11$

(b) $2x + 5y = 4$
$x + y = -1$

As shown in the last section, this false result means that the equations in the system have graphs that are parallel lines. The system is inconsistent and has no solution. ◀

Work Problem 2 at the side.

EXAMPLE 4 Use substitution to solve the system

$$2x + 3y = 8$$
$$-4x - 2y = 0.$$

The substitution method requires an equation solved for one of the variables. Let us choose the first equation of the system, $2x + 3y = 8$, and solve for x. Start by subtracting $3y$ on each side.

$$2x + 3y = 8$$
$$2x = 8 - 3y \quad \text{Subtract } 3y$$

Now divide both sides of this equation by 2.

$$x = \frac{8 - 3y}{2} \quad \text{Divide by 2}$$

Now substitute this value for x in the second equation of the system.

$$-4x - 2y = 0$$
$$-4\left(\frac{8 - 3y}{2}\right) - 2y = 0 \quad \text{Let } x = \frac{8 - 3y}{2}$$
$$-2(8 - 3y) - 2y = 0$$
$$-16 + 6y - 2y = 0 \quad \text{Distributive property}$$
$$-16 + 4y = 0$$
$$4y = 16$$
$$y = 4$$

Find x by letting $y = 4$ in $x = \frac{8 - 3y}{2}$.

$$x = \frac{8 - 3 \cdot 4}{2} \quad \text{Let } y = 4$$
$$x = \frac{8 - 12}{2}$$
$$x = \frac{-4}{2} = -2$$

The solution of the given system is $(-2, 4)$. Check this solution in both equations. ◀

Work Problem 3 at the side.

EXAMPLE 5 Use substitution to solve the system

$$2x = 4 - y \quad (1)$$
$$6 + 3y + 4x = 16 - x. \quad (2)$$

Start by simplifying the second equation by adding x and subtracting 6 on both sides. This gives the simplified system

$$2x = 4 - y \quad (1)$$
$$5x + 3y = 10. \quad (3)$$

ANSWERS
2. (a) $(3, 5)$ (b) $(-4, 9)$ (c) no solution
3. (a) $(3, -1)$ (b) $(-3, 2)$

Linear Systems

For the substitution method, one of the equations must be solved for either x or y. Since the coefficient of y in equation (1) is -1, avoid fractions by solving this equation for y.

$$2x = 4 - y \quad \text{(1)}$$
$$2x - 4 = -y$$
$$-2x + 4 = y$$

Now substitute $-2x + 4$ for y in equation (3).

$$5x + 3y = 10$$
$$5x + 3(-2x + 4) = 10$$
$$5x - 6x + 12 = 10 \quad \text{Distributive property}$$
$$-x + 12 = 10$$
$$-x = -2$$
$$x = 2$$

Since $y = -2x + 4$ and $x = 2$,

$$y = -2(2) + 4 = 0,$$

and the solution is $(2, 0)$. Check this solution by substitution in both equations of the given systems. ◀

Work Problem 4 at the side.

▶ When a system includes equations with fractions as coefficients, eliminate the fractions by multiplying both sides by a common denominator. Then solve the resulting system.

EXAMPLE 6   Solve the following system by any method.

$$3x + \frac{1}{4}y = 2 \quad \text{(1)}$$

$$\frac{1}{2}x + \frac{3}{4}y = \frac{-5}{2} \quad \text{(2)}$$

Begin by eliminating fractions. Clear equation (1) of fractions by multiplying both sides by 4.

$$4\left(3x + \frac{1}{4}y\right) = 4(2) \quad \text{Multiply by 4}$$

$$4(3x) + 4\left(\frac{1}{4}y\right) = 4(2) \quad \text{Distributive property}$$

$$12x + y = 8 \quad \text{(3)}$$

Now clear equation (2) of fractions by multiplying both sides by the common denominator 4.

$$4\left(\frac{1}{2}x + \frac{3}{4}y\right) = 4\left(\frac{-5}{2}\right) \quad \text{Multiply by 4}$$

$$4\left(\frac{1}{2}x\right) + 4\left(\frac{3}{4}y\right) = 4\left(\frac{-5}{2}\right) \quad \text{Distributive property}$$

$$2x + 3y = -10 \quad \text{(4)}$$

4. Solve each system by substitution. First simplify where necessary.

(a) $\quad x = 5 - 3y$
$\quad 2x + 3 = 5x - 4y + 14$

(b) $\;4x + 2x - y + 1 = 30$
$\qquad\qquad\qquad y = -4 + x$

(c) $\quad 5x - y = -14 + 2x + y$
$\quad 7x + 9y + 4 = 3x + 8y$

ANSWERS
4. (a) $(-1, 2)$ (b) $(5, 1)$ (c) $(-2, 4)$

5. Solve each system by any method. First clear all fractions.

(a) $\dfrac{2}{3}x + \dfrac{1}{2}y = 6$
$\dfrac{1}{2}x - \dfrac{3}{4}y = 0$

(b) $\dfrac{3}{5}x + \dfrac{1}{2}y = 7$
$\dfrac{7}{10}x - \dfrac{1}{5}y = 16$

The given system of equations has been simplified as follows.

$$12x + y = 8 \qquad (3)$$
$$2x + 3y = -10 \qquad (4)$$

Let us solve this system by the substitution method. Equation (3) can be solved for y by subtracting $12x$ from each side.

$$12x + y = 8$$
$$y = -12x + 8 \qquad \text{Subtract } 12x$$

Now substitute the result for y in equation (4).

$$2x + 3(-12x + 8) = -10 \qquad \text{Let } y = -12x + 8$$
$$2x - 36x + 24 = -10 \qquad \text{Distributive property}$$
$$-34x = -34 \qquad \text{Subtract } 24$$
$$x = 1 \qquad \text{Divide by } -34$$

Using $y = -12x + 8$ and $x = 1$ gives $y = -12(1) + 8 = -4$. The solution is $(1, -4)$. Check by substituting 1 for x and -4 for y in both of the original equations. Verify that the same solution is obtained if the addition method is used to solve the system of equations (3) and (4). ◀

Work Problem 5 at the side.

ANSWERS

5. (a) (6, 4) (b) (20, -10)

8.4 EXERCISES

Solve each system by the substitution method. Check each solution. See Examples 1–4.

1. $x + y = 6$
   $y = 2x$

2. $x + 3y = -11$
   $y = -4x$

3. $3x + 2y = 26$
   $x = y + 2$

4. $4x + 3y = -14$
   $x = y - 7$

5. $x + 5y = 3$
   $x - 2y = 10$

6. $5x + 2y = 14$
   $2x - y = 11$

7. $3x - 2y = 14$
   $2x + y = 0$

8. $2x - 5 = -y$
   $x + 3y = 0$

9. $x + y = 6$
   $x - y = 4$

10. $3x - 2y = 13$
    $x + y = 6$

11. $3x - y = 6$
    $y = 3x - 5$

12. $4x - y = 4$
    $y = 4x + 3$

13. $2x + 3y = 11$
    $2x - y = 7$

14. $3x + 4y = -10$
    $2x + y = -4$

15. $4x + y = 5$
    $5x + 3y = 1$

16. $5x + 2y = -19$
    $4x + y = -17$

17. $6x - 8y = 4$
    $3x = 4y + 2$

18. $12x + 18y = 12$
    $2x = 2 - 3y$

19. $4x + 5y = 5$
    $2x + 3y = 1$

20. $6x + 5y = 13$
    $3x + 2y = 4$

Solve each system by either the addition method or the substitution method. First simplify equations where necessary. Check each solution. See Example 5.

21. $x + 4y = 34$
    $y = 4x$

22. $3x - y = -14$
    $x = -2y$

414   Linear Systems

23. $4 + 4x - 3y = 34 + x$
    $4x = -y - 2 + 3x$

24. $5x - 4y = 42 - 8y - 2$
    $2x + y = x + 1$

25. $4x - 2y + 8 = 3x + 4y - 1$
    $3x + y = x + 8$

26. $5x - 4y - 8x - 2 = 6x + 3y - 3$
    $4x - y = -2y - 8$

27. $2x - 8y + 3y + 2 = 5y + 16$
    $8x - 2y = 4x + 28$

28. $7x - 9 + 2y - 8 = -3y + 4x + 13$
    $4y - 8x = -8 + 9x + 32$

29. $-2x + 3y = 12 + 2y$
    $2x - 5y + 4 = -8 - 4y$

30. $2x + 5y = 7 + 4y - x$
    $5x + 3y + 8 = 22 - x + y$

31. $y + 9 = 3x - 2y + 6$
    $5 - 3x + 24 = -2x + 4y + 3$

32. $5x - 2y = 16 + 4x - 10$
    $4x + 3y = 60 + 2x + y$

8.4 Exercises 415

Solve each system by the addition method or by the substitution method. First clear all fractions. Check each solution. See Example 6.

33. $x + \dfrac{1}{3}y = y - 2$

    $\dfrac{1}{4}x - y = x - y$

34. $\dfrac{5}{3}x + 2y = \dfrac{1}{3} + y$

    $2x - 3 + \dfrac{y}{3} = -2 + x$

35. $\dfrac{x}{6} + \dfrac{y}{6} = 1$

    $-\dfrac{1}{2}x - \dfrac{1}{3}y = -5$

36. $\dfrac{x}{2} - \dfrac{y}{3} = \dfrac{5}{6}$

    $\dfrac{x}{5} - \dfrac{y}{4} = \dfrac{1}{10}$

37. $\dfrac{x}{3} - \dfrac{3y}{4} = -\dfrac{1}{2}$

    $\dfrac{2x}{3} + \dfrac{y}{2} = 3$

38. $\dfrac{x}{5} + 2y = \dfrac{8}{5}$

    $\dfrac{3x}{5} + \dfrac{y}{2} = \dfrac{-7}{10}$

39. $\dfrac{x}{2} + \dfrac{y}{3} = \dfrac{7}{6}$

    $\dfrac{x}{4} - \dfrac{3y}{2} = \dfrac{9}{4}$

40. $\dfrac{5x}{2} - \dfrac{y}{3} = \dfrac{5}{6}$

    $\dfrac{4x}{3} + y = \dfrac{19}{3}$

8.5 APPLICATIONS OF LINEAR SYSTEMS

Many practical problems are more easily translated into equations if two variables are used. With two variables, a system of two equations is needed to find the desired solution. The examples in this section illustrate the method of solving word problems using two equations and two variables.

Recall from Chapter 3 the steps used in solving word problems. The steps presented there can be modified as follows to allow for two variables and two equations.

Solving a Word Problem with Two Variables

Step 1 Choose a variable to represent each of the unknown values that must be found. Write down what each variable is to represent.

Step 2 Translate the problem into two equations using both variables.

Step 3 Solve the system of two equations.

Step 4 Answer the question or questions asked in the problem.

Step 5 Check the solution in the words of the original problem.

▶ The first example shows how to use two variables to solve a problem about two unknown numbers.

EXAMPLE 1 The sum of two numbers is 63. Their difference is 19. Find the two numbers.

Step 1 Let x represent one number and y the other.

Step 2 Set up a system of equations from the information in the problem.

$x + y = 63$ The sum is 63
$x - y = 19$ The difference is 19

Step 3 Solve the system from Step 2.

Work Problem 1 at the side.

From Problem 1 at the side, $x = 41$ and $y = 22$.

Step 4 The numbers required in the problem are 41 and 22.

Step 5 Check: The sum of 41 and 22 is 63, and their difference is 19. The solution satisfies the conditions of the problem. ◀

Work Problem 2 at the side.

▶ The next example illustrates how to set up and solve a common type of word problem that involves two quantities and their costs.

EXAMPLE 2 Admission prices at a football game were $6 for adults and $2 for children. The total value of the tickets sold was $2528, and 454 tickets were sold. How many adults and how many children attended the game?

Objectives

Use linear systems to solve word problems about

▶ two numbers;

▶ money;

▶ mixtures;

▶ rate or speed using the distance formula.

1. Solve the system of equations

   $x + y = 63$
   $x - y = 19$.

2. (a) The sum of two numbers is 97. Their difference is 41. What are the numbers?

   (b) The sum of two numbers is 38. If twice the first is added to three times the second, the result is 99. Find the numbers.

ANSWER
1. $x = 41, y = 22$
2. (a) 69, 28 (b) 15, 23

3. The value of the tickets sold for a concert was $1850. The price for a regular ticket was $5.00, and student tickets were $3.50. A total of 400 tickets were sold.

   (a) Complete this table.

Kind of ticket	Number sold	Cost of each	Total value
Regular	r		
Student	s		
Totals			

   (b) Write a system of equations.

   (c) Solve the system.

Step 1 Let a represent the number of adults' tickets sold and c represent the number of children's tickets sold.

Step 2 The information given in the problem is summarized in the table. The entries in the "total value" column were found by multiplying the number of tickets sold by the price per ticket.

Kind of ticket	Number sold	Cost of each (in dollars)	Total value (in dollars)
Adult	a	6	$6a$
Child	c	2	$2c$
Total	454	—	2528

The total number of tickets sold was 454, so

$$a + c = 454$$

Since the total value was $2528, the right-hand column leads to

$$6a + 2c = 2528.$$

Step 3 These two equations give the following system.

$$a + c = 454 \quad (1)$$
$$6a + 2c = 2528 \quad (2)$$

Solve the system of equations with the addition method. First, multiply both sides of equation (1) by -2 to get $-2a - 2c = -908$. Then add the result to equation (2).

$$\begin{array}{r} -2a - 2c = -908 \\ 6a + 2c = 2528 \\ \hline 4a \quad\quad = 1620 \\ a \quad\quad = 405 \end{array}$$

Substitute 405 for a in equation (1) to get

$$a + c = 454 \quad (1)$$
$$\boxed{405} + c = 454 \quad \text{Let } a = 405$$
$$c = 49.$$

Step 4 There were 405 adults and 49 children at the game.

Step 5 Since 405 adults paid $6 each and 49 children paid $2 each, the value of tickets sold should be

$$405(6) + 49(2) = 2528,$$

or $2528. This result agrees with the given information. ◀

Work Problem 3 at the side.

In the rest of the examples try to identify each step in the solution of the problems.

▶ Mixture problems occur in many fields of application of mathematics. One important type of mixture problem comes up in the study of chemistry. This kind of problem can be solved as shown in the next example.

ANSWERS

3. (a)

Kind of ticket	Number sold	Cost of each	Total value
Regular	r	5	5r
Student	s	3.50	3.50s
Totals	400	—	1850

(b) $r + s = 400$
    $5r + 3.50s = 1850$

(c) 300 regular, 100 student

418   Linear Systems

EXAMPLE 3 A pharmacist needs 100 liters of 50% alcohol solution. She has on hand 30% alcohol solution and 80% alcohol solution, which she can mix. How many liters of each will be required to make the 100 liters of 50% alcohol solution?

A 30% solution means that 30% of the solution is alcohol and the rest is water. The information given above was used for the sketch in Figure 7.

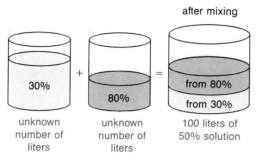

Figure 7

Let x represent the number of liters of 30% alcohol needed, and let y represent the number of liters of 80% alcohol. The information given in the problem is summarized in the table. (Each percent has been changed to its decimal form in the third column.)

Percent	Liters of solution	Liters of pure alcohol
30	x	$.30x$
80	y	$.80y$
50	100	$.50(100)$

The pharmacist will have $.30x$ liters of alcohol from the x liters of 30% solution and $.80y$ liters of alcohol from the y liters of 80% solution, for a total of $.30x + .80y$ liters of pure alcohol. In the mixture, she wants 100 liters of 50% solution. This 100 liters would contain $.50(100) = 50$ liters of pure alcohol. Since the amounts of pure alcohol must be equal,

$$.30x + .80y = 50.$$

The total number of liters is 100, or

$$x + y = 100.$$

These two equations give the system

$$.30x + .80y = 50$$
$$x + y = 100.$$

Work Problem 4 at the side.

From Problem 4 at the side, $x = 60$ and $y = 40$. The pharmacist should use 60 liters of the 30% solution and 40 liters of the 80% solution. Since $.30(60) + (.80)40 = 50$, this mix will give 100 liters of 50% solution, as required in the original problem. ◀

Work Problem 5 at the side.

4. Solve the system

$$.30x + .80y = 50$$
$$x + y = 100.$$

5. How many liters of 25% alcohol solution must be mixed with 12% solution to get 13 liters of 15% solution?

(a) Complete the table.

Percent	Liters	Liters of pure alcohol
25	x	$.25x$
12	y	
15	13	

(b) Write a system of equations and solve it.

(c) Joe needs 100 cc (cubic centimeters) of 20% salt solution for a chemistry experiment. The lab has on hand only 10% and 25% solutions. How much of each should he mix to get the desired amount of 20% solution?

ANSWERS

4. $x = 60, y = 40$
5. (a)

Percent	Liters	Liters of pure alcohol
25	x	$.25x$
12	y	$.12y$
15	13	$.15(13)$

(b) $x + y = 13$
$.25x + .12y = .15(13)$
3 liters of 25%, 10 liters of 12%

(c) $33\frac{1}{3}$ cc of 10%, $66\frac{2}{3}$ cc of 25%

8.5 Applications of Linear Systems

6. Use substitution to solve the system
$$4x + 4y = 400$$
$$x = 20 + y.$$

7. (a) In one hour Ann can row 2 miles against the current or 10 miles with the current. Find the speed of the current and Ann's speed in still water.

(b) Two cars that were 450 miles apart traveled toward each other. They met after 5 hours. If one car traveled twice as fast as the other, what were their speeds?

Complete this table.

	r	t	d
Faster car	x	5	
Slower car	y	5	

Write a system and solve it.

ANSWERS
6. $x = 60, y = 40$
7. (a) 4 miles per hour, 6 miles per hour

(b)

	r	t	d
Faster car	x	5	$5x$
Slower car	y	5	$5y$

$$5x + 5y = 450$$
$$x = 2y$$

$x = 60$ miles per hour, $y = 30$ miles per hour

▶ Word problems that use the distance formula, relating distance, rate, and time, often result in a system of two linear equations.

EXAMPLE 4 Two cars start from positions 400 miles apart and travel toward each other. They meet after 4 hours. (See Figure 8.) Find the average speed of each car if one travels 20 miles per hour faster than the other.

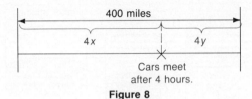
Cars meet after 4 hours.
Figure 8

Use the formula that relates distance, rate, and time, $d = rt$. Let x be the speed of the faster car and y the speed of the slower car. Since each car travels for 4 hours, the time, t, for each car is 4. This information is shown in the chart. The distance is found by using the formula $d = rt$ and the expressions already entered in the chart.

	r	t	d	
Faster car	x	4	$4x$	Find d from $d = rt$
Slower car	y	4	$4y$	

Since the total distance traveled by both cars is 400 miles,
$$4x + 4y = 400.$$
The faster car goes 20 miles per hour faster than the slower, giving
$$x = 20 + y.$$
This system of equations,
$$4x + 4y = 400$$
$$x = 20 + y,$$
can be solved by substitution.

Work Problem 6 at the side.

Problem 6 at the side gives the solution of the system: $x = 60$, $y = 40$. Thus, the speeds of the two cars were 40 miles per hour and 60 miles per hour. If each car travels for 4 hours, the total distance is
$$4(40) + 4(60) = 160 + 240 = 400$$
miles, as required. ◀

Work Problem 7 at the side.

The problems in this section also could be solved using only one variable, but for most of them the solution is simpler with two variables. Be careful: don't forget that two variables require two equations.

8.5 EXERCISES

Write a system of equations for each problem; then solve the system. Formulas are on the inside covers of the book. See Example 1.

1. The sum of two numbers is 52, and their difference is 34. Find the numbers.

2. Find two numbers whose sum is 56 and whose difference is 18.

3. A certain number is three times as large as a second number. Their sum is 96. What are the two numbers?

4. One number is five times as large as another. The difference of the numbers is 48. Find the numbers.

5. Two angles have a sum of 90°. Their difference is 20°. Find the angles.

6. The sum of two angles is 180°. One angle is 30° less than twice the other. Find the angles.

7. A rectangle is twice as long as it is wide. Its perimeter is 60 inches. Find the dimensions of the rectangle.

8. The perimeter of a triangle is 21 inches. If two sides are of equal length, and the third side is 3 inches longer than one of the equal sides, find the lengths of the three sides.

Write a system of equations for each problem; then solve the system. See Example 2.

9. The cashier at the Evergreen Ranch has some ten-dollar bills and some twenty-dollar bills. The total value of the money is $1480. If there is a total of 85 bills, how many of each type are there?

Kind of bill	Number of bills	Total value
$10	x	$10x$
$20	y	
Totals	85	$1480

10. A bank teller has 154 bills in $1 and $5 denominations. How many of each type of bill does he have if the total value of the money is $466?

Kind of bill	Number of bills	Total value
$1	x	
$5	y	
Totals	154	$466

11. A club secretary bought 8¢ and 10¢ pieces of candy to give to the members. She spent a total of $15.52. If she bought 170 pieces of candy, how many of each kind did she buy?

12. There were 311 tickets sold for a basketball game, some for students and some for nonstudents. Student tickets cost 25¢ each and nonstudent tickets cost 75¢ each. The total receipts were $108.75. How many of each type of tickets were sold?

13. A bank clerk has a total of 124 bills, both fives and tens. The total value of the money is $840. How many of each type of bill does he have?

14. A library buys a total of 54 books. Some cost $8 each and some cost $11 each. The total cost of the books is $492. How many of each type of book does the library buy?

15. Ms. Sullivan has $10,000 to invest, part at 10% and part at 14%. She wants the income from simple interest on the two investments to total $1100 yearly. How much should she invest at each rate?

16. Mr. Emerson has twice as much money invested at 14% as he has at 16%. If his yearly income from these investments is $880, how much does he have invested at each rate?

Write a system of equations for each problem; then solve the system. See Example 3.

17. A 90% antifreeze solution is to be mixed with a 75% solution to make 20 liters of a 78% solution. How many liters of 90% and 75% solutions should be used?

Liters of solution	Percent	Liters of pure antifreeze
x	90	$.90x$
y	75	
20	78	

18. A 40% potassium iodide solution is to be mixed with a 70% solution to get 60 liters of a 50% solution. How many liters of the 40% and 70% solutions will be needed?

Liters of solution	Percent	Liters of pure potassium iodide
x	40	
y	70	
60		

422 Linear Systems

19. How many liters of a 25% indicator solution should be mixed with a 55% solution to get 12 liters of a 45% solution?

20. A 60% solution of salt is to be mixed with an 80% solution to get 40 liters of a 65% solution. How many liters of the 60% and the 80% solutions will be needed?

21. A merchant wishes to mix coffee worth $6 per pound with coffee worth $3 per pound to get 90 pounds of a mixture worth $4 per pound. How many pounds of the $6 and the $3 coffee will be needed?

Pounds	Dollars per pound	Cost
x	6	$6x$
y	3	
90	4	

22. A grocer wishes to blend candy selling for $1.20 a pound with candy selling for $1.80 a pound to get a mixture that will be sold for $1.40 a pound. How many pounds of the $1.20 and the $1.80 candy should be used to get 30 pounds of the mixture?

Pounds	Dollars per pound	Cost
x	1.20	$1.20x$
y	1.80	
30		

23. How many barrels of olives worth $40 per barrel must be mixed with olives worth $60 per barrel to get 50 barrels of a mixture worth $48 per barrel?

24. A glue merchant wishes to mix some glue worth $70 per barrel with some glue worth $90 per barrel to get 80 barrels of a mixture worth $77.50 per barrel. How many barrels of each type should be used?

8.5 Exercises

Write a system of equations for each problem; then solve the system. See Example 4.

25. A boat takes 3 hours to go 24 miles upstream. It can go 36 miles downstream in the same time. Find the speed of the current and the speed of the boat in still water. Let x = the speed of the boat in still water and y = the speed of the current.

	d	r	t
Downstream	36	$x+y$	3
Upstream	24	$x-y$	3

26. It takes a boat $1\frac{1}{2}$ hours to go 12 miles downstream, and 6 hours to return. Find the speed of the boat in still water and the speed of the current.

	d	r	t
Downstream	12	$x+y$	$\frac{3}{2}$
Upstream	12	$x-y$	6

27. If a plane can travel 400 miles per hour into the wind and 540 miles per hour with the wind, find the speed of the wind and the speed of the plane in still air.

28. A small plane travels 100 miles per hour with the wind and 60 miles per hour against it. Find the speed of the wind and the speed of the plane in still air.

The next problems are "brain-busters."

29. At the beginning of a walk for charity, John and Harriet are 30 miles apart. If they leave at the same time and walk in the same direction, John overtakes Harriet in 60 hours. If they walk toward each other, they meet in 5 hours. What are their speeds?

30. Mr. Anderson left Farmersville in a plane at noon to travel to Exeter. Mr. Bentley left Exeter in his automobile at 2 P.M. to travel to Farmersville. It is 400 miles from Exeter to Farmersville. If the sum of their speeds was 120 miles per hour, and if they met at 4 P.M., find the speed of each.

31. The Smith family is coming to visit, and no one knows how many children they have. Janet, one of the girls, says she has as many brothers as sisters; her brother Steve says he has twice as many sisters as brothers. How many boys and how many girls are in the family?

32. In the Lopez family, the number of boys is one more than half the number of girls. One of the Lopez boys, Rico, says that he has one more sister than brothers. How many boys and girls are in the family?

8.6 SOLVING SYSTEMS OF LINEAR INEQUALITIES

Objective

▶ Solve systems of linear inequalities by graphing.

Graphing the solution of a linear inequality was discussed in Section 7.6. Let us review the method. To graph the solution of $x + 3y > 12$, first graph the line $x + 3y = 12$. (Find a few ordered pairs that satisfy the equation.) Because the points on the line do not satisfy the inequality, use a dashed line. Decide which side of the line should be shaded by choosing any test point not on the line, such as $(0, 0)$. Substitute 0 for x and 0 for y in the given inequality.

$$x + 3y > 12$$
$$0 + 0 > 12$$
$$0 > 12 \quad \text{False}$$

The test point does not satisfy the inequality, so shade the region on the side of the line that does not include $(0, 0)$, as shown in Figure 9.

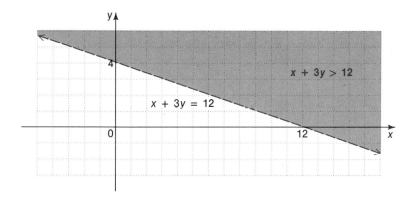

Figure 9

1. Graph each inequality.

    (a) $3x + 2y \geq 6$
    The line $3x + 2y = 6$ is graphed already.

    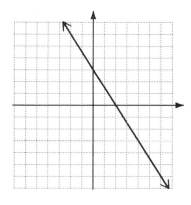

    (b) $2x - 5y < 10$

    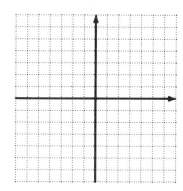

Work Problem 1 at the side.

▶ The same method is used to find the solution of a system of two linear inequalities, as shown in Examples 1 and 2. A **system of linear inequalities** contains two or more linear inequalities (and no other kinds of inequalities). The solution of a system of linear inequalities includes all points that make all inequalities of the system true at the same time.

EXAMPLE 1 Graph the solution of the system

$$3x + 2y \leq 6$$
$$2x - 5y \geq 10.$$

First graph the inequality $3x + 2y \leq 6$ using the steps described above. Then, on the same axes, graph the second inequality, $2x - 5y \geq 10$. The solution of the system is given by the overlap of the regions of the two graphs. This solution is the region in Figure 10 (on the next page) with the darkest shading; this region includes portions of both boundary lines. ◀

ANSWERS

1. (a)   (b)

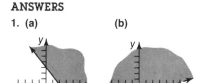

8.6 Solving Systems of Linear Inequalities

425

2. Graph the solution of the system

$x - 2y \leq 8$
$3x + y \geq 6$.

To get started, the graphs of $x - 2y = 8$ and $3x + y = 6$ are included.

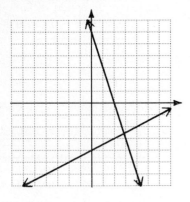

3. Graph the solution of the system

$x + 2y < 0$
$3x - 4y < 12$.

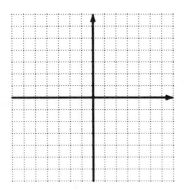

2.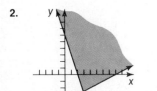

3.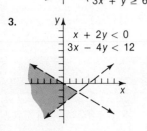

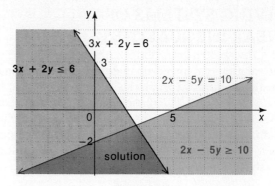

Figure 10

Work Problem 2 at the side.

EXAMPLE 2 Graph the solution of the system

$x - y > 5$
$2x + y < 2$.

Figure 11 shows the graphs of both $x - y > 5$ and $2x + y < 2$. Dashed lines show that the graphs of the inequalities do not include their boundary lines. The solution of the system is the region with the darkest shading. The solution does not include either boundary line. ◂

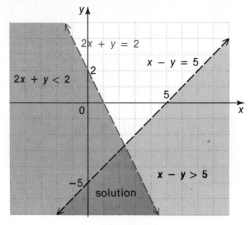

Figure 11

EXAMPLE 3 Graph the solution of the system

$4x - 3y \leq 8$
$x \geq 2$.

Recall that $x = 2$ is a vertical line through the point $(2, 0)$. The graph of the system is the region in Figure 12 with the darkest shading. ◂

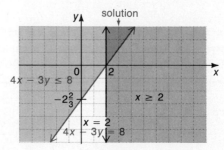

Figure 12

Work Problem 3 at the side.

426　　　　Linear Systems

8.6 EXERCISES

Graph the solution of each system of linear inequalities. See Examples 1–3.

1. $x + y \leq 6$
   $x - y \leq 1$

2. $x + y \geq 2$
   $x - y \leq 3$

3. $2x - 3y \leq 6$
   $x + y \geq -1$

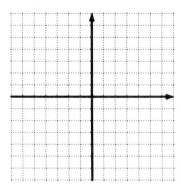 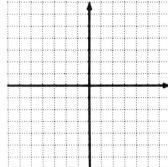 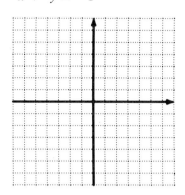

4. $4x + 5y \leq 20$
   $x - 2y \leq 5$

5. $x + 4y \leq 8$
   $2x - y \leq 4$

6. $3x + y \leq 6$
   $2x - y \leq 8$

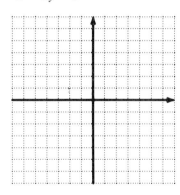 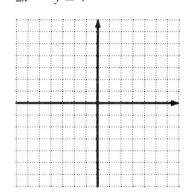 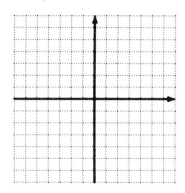

7. $x - 4y \leq 3$
   $x \geq 2y$

8. $2x + 3y \leq 6$
   $x - y \geq 5$

9. $x + 2y \leq 4$
   $x + 1 \geq y$

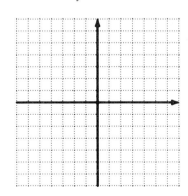 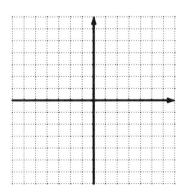 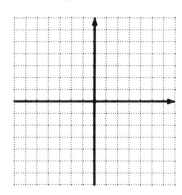

10. $y \leq 2x - 5$
    $x - 3y \leq 2$

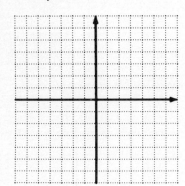

11. $4x + 3y \leq 6$
    $x - 2y \geq 4$

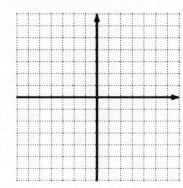

12. $3x - y \leq 4$
    $-6x + 2y \leq -10$

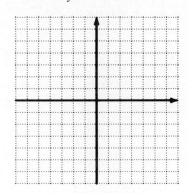

13. $x - 2y > 6$
    $2x + y > 4$

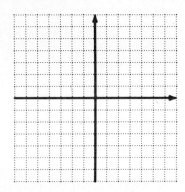

14. $3x + y < 4$
    $x + 2y > 2$

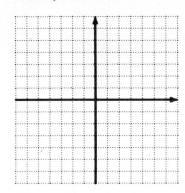

15. $x < 2y + 3$
    $x + y > 0$

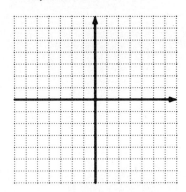

16. $2x + 3y < 6$
    $4x + 6y > 18$

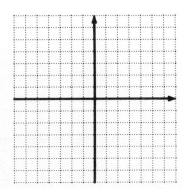

17. $x - 3y \leq 6$
    $x \geq -1$

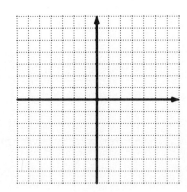

18. $2x + 5y \geq 10$
    $x \leq 4$

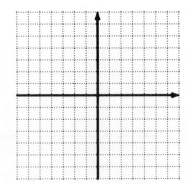

CHAPTER 8 REVIEW EXERCISES

(8.1) *Decide whether the given ordered pair is a solution of the given system.*

1. (3, 4)
   $4x - 2y = 4$
   $5x + y = 17$

2. (−2, 1)
   $5x + 3y = -7$
   $2x - 3y = -7$

3. (−5, 2)
   $x - 4y = -10$
   $2x + 3y = -4$

4. (6, 3)
   $3x + 8y = 42$
   $4x - 3y = 15$

5. (−1, −3)
   $x + 2y = -7$
   $-x + 3y = -8$

6. (2, 6)
   $3x - y = 0$
   $4x + 2y = 20$

Solve each system by graphing.

7. $x + y = 3$
   $2x - y = 3$

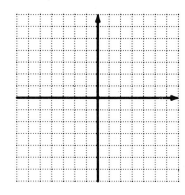

8. $x - 2y = -6$
   $2x + y = -2$

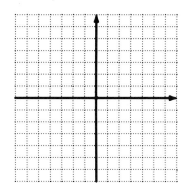

9. $2x + 3y = 1$
   $4x - y = -5$

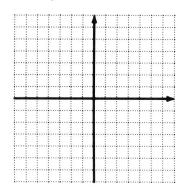

10. $x = 2y + 2$
    $2x - 4y = 4$

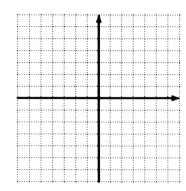

11. $y + 2 = 0$
    $3y = 6x$

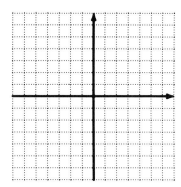

12. $2x + 5y = 10$
    $2x + 5y = 8$

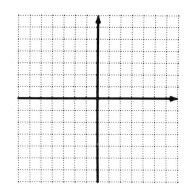

Chapter 8 Review Exercises **429**

(8.2) Solve each system by the addition method.

13. $x + y = 6$
    $2x + y = 8$

14. $3x - y = 13$
    $x - 2y = 1$

15. $5x + 4y = 7$
    $3x - 4y = 17$

16. $-4x + 3y = -7$
    $6x - 5y = 11$

17. $3x - 2y = 14$
    $5x - 4y = 24$

18. $2x + 6y = 18$
    $3x + 5y = 19$

19. $5x + 4y = 3$
    $7x + 5y = 3$

20. $3x + 5y = -1$
    $5x + 4y = 7$

(8.3)

21. $3x - 4y = 7$
    $6x - 8y = 14$

22. $2x + y = 5$
    $2x + y = 8$

23. $3x - 4y = -1$
    $-6x + 8y = 2$

24. $4x + y = 5$
    $-12x - 3y = -15$

(8.4) Solve each system by the substitution method.

25. $3x + y = 14$
    $x = 2y$

26. $2x - 5y = 5$
    $y = x + 2$

27. $4x + 5y = 35$
    $x + 1 = 2y$

28. $5x + y = -6$
    $x - 3y = 2$

430 Linear Systems

name   date   hour

Solve each system by any method. First simplify equations and clear of fractions where necessary.

29. $2x + 3y = 5$
$3x + 4y = 8$

30. $6x + 5y = 9$
$2x - 3y = 17$

31. $2x + y - x = 3y + 5$
$y + 2 = x - 5$

32. $5x - 3 + y = 4y + 8$
$2y + 1 = x - 3$

33. $\dfrac{x}{2} + \dfrac{y}{3} = \dfrac{-8}{3}$
$\dfrac{x}{4} + 2y = \dfrac{1}{2}$

34. $\dfrac{3x}{4} - \dfrac{y}{3} = \dfrac{7}{6}$
$\dfrac{x}{2} + \dfrac{2y}{3} = \dfrac{5}{3}$

35. $\dfrac{2x}{5} - \dfrac{y}{2} = \dfrac{7}{10}$
$\dfrac{x}{3} + y = 2$

36. $\dfrac{5x}{3} - \dfrac{y}{2} = -\dfrac{31}{3}$
$2x + \dfrac{y}{3} = \dfrac{26}{3}$

(8.5) *Solve each word problem by any method. Use two variables.*

37. The sum of two numbers is 42, and their difference is 14. Find the numbers.

38. One number is 2 more than twice as large as another. Their sum is 17. Find the numbers.

39. The perimeter of a rectangle is 40 meters. Its length is $1\frac{1}{2}$ times its width. Find the length and width of the rectangle.

40. A cashier has 20 bills, all of which are $10 or $20 bills. The total value of the money is $250. How many of each type does he have?

Chapter 8   Review Exercises   **431**

41. Ms. Branson has $18,000 to invest. She wants the total income from the money to be $2600. She can invest part of it at 12% and the rest at 16%. How much should she invest at each rate?

42. Candy that sells for $1.30 a pound is to be mixed with candy selling for $.90 a pound to get 50 pounds of a mix that will sell for $1 per pound. How much of each type should be used?

43. A 40% antifreeze solution is to be mixed with a 70% solution to get 60 liters of a 50% solution. How many liters of the 40% and 70% solutions will be needed?

44. A certain plane flying with the wind travels 540 miles in 2 hours. Later, flying against the same wind, the plane travels 690 miles in 3 hours. Find the speed of the plane in still air and the speed of the wind.

(8.6) *Graph the solution for each system of linear inequalities.*

45. $x + y \leq 3$
$2x \geq y$

46. $x + y \geq 4$
$x - y \leq 2$

47. $2x + y < 6$
$y - 2x < 6$

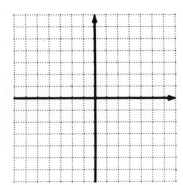
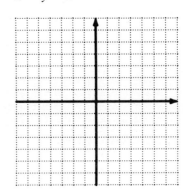
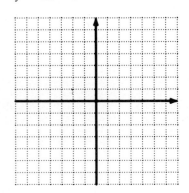

48. $x \geq 2$
$y \leq 5$

49. $y \geq 3x$
$2x + 3y \leq 4$

50. $y + 2 < 2x$
$x > 3$

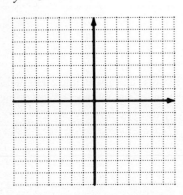
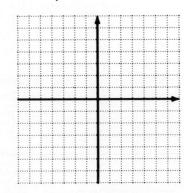
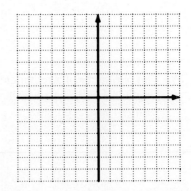

name _____ date _____ hour _____

CHAPTER 8   TEST

Solve each system by graphing.

1. $2x + y = 5$
$3x - y = 15$

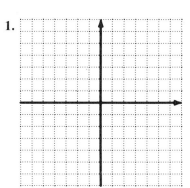

2. $3x + 2y = 8$
$5x + 4y = 10$

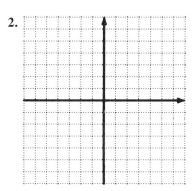

3. $x + 2y = 6$
$2x - y = 7$

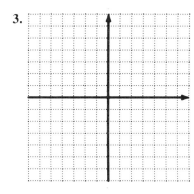

Solve each system by the addition method.

4. $2x - 5y = -13$
$3x + 5y = 43$

4. _____

5. $4x + 3y = 26$
$5x + 4y = 32$

5. _____

6. $6x + 5y = -13$
$3x + 2y = -4$

7. $4x + 5y = 8$
$-8x - 10y = -6$

8. $6x - 5y = 2$
$-2x + 3y = 2$

9. $\dfrac{6}{5}x - y = \dfrac{1}{5}$
$-\dfrac{2}{3}x + \dfrac{1}{6}y = \dfrac{1}{3}$

Solve each system by substitution.

10. $2x + y = 1$
$x = 8 + y$

11. $4x + 3y = 0$
$x + y = 2$

Solve each system by any method.

12. $8 + 3x - 4y = 14 - 3y$
$3x + y + 12 = 9x - y$

13. $\dfrac{x}{2} - \dfrac{y}{4} = -4$
$\dfrac{2x}{3} + \dfrac{5y}{4} = 1$

name _____ date _____ hour _____

14. The sum of two numbers is 39. If one number is doubled, the result is 3 less than the other. Find the numbers.

14. _____

15. The local record shop is having a sale. Some records cost $2.50 and some cost $3.75. Joe has exactly $20 to spend and wants to buy 6 records. How many can he buy at each price?

15. _____

16. A 40% solution of acid is to be mixed with a 60% solution to get 100 liters of a 45% solution. How many liters of each solution should be used?

16. _____

17. Two cars leave from the same place and travel in the same direction. One car travels one and one third times as fast as the other. After 3 hours, they are 45 miles apart. What is the speed of each car?

17. _____

Chapter 8 Test

Graph the solution of each system of inequalities.

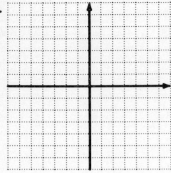

18. $2x + 7y \leq 14$
$x - y \geq 1$

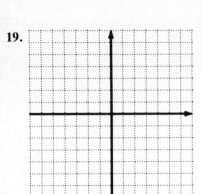

19. $2x - y \leq 6$
$4y + 12 \geq -3x$

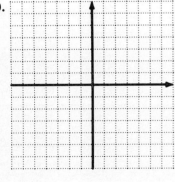

20. $3x - 5y < 15$
$y < 2$

CUMULATIVE REVIEW EXERCISES

Complete the given ordered pairs for each equation.

1. $4x + 7y = 28$ $(0, \underline{\ \ })$, $(\underline{\ \ }, 0)$, $(3, \underline{\ \ })$, $(\underline{\ \ }, 5)$

2. $6 + 2x = y$ $(0, \underline{\ \ })$, $(\underline{\ \ }, 0)$, $(-2, \underline{\ \ })$, $(\underline{\ \ }, 3)$

3. $x - 2y = 0$ $(0, \underline{\ \ })$, $(6, \underline{\ \ })$, $(\underline{\ \ }, -5)$, $(\underline{\ \ }, \frac{3}{2})$

4. $y + 3 = 0$ $(0, \underline{\ \ })$, $(2, \underline{\ \ })$, $(-4, \underline{\ \ })$, $(\frac{8}{3}, \underline{\ \ })$

Graph each linear equation.

5. $x - y = 4$

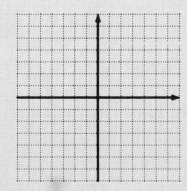

6. $3x + y = 6$

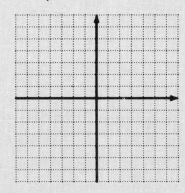

7. $y = 3x - 4$

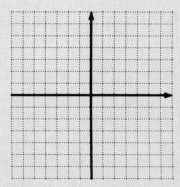

8. $y = 3x$

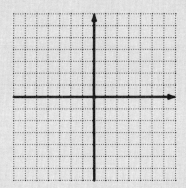

9. $x - 4 = 0$

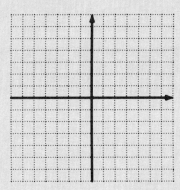

10. $y + 5 = 0$

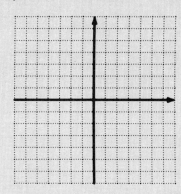

Find the slope of each of the following lines.

11. Through $(-5, 6)$ and $(1, -2)$

12. Through $(4, -1)$ and $(3, 5)$

13. $y = 4x - 3$

14. $x + 2y = 9$

Decide whether the lines in each of the following pairs are parallel, perpendicular, *or* neither.

15. $4x + y = 8$
    $8x + 2y = 3$

16. $3x - 2y = 4$
    $2x + 3y = 9$

17. $3x + 2y = 8$
    $6x - 4y = 1$

18. $x - 3y = 4$
    $3x + y = 2$

Write an equation for each line. Write answers in the form $ax + by = c$.

19. $m = -2$; $b = 3$

20. $m = \dfrac{4}{3}$; $b = 1$

21. $m = -\dfrac{5}{2}$; $b = -\dfrac{1}{4}$

22. $m = -\dfrac{3}{8}$; $b = -\dfrac{1}{6}$

23. Through $(2, -5)$; $m = 3$

24. Through $(1, -4)$; $m = \dfrac{2}{3}$

Graph each linear inequality.

25. $x - 3y \geq 6$

26. $2x + 3y < 12$

27. $4x - 5y \leq 20$

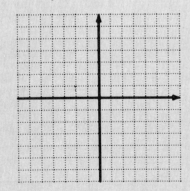
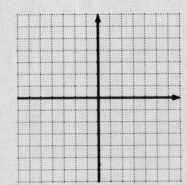
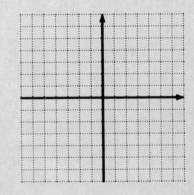

438 Linear Systems

28. $x \geq -3$

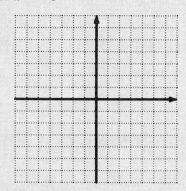

29. $2x < y$

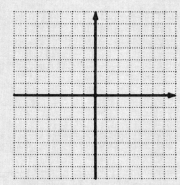

30. $3x + 2y \geq 0$

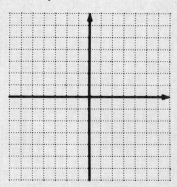

31. Is $(-6, 2)$ a solution for the following system?

$$3x + 2y = -14$$
$$4x + y = -26$$

Solve each system by any method.

32. $2x - y = -8$
$x + 2y = 11$

33. $4x + 5y = -8$
$3x + 4y = -7$

34. $3x + 5y = 1$
$x = y + 3$

35. $3x + 4y = 2$
$6x + 8y = 1$

36. $2x - 5y = 7$
$-4x + 10y = -14$

37. $y = 2x$
$4x + y = 3$

Solve each word problem by any method. Use two variables.

38. Two numbers have a sum of 24 and a difference of 6. Find the numbers.

Cumulative Review Exercises **439**

39. A rectangle has a perimeter of 40 meters. The length is 2 meters more than the width. Find the width of the rectangle.

Let x represent the width.

Length = _____

$P = 2L + 2W$

x

40. The cashier at a small motel has 17 bills, all of which are fives and tens. The value of the money is $125. How many of each type of bill are there?

Type of bill	Number of bills	Value of bills
Fives	x	
Tens	y	
Totals		

41. The manager of a small pension fund has $75,000 to invest. She wants to invest part in a bond fund paying 10% and the balance in a stock fund paying 8%. How much should be invested at each rate to produce $6600 in annual interest?

Graph each system of inequalities.

42. $y < x - 1$
$x > y + 2$

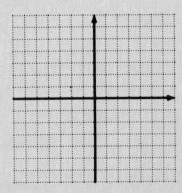

43. $x - y \leq 2$
$2x + 3y \geq 6$

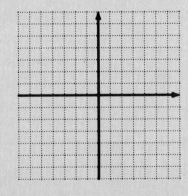

44. $y \leq 2x$
$x \geq -1$

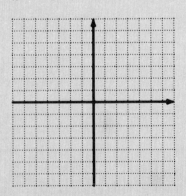

Roots and Radicals

9

9.1 FINDING ROOTS

The *square* of a number is found by multiplying the number by itself.

If $a = 7$, then $a^2 = 7 \cdot 7 = 49$.
If $a = -5$, then $a^2 = (-5)(-5) = 25$.
If $a = -\frac{1}{2}$, then $a^2 = \left(-\frac{1}{2}\right)\left(-\frac{1}{2}\right) = \frac{1}{4}$.

In this chapter, the opposite problem is considered.

If $a^2 = 49$, then $a = ?$
If $a^2 = 100$, then $a = ?$
If $a^2 = 25$, then $a = ?$

▶ Finding a requires a number that can be multiplied by itself to result in the given number. The number a is called a **square root** of the number a^2.

EXAMPLE 1 Find a square root of 49 by thinking of a number that multiplied by itself gives 49. One square root of 49 is 7, since $7 \cdot 7 = 49$. Another square root of 49 is -7, since $(-7)(-7) = 49$. The number 49 has two square roots, 7 and -7. One is positive and one is negative. ◀

Work Problem 1 at the side.

All numbers with integer square roots are called **perfect squares**. The first 100 perfect squares are listed in Table 2 in the back of the book.

The positive square root of a number is written with the symbol $\sqrt{}$. For example, the positive square root of 121 is 11, written

$$\sqrt{121} = 11.$$

The symbol $-\sqrt{}$ is used for the negative square root of a number. For example, the negative square root of 121 is -11, written

$$-\sqrt{121} = -11.$$

The symbol $\sqrt{}$ is called a **radical sign** and, used alone, always represents the positive square root (except that $\sqrt{0} = 0$). The number inside the radical sign is called the **radicand** and the entire expression, radical sign and radicand, is called a **radical**. An algebraic expression containing a radical is called a **radical expression.**

Objectives

▶ Find square roots.
▶ Decide whether a given root is rational or irrational.
▶ Find decimal approximations for irrational square roots.
▶ Use the Pythagorean formula.
▶ Find higher roots.

1. Find all square roots.

   (a) 100

   (b) 25

   (c) 36

   (d) 64

ANSWERS
1. (a) 10, −10 (b) 5, −5
   (c) 6, −6 (d) 8, −8

9.1 Finding Roots   441

2. Find each square root.

(a) $\sqrt{16}$

(b) $-\sqrt{169}$

(c) $-\sqrt{225}$

(d) $\sqrt{729}$

(e) $\sqrt{\dfrac{36}{25}}$

3. Write *rational* or *irrational* for each square root.

(a) $\sqrt{9}$

(b) $\sqrt{7}$

(c) $\sqrt{16}$

(d) $\sqrt{72}$

If a is a nonnegative real number, then
$\sqrt{a}$ is the nonnegative square root of a,
$-\sqrt{a}$ is the negative square root of a.
Also, $\sqrt{0} = 0$, and for nonnegative a,
$$\sqrt{a} \cdot \sqrt{a} = (\sqrt{a})^2 = a.$$

EXAMPLE 2 Find each square root.

(a) $\sqrt{144}$

The radical $\sqrt{144}$ represents the positive square root of 144. Think of a positive number whose square is 144. The number is 12 (see the table of perfect squares), and
$$\sqrt{144} = 12.$$

(b) $-\sqrt{1024}$

This symbol represents the negative square root of 1024. From the table,
$$-\sqrt{1024} = -32.$$

(c) $-\sqrt{\dfrac{16}{49}} = -\dfrac{4}{7}$ ◂

Work Problem 2 at the side.

▶ A number that is not a perfect square has a square root that is not a rational number. For example, $\sqrt{5}$ is not a rational number, because it cannot be written as the ratio of two integers. However, $\sqrt{5}$ is a real number and corresponds to a point on the number line. As mentioned in Chapter 2, a real number that is not rational is called an **irrational number**. The number $\sqrt{5}$ is irrational. Many square roots of integers are irrational.

Not every number has a *real number* square root. For example, there is no real number that can be squared to get -36. (The square of a real number can never be negative.) Because of this, $\sqrt{-36}$ is not a real number.

If a is a negative real number, $\sqrt{a}$ is not a real number.

EXAMPLE 3 Tell whether each square root is rational or irrational.

(a) $\sqrt{17}$

Since 17 is not a perfect square, $\sqrt{17}$ is irrational.

(b) $\sqrt{64}$

The number 64 is a perfect square, 8^2, so $\sqrt{64} = 8$, a rational number. ◂

Work Problem 3 at the side.

ANSWERS
2. (a) 4 (b) -13 (c) -15
   (d) 27 (e) $\dfrac{6}{5}$
3. (a) rational (3) (b) irrational
   (c) rational (4) (d) irrational

Not all irrational numbers are square roots of integers. For example, π (approximately 3.14159) is an irrational number that is not a square root of any integer.

▶ Even if a number is irrational, a decimal that approximates the number can be found by using a table or certain calculators. For square roots, the square root table can be used, as shown in Example 4.

EXAMPLE 4 Find a decimal approximation for each square root. Round answers to the nearest thousandth.

(a) $\sqrt{11}$

Using the square root key of a calculator gives $3.31662479 \approx 3.317$, where $\approx$ means "is approximately equal to." To use the square root table, find 11 at the left. The approximate square root is given in the column having $\sqrt{n}$ at the top. You should find that

$$\sqrt{11} \approx 3.317.$$

(b) $\sqrt{39} \approx 6.245$. Use the table or a calculator.

(c) $\sqrt{740}$

There is no 740 in the "n" column of the square root table. However, $740 = 74 \times 10$, so $\sqrt{740}$ can be found in the "$\sqrt{10n}$" column. Using the table or a calculator,

$$\sqrt{740} \approx 27.203. \blacktriangleleft$$

Work Problem 4 at the side.

▶ One application of square roots comes from the Pythagorean formula. Recall from Section 5.6 that by this formula if c is the length of the longest side of a right triangle, and a and b are the lengths of the two shorter sides, then

$$a^2 + b^2 = c^2.$$

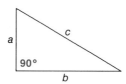

EXAMPLE 5 Find the third side of each right triangle with sides a, b, and c, where c is the longest side.

(a) $a = 3, b = 4$

Use the formula to find c^2 first.

$$c^2 = a^2 + b^2$$
$$c^2 = 3^2 + 4^2$$
$$= 9 + 16 = 25$$

Now find the positive square root of 25 to get c.

$$c = \sqrt{25} = 5$$

(Although -5 is also a square root of 25, the length of a side of a triangle must be a positive number.)

4. Find a decimal approximation for each square root.

(a) $\sqrt{28}$

(b) $\sqrt{63}$

(c) $\sqrt{190}$

(d) $\sqrt{1000}$

ANSWERS
4. (a) 5.292 (b) 7.937 (c) 13.784
   (d) 31.623

9.1 Finding Roots

5. Find the missing side in each right triangle.

   (a) $a = 7, b = 24$

   (b) $c = 15, b = 13$

   (c) $c = 11, a = 8$

6. Find each cube root

   (a) $\sqrt[3]{27}$

   (b) $-\sqrt[3]{27}$

   (c) $\sqrt[3]{-27}$

7. Find each root. You may have to use trial and error.

   (a) $\sqrt[4]{81}$

   (b) $-\sqrt[4]{256}$

   (c) $\sqrt[5]{-243}$

   (d) $-\sqrt[5]{243}$

ANSWERS
5. (a) 25 (b) $\sqrt{56} \approx 7.483$
   (c) $\sqrt{57} \approx 7.550$
6. (a) 3 (b) -3 (c) -3
7. (a) 3 (b) -4 (c) -3 (d) -3

(b) $c = 9, b = 5$

Substitute the given values in the formula, $c^2 = a^2 + b^2$. Then solve for a^2.

$$9^2 = a^2 + 5^2$$
$$81 = a^2 + 25$$
$$56 = a^2$$

Use the table or a calculator to find $a = \sqrt{56} \approx 7.483$. ◀

Work Problem 5 at the side.

Finding the square root of a number is the inverse of squaring a number. In a similar way, there are inverses to finding the cube of a number, or finding the fourth or higher power of a number. These inverses are called finding the **cube root**, written $\sqrt[3]{a}$, the **fourth root**, written $\sqrt[4]{a}$, and so on. In $\sqrt[n]{a}$, the number n is the **index** or **order** of the radical. It would be possible to write $\sqrt[2]{a}$ instead of $\sqrt{a}$, but the simpler symbol $\sqrt{a}$ is customary since the square root is the most commonly used root.

Table 1, Selected Powers of Numbers, may help you find roots; this table is given at the back of the book. Some calculators also can be used to find these roots.

EXAMPLE 6 Find each cube root.

(a) $\sqrt[3]{8}$

Look for a number that can be cubed to give 8. Since $2^3 = 8$, it follows that $\sqrt[3]{8} = 2$.

(b) $\sqrt[3]{-8}$

$\sqrt[3]{-8} = -2$ because $(-2)^3 = -8$. ◀

As these examples suggest, the cube root of a positive number is positive, and the cube root of a negative number is negative.

Work Problem 6 at the side.

When the index of the radical is even (square root, fourth root, sixth root, and so on), the radicand must be nonnegative to obtain a real number root. Also, for even indexes the symbols $\sqrt{}$, $\sqrt[4]{}$, $\sqrt[6]{}$, and so on are used for the *nonnegative roots*.

EXAMPLE 7 Find each root.

(a) $\sqrt[4]{16}$

$\sqrt[4]{16} = 2$ because 2 is positive and $2^4 = 16$. Also, $-\sqrt[4]{16} = -2$, but there is no real number that equals $\sqrt[4]{-16}$.

(b) $-\sqrt[5]{32}$

First find $\sqrt[5]{32}$. Since 2 is the number whose fifth power is 32,

$$\sqrt[5]{32} = 2.$$

If $\sqrt[5]{32} = 2$, then $-\sqrt[5]{32} = -2.$ ◀

Work Problem 7 at the side.

Roots and Radicals

9.1 EXERCISES

Find all square roots of each number. See Example 1.

1. 9
2. 16
3. 121
4. $\dfrac{196}{25}$

5. $\dfrac{400}{81}$
6. $\dfrac{900}{49}$
7. 625
8. 961

9. 1521
10. 2209
11. 3969
12. 4624

Find each root that exists. See Example 2.

13. $\sqrt{4}$
14. $\sqrt{9}$
15. $\sqrt{25}$
16. $\sqrt{36}$

17. $-\sqrt{64}$
18. $-\sqrt{100}$
19. $\sqrt{169}$
20. $\sqrt{196}$

21. $\sqrt{900}$
22. $\sqrt{1600}$
23. $-\sqrt{1681}$
24. $-\sqrt{2116}$

25. $\sqrt{\dfrac{36}{49}}$
26. $\sqrt{\dfrac{121}{144}}$
27. $-\sqrt{\dfrac{100}{169}}$

28. $-\sqrt{5625}$
29. $\sqrt{-9}$
30. $\sqrt{-25}$

Write rational or irrational for each number. If a number is rational, give its exact value. If a number is irrational, give a decimal approximation for the square root. See Examples 3 and 4.

31. $\sqrt{16}$

32. $\sqrt{81}$

33. $\sqrt{15}$

34. $\sqrt{31}$

35. $\sqrt{47}$

36. $\sqrt{53}$

37. $-\sqrt{121}$

38. $-\sqrt{144}$

39. $\sqrt{110}$

40. $\sqrt{170}$

41. $\sqrt{400}$

42. $\sqrt{900}$

43. $-\sqrt{200}$

44. $-\sqrt{260}$

45. $\sqrt{570}$

46. $\sqrt{690}$

Find the missing side of each right triangle with sides a, b, and c, where c is the longest side. See Example 5.

47. $a = 6, b = 8$

48. $a = 5, b = 12$

49. $c = 17, a = 8$

50. $c = 26, b = 10$

51. $a = 10, b = 8$

52. $a = 9, b = 7$

53. $c = 12, b = 7$

54. $c = 8, a = 3$

Find each of the following roots that are real numbers. Use Table 1 in the back of the book, if necessary. See Examples 6 and 7.

55. $\sqrt[3]{1000}$

56. $\sqrt[3]{8}$

57. $\sqrt[3]{125}$

58. $\sqrt[3]{216}$

59. $-\sqrt[3]{8}$

60. $-\sqrt[3]{64}$

61. $\sqrt[3]{-8}$

62. $\sqrt[3]{-27}$

63. $\sqrt[4]{1}$

64. $-\sqrt[4]{16}$

65. $\sqrt[4]{-16}$

66. $-\sqrt[4]{-625}$

67. $-\sqrt[4]{81}$

68. $-\sqrt[4]{256}$

69. $\sqrt[5]{-32}$

70. $-\sqrt[5]{32}$

Use the $\sqrt{}$ key on your calculator to find approximations for the following roots. Round answers to the nearest thousandth.

71. $\sqrt[4]{27}$

72. $-\sqrt[4]{28}$

73. $\sqrt[4]{1.42}$

74. $\sqrt[4]{2.04}$

75. $\sqrt[4]{265.3}$

76. $-\sqrt[4]{57.68}$

9.2 PRODUCTS AND QUOTIENTS OF RADICALS

Objectives
1. Multiply radicals.
2. Simplify radical expressions.
3. Simplify radical quotients.
4. Use product and quotient rules to simplify higher roots.

1 Several useful rules for finding products and quotients of radicals are developed in this section. To get a rule for products, recall that

$$\sqrt{4} \cdot \sqrt{9} = 2 \cdot 3 = 6 \quad \text{and} \quad \sqrt{4 \cdot 9} = \sqrt{36} = 6,$$

showing that

$$\sqrt{4} \cdot \sqrt{9} = \sqrt{4 \cdot 9}.$$

This result illustrates the *product rule for radicals*.

> **Product Rule for Radicals**
> For nonnegative real numbers a and b,
> $$\sqrt{a} \cdot \sqrt{b} = \sqrt{a \cdot b}.$$
> The product of two radicals is the radical of the product.

(Recall that the square root of a negative number is not a real number.)

EXAMPLE 1 Use the product rule for radicals to find each product.
(a) $\sqrt{2} \cdot \sqrt{3} = \sqrt{2 \cdot 3} = \sqrt{6}$
(b) $\sqrt{7} \cdot \sqrt{5} = \sqrt{35}$
(c) $\sqrt{11} \cdot \sqrt{a} = \sqrt{11a}$ Assume $a > 0$ ◀

Work Problem 1 at the side.

2 A very important use of the product rule is in simplifying radical expressions: as a first step, a radical expression is **simplified** when no perfect square factor remains under the radical sign. Example 2 shows how this is done.

EXAMPLE 2 Simplify each radical.
(a) $\sqrt{20}$

Since 20 has a perfect square factor of 4,

$$\sqrt{20} = \sqrt{4 \cdot 5} \qquad \text{4 is a perfect square}$$
$$= \sqrt{4} \cdot \sqrt{5} \qquad \text{Product rule}$$
$$= 2\sqrt{5}. \qquad \sqrt{4} = 2$$

Thus, $\sqrt{20} = 2\sqrt{5}$. Since 5 has no perfect square factor (other than 1), $2\sqrt{5}$ is called the **simplified form** of $\sqrt{20}$.

(b) $\sqrt{72}$

Look down the list of perfect squares in the square root table. Find the largest of these numbers that is a factor of 72. The largest is 36, so

$$\sqrt{72} = \sqrt{36 \cdot 2} \qquad \text{36 is a perfect square}$$
$$= \sqrt{36} \cdot \sqrt{2}$$
$$= 6\sqrt{2}.$$

1. Use the product rule for radicals to find each product.

(a) $\sqrt{6} \cdot \sqrt{11}$

(b) $\sqrt{2} \cdot \sqrt{7}$

(c) $\sqrt{17} \cdot \sqrt{3}$

(d) $\sqrt{10} \cdot \sqrt{r}, \quad r > 0$

ANSWERS
1. (a) $\sqrt{66}$ (b) $\sqrt{14}$ (c) $\sqrt{51}$
   (d) $\sqrt{10r}$

2. Simplify each radical.

   (a) $\sqrt{8}$

   (b) $\sqrt{27}$

   (c) $\sqrt{50}$

   (d) $\sqrt{60}$

3. Find each product and simplify.

   (a) $\sqrt{3} \cdot \sqrt{15}$

   (b) $\sqrt{7} \cdot \sqrt{14}$

   (c) $\sqrt{10} \cdot \sqrt{50}$

   (d) $\sqrt{12} \cdot \sqrt{12}$

(c) $\sqrt{300} = \sqrt{100 \cdot 3}$ 100 is a perfect square
$= \sqrt{100} \cdot \sqrt{3}$
$= 10\sqrt{3}$

(d) $\sqrt{15}$

The number 15 has no perfect square factors, so $\sqrt{15}$ cannot be simplified further. ◀

Work Problem 2 at the side.

Sometimes the product rule can be used to simplify an answer, as Example 3 shows.

EXAMPLE 3 Find each product and simplify.

(a) $\sqrt{9} \cdot \sqrt{75} = 3\sqrt{75}$ $\sqrt{9} = 3$
$= 3\sqrt{25 \cdot 3}$
$= 3\sqrt{25} \cdot \sqrt{3}$ Product rule
$= 3 \cdot 5\sqrt{3}$
$= 15\sqrt{3}$

(b) $\sqrt{8} \cdot \sqrt{12} = \sqrt{8 \cdot 12}$ Product rule
$= \sqrt{96}$
$= \sqrt{16 \cdot 6}$ 16 is a perfect square
$= \sqrt{16} \cdot \sqrt{6}$ Product rule
$= 4\sqrt{6}$ ◀

Work Problem 3 at the side.

▶ The quotient rule for radicals is very similar to the product rule:

> **Quotient Rule for Radicals**
> If a and b are nonnegative real numbers and b is not 0,
> $$\frac{\sqrt{a}}{\sqrt{b}} = \sqrt{\frac{a}{b}}.$$
> The quotient of radicals is the radical of the quotient.

EXAMPLE 4 Use the quotient rule to simplify each radical.

(a) $\sqrt{\dfrac{25}{9}} = \dfrac{\sqrt{25}}{\sqrt{9}} = \dfrac{5}{3}$ Quotient rule

(b) $\sqrt{\dfrac{144}{49}} = \dfrac{\sqrt{144}}{\sqrt{49}} = \dfrac{12}{7}$ Quotient rule

(c) $\sqrt{\dfrac{3}{4}} = \dfrac{\sqrt{3}}{\sqrt{4}} = \dfrac{\sqrt{3}}{2}$ ◀

ANSWERS
2. (a) $2\sqrt{2}$ (b) $3\sqrt{3}$ (c) $5\sqrt{2}$
   (d) $2\sqrt{15}$
3. (a) $3\sqrt{5}$ (b) $7\sqrt{2}$ (c) $10\sqrt{5}$
   (d) 12

Roots and Radicals

EXAMPLE 5 $\dfrac{27\sqrt{15}}{9\sqrt{3}} = \dfrac{27}{9} \cdot \dfrac{\sqrt{15}}{\sqrt{3}}$

$\phantom{\dfrac{27\sqrt{15}}{9\sqrt{3}}} = \dfrac{27}{9} \cdot \sqrt{\dfrac{15}{3}}$ Quotient rule

$\phantom{\dfrac{27\sqrt{15}}{9\sqrt{3}}} = 3\sqrt{5}$ ◀

Work Problem 4 at the side.

Some problems require both the product and the quotient rules, as Example 6 shows.

EXAMPLE 6 Simplify $\sqrt{\dfrac{3}{5}} \cdot \sqrt{\dfrac{4}{5}}$.

Use the product and quotient rules.

$\sqrt{\dfrac{3}{5}} \cdot \sqrt{\dfrac{4}{5}} = \dfrac{\sqrt{3}}{\sqrt{5}} \cdot \dfrac{\sqrt{4}}{\sqrt{5}}$ Quotient rule

$\phantom{\sqrt{\dfrac{3}{5}} \cdot \sqrt{\dfrac{4}{5}}} = \dfrac{\sqrt{3} \cdot \sqrt{4}}{\sqrt{5} \cdot \sqrt{5}}$

$\phantom{\sqrt{\dfrac{3}{5}} \cdot \sqrt{\dfrac{4}{5}}} = \dfrac{\sqrt{3} \cdot 2}{\sqrt{25}}$ Product rule

$\phantom{\sqrt{\dfrac{3}{5}} \cdot \sqrt{\dfrac{4}{5}}} = \dfrac{2\sqrt{3}}{5}$ ◀

Work Problem 5 at the side.

Finally, the properties of this section are also valid when variables appear under the radical sign, as long as all the variables represent only nonnegative real numbers. For example, $\sqrt{5^2} = 5$, but $\sqrt{(-5)^2} = \sqrt{25} \neq -5$. That is,

> for a real number a,
> $$\sqrt{a^2} = a$$
> only if a is nonnegative.

EXAMPLE 7 Simplify each radical. Assume all variables represent positive real numbers.

(a) $\sqrt{25m^4} = \sqrt{25} \cdot \sqrt{m^4}$
$\phantom{\sqrt{25m^4}} = 5m^2$ Product rule

(b) $\sqrt{64p^{10}} = 8p^5$

(c) $\sqrt{r^9} = \sqrt{r^8 \cdot r}$
$\phantom{\sqrt{r^9}} = \sqrt{r^8} \cdot \sqrt{r} = r^4\sqrt{r}$

(d) $\sqrt{\dfrac{5}{x^2}} = \dfrac{\sqrt{5}}{\sqrt{x^2}} = \dfrac{\sqrt{5}}{x}$ Quotient rule ◀

---

4. Use the quotient rule to simplify each radical.

(a) $\sqrt{\dfrac{81}{16}}$

(b) $\sqrt{\dfrac{100}{9}}$

(c) $\sqrt{\dfrac{10}{49}}$

(d) $\dfrac{8\sqrt{50}}{4\sqrt{5}}$

5. Multiply and then simplify each product.

(a) $\sqrt{\dfrac{5}{6}} \cdot \sqrt{120}$

(b) $\sqrt{\dfrac{3}{8}} \cdot \sqrt{\dfrac{7}{2}}$

ANSWERS

4. (a) $\dfrac{9}{4}$ (b) $\dfrac{10}{3}$ (c) $\dfrac{\sqrt{10}}{7}$
   (d) $2\sqrt{10}$

5. (a) 10 (b) $\dfrac{\sqrt{21}}{4}$

9.2 Products and Quotients of Radicals

6. Simplify each radical. Assume all variables represent positive real numbers.

   (a) $\sqrt{36y^6}$

   (b) $\sqrt{100p^8}$

   (c) $\sqrt{a^5}$

   (d) $\sqrt{\dfrac{7}{p^2}}$

7. Simplify each radical.

   (a) $\sqrt[3]{16}$

   (b) $\sqrt[3]{81}$

   (c) $\sqrt[4]{48}$

   (d) $\sqrt[4]{128}$

   (e) $\sqrt[4]{\dfrac{256}{81}}$

Work Problem 6 at the side.

▶ The product rule and the quotient rule for radicals also work for other roots, as in the next example.

EXAMPLE 8 Simplify each radical.

(a) $\sqrt[3]{32} = \sqrt[3]{8 \cdot 4}$ 8 is a perfect cube
$= \sqrt[3]{8} \cdot \sqrt[3]{4}$ Product rule
$= 2\sqrt[3]{4}$

(b) $\sqrt[3]{108} = \sqrt[3]{27 \cdot 4}$
$= \sqrt[3]{27} \cdot \sqrt[3]{4}$ Product rule
$= 3\sqrt[3]{4}$

(c) $\sqrt[3]{\dfrac{8}{125}} = \dfrac{\sqrt[3]{8}}{\sqrt[3]{125}}$ Quotient rule
$= \dfrac{2}{5}$

(d) $\sqrt[4]{32} = \sqrt[4]{16} \cdot \sqrt[4]{2}$ Product rule
$= 2\sqrt[4]{2}$

(e) $\sqrt[4]{\dfrac{16}{625}} = \dfrac{\sqrt[4]{16}}{\sqrt[4]{625}}$ Quotient rule
$= \dfrac{2}{5}$ ◀

Work Problem 7 at the side.

ANSWERS
6. (a) $6y^3$ (b) $10p^4$ (c) $a^2\sqrt{a}$
   (d) $\dfrac{\sqrt{7}}{p}$
7. (a) $2\sqrt[3]{2}$ (b) $3\sqrt[3]{3}$ (c) $2\sqrt[4]{3}$
   (d) $2\sqrt[4]{8}$ (e) $\dfrac{4}{3}$

name　　　　　　　　　　　　　　　　date　　　　　hour

9.2 EXERCISES

Use the product rule to simplify each radical or radical expression. See Examples 1–3.

1. $\sqrt{8} \cdot \sqrt{2}$
2. $\sqrt{27} \cdot \sqrt{3}$
3. $\sqrt{6} \cdot \sqrt{6}$

4. $\sqrt{11} \cdot \sqrt{11}$
5. $\sqrt{21} \cdot \sqrt{21}$
6. $\sqrt{17} \cdot \sqrt{17}$

7. $\sqrt{3} \cdot \sqrt{7}$
8. $\sqrt{2} \cdot \sqrt{5}$
9. $\sqrt{27}$

10. $\sqrt{45}$
11. $\sqrt{28}$
12. $\sqrt{40}$

13. $\sqrt{18}$
14. $\sqrt{75}$
15. $\sqrt{48}$

16. $\sqrt{80}$
17. $\sqrt{125}$
18. $\sqrt{150}$

19. $\sqrt{700}$
20. $\sqrt{1100}$
21. $10\sqrt{27}$

9.2 Exercises　　451

22. $4\sqrt{8}$

23. $2\sqrt{20}$

24. $5\sqrt{80}$

25. $\sqrt{27} \cdot \sqrt{48}$

26. $\sqrt{75} \cdot \sqrt{27}$

27. $\sqrt{50} \cdot \sqrt{72}$

28. $\sqrt{98} \cdot \sqrt{8}$

29. $\sqrt{7} \cdot \sqrt{21}$

30. $\sqrt{12} \cdot \sqrt{48}$

31. $\sqrt{15} \cdot \sqrt{45}$

32. $\sqrt{20} \cdot \sqrt{45}$

33. $\sqrt{80} \cdot \sqrt{15}$

34. $\sqrt{60} \cdot \sqrt{12}$

35. $\sqrt{50} \cdot \sqrt{20}$

36. $\sqrt{72} \cdot \sqrt{12}$

Use the quotient rule and the product rule, as necessary, to simplify each radical or radical expression. See Examples 4–6.

37. $\sqrt{\dfrac{100}{9}}$

38. $\sqrt{\dfrac{225}{16}}$

39. $\sqrt{\dfrac{36}{49}}$

40. $\sqrt{\dfrac{256}{9}}$

41. $\sqrt{\dfrac{5}{16}}$

42. $\sqrt{\dfrac{11}{25}}$

43. $\sqrt{\dfrac{30}{49}}$

44. $\sqrt{\dfrac{10}{121}}$

name   date   hour

45. $\sqrt{\dfrac{1}{5}} \cdot \sqrt{\dfrac{4}{5}}$ **46.** $\sqrt{\dfrac{2}{3}} \cdot \sqrt{\dfrac{2}{27}}$ **47.** $\sqrt{\dfrac{2}{5}} \cdot \sqrt{\dfrac{8}{125}}$

48. $\sqrt{\dfrac{3}{8}} \cdot \sqrt{\dfrac{3}{2}}$ **49.** $\dfrac{\sqrt{75}}{\sqrt{3}}$ **50.** $\dfrac{\sqrt{200}}{\sqrt{2}}$

51. $\dfrac{\sqrt{48}}{\sqrt{3}}$ **52.** $\dfrac{\sqrt{72}}{\sqrt{8}}$ **53.** $\dfrac{15\sqrt{10}}{5\sqrt{2}}$

54. $\dfrac{18\sqrt{20}}{2\sqrt{10}}$ **55.** $\dfrac{25\sqrt{50}}{5\sqrt{5}}$ **56.** $\dfrac{26\sqrt{10}}{13\sqrt{5}}$

Simplify each radical or radical expression. Assume that all variables represent positive real numbers. See Example 7.

57. $\sqrt{y} \cdot \sqrt{y}$ **58.** $\sqrt{m} \cdot \sqrt{m}$ **59.** $\sqrt{x} \cdot \sqrt{z}$ **60.** $\sqrt{p} \cdot \sqrt{q}$

61. $\sqrt{x^2}$ **62.** $\sqrt{y^2}$ **63.** $\sqrt{x^4}$ **64.** $\sqrt{y^4}$

9.2 Exercises

65. $\sqrt{x^2y^4}$

66. $\sqrt{x^4y^8}$

67. $\sqrt{x^3}$

68. $\sqrt{y^3}$

69. $\sqrt{\dfrac{16}{x^2}}$

70. $\sqrt{\dfrac{100}{m^4}}$

71. $\sqrt{\dfrac{11}{r^4}}$

72. $\sqrt{\dfrac{23}{y^6}}$

Simplify each radical. See Example 8.

73. $\sqrt[3]{40}$

74. $\sqrt[3]{48}$

75. $\sqrt[3]{54}$

76. $\sqrt[3]{135}$

77. $\sqrt[3]{128}$

78. $\sqrt[3]{192}$

79. $\sqrt[4]{80}$

80. $\sqrt[4]{243}$

81. $\sqrt[3]{\dfrac{8}{27}}$

82. $\sqrt[4]{\dfrac{1}{256}}$

83. $\sqrt[4]{\dfrac{10{,}000}{81}}$

84. $\sqrt[3]{\dfrac{216}{125}}$

9.3 ADDITION AND SUBTRACTION OF RADICALS

Objectives
1. Add and subtract radical expressions.
2. Simplify radical expressions before adding or subtracting.
3. Simplify radical expressions with multiplication.

▶ Add or subtract radical expressions by using the distributive property. For example,
$$8\sqrt{3} + 6\sqrt{3} = (8 + 6)\sqrt{3} = 14\sqrt{3}.$$
Also,
$$2\sqrt{11} - 7\sqrt{11} = -5\sqrt{11}.$$

Only **like radicals,** those that are multiples of the same root of the same number, can be combined in this way.

EXAMPLE 1 Add or subtract, as indicated.
(a) $3\sqrt{6} + 5\sqrt{6} = (3 + 5)\sqrt{6} = 8\sqrt{6}$ Distributive property
(b) $5\sqrt{10} - 7\sqrt{10} = (5 - 7)\sqrt{10} = -2\sqrt{10}$
(c) $\sqrt[3]{5} + \sqrt[3]{5} = 1\sqrt[3]{5} + 1\sqrt[3]{5} = (1 + 1)\sqrt[3]{5} = 2\sqrt[3]{5}$
(d) $\sqrt[4]{7} + 2\sqrt[4]{7} = 1\sqrt[4]{7} + 2\sqrt[4]{7} = 3\sqrt[4]{7}$
(e) $\sqrt{3} + \sqrt{7}$ cannot be further simplified. ◀

Work Problem 1 at the side.

▶ Sometimes each radical expression in a sum or difference must be simplified first. Doing this might cause like radicals to appear, which then can be added or subtracted.

EXAMPLE 2 Simplify as much as possible.
(a) $3\sqrt{2} + \sqrt{8} = 3\sqrt{2} + \sqrt{4 \cdot 2}$ Simplify $\sqrt{8}$
$= 3\sqrt{2} + \sqrt{4} \cdot \sqrt{2}$
$= 3\sqrt{2} + 2\sqrt{2}$
$= 5\sqrt{2}$

(b) $\sqrt{18} - \sqrt{27} = \sqrt{9 \cdot 2} - \sqrt{9 \cdot 3}$ Simplify $\sqrt{18}$ and $\sqrt{27}$
$= \sqrt{9} \cdot \sqrt{2} - \sqrt{9} \cdot \sqrt{3}$
$= 3\sqrt{2} - 3\sqrt{3}$

Since $\sqrt{2}$ and $\sqrt{3}$ are unlike radicals, this difference cannot be simplified further.

(c) $2\sqrt{12} + 3\sqrt{75} = 2(\sqrt{4} \cdot \sqrt{3}) + 3(\sqrt{25} \cdot \sqrt{3})$
$= 2(2\sqrt{3}) + 3(5\sqrt{3})$
$= 4\sqrt{3} + 15\sqrt{3}$
$= 19\sqrt{3}$

(d) $3\sqrt[3]{16} + 5\sqrt[3]{2} = 3(\sqrt[3]{8} \cdot \sqrt[3]{2}) + 5\sqrt[3]{2}$
$= 3(2\sqrt[3]{2}) + 5\sqrt[3]{2}$
$= 6\sqrt[3]{2} + 5\sqrt[3]{2}$
$= 11\sqrt[3]{2}$ ◀

1. Add or subtract, as indicated.
   (a) $8\sqrt{5} + 2\sqrt{5}$
   (b) $-4\sqrt{3} + 9\sqrt{3}$
   (c) $12\sqrt[3]{11} - 2\sqrt[3]{11}$
   (d) $\sqrt[4]{15} + \sqrt[4]{15}$
   (e) $2\sqrt{7} + 2\sqrt{10}$

ANSWERS
1. (a) $10\sqrt{5}$ (b) $5\sqrt{3}$ (c) $10\sqrt[3]{11}$
   (d) $2\sqrt[4]{15}$ (e) cannot be simplified

9.3 Addition and Subtraction of Radicals

2. Simplify as much as possible.

(a) $\sqrt{8} + 4\sqrt{2}$

(b) $\sqrt{27} + \sqrt{12}$

(c) $5\sqrt{200} - 6\sqrt{18}$

(b) $8\sqrt[3]{40} + 7\sqrt[3]{135}$

3. Simplify as much as possible. Assume all variables represent nonnegative real numbers.

(a) $\sqrt{7} \cdot \sqrt{21} + 2\sqrt{27}$

(b) $\sqrt{3} \cdot \sqrt{48} + 5\sqrt{3}$

(c) $\sqrt{3r} \cdot \sqrt{6} + \sqrt{8r}$

(d) $6\sqrt{5m} - 9\sqrt{m} \cdot \sqrt{125}$

(e) $\sqrt[3]{5a^3} + \sqrt[3]{4a} \cdot \sqrt[3]{10a^2}$

Work Problem 2 at the side.

3 Some radical expressions require both multiplication and addition (or subtraction). The order of operations presented earlier still applies.

EXAMPLE 3 Simplify each expression. Assume all variables represent nonnegative real numbers.

(a) $\sqrt{5} \cdot \sqrt{15} + 4\sqrt{3} = \sqrt{5 \cdot 15} + 4\sqrt{3}$
$= \sqrt{75} + 4\sqrt{3}$
$= \sqrt{25 \cdot 3} + 4\sqrt{3}$
$= \sqrt{25} \cdot \sqrt{3} + 4\sqrt{3}$
$= 5\sqrt{3} + 4\sqrt{3}$
$= 9\sqrt{3}$

(b) $\sqrt{2} \cdot \sqrt{6k} + \sqrt{27k} = \sqrt{12k} + \sqrt{27k}$
$= \sqrt{4 \cdot 3k} + \sqrt{9 \cdot 3k}$
$= \sqrt{4} \cdot \sqrt{3k} + \sqrt{9} \cdot \sqrt{3k}$
$= 2\sqrt{3k} + 3\sqrt{3k}$
$= 5\sqrt{3k}$

(c) $\sqrt[3]{2} \cdot \sqrt[3]{16m^3} - \sqrt[3]{108m^3} = \sqrt[3]{32m^3} - \sqrt[3]{(27m^3)4}$
$= \sqrt[3]{(8m^3)4} - \sqrt[3]{(27m^3)4}$
$= 2m\sqrt[3]{4} - 3m\sqrt[3]{4}$
$= -m\sqrt[3]{4}$ ◀

Work Problem 3 at the side.

Remember:

> a sum or difference of radicals can be simplified only if the radicals are **like radicals.**

For example,
$$\sqrt{5} + 3\sqrt{5} = 4\sqrt{5},$$
but $\sqrt{5} + 5\sqrt{3}$ cannot be simplified further.

ANSWERS
2. (a) $6\sqrt{2}$ (b) $5\sqrt{3}$ (c) $32\sqrt{2}$
   (d) $37\sqrt[3]{5}$
3. (a) $13\sqrt{3}$ (b) $12 + 5\sqrt{3}$
   (c) $5\sqrt{2r}$ (d) $-39\sqrt{5m}$
   (e) $3a\sqrt[3]{5}$

9.3 EXERCISES

Simplify and add or subtract wherever possible. See Examples 1–2.

1. $2\sqrt{3} + 5\sqrt{3}$

2. $6\sqrt{5} + 8\sqrt{5}$

3. $4\sqrt{7} - 9\sqrt{7}$

4. $6\sqrt{2} - 8\sqrt{2}$

5. $\sqrt{6} + \sqrt{6}$

6. $\sqrt{11} + \sqrt{11}$

7. $\sqrt{17} + 2\sqrt{17}$

8. $3\sqrt{19} + \sqrt{19}$

9. $5\sqrt{7} - \sqrt{7}$

10. $12\sqrt{14} - \sqrt{14}$

11. $\sqrt{45} + 2\sqrt{20}$

12. $\sqrt{24} + 5\sqrt{54}$

13. $-\sqrt{12} + \sqrt{75}$

14. $2\sqrt{27} - \sqrt{300}$

15. $5\sqrt{72} - 2\sqrt{50}$

16. $6\sqrt{18} - 4\sqrt{32}$

17. $-5\sqrt{32} + \sqrt{98}$

18. $4\sqrt{75} + 3\sqrt{12}$

19. $5\sqrt{7} - 2\sqrt{28} + 6\sqrt{63}$

20. $3\sqrt{11} + 5\sqrt{44} - 3\sqrt{99}$

21. $6\sqrt{5} + 3\sqrt{20} - 8\sqrt{45}$

22. $7\sqrt{3} + 2\sqrt{12} - 5\sqrt{27}$

23. $2\sqrt{8} - 5\sqrt{32} + 2\sqrt{48}$

24. $5\sqrt{72} - 3\sqrt{48} - 4\sqrt{128}$

25. $4\sqrt{50} + 3\sqrt{12} + 5\sqrt{45}$

26. $6\sqrt{18} + 2\sqrt{48} - 6\sqrt{28}$

27. $\frac{1}{4}\sqrt{288} - \frac{1}{6}\sqrt{72}$

28. $\frac{2}{3}\sqrt{27} - \frac{3}{4}\sqrt{48}$

29. $\frac{3}{5}\sqrt{75} - \frac{2}{3}\sqrt{45}$

30. $\frac{5}{8}\sqrt{128} - \frac{3}{4}\sqrt{160}$

Simplify each expression. Assume that all variables represent nonnegative real numbers. See Example 3.

31. $\sqrt{3} \cdot \sqrt{7} + 2\sqrt{21}$
32. $\sqrt{13} \cdot \sqrt{2} + 3\sqrt{26}$

33. $\sqrt{6} \cdot \sqrt{2} + 3\sqrt{3}$
34. $4\sqrt{15} \cdot \sqrt{3} - 2\sqrt{5}$

35. $4\sqrt[3]{16} - 3\sqrt[3]{54}$
36. $5\sqrt[3]{128} + 3\sqrt[3]{250}$

37. $3\sqrt[3]{24} + 6\sqrt[3]{81}$
38. $2\sqrt[4]{48} - \sqrt[4]{243}$

39. $5\sqrt[4]{32} + 2\sqrt[4]{32} \cdot \sqrt[4]{4}$
40. $8\sqrt[3]{48} + 10\sqrt[3]{3} \cdot \sqrt[3]{18}$

41. $\sqrt{9x} + \sqrt{49x} - \sqrt{16x}$
42. $\sqrt{4a} - \sqrt{16a} + \sqrt{9a}$
43. $\sqrt{4a} + 6\sqrt{a} + \sqrt{25a}$

44. $\sqrt{6x^2} + x\sqrt{54}$
45. $\sqrt{75x^2} + x\sqrt{300}$
46. $\sqrt{20y^2} - 3y\sqrt{5}$

47. $3\sqrt{8x^2} - 4x\sqrt{2}$
48. $6r\sqrt{27r^2s} + 3r^2\sqrt{3s}$
49. $6\sqrt[3]{8p^2} - 2\sqrt[3]{27p^2}$

50. $5\sqrt[4]{m^3} + 8\sqrt[4]{16m^3}$
51. $5\sqrt[4]{m^4} + 3\sqrt[4]{81m^4}$
52. $10\sqrt[3]{4m^3} - 3m\sqrt[3]{32}$

9.4 RATIONALIZING THE DENOMINATOR

Objectives

▶ Rationalize denominators with square roots.
▶ Simplify expressions by rationalizing the denominator.
▶ Rationalize denominators with cube roots.

▶ Decimal approximations for radicals were found in the first section of this chapter. For more complicated radical expressions, it is easier to find these decimals if the denominators do not contain any radicals. For example, the radical in the denominator of

$$\frac{\sqrt{3}}{\sqrt{2}}$$

can be eliminated by multiplying the numerator and the denominator by $\sqrt{2}$.

$$\frac{\sqrt{3}}{\sqrt{2}} = \frac{\sqrt{3} \cdot \sqrt{2}}{\sqrt{2} \cdot \sqrt{2}} = \frac{\sqrt{6}}{2} \qquad \sqrt{2} \cdot \sqrt{2} = \sqrt{4} = 2$$

This process of changing the denominator from a radical (irrational number) to a rational number is called **rationalizing the denominator.** The value of the radical expression is not changed; only the form is changed.

EXAMPLE 1 Rationalize each denominator.

(a) $\frac{9}{\sqrt{6}}$

Multiply both numerator and denominator by $\sqrt{6}$.

$$\frac{9}{\sqrt{6}} = \frac{9 \cdot \sqrt{6}}{\sqrt{6} \cdot \sqrt{6}}$$

$$= \frac{9\sqrt{6}}{6} \qquad \sqrt{6} \cdot \sqrt{6} = 6$$

$$= \frac{3\sqrt{6}}{2}$$

(b) $\frac{12}{\sqrt{8}}$

The denominator could be rationalized here by multiplying by $\sqrt{8}$. However, the result can be found faster by multiplying numerator and denominator by $\sqrt{2}$. This is because $\sqrt{8} \cdot \sqrt{2} = \sqrt{16} = 4$, a rational number.

$$\frac{12}{\sqrt{8}} = \frac{12 \cdot \sqrt{2}}{\sqrt{8} \cdot \sqrt{2}} \qquad \text{Multiply by } \sqrt{2}$$

$$= \frac{12\sqrt{2}}{\sqrt{16}} \qquad \text{Product rule}$$

$$= \frac{12\sqrt{2}}{4} \qquad \sqrt{16} = 4$$

$$= 3\sqrt{2} \blacktriangleleft$$

Work Problem 1 at the side.

▶ Radicals are considered simplified only if any denominators are rationalized, as shown in Examples 2–4.

1. Rationalize each denominator.

(a) $\frac{3}{\sqrt{5}}$

(b) $\frac{-6}{\sqrt{11}}$

(c) $\frac{\sqrt{7}}{\sqrt{2}}$

(d) $\frac{20}{\sqrt{18}}$

(e) $\frac{-12}{\sqrt{32}}$

ANSWERS

1. (a) $\frac{3\sqrt{5}}{5}$ (b) $\frac{-6\sqrt{11}}{11}$ (c) $\frac{\sqrt{14}}{2}$
(d) $\frac{10\sqrt{2}}{3}$ (e) $\frac{-3\sqrt{2}}{2}$

9.4 Rationalizing the Denominator

2. Simplify by rationalizing each denominator.

   (a) $\sqrt{\dfrac{1}{7}}$

   (b) $\sqrt{\dfrac{3}{2}}$

   (c) $\sqrt{\dfrac{5}{18}}$

   (d) $\sqrt{\dfrac{16}{11}}$

3. Simplify.

   (a) $\sqrt{\dfrac{1}{2}} \cdot \sqrt{\dfrac{5}{6}}$

   (b) $\sqrt{\dfrac{1}{10}} \cdot \sqrt{20}$

   (c) $\sqrt{\dfrac{5}{8}} \cdot \sqrt{\dfrac{24}{10}}$

ANSWERS

2. (a) $\dfrac{\sqrt{7}}{7}$ (b) $\dfrac{\sqrt{6}}{2}$ (c) $\dfrac{\sqrt{10}}{6}$
   (d) $\dfrac{4\sqrt{11}}{11}$

3. (a) $\dfrac{\sqrt{15}}{6}$ (b) $\sqrt{2}$ (c) $\dfrac{\sqrt{6}}{2}$

460

EXAMPLE 2 Simplify $\sqrt{\dfrac{27}{5}}$ by rationalizing the denominator.

First use the quotient rule for radicals.

$$\sqrt{\dfrac{27}{5}} = \dfrac{\sqrt{27}}{\sqrt{5}}$$

Now multiply both numerator and denominator by $\sqrt{5}$.

$$\dfrac{\sqrt{27}}{\sqrt{5}} = \dfrac{\sqrt{27} \cdot \sqrt{5}}{\sqrt{5} \cdot \sqrt{5}}$$

$$= \dfrac{\sqrt{9 \cdot 3} \cdot \sqrt{5}}{5}$$

$$= \dfrac{\sqrt{9} \cdot \sqrt{3} \cdot \sqrt{5}}{5}$$

$$= \dfrac{3 \cdot \sqrt{3} \cdot \sqrt{5}}{5}$$

$$= \dfrac{3\sqrt{15}}{5} \blacktriangleleft$$

Work Problem 2 at the side.

EXAMPLE 3 Simplify $\sqrt{\dfrac{5}{8}} \cdot \sqrt{\dfrac{1}{6}}$.

Use both the quotient rule and the product rule.

$$\sqrt{\dfrac{5}{8}} \cdot \sqrt{\dfrac{1}{6}} = \sqrt{\dfrac{5}{8} \cdot \dfrac{1}{6}}$$

$$= \sqrt{\dfrac{5}{48}} = \dfrac{\sqrt{5}}{\sqrt{48}}$$

Now rationalize the denominator by multiplying by $\sqrt{3}$ (since $\sqrt{48} \cdot \sqrt{3} = \sqrt{48 \cdot 3} = \sqrt{144} = 12$).

$$\dfrac{\sqrt{5}}{\sqrt{48}} = \dfrac{\sqrt{5} \cdot \sqrt{3}}{\sqrt{48} \cdot \sqrt{3}}$$

$$= \dfrac{\sqrt{15}}{\sqrt{144}} = \dfrac{\sqrt{15}}{12} \blacktriangleleft$$

Work Problem 3 at the side.

EXAMPLE 4 Rationalize the denominator of $\dfrac{\sqrt{4x}}{\sqrt{y}}$. Assume that x and y are positive real numbers.

Multiply numerator and denominator by $\sqrt{y}$.

$$\dfrac{\sqrt{4x}}{\sqrt{y}} = \dfrac{\sqrt{4x} \cdot \sqrt{y}}{\sqrt{y} \cdot \sqrt{y}} = \dfrac{\sqrt{4xy}}{y} = \dfrac{2\sqrt{xy}}{y} \blacktriangleleft$$

Roots and Radicals

Work Problem 4 at the side.

3 Rationalize a denominator with a cube root by changing the radicand in the denominator to a perfect cube, as shown in the next example.

EXAMPLE 5 Rationalize each denominator.

(a) $\sqrt[3]{\dfrac{3}{2}}$

Multiply the numerator and the denominator by enough factors of 2 to make the denominator a perfect cube. This will eliminate the radical in the denominator. Here, multiply by $\sqrt[3]{2^2}$.

$$\sqrt[3]{\dfrac{3}{2}} = \dfrac{\sqrt[3]{3}}{\sqrt[3]{2}}$$
$$= \dfrac{\sqrt[3]{3} \cdot \sqrt[3]{2^2}}{\sqrt[3]{2} \cdot \sqrt[3]{2^2}}$$
$$= \dfrac{\sqrt[3]{3 \cdot 2^2}}{\sqrt[3]{2^3}}$$
$$= \dfrac{\sqrt[3]{12}}{2} \qquad \sqrt[3]{2^3} = \sqrt[3]{8} = 2$$

(b) $\dfrac{\sqrt[3]{3}}{\sqrt[3]{4}}$

Since $4 \cdot 2 = 2^2 \cdot 2 = 2^3$, multiply numerator and denominator by $\sqrt[3]{2}$.

$$\dfrac{\sqrt[3]{3}}{\sqrt[3]{4}} = \dfrac{\sqrt[3]{3} \cdot \sqrt[3]{2}}{\sqrt[3]{4} \cdot \sqrt[3]{2}}$$
$$= \dfrac{\sqrt[3]{6}}{\sqrt[3]{8}}$$
$$= \dfrac{\sqrt[3]{6}}{2} \blacktriangleleft$$

Work Problem 5 at the side.

---

4. Rationalize each denominator. Assume that all variables represent positive numbers.

(a) $\dfrac{\sqrt{2r}}{\sqrt{a}}$

(b) $\sqrt{\dfrac{z^2 x}{9m}}$

(c) $\sqrt{\dfrac{7r^2 s^2}{m}}$

5. Rationalize each denominator.

(a) $\sqrt[3]{\dfrac{1}{3}}$

(b) $\sqrt[3]{\dfrac{1}{16}}$

(c) $\sqrt[3]{\dfrac{5}{2}}$

(d) $\sqrt[3]{\dfrac{25}{36}}$

ANSWERS

4. (a) $\dfrac{\sqrt{2ra}}{a}$ (b) $\dfrac{z\sqrt{xm}}{3m}$ (c) $\dfrac{rs\sqrt{7m}}{m}$

5. (a) $\dfrac{\sqrt[3]{9}}{3}$ (b) $\dfrac{\sqrt[3]{4}}{4}$ (c) $\dfrac{\sqrt[3]{20}}{2}$

(d) $\dfrac{\sqrt[3]{150}}{6}$

9.4 EXERCISES

Perform the indicated operations, and write all answers in simplest form. Rationalize all denominators. See Examples 1–3.

1. $\dfrac{6}{\sqrt{5}}$

2. $\dfrac{4}{\sqrt{2}}$

3. $\dfrac{5}{\sqrt{5}}$

4. $\dfrac{15}{\sqrt{15}}$

5. $\dfrac{3}{\sqrt{7}}$

6. $\dfrac{12}{\sqrt{10}}$

7. $\dfrac{8\sqrt{3}}{\sqrt{5}}$

8. $\dfrac{9\sqrt{6}}{\sqrt{5}}$

9. $\dfrac{12\sqrt{10}}{8\sqrt{3}}$

10. $\dfrac{9\sqrt{15}}{6\sqrt{2}}$

11. $\dfrac{8}{\sqrt{27}}$

12. $\dfrac{12}{\sqrt{18}}$

13. $\dfrac{3}{\sqrt{50}}$

14. $\dfrac{5}{\sqrt{75}}$

15. $\dfrac{12}{\sqrt{72}}$

16. $\dfrac{21}{\sqrt{45}}$

17. $\dfrac{9}{\sqrt{32}}$

18. $\dfrac{50}{\sqrt{125}}$

19. $\dfrac{\sqrt{8}}{\sqrt{2}}$

20. $\dfrac{\sqrt{27}}{\sqrt{3}}$

21. $\dfrac{\sqrt{10}}{\sqrt{5}}$

22. $\dfrac{\sqrt{6}}{\sqrt{3}}$

23. $\dfrac{\sqrt{40}}{\sqrt{3}}$

24. $\dfrac{\sqrt{5}}{\sqrt{8}}$

25. $\sqrt{\dfrac{1}{2}}$

26. $\sqrt{\dfrac{1}{8}}$

27. $\sqrt{\dfrac{10}{7}}$

28. $\sqrt{\dfrac{2}{3}}$

29. $\sqrt{\dfrac{9}{5}}$

30. $\sqrt{\dfrac{16}{7}}$

31. $\sqrt{\dfrac{7}{5}} \cdot \sqrt{\dfrac{5}{3}}$

32. $\sqrt{\dfrac{1}{3}} \cdot \sqrt{\dfrac{2}{5}}$

33. $\sqrt{\dfrac{3}{4}} \cdot \sqrt{\dfrac{1}{5}}$ 34. $\sqrt{\dfrac{1}{10}} \cdot \sqrt{\dfrac{10}{3}}$ 35. $\sqrt{\dfrac{21}{7}} \cdot \sqrt{\dfrac{21}{8}}$ 36. $\sqrt{\dfrac{1}{11}} \cdot \sqrt{\dfrac{33}{16}}$

37. $\sqrt{\dfrac{2}{5}} \cdot \sqrt{\dfrac{3}{10}}$ 38. $\sqrt{\dfrac{9}{8}} \cdot \sqrt{\dfrac{7}{16}}$ 39. $\sqrt{\dfrac{5}{8}} \cdot \sqrt{\dfrac{5}{6}}$

Perform the indicated operations, and write all answers in simplest form. Rationalize all denominators. Assume that all variables represent positive real numbers. See Example 4.

40. $\sqrt{\dfrac{5}{x}}$ 41. $\sqrt{\dfrac{6}{p}}$ 42. $\sqrt{\dfrac{4r^3}{s}}$

43. $\sqrt{\dfrac{6p^3}{3m}}$ 44. $\sqrt{\dfrac{a^3 b}{6}}$ 45. $\sqrt{\dfrac{x^2}{4y}}$

46. $\sqrt{\dfrac{m^2 n}{2}}$ 47. $\sqrt{\dfrac{9a^2 r}{5}}$ 48. $\sqrt{\dfrac{2x^2 z^4}{3y}}$

Rationalize the denominators of the following cube roots. See Example 5.

49. $\sqrt[3]{\dfrac{1}{2}}$ 50. $\sqrt[3]{\dfrac{1}{4}}$ 51. $\sqrt[3]{\dfrac{1}{32}}$ 52. $\sqrt[3]{\dfrac{1}{5}}$

53. $\sqrt[3]{\dfrac{1}{11}}$ 54. $\sqrt[3]{\dfrac{3}{2}}$ 55. $\sqrt[3]{\dfrac{2}{5}}$ 56. $\sqrt[3]{\dfrac{4}{9}}$

57. $\sqrt[3]{\dfrac{3}{4}}$ 58. $\sqrt[3]{\dfrac{3}{25}}$ 59. $\sqrt[3]{\dfrac{7}{36}}$ 60. $\sqrt[3]{\dfrac{11}{49}}$

9.5 SIMPLIFYING RADICAL EXPRESSIONS

It can be difficult to decide on the "simplest" form of a radical. In this book, a radical expression is simplified when the following five conditions are satisfied. Although the rules are illustrated with square roots, they hold true for higher roots as well.

Simplifying Radical Expressions

1. If a radical represents a rational number, then that rational number should be used in place of the radical.

   For example, $\sqrt{49}$ is simplified by writing 7; $\sqrt{64}$ by writing 8; $\sqrt{\frac{169}{9}}$ by writing $\frac{13}{3}$.

2. If a radical expression contains products of radicals, the product rule for radicals, $\sqrt{x} \cdot \sqrt{y} = \sqrt{xy}$, should be used to get a single radical.

   For example, $\sqrt{3} \cdot \sqrt{2}$ is simplified to $\sqrt{6}$; $\sqrt{5} \cdot \sqrt{x}$ to $\sqrt{5x}$.

3. If a radicand has a factor that is a perfect square, the radical should be expressed as the product of the positive square root of the perfect square and the remaining radical factor.

   For example, $\sqrt{20}$ is simplified to $\sqrt{20} = \sqrt{4 \cdot 5} = \sqrt{4} \cdot \sqrt{5} = 2\sqrt{5}$; $\sqrt{75}$ to $5\sqrt{3}$.

4. If a radical expression contains sums or differences of radicals, the distributive property should be used to combine like radicals.

   For example, $3\sqrt{2} + 4\sqrt{2}$ is combined as $7\sqrt{2}$, but $3\sqrt{2} + 4\sqrt{3}$ cannot be further combined.

5. Any radicals in a denominator should be changed to rational numbers.

   For example, $\frac{5}{\sqrt{3}}$ is rationalized as $\frac{5}{\sqrt{3}} = \frac{5\sqrt{3}}{\sqrt{3} \cdot \sqrt{3}} = \frac{5\sqrt{3}}{3}$.

Objectives

▶ 1 Simplify radical expressions with sums.

▶ 2 Simplify radical expressions with products.

▶ 3 Simplify radical expressions with quotients.

▶ 4 Write radical expressions with quotients in lowest terms.

1. Simplify each radical expression.

   (a) $\sqrt{36} + \sqrt{25}$

   (b) $3\sqrt{3} + 2\sqrt{27}$

   (c) $4\sqrt{8} - 2\sqrt{32}$

   (d) $2\sqrt{12} - 5\sqrt{48}$

EXAMPLE 1 Simplify each of the following.

(a) $\sqrt{16} + \sqrt{9}$

   Here $\sqrt{16} + \sqrt{9} = 4 + 3 = 7$.

(b) $5\sqrt{2} + 2\sqrt{18}$

   First simplify $\sqrt{18}$.

   $5\sqrt{2} + 2\sqrt{18} = 5\sqrt{2} + 2(\sqrt{9} \cdot \sqrt{2})$
   $= 5\sqrt{2} + 2(3\sqrt{2})$
   $= 5\sqrt{2} + 6\sqrt{2}$
   $= 11\sqrt{2}$ ◀

Work Problem 1 at the side.

ANSWERS
1. (a) 11 (b) $9\sqrt{3}$ (c) 0
   (d) $-16\sqrt{3}$

2. Find each product. Simplify the answers.

(a) $\sqrt{7}(\sqrt{2} + \sqrt{5})$

(b) $\sqrt{2}(\sqrt{8} + \sqrt{20})$

(c) $(\sqrt{2} + 5\sqrt{3}) \cdot (\sqrt{3} - 2\sqrt{2})$

(d) $(2\sqrt{7} + \sqrt{10}) \cdot (3\sqrt{7} - 2\sqrt{10})$

(e) $(\sqrt{2} - \sqrt{5}) \cdot (\sqrt{10} + \sqrt{2})$

▶ The next examples show how to simplify radical expressions with products.

EXAMPLE 2 Simplify $\sqrt{5}(\sqrt{8} - \sqrt{32})$.

Start by simplifying $\sqrt{8}$ and $\sqrt{32}$.

$$\sqrt{8} = \boxed{2\sqrt{2}} \quad \text{and} \quad \sqrt{32} = \boxed{4\sqrt{2}}$$

Now simplify inside the parentheses.

$$\sqrt{5}(\sqrt{8} - \sqrt{32}) = \sqrt{5}(\boxed{2\sqrt{2} - 4\sqrt{2}})$$
$$= \sqrt{5}(\boxed{-2\sqrt{2}})$$
$$= -2\sqrt{5 \cdot 2}$$
$$= -2\sqrt{10} \quad \blacktriangleleft$$

EXAMPLE 3 Simplify each product.

(a) $(\sqrt{3} + 2\sqrt{5})(\sqrt{3} - 4\sqrt{5})$

The products of these sums of radicals can be found in the same way that we found the product of binomials in Chapter 4. The pattern of multiplication is the same, using the FOIL method.

$$(\sqrt{3} + 2\sqrt{5})(\sqrt{3} - 4\sqrt{5})$$
$$= \underbrace{\sqrt{3} \cdot \sqrt{3}}_{\text{First}} + \underbrace{\sqrt{3} \cdot (-4\sqrt{5})}_{\text{Outside}}$$
$$+ \underbrace{2\sqrt{5}(\sqrt{3})}_{\text{Inside}} + \underbrace{2\sqrt{5}(-4\sqrt{5})}_{\text{Last}}$$
$$= 3 - 4\sqrt{15} + 2\sqrt{15} - 8 \cdot 5$$
$$= 3 - 2\sqrt{15} - 40$$
$$= -37 - 2\sqrt{15}$$

(b) $(\sqrt{3} + \sqrt{21})(\sqrt{3} - \sqrt{7})$
$$= \sqrt{3}(\sqrt{3}) + \sqrt{3}(-\sqrt{7}) + \sqrt{21}(\sqrt{3})$$
$$+ \sqrt{21}(-\sqrt{7})$$
$$= 3 - \sqrt{21} + \sqrt{63} - \sqrt{147}$$
$$= 3 - \sqrt{21} + \sqrt{9} \cdot \sqrt{7} - \sqrt{49} \cdot \sqrt{3}$$
$$= 3 - \sqrt{21} + 3\sqrt{7} - 7\sqrt{3}$$

Since there are no like radicals, no terms may be combined. ◀

Work Problem 2 at the side.

Since radicals represent real numbers, the special products of binomials discussed in Chapters 4 and 5 can be used to find products of radicals. Example 4 uses the rule for the product that gives a difference of two squares,

$$(a + b)(a - b) = a^2 - b^2.$$

ANSWERS
2. (a) $\sqrt{14} + \sqrt{35}$ (b) $4 + 2\sqrt{10}$
(c) $11 - 9\sqrt{6}$ (d) $22 - \sqrt{70}$
(e) $2\sqrt{5} + 2 - 5\sqrt{2} - \sqrt{10}$

EXAMPLE 4 Find each product.

(a) $(4 + \sqrt{3})(4 - \sqrt{3})$

Follow the pattern given above. Let $a = 4$ and $b = \sqrt{3}$.

$$(4 + \sqrt{3})(4 - \sqrt{3}) = (4)^2 - (\sqrt{3})^2$$
$$= 16 - 3 \qquad 4^2 = 16, (\sqrt{3})^2 = 3$$
$$= 13$$

(b) $(\sqrt{12} - \sqrt{6})(\sqrt{12} + \sqrt{6}) = (\sqrt{12})^2 - (\sqrt{6})^2$
$$= 12 - 6$$
$$= 6 \blacktriangleleft$$

Work Problem 3 at the side.

Both products in Example 4 resulted in rational numbers. The expressions in those products, such as $4 + \sqrt{3}$ and $4 - \sqrt{3}$, are called **conjugates** of each other.

▶ Products of radicals similar to those in Example 4 can be used to rationalize the denominators in more complicated quotients, such as

$$\frac{2}{4 - \sqrt{3}}.$$

By Example 4(a), if this denominator, $4 - \sqrt{3}$, is multiplied by $4 + \sqrt{3}$, then the product $(4 - \sqrt{3})(4 + \sqrt{3})$ is the rational number 13. Multiplying numerator and denominator of the quotient by $4 + \sqrt{3}$ gives

$$\frac{2}{4 - \sqrt{3}} = \frac{2(4 + \sqrt{3})}{(4 - \sqrt{3})(4 + \sqrt{3})}$$
$$= \frac{2(4 + \sqrt{3})}{13}.$$

The denominator now has been rationalized; it contains no radical signs.

EXAMPLE 5 Rationalize the denominator in the quotient

$$\frac{5}{3 + \sqrt{5}}.$$

The radical in the denominator can be eliminated by multiplying both numerator and denominator by $3 - \sqrt{5}$.

$$\frac{5}{3 + \sqrt{5}} = \frac{5(3 - \sqrt{5})}{(3 + \sqrt{5})(3 - \sqrt{5})}$$
$$= \frac{5(3 - \sqrt{5})}{3^2 - (\sqrt{5})^2}$$
$$= \frac{5(3 - \sqrt{5})}{9 - 5}$$
$$= \frac{5(3 - \sqrt{5})}{4} \blacktriangleleft$$

---

3. Find each product. Simplify the answers.

(a) $(3 + \sqrt{5})(3 - \sqrt{5})$

(b) $(\sqrt{3} - 2)(\sqrt{3} + 2)$

(c) $(\sqrt{5} + \sqrt{3})(\sqrt{5} - \sqrt{3})$

(d) $(\sqrt{11} + \sqrt{14}) \cdot (\sqrt{11} - \sqrt{14})$

ANSWERS
3. (a) 4 (b) −1 (c) 2 (d) −3

4. Rationalize each denominator.

   (a) $\dfrac{5}{4 + \sqrt{2}}$

   (b) $\dfrac{3}{2 - \sqrt{5}}$

   (c) $\dfrac{1}{6 + \sqrt{3}}$

5. Write each quotient in lowest terms.

   (a) $\dfrac{4 + 8\sqrt{2}}{2}$

   (b) $\dfrac{5\sqrt{3} - 15}{10}$

   (c) $\dfrac{8\sqrt{5} + 12}{16}$

EXAMPLE 6 Simplify $\dfrac{6 + \sqrt{2}}{\sqrt{2} - 5}$.

Multiply numerator and denominator by $\sqrt{2} + 5$.

$$\dfrac{6 + \sqrt{2}}{\sqrt{2} - 5} = \dfrac{(6 + \sqrt{2})(\sqrt{2} + 5)}{(\sqrt{2} - 5)(\sqrt{2} + 5)}$$

$$= \dfrac{6\sqrt{2} + 30 + 2 + 5\sqrt{2}}{2 - 25}$$

$$= \dfrac{11\sqrt{2} + 32}{-23} = -\dfrac{11\sqrt{2} + 32}{23} \blacktriangleleft$$

Work Problem 4 at the side.

The final example shows how to write certain quotients in lowest terms.

EXAMPLE 7 Write $\dfrac{3\sqrt{3} + 9}{12}$ in lowest terms.

Factor the numerator, and then divide numerator and denominator by any common factors.

$$\dfrac{3\sqrt{3} + 9}{12} = \dfrac{3(\sqrt{3} + 3)}{12} = \dfrac{\sqrt{3} + 3}{4} \blacktriangleleft$$

Work Problem 5 at the side.

ANSWERS

4. (a) $\dfrac{5(4 - \sqrt{2})}{14}$ (b) $-3(2 + \sqrt{5})$

   (c) $\dfrac{6 - \sqrt{3}}{33}$

5. (a) $2 + 4\sqrt{2}$ (b) $\dfrac{\sqrt{3} - 3}{2}$

   (c) $\dfrac{2\sqrt{5} + 3}{4}$

9.5 EXERCISES

Simplify each expression. Use the five rules given in the text. See Examples 1–4.

1. $3\sqrt{5} + 8\sqrt{45}$

2. $6\sqrt{2} + 4\sqrt{18}$

3. $9\sqrt{50} - 4\sqrt{72}$

4. $3\sqrt{80} - 5\sqrt{45}$

5. $\sqrt{2}(\sqrt{8} - \sqrt{32})$

6. $\sqrt{3}(\sqrt{27} - \sqrt{3})$

7. $\sqrt{5}(\sqrt{3} + \sqrt{7})$

8. $\sqrt{7}(\sqrt{10} - \sqrt{3})$

9. $2\sqrt{5}(\sqrt{2} + \sqrt{5})$

10. $3\sqrt{7}(2\sqrt{7} - 4\sqrt{5})$

11. $-\sqrt{14} \cdot \sqrt{2} - \sqrt{28}$

12. $\sqrt{6} \cdot \sqrt{3} - 2\sqrt{50}$

13. $(2\sqrt{6} + 3)(3\sqrt{6} - 5)$

14. $(4\sqrt{5} - 2)(2\sqrt{5} + 3)$

15. $(5\sqrt{7} - 2\sqrt{3})(3\sqrt{7} + 3\sqrt{3})$

16. $(2\sqrt{10} + 5\sqrt{2})(3\sqrt{10} - 4\sqrt{2})$

9.5 Exercises

17. $(3\sqrt{2} + 4)(3\sqrt{2} + 4)$

18. $(4\sqrt{5} - 1)(4\sqrt{5} - 1)$

19. $(2\sqrt{7} - 3)^2$

20. $(3\sqrt{5} + 5)^2$

21. $(3 - \sqrt{2})(3 + \sqrt{2})$

22. $(7 - \sqrt{5})(7 + \sqrt{5})$

23. $(2 + \sqrt{8})(2 - \sqrt{8})$

24. $(3 + \sqrt{11})(3 - \sqrt{11})$

25. $(\sqrt{6} - \sqrt{5})(\sqrt{6} + \sqrt{5})$

26. $(\sqrt{11} + \sqrt{10})(\sqrt{11} - \sqrt{10})$

27. $(\sqrt{2} + \sqrt{3})(\sqrt{6} - \sqrt{2})$

28. $(\sqrt{3} + \sqrt{5})(\sqrt{15} - \sqrt{5})$

29. $(\sqrt{8} - \sqrt{2})(\sqrt{2} + \sqrt{4})$

30. $(\sqrt{6} - \sqrt{3})(\sqrt{3} + \sqrt{18})$

31. $(\sqrt{5} + \sqrt{30})(\sqrt{6} + \sqrt{3})$

32. $(\sqrt{10} - \sqrt{20})(\sqrt{2} - \sqrt{5})$

name date hour

Write each quotient in lowest terms. See Example 7.

33. $\dfrac{5\sqrt{7} - 10}{5}$

34. $\dfrac{6\sqrt{5} - 9}{3}$

35. $\dfrac{2\sqrt{3} + 10}{8}$

36. $\dfrac{4\sqrt{6} + 6}{10}$

37. $\dfrac{12 - 2\sqrt{10}}{4}$

38. $\dfrac{9 - 6\sqrt{2}}{12}$

Rationalize the denominators. See Examples 5 and 6.

39. $\dfrac{1}{3 + \sqrt{2}}$

40. $\dfrac{1}{4 - \sqrt{3}}$

41. $\dfrac{5}{2 + \sqrt{5}}$

42. $\dfrac{6}{3 + \sqrt{7}}$

43. $\dfrac{7}{2 - \sqrt{11}}$

44. $\dfrac{38}{5 - \sqrt{6}}$

9.5 Exercises

45. $\dfrac{\sqrt{2}}{1 + \sqrt{2}}$

46. $\dfrac{\sqrt{7}}{2 - \sqrt{7}}$

47. $\dfrac{\sqrt{5}}{1 - \sqrt{5}}$

48. $\dfrac{\sqrt{3}}{2 + \sqrt{3}}$

49. $\dfrac{\sqrt{12}}{\sqrt{3} + 1}$

50. $\dfrac{\sqrt{18}}{\sqrt{2} - 1}$

51. $\dfrac{2\sqrt{3}}{\sqrt{3} + 5}$

52. $\dfrac{\sqrt{12}}{2 - \sqrt{10}}$

53. $\dfrac{\sqrt{2} + 3}{\sqrt{3} - 1}$

54. $\dfrac{\sqrt{5} + 2}{2 - \sqrt{3}}$

55. $\dfrac{6 - \sqrt{5}}{\sqrt{2} + 2}$

56. $\dfrac{3 + \sqrt{2}}{\sqrt{2} + 1}$

9.6 EQUATIONS WITH RADICALS

The addition and multiplication properties are not enough to solve an equation with radicals such as

$$\sqrt{x + 1} = 3.$$

▶ Solving equations with square roots requires a new property, the *squaring property*.

Squaring Property of Equality
If both sides of a given equation are squared, all solutions of the original equation are *among* the solutions of the squared equation.

Be very careful with the squaring property: Using this property can give a new equation with *more* solutions than the original equation. For example, starting with the equation $y = 4$ and squaring both sides gives

$$y^2 = 4^2, \quad \text{or} \quad y^2 = 16.$$

This last equation, $y^2 = 16$, has *two* solutions, 4 or -4, while the original equation, $y = 4$, has only *one* solution, 4. Because of this possibility,

all proposed solutions from the squared equation must be checked in the original equation.

EXAMPLE 1 Solve the equation $\sqrt{p + 1} = 3$.

Use the squaring property of equality to square both sides of the equation.

$$(\sqrt{p + 1})^2 = 3^2$$
$$p + 1 = 9$$
$$p = 8 \quad \text{Subtract 1}$$

Now check this answer in the original equation.

$$\sqrt{p + 1} = 3$$
$$\sqrt{8 + 1} = 3 \quad \text{Let } p = 8$$
$$\sqrt{9} = 3$$
$$3 = 3 \quad \text{True}$$

Since this statement is true, 8 is the solution of $\sqrt{p + 1} = 3$. In this case the squared equation had just one solution, which also satisfied the original equation. ◀

Work Problem 1 at the side.

EXAMPLE 2 Solve $3\sqrt{x} = \sqrt{x + 8}$.

Squaring both sides gives

$$(3\sqrt{x})^2 = (\sqrt{x + 8})^2$$
$$3^2(\sqrt{x})^2 = (\sqrt{x + 8})^2$$
$$9x = x + 8$$
$$8x = 8 \quad \text{Subtract } x$$
$$x = 1. \quad \text{Divide by 8}$$

Objectives

▶ Solve equations with radicals.

▶ Identify equations with no solutions.

▶ Solve equations by squaring a binomial.

1. Solve each equation.

   (a) $\sqrt{k} = 3$

   (b) $\sqrt{m - 2} = 4$

   (c) $\sqrt{9 - y} = 4$

ANSWERS
1. (a) 9 (b) 18 (c) -7

9.6 Equations with Radicals   473

2. Solve each equation.

   (a) $\sqrt{2k+1} = 5$

   (b) $\sqrt{3x+9} = 2\sqrt{x}$

   (c) $5\sqrt{a} = \sqrt{20a+5}$

3. Solve each equation that has a solution. (*Hint:* In (c) subtract 4 from both sides.)

   (a) $\sqrt{m} = -5$

   (b) $\sqrt{x} = 2$

   (c) $\sqrt{y} + 4 = 0$

   (d) $m = \sqrt{m^2 - 4m - 16}$

Check this proposed solution.

$$3\sqrt{x} = \sqrt{x+8}$$
$$3\sqrt{1} = \sqrt{1+8} \quad \text{Let } x = 1$$
$$3(1) = \sqrt{9}$$
$$3 = 3 \quad \text{True}$$

The solution of $3\sqrt{x} = \sqrt{x+8}$ is the number 1. ◀

Work Problem 2 at the side.

▶ Not all equations with radicals have a solution, as shown by the equations in Examples 3 and 4.

EXAMPLE 3 Solve the equation $\sqrt{y} = -3$.

Square both sides.

$$(\sqrt{y})^2 = (-3)^2$$
$$y = 9$$

Check this proposed answer in the original equation.

$$\sqrt{y} = -3$$
$$\sqrt{9} = -3 \quad \text{Let } y = 9$$
$$3 = -3 \quad \text{False}$$

Since the statement $3 = -3$ is false, the number 9 is not a solution of the given equation, and $\sqrt{y} = -3$ has no solution. Since $\sqrt{y}$ represents the *nonnegative* square root of y, we might have seen immediately that there is no solution. ◀

EXAMPLE 4 Solve $a = \sqrt{a^2 + 5a + 10}$.

Square both sides.

$$(a)^2 = (\sqrt{a^2 + 5a + 10})^2$$
$$a^2 = a^2 + 5a + 10$$
$$0 = 5a + 10 \quad \text{Subtract } a^2$$
$$-10 = 5a \quad \text{Subtract 10}$$
$$a = -2 \quad \text{Divide by 5}$$

Check this proposed solution in the original equation.

$$a = \sqrt{a^2 + 5a + 10}$$
$$-2 = \sqrt{(-2)^2 + 5(-2) + 10} \quad \text{Let } a = -2$$
$$-2 = \sqrt{4 - 10 + 10}$$
$$-2 = 2 \quad \text{False}$$

Since $a = -2$ leads to a false result, the equation has no solution. ◀

Work Problem 3 at the side.

▶ The next examples use the fact that

$$(a+b)^2 = a^2 + 2ab + b^2.$$

ANSWERS

2. (a) 12 (b) 9 (c) 1
3. (a) no solution (b) 4
   (c) no solution (d) no solution

By this pattern, for example,
$$(y - 3)^2 = y^2 - 2(y)(3) + (-3)^2$$
$$= y^2 - 6y + 9.$$

Work Problem 4 at the side.

EXAMPLE 5 Solve the equation $\sqrt{2y - 3} = y - 3$.

Square both sides, using the result above on the right side of the equation.
$$(\sqrt{2y - 3})^2 = (y - 3)^2$$
$$2y - 3 = y^2 - 6y + 9$$

This equation is quadratic because of the y^2 term. Solving this equation requires that one side be equal to 0. Subtract $2y$ and add 3, getting
$$0 = y^2 - 8y + 12.$$
Solve this equation by factoring.
$$0 = (y - 6)(y - 2)$$
Set each factor equal to 0.
$$y - 6 = 0 \quad \text{or} \quad y - 2 = 0$$
$$y = 6 \quad \text{or} \quad y = 2$$

Check both of these proposed solutions in the original equation.

If $y = 6$,
$$\sqrt{2y - 3} = y - 3$$
$$\sqrt{2(6) - 3} = 6 - 3$$
$$\sqrt{12 - 3} = 3$$
$$\sqrt{9} = 3$$
$$3 = 3. \quad \text{True}$$

If $y = 2$,
$$\sqrt{2y - 3} = y - 3$$
$$\sqrt{2(2) - 3} = 2 - 3$$
$$\sqrt{4 - 3} = -1$$
$$\sqrt{1} = -1$$
$$1 = -1. \quad \text{False}$$

Only 6 is a valid solution of the equation. ◀

Work Problem 5 at the side.

Sometimes it is necessary to write an equation in a different form before squaring both sides. The next example shows why.

EXAMPLE 6 Solve the equation $\sqrt{x} + 1 = 2x$.

Squaring both sides gives
$$(\sqrt{x} + 1)^2 = (2x)^2$$
$$x + 2\sqrt{x} + 1 = 4x^2,$$

an equation that is more complicated, and still contains a radical. It would be better instead to rewrite the original equation so that the radical is alone on one side of the equals sign. Get the radical alone by subtracting 1 from both sides to get
$$\sqrt{x} = 2x - 1.$$

4. Square each expression.

   (a) $m - 5$

   (b) $2k - 5$

   (c) $3m - 2p$

   (d) $4z + 9y$

5. Solve each equation.

   (a) $\sqrt{6w + 6} = w + 1$

   (b) $x + 3 = \sqrt{12x + 1}$

   (c) $2u - 1 = \sqrt{10u + 9}$

ANSWERS
4. (a) $m^2 - 10m + 25$
   (b) $4k^2 - 20k + 25$
   (c) $9m^2 - 12mp + 4p^2$
   (d) $16z^2 + 72zy + 81y^2$
5. (a) 5, −1 (b) 4, 2 (c) 4

6. Solve each equation.

   (a) $\sqrt{x} - 3 = x - 15$

   (b) $\sqrt{z + 5} + 2 = z + 5$

Now square both sides.
$$(\sqrt{x})^2 = (2x - 1)^2$$
$$x = 4x^2 - 4x + 1$$

Subtract x from both sides.
$$0 = 4x^2 - 5x + 1$$

This quadratic equation can be solved by factoring.
$$0 = (4x - 1)(x - 1)$$
$$4x - 1 = 0 \quad \text{or} \quad x - 1 = 0$$
$$x = \frac{1}{4} \quad \text{or} \quad x = 1$$

Both of these proposed solutions must be checked in the original equation.

If $x = \frac{1}{4}$,

$\sqrt{x} + 1 = 2x$

$\sqrt{\frac{1}{4}} + 1 = 2\left(\frac{1}{4}\right)$

$\frac{1}{2} + 1 = \frac{1}{2}.$ False

If $x = 1$,

$\sqrt{x} + 1 = 2x$

$\sqrt{1} + 1 = 2(1)$

$2 = 2.$ True

The only solution to the original equation is 1. ◀

Work Problem 6 at the side.

Here is a summary of the steps to use when solving an equation with radicals.

Solving an Equation with Radicals

Step 1 Arrange the terms so that there is no more than one radical on each side of the equation.

Step 2 Square both sides.

Step 3 Combine like terms.

Step 4 If there is still a term with a radical, repeat Steps 1–3.

Step 5 Solve the equation.

Step 6 Check all solutions from Step 5 in the original equation.

ANSWERS

6. (a) 16 (b) −1

9.6 EXERCISES

Find all solutions for each equation. See Examples 1–4.

1. $\sqrt{x} = 2$

2. $\sqrt{m} = 5$

3. $\sqrt{y + 3} = 2$

4. $\sqrt{z + 1} = 5$

5. $\sqrt{t - 3} = 2$

6. $\sqrt{r + 5} = 4$

7. $\sqrt{n + 8} = 1$

8. $\sqrt{k + 10} = 2$

9. $\sqrt{m + 5} = 0$

10. $\sqrt{y - 4} = 0$

11. $\sqrt{z + 5} = -2$

12. $\sqrt{t - 3} = -2$

13. $\sqrt{k} - 2 = 5$

14. $\sqrt{p} - 3 = 7$

15. $\sqrt{y} + 4 = 2$

16. $\sqrt{m} + 6 = 5$

17. $\sqrt{5t - 9} = 2\sqrt{t}$

18. $\sqrt{3n + 4} = 2\sqrt{n}$

19. $3\sqrt{r} = \sqrt{8r + 16}$

20. $2\sqrt{r} = \sqrt{3r + 9}$

21. $\sqrt{5y - 5} = \sqrt{4y + 1}$

22. $\sqrt{2x + 2} = \sqrt{3x - 5}$

23. $\sqrt{x + 2} = \sqrt{2x - 5}$

24. $\sqrt{3m + 3} = \sqrt{5m - 1}$

25. $p = \sqrt{p^2 - 3p - 12}$

26. $k = \sqrt{k^2 - 2k + 10}$

27. $2r = \sqrt{4r^2 + 5r - 30}$

28. $3t = \sqrt{9t^2 - 6t + 12}$

Find all solutions for each equation. See Examples 5 and 6. Remember that $(a + b)^2 = a^2 + 2ab + b^2$ *and* $(\sqrt{a})^2 = a$.

29. $\sqrt{2x + 1} = x - 7$

30. $\sqrt{5x + 1} = x + 1$

31. $\sqrt{3x + 10} = 2x - 5$

32. $\sqrt{4x + 13} = 2x - 1$

33. $\sqrt{x + 1} - 1 = x$

34. $\sqrt{3x + 3} + 5 = x$

35. $\sqrt{4x + 5} - 2 = 2x - 7$

36. $\sqrt{6x + 7} - 1 = x + 1$

37. $3\sqrt{x + 13} = x + 9$

38. $2\sqrt{x + 7} = x - 1$

39. $\sqrt{4x} - x + 3 = 0$

40. $\sqrt{2x} - x + 4 = 0$

9.6 Exercises

41. $\sqrt{3x-4} = x-10$

42. $\sqrt{x+9} = x+3$

Solve each problem.

43. The square root of the sum of a number and 4 has the value 5. Find the number.

44. A certain number is the same as the square root of the product of 8 and the number. Find the number.

45. Three times the square root of 2 equals the square root of the sum of some number and 10. Find the number.

46. The negative square root of a number equals that number decreased by 2. Find the number.

47. Police sometimes use the following procedure to estimate the speed at which a car was traveling at the time of an accident. A police officer drives the car involved in the accident under conditions similar to those during which the accident took place and then skids to a stop. If the car is driven at 30 miles per hour, then the speed at the time of the accident is given by

$$s = 30\sqrt{\frac{a}{p}},$$

where a is the length of the skid marks left at the time of the accident and p is the length of the skid marks in the police test. Find s for the following values of a and p.

(a) $a = 900$ feet; $p = 100$ feet

(b) $a = 400$ feet; $p = 25$ feet

(c) $a = 80$ feet; $p = 20$ feet

(d) $a = 120$ feet; $p = 30$ feet

CHAPTER 9 REVIEW EXERCISES

(9.1) *Find all square roots of each number.*

1. 49
2. 81
3. 196
4. 121

5. 225
6. 729
7. 2401
8. 7569

Find each root that exists.

9. $\sqrt{16}$
10. $-\sqrt{36}$
11. $\sqrt{400}$
12. $-\sqrt{2704}$

13. $\sqrt{-8100}$
14. $-\sqrt{4225}$
15. $\sqrt[3]{27}$
16. $\sqrt[3]{-64}$

17. $-\sqrt[3]{8}$
18. $\sqrt[4]{16}$
19. $-\sqrt[4]{16}$
20. $\sqrt[4]{-16}$

21. $\sqrt[5]{32}$
22. $\sqrt[5]{-243}$
23. $-\sqrt[5]{32}$
24. $\sqrt[6]{64}$

Write rational *or* irrational *for each number. If a number is rational, give its exact value. If a number is irrational, give a decimal approximation for the number. Round approximations to the nearest thousandth.*

25. $\sqrt{15}$
26. $\sqrt{64}$
27. $-\sqrt{169}$
28. $-\sqrt{170}$

(9.2) *Use the product rule to simplify each expression.*

29. $\sqrt{2} \cdot \sqrt{5}$
30. $\sqrt{12} \cdot \sqrt{3}$
31. $\sqrt{5} \cdot \sqrt{15}$

32. $\sqrt{8} \cdot \sqrt{8}$
33. $\sqrt{27}$
34. $\sqrt{48}$

35. $\sqrt{160}$
36. $\sqrt{320}$
37. $\sqrt{98} \cdot \sqrt{50}$

38. $\sqrt{12} \cdot \sqrt{27}$
39. $\sqrt{32} \cdot \sqrt{48}$
40. $\sqrt{50} \cdot \sqrt{125}$

Use the product rule and the quotient rule, as necessary, to simplify each expression.

41. $\sqrt{\dfrac{9}{4}}$
42. $\sqrt{\dfrac{121}{400}}$
43. $\sqrt{\dfrac{3}{25}}$

Chapter 9 Review Exercises

44. $\sqrt{\dfrac{10}{169}}$

45. $\sqrt{\dfrac{1}{6}} \cdot \sqrt{\dfrac{5}{6}}$

46. $\sqrt{\dfrac{2}{5}} \cdot \sqrt{\dfrac{2}{45}}$

47. $\dfrac{3\sqrt{10}}{\sqrt{2}}$

48. $\dfrac{24\sqrt{12}}{6\sqrt{3}}$

49. $\dfrac{12\sqrt{75}}{4\sqrt{3}}$

Simplify each expression. Assume that all variables represent positive real numbers.

50. $\sqrt{p} \cdot \sqrt{p}$

51. $\sqrt{k} \cdot \sqrt{m}$

52. $\sqrt{y^2}$

53. $\sqrt{r^{18}}$

54. $\sqrt{x^{10}y^{16}}$

55. $\sqrt{x^9}$

56. $\sqrt{\dfrac{36}{p^2}}$

57. $\sqrt{\dfrac{13}{k^4}}$

58. $\sqrt{\dfrac{100}{y^{10}}}$

59. $\sqrt[3]{54}$

60. $\sqrt[3]{250}$

61. $\sqrt[3]{\dfrac{5}{8}}$

62. $\sqrt[3]{\dfrac{6}{125}}$

63. $\sqrt[3]{8x^2}$

64. $\sqrt[3]{27x^4}$

(9.3) Simplify and combine terms wherever possible.

65. $\sqrt{7} + \sqrt{7}$

66. $3\sqrt{2} + 5\sqrt{2}$

67. $3\sqrt{75} + 2\sqrt{27}$

68. $4\sqrt{12} + \sqrt{48}$

69. $4\sqrt{24} - 3\sqrt{54} + \sqrt{6}$

70. $2\sqrt{7} - 4\sqrt{28} + 3\sqrt{63}$

71. $\dfrac{2}{5}\sqrt{75} - \dfrac{3}{4}\sqrt{160}$

72. $\dfrac{1}{3}\sqrt{18} + \dfrac{1}{4}\sqrt{32}$

73. $\sqrt{15} \cdot \sqrt{2} + 5\sqrt{30}$

74. $\sqrt{5} \cdot \sqrt{3} + 4\sqrt{15} - \sqrt{3}$

75. $4\sqrt[4]{16} + 3\sqrt[4]{32} \cdot \sqrt[4]{2}$

76. $3\sqrt[3]{54} - 2\sqrt[3]{16}$

Simplify each expression. Assume that all variables represent nonnegative real numbers.

77. $\sqrt{4x} + \sqrt{36x} + \sqrt{9x}$

78. $\sqrt{16p} + 3\sqrt{p} - \sqrt{49p}$

79. $\sqrt{20m^2} - m\sqrt{45}$

80. $3k\sqrt{8k^2n} + 5k^2\sqrt{2n}$

(9.4) Perform the indicated operations, and write all answers in simplest form. Rationalize all denominators. Assume that all variables represent positive real numbers.

81. $\dfrac{10}{\sqrt{3}}$

82. $\dfrac{5}{\sqrt{2}}$

83. $\dfrac{3\sqrt{2}}{\sqrt{5}}$

84. $\dfrac{8}{\sqrt{8}}$

85. $\dfrac{12}{\sqrt{24}}$

86. $\dfrac{\sqrt{2}}{\sqrt{15}}$

87. $\sqrt{\dfrac{3}{5}}$

88. $\sqrt{\dfrac{5}{14}} \cdot \sqrt{28}$

89. $\sqrt{\dfrac{2}{3}} \cdot \sqrt{\dfrac{1}{5}}$

90. $\sqrt{\dfrac{1}{6}} \cdot \sqrt{\dfrac{18}{5}}$

91. $\sqrt[3]{\dfrac{1}{3}}$

92. $\sqrt[3]{\dfrac{5}{4}}$

93. $\sqrt{\dfrac{7}{x}}$

94. $\sqrt{\dfrac{9m^3}{2p}}$

95. $\sqrt{\dfrac{r^3 t}{5w^2}}$

(9.5) Simplify each expression.

96. $2\sqrt{6} + 4\sqrt{54}$

97. $-\sqrt{3}(\sqrt{5} - \sqrt{27})$

98. $3\sqrt{2}(\sqrt{3} + 2\sqrt{2})$

99. $(2\sqrt{3} - 4)(5\sqrt{3} + 2)$

100. $(5\sqrt{7} + 2)^2$

101. $(\sqrt{5} - \sqrt{7})(\sqrt{5} + \sqrt{7})$

102. $(2\sqrt{3} + 5)(2\sqrt{3} - 5)$

103. $(\sqrt{7} + 2\sqrt{6})(\sqrt{12} - \sqrt{2})$

Rationalize the denominators.

104. $\dfrac{1}{2 + \sqrt{5}}$

105. $\dfrac{2}{\sqrt{2} - 3}$

106. $\dfrac{\sqrt{8}}{\sqrt{2} + 6}$

107. $\dfrac{\sqrt{3}}{1 + \sqrt{3}}$

108. $\dfrac{\sqrt{5} - 1}{\sqrt{2} + 3}$

109. $\dfrac{2 + \sqrt{6}}{\sqrt{3} - 1}$

(9.6) *Find all solutions for each equation.*

110. $\sqrt{m} = 5$

111. $\sqrt{p} = -2$

112. $\sqrt{k + 1} = 10$

113. $\sqrt{y - 8} = 0$

114. $\sqrt{r} - 3 = 8$

115. $\sqrt{x} + 2 = 1$

116. $\sqrt{5m + 4} = 3\sqrt{m}$

117. $\sqrt{2p + 3} = \sqrt{5p - 3}$

118. $\sqrt{4y + 1} = y - 1$

119. $\sqrt{-2k - 4} = k + 2$

120. $\sqrt{2 - x} + 3 = x + 7$

121. $\sqrt{x} - x + 2 = 0$

name _____ date _____ hour _____

CHAPTER 9 TEST

Find the indicated root. Use the square root table or a calculator to give a decimal approximation if necessary.

1. $\sqrt{100}$

2. $\sqrt{77}$

3. $\sqrt{190}$

4. $\sqrt[3]{-27}$

5. $\sqrt[4]{625}$

Simplify where possible.

6. $\sqrt{\dfrac{64}{169}}$

7. $\sqrt{50}$

8. $\sqrt{\dfrac{3}{8}} \cdot \sqrt{\dfrac{32}{27}}$

9. $\dfrac{20\sqrt{18}}{5\sqrt{3}}$

10. $\sqrt[3]{32}$

11. $\sqrt{12} + \sqrt{48}$

12. $\sqrt{20} - \sqrt{45}$

13. $3\sqrt{6} + \sqrt{14}$

14. $3\sqrt{28} + \sqrt{63}$

15. $3\sqrt{27x} - 4\sqrt{48x}$

1. _____
2. _____
3. _____
4. _____
5. _____
6. _____
7. _____
8. _____
9. _____
10. _____
11. _____
12. _____
13. _____
14. _____
15. _____

16. _____

17. _____

18. _____

19. _____

20. _____

21. _____

22. _____

23. _____

24. _____

25. _____

26. _____

27. _____

28. _____

29. _____

30. _____

16. $\sqrt[3]{32x^2y^3}$

17. $(6 - \sqrt{5})(6 + \sqrt{5})$

18. $(2 - \sqrt{7})(3\sqrt{2} + 1)$

19. $(1 - \sqrt{3})^2$

20. $(\sqrt{5} + \sqrt{6})^2$

Rationalize each denominator.

21. $\dfrac{4}{\sqrt{3}}$

22. $\dfrac{3\sqrt{2}}{\sqrt{6}}$

23. $\sqrt[3]{\dfrac{5}{9}}$

24. $\dfrac{-3}{4 - \sqrt{3}}$

25. $\dfrac{\sqrt{2} + 1}{3 - \sqrt{7}}$

Solve each equation.

26. $\sqrt{m} + 5 = 0$

27. $\sqrt{k + 2} = 5$

28. $\sqrt{2y + 8} = 2\sqrt{y}$

29. $6\sqrt{k} - 3 = k + 2$

30. $\sqrt{y + 3} = y + 1$

Quadratic Equations

10

10.1 SOLVING QUADRATIC EQUATIONS BY THE SQUARE ROOT METHOD

Recall that a *quadratic equation* is an equation that can be written in the form

$$ax^2 + bx + c = 0,$$

for real numbers a, b, and c, with $a \neq 0$. In Chapter 5 these equations were solved by factoring. As mentioned there, not all quadratic equations can easily be solved by factoring (for example, $x^2 - x + 1 = 0$ cannot). Other ways to solve quadratic equations are shown in this chapter. For example, the quadratic equation

$$(x - 3)^2 = 16,$$

with the square of a binomial equal to a number, can be solved with square roots.

▶ The *square root property of equations* justifies taking square roots of both sides of an equation.

> **Square Root Property of Equations**
> If b is a positive number and if $a^2 = b$, then
> $$a = \sqrt{b} \quad \text{or} \quad a = -\sqrt{b}.$$

EXAMPLE 1 Solve each equation. Write radicals in simplified form.

(a) $x^2 = 16$

By the square root property, if $x^2 = 16$, then

$$x = \sqrt{16} = 4 \quad \text{or} \quad x = -\sqrt{16} = -4.$$

Check each solution by substituting back in the original equation.

(b) $z^2 = 5$

The solutions are $z = \sqrt{5}$ or $z = -\sqrt{5}$.

(c) $m^2 = 8$

Use the square root property to get $m = \sqrt{8}$ or $m = -\sqrt{8}$. Simplify $\sqrt{8}$ as $\sqrt{8} = 2\sqrt{2}$, so $m = 2\sqrt{2}$ or $m = -2\sqrt{2}$.

(d) $y^2 = -4$

Since -4 is a negative number and since the square of a real number cannot be negative, there is no real number solution for this equation. (The square root property cannot be used because of the requirement that b must be positive.) ◀

Work Problem 1 at the side.

Objectives

▶ Solve equations of the form $x^2 = $ a number.

▶ Solve equations of the form $(ax + b)^2 = $ a number.

1. Solve each equation. Write radicals in simplified form.

   (a) $k^2 = 49$

   (b) $b^2 = 11$

   (c) $c^2 = 12$

   (d) $x^2 = -9$

ANSWERS
1. (a) 7, −7 (b) $\sqrt{11}$, $-\sqrt{11}$
   (c) $2\sqrt{3}$, $-2\sqrt{3}$ (d) no real number solution

2. Solve each equation.

(a) $(m + 2)^2 = 36$

(b) $(p - 4)^2 = 3$

3. Solve each equation.

(a) $(2x - 5)^2 = 18$

(b) $(5m + 1)^2 = 7$

(c) $(9k - 2)^2 = 48$

4. Solve each equation.

(a) $p^2 = -9$

(b) $(x + 8)^2 = -25$

(c) $(7z - 1)^2 = -1$

ANSWERS

2. (a) $4, -8$ (b) $4 + \sqrt{3}, 4 - \sqrt{3}$
3. (a) $\dfrac{5 + 3\sqrt{2}}{2}, \dfrac{5 - 3\sqrt{2}}{2}$
   (b) $\dfrac{-1 + \sqrt{7}}{5}, \dfrac{-1 - \sqrt{7}}{5}$
   (c) $\dfrac{2 + 4\sqrt{3}}{9}, \dfrac{2 - 4\sqrt{3}}{9}$
4. (a) no real number solution
   (b) no real number solution
   (c) no real number solution

488   Quadratic Equations

▶ The equation $(x - 3)^2 = 16$ also can be solved with the square root property of equations. If $(x - 3)^2 = 16$, then

$$x - 3 = 4 \quad \text{or} \quad x - 3 = -4. \quad \text{Square root property}$$

Solve each of the last two equations to get

$$x - 3 = 4 \quad \text{or} \quad x - 3 = -4$$
$$x = 7 \quad \text{or} \quad x = -1. \quad \text{Add 3}$$

Check both answers in the original equation.

$$(7 - 3)^2 = 4^2 = 16 \quad \text{and} \quad (-1 - 3)^2 = (-4)^2 = 16$$

Both 7 and -1 are solutions.

EXAMPLE 2   Solve $(x - 1)^2 = 6$.

By the square root property of equations,

$$x - 1 = \sqrt{6} \quad \text{or} \quad x - 1 = -\sqrt{6} \quad \text{Square root property}$$
$$x = 1 + \sqrt{6} \quad \text{or} \quad x = 1 - \sqrt{6}. \quad \text{Add 1}$$

Check: $(1 + \sqrt{6} - 1)^2 = (\sqrt{6})^2 = 6;$
$(1 - \sqrt{6} - 1)^2 = (-\sqrt{6})^2 = 6.$

The solutions are $1 + \sqrt{6}$ and $1 - \sqrt{6}$. ◀

Work Problem 2 at the side.

EXAMPLE 3   Solve the equation $(3r - 2)^2 = 27$.

The square root property gives

$$3r - 2 = \sqrt{27} \quad \text{or} \quad 3r - 2 = -\sqrt{27}.$$

Now simplify the radical: $\sqrt{27} = \sqrt{9 \cdot 3} = \sqrt{9} \cdot \sqrt{3} = 3\sqrt{3}$, so

$$3r - 2 = 3\sqrt{3} \quad \text{or} \quad 3r - 2 = -3\sqrt{3}$$
$$3r = 2 + 3\sqrt{3} \quad \text{or} \quad 3r = 2 - 3\sqrt{3} \quad \text{Add 2}$$
$$r = \dfrac{2 + 3\sqrt{3}}{3} \quad \text{or} \quad r = \dfrac{2 - 3\sqrt{3}}{3}. \quad \text{Divide by 3}$$

The solutions are

$$\dfrac{2 + 3\sqrt{3}}{3} \quad \text{and} \quad \dfrac{2 - 3\sqrt{3}}{3}. \quad ◀$$

Work Problem 3 at the side.

EXAMPLE 4   Solve $(x + 3)^2 = -9$.

The square root of -9 is not a real number. There is no real number solution for this equation. ◀

Work Problem 4 at the side.

name　　　　　　　　　　　　　　　　　　　　date　　　　　hour

10.1 EXERCISES

Solve each equation by using the square root property. Express all radicals in simplest form. See Example 1.

1. $x^2 = 25$
2. $y^2 = 100$
3. $x^2 = 64$
4. $z^2 = 81$

5. $m^2 = 13$
6. $x^2 = 7$
7. $p^2 = 2$
8. $q^2 = 6$

9. $x^2 = 24$
10. $t^2 = 27$
11. $m^2 = \dfrac{169}{9}$
12. $r^2 = \dfrac{144}{25}$

13. $k^2 = \dfrac{9}{16}$
14. $q^2 = \dfrac{225}{16}$
15. $r^2 = 1.21$
16. $k^2 = 1.96$

17. $k^2 = 2.56$
18. $z^2 = 9.61$
19. $r^2 = 77.44$
20. $y^2 = 43.56$

Solve each equation by using the square root property. Express all radicals in simplest form. See Examples 2–4.

21. $(x - 2)^2 = 16$
22. $(r + 4)^2 = 25$
23. $(a + 4)^2 = 10$

24. $(r - 3)^2 = 15$
25. $(x - 1)^2 = 32$
26. $(y + 5)^2 = 28$

27. $(2m - 1)^2 = 9$
28. $(3y - 7)^2 = 4$
29. $(3z + 5)^2 = 9$

30. $(2y - 7)^2 = 49$ **31.** $(2a - 5)^2 = 30$ **32.** $(2y + 3)^2 = 45$

33. $(3p - 1)^2 = 18$ **34.** $(5r - 6)^2 = 75$ **35.** $(2k - 5)^2 = 98$

36. $(4x - 1)^2 = 48$ **37.** $(3m + 4)^2 = 8$ **38.** $(2p - 5)^2 = 180$

Solve each equation by using the square root property. Round solutions to the nearest hundredth.

39. $(m - 1.9)^2 = 7.3$ **40.** $(y + 2.6)^2 = 4.8$ **41.** $(2.11p + 3.42)^2 = 9.58$

42. $(1.71m - 6.20)^2 = 5.41$ **43.** $(6.95x - 1.72)^2 = 86.4$ **44.** $(3.76y + 4.19)^2 = 72.8$

45. This exercise uses the formula for the distance traveled by a falling object:

$$d = 16t^2,$$

where d is the distance the object falls in t seconds. One expert at marksmanship can hold a silver dollar above his head, drop it, draw his gun, and shoot the coin as it passes waist level. If the coin falls about 4 feet, estimate the time that elapses between the dropping of the coin and the shot.

46. The illumination produced by a light source depends on the distance from the source. For a particular light source, this relationship can be expressed as

$$d^2 = \frac{4050}{I},$$

where d is the distance from the source (in meters) and I is the amount of illumination in footcandles. How far from the source is the illumination equal to 50 footcandles?

10.2 SOLVING QUADRATIC EQUATIONS BY COMPLETING THE SQUARE

Objectives

▶ 1 Solve quadratic equations by completing the square when the coefficient of the squared term is 1.

▶ 2 Solve quadratic equations by completing the square when the coefficient of the squared term is not 1.

▶ 3 Simplify an equation before solving.

▶ 1 The properties studied so far are not enough to solve the equation

$$x^2 + 6x + 7 = 0.$$

For a method of solving this equation, recall the method from the preceding section for solving equations of the type

$$(x + 3)^2 = 2.$$

If the equation $x^2 + 6x + 7 = 0$ could be rewritten in a form like $(x + 3)^2 = 2$, it could be solved with the square root property. The next example shows how to rewrite the equation so it can be solved.

Work Problem 1 at the side.

EXAMPLE 1 Solve $x^2 + 6x + 7 = 0$.

Start by subtracting 7 from both sides of the equation.

$$x^2 + 6x = -7$$

The quantity on the left-hand side of $x^2 + 6x = -7$ must be made into a perfect square trinomial. The expression $x^2 + 6x + 9$ is a perfect square, since

$$x^2 + 6x + 9 = (x + 3)^2.$$

Therefore, if 9 is added to both sides, the equation will have a perfect square trinomial on the left-hand side, as needed.

$$x^2 + 6x \boxed{+ 9} = -7 \boxed{+ 9} \quad \text{Add 9}$$
$$(x + 3)^2 = 2 \quad \text{Factor}$$

Now use the square root property to complete the solution.

$$x + 3 = \sqrt{2} \quad \text{or} \quad x + 3 = -\sqrt{2}$$
$$x = -3 + \sqrt{2} \quad \text{or} \quad x = -3 - \sqrt{2}$$

The solutions of the original equation are $-3 + \sqrt{2}$ and $-3 - \sqrt{2}$. Check by substituting $-3 + \sqrt{2}$ and $-3 - \sqrt{2}$ for x in the original equation. ◀

The process of changing the form of the equation in Example 1 from

$$x^2 + 6x + 7 = 0 \quad \text{to} \quad (x + 3)^2 = 2$$

is called **completing the square**. When completing the square, only the *form* of the equation is changed. To see this, simplify $(x + 3)^2 = 2$; the result will be $x^2 + 6x + 7 = 0$.

EXAMPLE 2 Find the solutions of the quadratic equation

$$m^2 - 8m = 5.$$

A suitable number must be added to both sides to make the left side a perfect square. Find this number as follows: recall from Chapter 4 that

$$(m + a)^2 = m^2 + 2am + a^2.$$

In the equations of this chapter, the value of $2a$ is known, and a^2 must be found. From the middle term, $2am$, find a^2 by taking half the coefficient of m [since $\frac{1}{2}(2a) = a$] and squaring the result. Add the result, a^2, to both sides.

1. As a review, factor each of these perfect square trinomials.

   (a) $x^2 + 6x + 9$

   (b) $q^2 - 20q + 100$

   (c) $9m^2 - 30m + 25$

   (d) $100r^2 - 60m + 9$

ANSWERS
1. (a) $(x + 3)^2$ (b) $(q - 10)^2$
   (c) $(3m - 5)^2$ (d) $(10r - 3)^2$

10.2 Solving Quadratic Equations by Completing the Square

2. Solve by completing the square.

   (a) $x^2 - 14x = -40$

   (b) $a^2 + 3a = 1$

   (c) $z^2 + 5z + 3 = 0$

In the equation $m^2 - 8m = 5$, the coefficient of m is -8, and half of -8 is -4. Squaring -4 gives 16, which is the number to be added to both sides.

$$m^2 - 8m + 16 = 5 + 16 \qquad (1) \qquad \text{Add 16}$$

The trinomial $m^2 - 8m + 16$ is a perfect square trinomial. Factor this trinomial to get

$$m^2 - 8m + 16 = (m - 4)^2.$$

Equation (1) becomes

$$(m - 4)^2 = 21.$$

Now use the square root property.

$$m - 4 = \sqrt{21} \qquad \text{or} \qquad m - 4 = -\sqrt{21}$$
$$m = 4 + \sqrt{21} \qquad \text{or} \qquad m = 4 - \sqrt{21}$$

The solutions are

$$4 + \sqrt{21} \quad \text{and} \quad 4 - \sqrt{21}. \blacktriangleleft$$

Work Problem 2 at the side.

▶ The process of completing the square discussed above requires the coefficient of the squared term to be 1. For an equation of the form $ax^2 + bx + c = 0$, get a coefficient of 1 on x^2 by first dividing both sides of the equation by a. The next examples illustrate this approach.

EXAMPLE 3 Solve $4y^2 + 16y = 9$.

Before completing the square, the coefficient of the squared term must be 1. Here the coefficient of y^2 is 4. Make the coefficient 1 by dividing both sides of the equation by 4.

$$y^2 + 4y = \frac{9}{4} \qquad \text{Divide by 4}$$

Now take half the coefficient of y, or $(\frac{1}{2})(4) = 2$, and square the result: $2^2 = 4$. Add 4 to both sides of the equation, and perform the addition on the right-hand side.

$$y^2 + 4y + 4 = \frac{9}{4} + 4 \qquad \text{Add 4}$$

$$y^2 + 4y + 4 = \frac{25}{4}$$

Factor on the left.

$$(y + 2)^2 = \frac{25}{4}$$

Use the square root property of equations and solve for y.

$$y + 2 = \frac{5}{2} \qquad \text{or} \qquad y + 2 = -\frac{5}{2} \qquad \text{Square root property}$$

$$y = -2 + \frac{5}{2} \qquad \text{or} \qquad y = -2 - \frac{5}{2} \qquad \text{Subtract 2}$$

$$y = \frac{1}{2} \qquad \text{or} \qquad y = -\frac{9}{2}$$

The two solutions are $\frac{1}{2}$ and $-\frac{9}{2}$. ◀

ANSWERS

2. (a) 4, 10 (b) $\dfrac{-3 + \sqrt{13}}{2}$, $\dfrac{-3 - \sqrt{13}}{2}$

   (c) $\dfrac{-5 + \sqrt{13}}{2}$, $\dfrac{-5 - \sqrt{13}}{2}$

Quadratic Equations

Work Problem 3 at the side.

EXAMPLE 4 Solve $2x^2 - 7x = 9$.

Divide both sides of the equation by 2 to get a coefficient of 1 for the x^2 term.

$$x^2 - \frac{7}{2}x = \frac{9}{2} \qquad \text{Divide by 2}$$

Now take half the coefficient of x and square it. Half of $-\frac{7}{2}$ is $-\frac{7}{4}$, and $(-\frac{7}{4})^2 = \frac{49}{16}$. Add $\frac{49}{16}$ to both sides of the equation, and write the left side as a perfect square.

$$x^2 - \frac{7}{2}x + \frac{49}{16} = \frac{9}{2} + \frac{49}{16} \qquad \text{Add } \frac{49}{16}$$

$$\left(x - \frac{7}{4}\right)^2 = \frac{121}{16} \qquad \text{Factor}$$

Use the square root property.

$$x - \frac{7}{4} = \sqrt{\frac{121}{16}} \qquad \text{or} \qquad x - \frac{7}{4} = -\sqrt{\frac{121}{16}}$$

Since $\sqrt{\frac{121}{16}} = \frac{11}{4}$,

$$x - \frac{7}{4} = \frac{11}{4} \qquad \text{or} \qquad x - \frac{7}{4} = -\frac{11}{4}$$

$$x = \frac{18}{4} \qquad \text{or} \qquad x = -\frac{4}{4} \qquad \text{Add } \frac{7}{4}$$

$$x = \frac{9}{2} \qquad \text{or} \qquad x = -1.$$

The solutions are $\frac{9}{2}$ and -1. ◀

Work Problem 4 at the side.

EXAMPLE 5 Solve $4p^2 + 8p + 5 = 0$.

First divide both sides by 4 to get the coefficient 1 for the p^2 term.

$$p^2 + 2p + \frac{5}{4} = 0 \qquad \text{Divide each side by 4}$$

Subtract $\frac{5}{4}$ from each side, which gives

$$p^2 + 2p = -\frac{5}{4}. \qquad \text{Subtract } \frac{5}{4} \text{ from each side}$$

The coefficient of p is 2. Take half of 2, square the result, and add this square to both sides. The left-hand side can then be written as a perfect square.

$$p^2 + 2p + 1 = -\frac{5}{4} + 1 \qquad \text{Add 1 on each side}$$

$$(p + 1)^2 = -\frac{1}{4} \qquad \text{Factor}$$

The square root of $-\frac{1}{4}$ is not a real number, so the square root property does not apply. This equation has no real number solutions. ◀

10.2 Solving Quadratic Equations by Completing the Square

---

3. Solve by completing the square.

   (a) $9m^2 + 18m + 5 = 0$

   (b) $4k^2 - 24k + 11 = 0$

4. Solve by completing the square.

   (a) $3x^2 + 5x - 2 = 0$

   (b) $2p^2 - 3p = 1$

ANSWERS

3. (a) $-\frac{1}{3}, -\frac{5}{3}$ (b) $\frac{11}{2}, \frac{1}{2}$

4. (a) $-2, \frac{1}{3}$ (b) $\frac{3 + \sqrt{17}}{4}$, $\frac{3 - \sqrt{17}}{4}$

5. Solve by completing the square.

(a) $a^2 + 3a + 10 = 0$

(b) $5v^2 + 3v + 1 = 0$

6. Solve each equation.

(a) $r^2 + 2 = 3r$

(b) $m^2 - 6 = m$

(c) $3p^2 + 7p = p^2 + 2p + 3$

ANSWERS
5. (a) no real number solution
   (b) no real number solution
6. (a) 2, 1 (b) 3, −2 (c) $-3, \frac{1}{2}$

494

Work Problem 5 at the side.

The steps in solving a quadratic equation by completing the square are summarized below.

> Use *completing the square* to solve the quadratic equation $ax^2 + bx + c = 0$ as follows.
>
> *Step 1* If the coefficient of the squared term is 1, proceed to Step 2. If the coefficient of the squared term is not 1 but some other nonzero number, divide both sides of the equation by this coefficient. This gives an equation that has 1 as coefficient of the squared term.
>
> *Step 2* Make sure that all terms with variables are on one side of the equals sign and that all constants are on the other side.
>
> *Step 3* Take half the coefficient of x and square the result. Add the square to both sides of the equation. By factoring, the side containing the variables can now be written as a perfect square.
>
> *Step 4* Apply the square root property of equations.

The next example shows how to simplify an equation before solving it.

EXAMPLE 6 Solve $(m - 2)(m + 1) = 4$.

Start by multiplying on the left. Use the FOIL method.

$$(m - 2)(m + 1) = 4$$
$$m^2 - m - 2 = 4 \quad \text{Use FOIL}$$
$$m^2 - m = 6 \quad \text{Add 2}$$

Now complete the square. Half of -1 is $-\frac{1}{2}$, and $(-\frac{1}{2})^2 = \frac{1}{4}$. Add $\frac{1}{4}$ on each side.

$$m^2 - m + \frac{1}{4} = 6 + \frac{1}{4} \quad \text{Add } \frac{1}{4}$$

Factor on the left, and add on the right.

$$\left(m - \frac{1}{2}\right)^2 = \frac{25}{4}$$

Complete the solution.

$$m - \frac{1}{2} = \frac{5}{2} \quad \text{or} \quad m - \frac{1}{2} = -\frac{5}{2} \quad \text{Square root property}$$
$$m = \frac{1}{2} + \frac{5}{2} \quad \text{or} \quad m = \frac{1}{2} - \frac{5}{2} \quad \text{Add } \frac{1}{2}$$
$$m = 3 \quad \text{or} \quad m = -2$$

Check each of these solutions by substituting it in the original equation. ◀

Work Problem 6 at the side.

Quadratic Equations

10.2 EXERCISES

Find the number that should be added to each expression to make it a perfect square. See Example 2.

1. $x^2 + 2x$
2. $y^2 - 4y$
3. $x^2 + 18x$

4. $m^2 - 6m$
5. $x^2 + 3x$
6. $r^2 + 7r$

7. $y^2 + 5y$
8. $q^2 - 9q$

Solve each equation by completing the square. See Examples 1–6.

9. $x^2 + 4x = -3$
10. $a^2 + 2a = 5$
11. $m^2 + 4m = -1$

12. $z^2 + 6z = -8$
13. $q^2 - 8q = -16$
14. $x^2 - 6x + 1 = 0$

15. $b^2 - 2b - 2 = 0$
16. $c^2 + 3c = 2$
17. $k^2 + 5k - 3 = 0$

10.2 Exercises 495

18. $2m^2 + 4m = -7$

19. $4y^2 + 4y - 3 = 0$

20. $-x^2 + 6x = 4$

21. $2m^2 - 4m - 5 = 0$

22. $-x^2 + 4 = 2x$

23. $m^2 - 4m + 8 = 6m$

24. $2z^2 = 8z + 5 - 4z^2$

25. $3r^2 - 2 = 6r + 3$

26. $4p - 3 = p^2 + 2p$

27. $(x + 1)(x + 3) = 2$

28. $(x - 3)(x + 1) = 1$

29. $(r - 2)(r + 2) = 5$

30. A rule for estimating the number of board feet of lumber that can be cut from a log of a certain length depends on the diameter of the log. The diameter d required to get 9 board feet is found from the equation

$$\left(\frac{d - 4}{4}\right)^2 = 9.$$

Solve this equation for d. Are both answers reasonable?

10.3 SOLVING QUADRATIC EQUATIONS BY THE QUADRATIC FORMULA

Completing the square can be used to solve any quadratic equation, but the method often is not very handy. This section introduces a general formula, the *quadratic formula,* that gives the solution for any quadratic equation.

Get the quadratic formula by starting with the general form of a quadratic equation,
$$ax^2 + bx + c = 0, \quad a \neq 0.$$

The restriction $a \neq 0$ is important in order to make sure that the equation is quadratic. If $a = 0$, then the equation becomes $0x^2 + bx + c = 0$, or $bx + c = 0$, which is a linear, not a quadratic equation.

▶ The first step in solving a quadratic equation by this new method is to identify the letters a, b, and c in the general form of the quadratic equation.

EXAMPLE 1 Match the coefficients of each of the following quadratic equations with the letters a, b, and c of the general quadratic equation
$$ax^2 + bx + c = 0.$$

(a) $2x^2 + 3x - 5 = 0$

In this example $a = 2$, $b = 3$, and $c = -5$.

(b) $-x^2 + 2 = 6x$

First rewrite the equation with 0 on one side to match the general form $ax^2 + bx + c = 0$.
$$-x^2 + 2 = 6x$$
$$-x^2 - 6x + 2 = 0 \quad \text{Subtract } 6x$$

Now identify $a = -1$, $b = -6$, and $c = 2$.

(c) $2(x + 3)(x - 1) = 5$

Multiply on the left.
$$2(x + 3)(x - 1) = 2(x^2 + 2x - 3)$$
$$= 2x^2 + 4x - 6 \quad \text{Distributive property}$$

The equation becomes
$$2x^2 + 4x - 6 = 5,$$
or $2x^2 + 4x - 11 = 0$, so that $a = 2$, $b = 4$, and $c = -11$. ◀

Work Problem 1 at the side.

The next steps show how to obtain the formula for solving a quadratic equation once a, b, and c are known. First, get the coefficient 1 for the x^2 term by dividing both sides by a.
$$x^2 + \frac{b}{a}x + \frac{c}{a} = 0 \quad \text{Divide by } a$$

Next, subtract $\frac{c}{a}$ from each side.
$$x^2 + \frac{b}{a}x = -\frac{c}{a} \quad \text{Subtract } \frac{c}{a}$$

Objectives

▶ 1 Identify the letters a, b, and c in a quadratic equation.

▶ 2 Use the quadratic formula to solve quadratic equations.

▶ 3 Solve quadratic equations with only one solution.

▶ 4 Solve quadratic equations with fractions.

1. Match the coefficients of each of the following quadratic equations with the letters a, b, and c of the general quadratic equation $ax^2 + bx + c = 0$.

   (a) $5x^2 + 2x - 1 = 0$

   (b) $3m^2 = m - 2$

   (c) $p(p + 5) = 4$

   (d) $3(m - 2)(m + 1) = 7$

ANSWERS
1. (a) $a = 5$, $b = 2$, $c = -1$
   (b) $a = 3$, $b = -1$, $c = 2$
   (c) $a = 1$, $b = 5$, $c = -4$
   (d) $a = 3$, $b = -3$, $c = -13$

10.3 Solving Quadratic Equations by the Quadratic Formula

2. Complete these steps to simplify the right side of the equation.

(a) $-\dfrac{c}{a} + \dfrac{b^2}{4a^2} = \dfrac{b^2}{4a^2} - \dfrac{?}{\underline{}}$

(b) $\dfrac{b^2}{4a^2} - \dfrac{c}{a} = \dfrac{b^2}{4a^2} - \dfrac{?}{4a^2}$

(c) $\dfrac{b^2}{4a^2} - \dfrac{4ac}{4a^2} = \dfrac{?}{4a^2}$

ANSWERS

2. (a) $\dfrac{c}{a}$ (b) $4ac$ (c) $b^2 - 4ac$

To complete the square on the left, take half the coefficient of x, or $\dfrac{1}{2} \cdot \dfrac{b}{a} = \dfrac{b}{2a}$. Square $\dfrac{b}{2a}$ to get $\dfrac{b^2}{4a^2}$. Add $\dfrac{b^2}{4a^2}$ to both sides of the equation.

$$x^2 + \dfrac{b}{a}x + \dfrac{b^2}{4a^2} = -\dfrac{c}{a} + \dfrac{b^2}{4a^2} \quad \text{Add } \dfrac{b^2}{4a^2}$$

Rewrite the left-hand side as a perfect square.

$$\left(x + \dfrac{b}{2a}\right)^2 = -\dfrac{c}{a} + \dfrac{b^2}{4a^2} \quad \text{Factor}$$

Now simplify the right-hand side of the equation.

Work Problem 2 at the side.

From the result of Problem 2 at the side,

$$\left(x + \dfrac{b}{2a}\right)^2 = \dfrac{b^2 - 4ac}{4a^2}.$$

Apply the square root property.

$$x + \dfrac{b}{2a} = \sqrt{\dfrac{b^2 - 4ac}{4a^2}} \quad \text{or} \quad x + \dfrac{b}{2a} = -\sqrt{\dfrac{b^2 - 4ac}{4a^2}}$$

Simplify the radical.

$$\sqrt{\dfrac{b^2 - 4ac}{4a^2}} = \dfrac{\sqrt{b^2 - 4ac}}{\sqrt{4a^2}} = \dfrac{\sqrt{b^2 - 4ac}}{2a}$$

Write the solutions as follows.

$$x + \dfrac{b}{2a} = \dfrac{\sqrt{b^2 - 4ac}}{2a} \quad \text{or} \quad x + \dfrac{b}{2a} = \dfrac{-\sqrt{b^2 - 4ac}}{2a}$$

$$x = \dfrac{-b}{2a} + \dfrac{\sqrt{b^2 - 4ac}}{2a} \quad \text{or} \quad x = \dfrac{-b}{2a} - \dfrac{\sqrt{b^2 - 4ac}}{2a}$$

$$x = \dfrac{-b + \sqrt{b^2 - 4ac}}{2a} \quad \text{or} \quad x = \dfrac{-b - \sqrt{b^2 - 4ac}}{2a}$$

The solutions of the general quadratic equation $ax^2 + bx + c = 0$ (with $a \neq 0$) are

$$\dfrac{-b + \sqrt{b^2 - 4ac}}{2a} \quad \text{and} \quad \dfrac{-b - \sqrt{b^2 - 4ac}}{2a}.$$

For convenience, the solutions are often expressed in compact form by using the symbol ± (read "plus or minus"). The result is called the **quadratic formula,** a key result that should be memorized.

> The solutions of the quadratic equation $ax^2 + bx + c = 0$ are given by the **quadratic formula,**
>
> $$x = \dfrac{-b \pm \sqrt{b^2 - 4ac}}{2a}, \quad a \neq 0.$$

▶ The next example shows how to use the quadratic formula.

EXAMPLE 2 Use the quadratic formula to solve $2x^2 - 7x - 9 = 0$.

Match the coefficients of the variables with those of the general quadratic equation

$$ax^2 + bx + c = 0.$$

Here, $a = 2$, $b = -7$, and $c = -9$. Substitute these numbers into the quadratic formula and simplify the result.

$$x = \frac{-b \pm \sqrt{b^2 - 4ac}}{2a}$$

$$= \frac{-(-7) \pm \sqrt{(-7)^2 - 4(2)(-9)}}{2(2)} \quad \text{Let } a = 2, b = -7, c = -9$$

$$= \frac{7 \pm \sqrt{49 + 72}}{4} = \frac{7 \pm \sqrt{121}}{4}$$

Since $\sqrt{121} = 11$,

$$x = \frac{7 \pm 11}{4}.$$

Write the two separate solutions by first using the plus sign,

$$x = \frac{7 + 11}{4} = \frac{18}{4} = \frac{9}{2},$$

and then using the minus sign,

$$x = \frac{7 - 11}{4} = -\frac{4}{4} = -1.$$

The solutions of $2x^2 - 7x - 9 = 0$ are $\frac{9}{2}$ and -1. Check by substituting back in the original equation. ◀

Work Problem 3 at the side.

EXAMPLE 3 Solve $x^2 = 2x + 1$.

Find a, b, and c by rewriting the equation to get 0 on one side. Add $-2x - 1$ to both sides of the equation to get

$$x^2 - 2x - 1 = 0.$$

Then $a = 1$, $b = -2$, and $c = -1$. The solution is found by substituting these values into the quadratic formula.

$$x = \frac{-b \pm \sqrt{b^2 - 4ac}}{2a}$$

$$= \frac{-(-2) \pm \sqrt{(-2)^2 - 4(1)(-1)}}{2(1)} \quad \text{Let } a = 1, b = -2, c = -1$$

$$= \frac{2 \pm \sqrt{4 + 4}}{2} = \frac{2 \pm \sqrt{8}}{2}$$

Since $\sqrt{8} = \sqrt{4 \cdot 2} = \sqrt{4} \cdot \sqrt{2} = 2\sqrt{2}$,

$$x = \frac{2 \pm 2\sqrt{2}}{2}.$$

3. Solve by using the quadratic formula.

(a) $2x^2 + 3x - 5 = 0$

(b) $6p^2 + p = 1$

ANSWERS

3. (a) $1, -\frac{5}{2}$ (b) $-\frac{1}{2}, \frac{1}{3}$

10.3 Solving Quadratic Equations by the Quadratic Formula

4. Solve by using the quadratic formula.

   (a) $-y^2 = 8y + 1$

   (b) $4m^2 - 12m + 2 = 0$

   (c) $81k^2 - 36k + 4 = 0$

5. Solve each equation.

   (a) $9y^2 - 12y + 4 = 0$

   (b) $p^2 - 6p + 9 = 0$

   (c) $16m^2 - 24m + 9 = 0$

6. Solve by using the quadratic formula.

   (a) $m^2 = \frac{2}{3}m + \frac{4}{9}$

   (b) $x^2 - \frac{4}{3}x + \frac{2}{3} = 0$

Write these solutions in lowest terms by factoring $2 \pm 2\sqrt{2}$ as $2(1 \pm \sqrt{2})$ to get

$$x = \frac{2(1 \pm \sqrt{2})}{2} = 1 \pm \sqrt{2}.$$

The two solutions of the original equation are

$$1 + \sqrt{2} \quad \text{and} \quad 1 - \sqrt{2}. \blacktriangleleft$$

Work Problem 4 at the side.

▶ If the radicand in the quadratic formula, $b^2 - 4ac$, equals 0, then $-b$ and 0 will be added and subtracted, giving just $-b$ in the numerator. This produces only one rational number solution for the equation. Also, this means that the trinomial $ax^2 + bx + c$ is a perfect square.

EXAMPLE 4 Solve $4x^2 + 25 = 20x$.

Write the equation as

$$4x^2 - 20x + 25 = 0. \quad \text{Subtract } 20x$$

Here, $a = 4$, $b = -20$, and $c = 25$. By the quadratic formula,

$$x = \frac{-(-20) \pm \sqrt{(-20)^2 - 400}}{8} = \frac{20 \pm 0}{8} = \frac{5}{2}.$$

Since there is just one solution, the trinomial $4x^2 - 20x + 25$ is a perfect square. ◀

Work Problem 5 at the side.

▶ The final example shows how to solve quadratic equations with fractions.

EXAMPLE 5 Solve the equation

$$\frac{1}{10}t^2 = \frac{2}{5}t - \frac{1}{2}.$$

Eliminate the denominators by multiplying both sides of the equation by the common denominator, 10.

$$10\left(\frac{1}{10}t^2\right) = 10\left(\frac{2}{5}t - \frac{1}{2}\right)$$

$$t^2 = 4t - 5$$

Add $-4t$ and 5 to both sides of the equation to get

$$t^2 - 4t + 5 = 0.$$

From this form, identify $a = 1$, $b = -4$, and $c = 5$. Use the quadratic formula to complete the solution.

$$t = \frac{4 \pm \sqrt{4^2 - 4(1)(5)}}{2(1)} = \frac{4 \pm \sqrt{16 - 20}}{2} = \frac{4 \pm \sqrt{-4}}{2}$$

The radical $\sqrt{-4}$ is not a real number, so the equation has no real number solution. ◀

Work Problem 6 at the side.

ANSWERS

4. (a) $-4 + \sqrt{15}, -4 - \sqrt{15}$
   (b) $\frac{3 + \sqrt{7}}{2}, \frac{3 - \sqrt{7}}{2}$ (c) $\frac{2}{9}$

5. (a) $\frac{2}{3}$ (b) 3 (c) $\frac{3}{4}$

6. (a) $\frac{1 + \sqrt{5}}{3}, \frac{1 - \sqrt{5}}{3}$
   (b) no real number solution

500 Quadratic Equations

10.3 EXERCISES

For each equation, identify the letters a, b, and c of the general quadratic equation, $ax^2 + bx + c = 0$. Do not solve. See Example 1.

1. $3x^2 + 4x - 8 = 0$
   a = ____
   b = ____
   c = ____

2. $9x^2 + 2x - 3 = 0$
   a = ____
   b = ____
   c = ____

3. $2x^2 = 3x - 2$
   a = ____
   b = ____
   c = ____

4. $9x^2 - 2 = 4x$
   a = ____
   b = ____
   c = ____

5. $3x^2 - 8x = 0$
   a = ____
   b = ____
   c = ____

6. $5x^2 = 2x$
   a = ____
   b = ____
   c = ____

7. $(x - 3)(x + 4) = 0$
   a = ____
   b = ____
   c = ____

8. $(x + 6)^2 = 3$
   a = ____
   b = ____
   c = ____

9. $9(x - 1)(x + 2) = 8$
   a = ____
   b = ____
   c = ____

10. $(3x - 1)(2x + 5) = x(x - 1)$
    a = ____
    b = ____
    c = ____

Use the quadratic formula to solve each equation. Write all radicals in simplified form. Write answers in lowest terms. See Examples 2–4.

11. $p^2 + 2p - 2 = 0$

12. $6k^2 + 6k + 1 = 0$

13. $y^2 + 4y + 4 = 0$

14. $3r^2 - 5r + 1 = 0$

15. $z^2 = 13 - 12z$

16. $x^2 = 8x + 9$

17. $2w^2 + 12w + 5 = 0$

18. $k^2 = 20k - 19$

19. $5x^2 + 4x - 1 = 0$

20. $5n^2 + n - 1 = 0$

21. $2z^2 = 3z + 5$

22. $7r - 2r^2 + 30 = 0$

23. $z^2 + 6z + 9 = 0$

24. $x^2 - 2x + 1 = 0$

25. $4p^2 - 12p + 9 = 0$

26. $9r^2 + 6r + 1 = 0$

27. $5m^2 + 5m = 0$

28. $4y^2 - 8y = 0$

29. $6p^2 = 10p$

30. $3r^2 = 16r$

31. $m^2 - 20 = 0$

name date hour

32. $k^2 - 5 = 0$

33. $9r^2 - 16 = 0$

34. $4y^2 - 25 = 0$

35. $2x^2 + 2x + 4 = 4 - 2x$

36. $3x^2 - 4x + 3 = 8x - 1$

37. $2x^2 + x + 7 = 0$

38. $x^2 + x + 1 = 0$

39. $2x^2 = 3x - 2$

40. $x^2 = 5x - 20$

Use the quadratic formula to solve each equation. See Example 5.

41. $\dfrac{1}{2}x^2 = 1 - \dfrac{1}{6}x$

42. $\dfrac{3}{2}r^2 - r = \dfrac{4}{3}$

43. $\dfrac{2}{3}m^2 - \dfrac{4}{9}m - \dfrac{1}{3} = 0$

44. $\dfrac{3}{5}x - \dfrac{2}{5}x^2 = -1$

45. $\dfrac{r^2}{2} = r + \dfrac{1}{2}$

46. $\dfrac{m^2}{4} + \dfrac{3m}{2} + 1 = 0$

47. $k^2 = \dfrac{2k}{3} + \dfrac{2}{9}$

48. $\dfrac{2y^2}{7} + \dfrac{10}{7}y + 1 = 0$

49. $\dfrac{m^2}{2} = \dfrac{m}{2} - 1$

50. $\dfrac{3k^2}{8} - k = -\dfrac{17}{24}$

51. $2 = \dfrac{4}{x} - \dfrac{3}{x^2}$

52. $y^2 + \dfrac{8}{3}y = -\dfrac{7}{3}$

53. The time t in seconds for a car to skid 48 feet is given (approximately) by

$$48 = 64t - 16t^2.$$

Solve this equation for t. Are both answers reasonable?

54. The distance in feet an object falls in t seconds is given by

$$d = 16t^2.$$

Let $d = 400$ and solve for t. Are both answers reasonable?

SUPPLEMENTARY EXERCISES ON QUADRATIC EQUATIONS

Four methods have now been introduced for solving quadratic equations written in the form $ax^2 + bx + c = 0$. The chart below shows some advantages and some disadvantages of each method.

Method	Advantages	Disadvantages
1. Factoring	Usually the fastest method.	Not all equations are easily factorable. Some factorable equations are hard to factor.
2. Completing the square	Can always be used (also, the procedure is useful in other areas of mathematics).	Requires more steps than other methods.
3. Quadratic formula	Can always be used.	More difficult than factoring because of the $\sqrt{b^2 - 4ac}$ expression.
4. Square root method	Simplest method for solving equations of the form $(x + a)^2 = b$.	Few equations are given in this form.

Solve each quadratic equation by the best method.

1. $y^2 + 3y + 1 = 0$
2. $r^2 = 25$
3. $m^2 = 169$
4. $m^2 = 400$
5. $k^2 = \dfrac{49}{81}$
6. $p^2 = \dfrac{100}{81}$
7. $z^2 = 1.21$
8. $m^2 = 4.41$
9. $x^2 + 3x + 2 = 0$
10. $m^2 - 4m + 3 = 0$
11. $p^2 + 3p - 2 = 0$
12. $z^2 - 9z + 20 = 0$
13. $(2p - 1)^2 = 10$
14. $(3r - 2)^2 = 9$
15. $(3r + 2)^2 = 5$
16. $(5m + 1)^2 = 36$
17. $(y + 6)^2 = 121$
18. $(y - 2)^2 = 196$

19. $(7m - 1)^2 = 32$

20. $(3k - 7)^2 = 24$

21. $(5r - 7)^2 = -1$

22. $(5y - 3)^2 = -8$

23. $2x^2 - x = 1$

24. $2m^2 = 3m + 2$

25. $2p^2 + 1 = p$

26. $3r^2 + 8r = 3$

27. $8m^2 = 2m + 15$

28. $10m^2 + m = 2$

29. $3p^2 + 5p + 1 = 0$

30. $m^2 - 2m - 1 = 0$

31. $2q^2 + 3q = 1$

32. $y^2 + 6y + 4 = 0$

33. $5k^2 + 8 = 22k$

34. $z^2 + 12z + 25 = 0$

35. $3z^2 = 4z + 1$

36. $2m^2 = 2m + 1$

37. $4x^2 = 5x - 1$

38. $z^2 - 11z + 2 = 0$

39. $4x^2 + 5x = 1$

40. $9m^2 + 12m = 7$

41. $15t^2 + 58t + 48 = 0$

42. $4r^2 + 14r + 11 = 0$

43. $p^2 + 5p + 5 = 0$

44. $9k^2 = 48k + 64$

45. $k^2 - 3k - 3 = 0$

46. $\frac{1}{5}z^2 = z + \frac{14}{5}$

47. $3m^2 - 4m = 4$

48. $\frac{1}{2}r^2 = r + \frac{15}{2}$

49. $5k^2 + 17k = 12$

50. $m^2 = \frac{5}{12}m + \frac{1}{6}$

51. $2x^2 - 5x + 1 = 0$

52. $z^2 - z + 3 = 0$

53. $k^2 + \frac{4}{15}k = \frac{4}{15}$

54. $4m^2 - 11m + 10 = 0$

10.4 GRAPHING QUADRATIC EQUATIONS IN TWO VARIABLES

▶ As described in Chapter 7, the graph of a straight line shows all the solutions of a linear equation in two variables, such as $x + y = 4$, or $3x - 2y = 6$. Quadratic equations in two variables, of the form $y = ax^2 + bx + c$, are graphed in this section. Perhaps the simplest such quadratic equation is

$$y = x^2$$

(or $y = 1x^2 + 0x + 0$). The graph of this equation cannot be a straight line since only linear equations of the form

$$ax + by + c = 0$$

have graphs that are straight lines. However, $y = x^2$ can be graphed in much the same way that straight lines were graphed, by finding ordered pairs that satisfy the equation $y = x^2$.

EXAMPLE 1 Graph $y = x^2$.

Select several values for x; then find the corresponding y-values. For example, selecting $x = 2$ gives

$$y = 2^2 = 4,$$

with the point $(2, 4)$ on the graph of $y = x^2$. (Recall that in an ordered pair such as $(2, 4)$, the x-value comes first and the y-value second.)

Work Problem 1 at the side.

If the points from Problem 1 at the side are plotted on a coordinate system and a smooth curve drawn through them, the graph is as shown in Figure 1. ◀

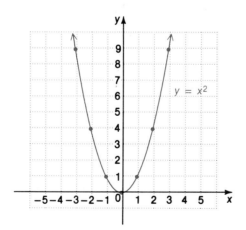

Figure 1

The curve in Figure 1 is called a **parabola.** The point $(0, 0)$, the lowest point on this graph, is called the **vertex** of the parabola.

Every expression of the form

$$y = ax^2 + bx + c,$$

with $a \neq 0$, has a graph that is a parabola. Because of its many useful properties, the parabola occurs frequently in real-life applications. For

Objectives

▶ Graph quadratic equations.

▶ Find the vertex of a parabola.

1. Complete the following ordered pairs for $y = x^2$.

x	y	(x, y)
3	___	(3,)
2	4	(2, 4)
1	___	(1,)
0	___	(0,)
−1	___	(−1,)
−2	___	(−2,)
−3	___	(−3,)

ANSWERS

1. (3, 9), (1, 1), (0, 0), (−1, 1), (−2, 4), (−3, 9)

10.4 Graphing Quadratic Equations in Two Variables

2. Graph each parabola.

(a) $y = -\dfrac{1}{2}x^2$

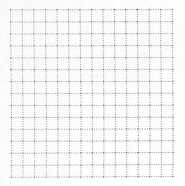

(b) $y = 3x^2$

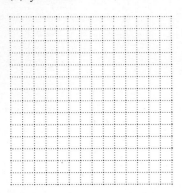

ANSWERS

2. (a)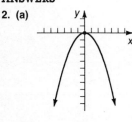

(b)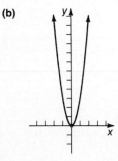

example, if an object is thrown into the air, the path that the object follows is a parabola (ignoring wind resistance). The cross sections of radar, spotlight, and telescope reflectors also form parabolas.

EXAMPLE 2 Graph the parabola $y = -x^2$.

We could select values for x and then find the corresponding y-values. But for a given x-value, the y-value will be the negative of the corresponding y-value of the parabola $y = x^2$ discussed above. Because of this, the new parabola has the same shape as the one in the preceding figure but is turned in the opposite direction, so that it opens downward, as shown in Figure 2. Here the vertex, $(0, 0)$, is the *highest* point on the graph. ◀

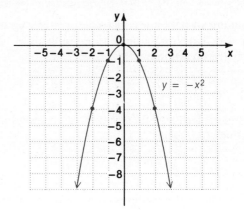

Figure 2

EXAMPLE 3 Graph the parabola $y = \tfrac{1}{2}x^2$.

For any value of x that might be chosen, the value of y will be $\tfrac{1}{2}$ the corresponding value of y from $y = x^2$. For this reason, the graph of $y = (\tfrac{1}{2})x^2$ will be *broader* than the graph of $y = x^2$. As shown in Figure 3, the graph opens upward and has $(0, 0)$ as vertex. ◀

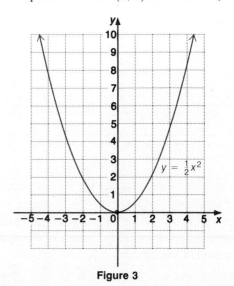

Figure 3

Work Problem 2 at the side.

▶ The next examples show how to find the vertex of a parabola from its equation.

EXAMPLE 4   Graph the parabola $y = (x - 2)^2$.

The vertex of this parabola is the point at which y has its largest or smallest value. Since $(x - 2)^2$ is always nonnegative, the value of y will be nonnegative. The smallest y-value, the y-value of the vertex, will be 0. Find the value of x at the vertex by letting $y = 0$ and solving for x.

$$y = (x - 2)^2$$
$$0 = (x - 2)^2 \quad \text{Let } y = 0$$
$$0 = x - 2$$
$$x = 2$$

The vertex is $(2, 0)$.

Find some other points on the graph by selecting values for x and finding the corresponding values of y. For example, if $x = -1$, then

$$y = (-1 - 2)^2 = (-3)^2 = 9.$$

Work Problem 3 at the side.

Plotting the points from Problem 3 gives the graph shown in Figure 4. This parabola has the same shape as $y = x^2$, but is shifted two units to the right, with vertex $(2, 0)$. ◀

Work Problem 4 at the side.

EXAMPLE 5   Graph the parabola $y = x^2 - 3$.

The smallest value of y will occur when $x = 0$, so that $y = 0 - 3$, or $y = -3$, making the vertex of this parabola $(0, -3)$. The graph is shown in Figure 5. This time the graph is shifted three units downward as compared with the graph of $y = x^2$. ◀

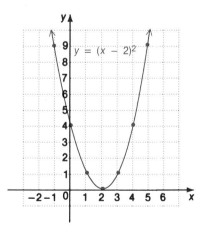

Figure 4

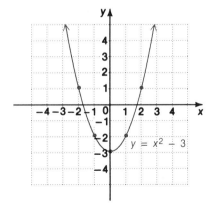

Figure 5

Example 4 showed a parabola shifted to the right and Example 5 showed a parabola shifted downward. The next example shows a combination of these two kinds of shifts.

3. Complete each ordered pair for $y = (x - 2)^2$.

(−1,)   (3,)
(0,)   (4,)
(1,)   (5,)
(2,)

4. Graph each parabola and identify the vertex. First draw the axes. (*Hint:* Place the x-axes near the bottom of the grid.)

(a) $y = (x - 3)^2$

(b) $y = (x + 3)^2$

ANSWERS

3. (−1, 9), (3, 1), (0, 4), (4, 4), (1, 1), (5, 9), (2, 0)

4.
(a)   (b)

(3, 0)   (−3, 0)

10.4   Graphing Quadratic Equations in Two Variables

509

5. Graph each parabola. Identify the vertex.

(a) $y = x^2 + 2$

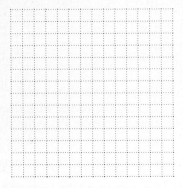

(b) $y = (x - 1)^2 + 3$

EXAMPLE 6 Graph $y = (x - 2)^2 - 3$.

As shown in Figure 6, the vertex is $(2, -3)$. Thus, the graph is the same shape as that of $y = x^2$, but is shifted two units to the right and three units downward. See Examples 4 and 5 to see why. ◀

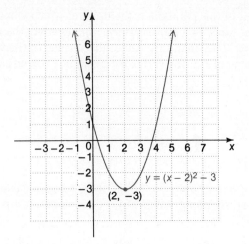

Figure 6

Work Problem 5 at the side.

These results on graphing parabolas are summarized below.

> The graph of
> $$y = a(x - h)^2 + k$$
> is a parabola with vertex at (h, k).
> The parabola opens upward if $a > 0$ and downward if $a < 0$.
> The parabola is broader than $y = x^2$ if $0 < |a| < 1$, and the parabola is narrower than $y = x^2$ if $|a| > 1$.

ANSWERS

5. (a)

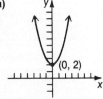

(b)

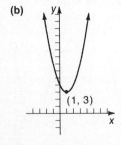

Quadratic Equations

10.4 EXERCISES

Sketch the graph of each equation. Identify each vertex. See Examples 1–6.

1. $y = 2x^2$

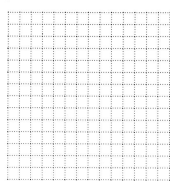

2. $y = -2x^2$

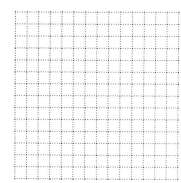

3. $y = (x + 1)^2$

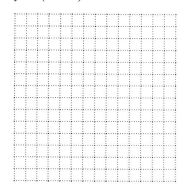

4. $y = (x - 2)^2$

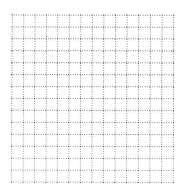

5. $y = -(x + 1)^2$

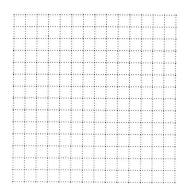

6. $y = -(x - 2)^2$

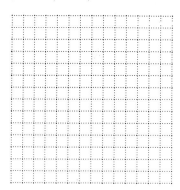

7. $y = x^2 + 1$

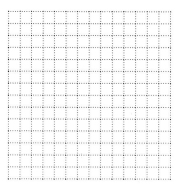

8. $y = -x^2 - 2$

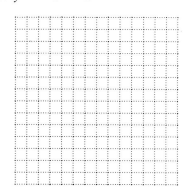

9. $y = 2 - x^2$

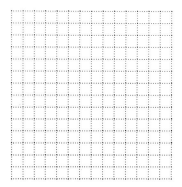

10.4 Exercises 511

10. $y = -4 + x^2$

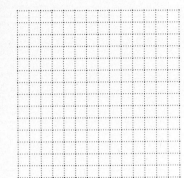

11. $y = \dfrac{1}{2}x^2 + 2$

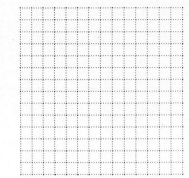

12. $y = -\dfrac{1}{3}x^2 + 1$

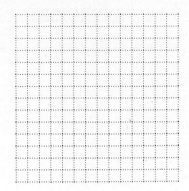

13. $y = (x + 1)^2 + 2$

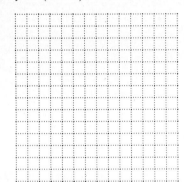

14. $y = (x - 2)^2 - 1$

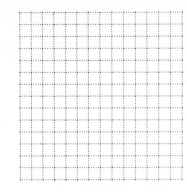

15. $y = (x - 4)^2 - 1$

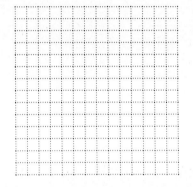

16. $y = (x + 3)^2 + 3$

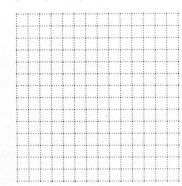

17. $y = 1 - (x + 2)^2$

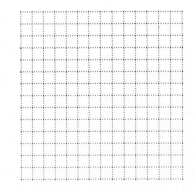

18. $y = 4 - (x - 3)^2$

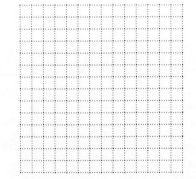

name date hour

CHAPTER 10   REVIEW EXERCISES

(10.1)   *Solve each equation by using the square root property. Express all radicals in simplest form.*

1. $y^2 = 49$

2. $x^2 = 15$

3. $m^2 = 48$

4. $(k + 2)^2 = 9$

5. $(r - 3)^2 = 7$

6. $(2p + 1)^2 = 11$

7. $(3k + 2)^2 = 12$

8. $\left(x + \dfrac{1}{2}\right)^2 = \dfrac{3}{4}$

(10.2)   *Solve each equation by completing the square.*

9. $m^2 + 6m + 5 = 0$

10. $p^2 + 4p = 7$

11. $-x^2 + 5 = 2x$

Chapter 10   Review Exercises   **513**

12. $2y^2 + 8y = 3$ **13.** $5k^2 - 3k - 2 = 0$ **14.** $(4a + 1)(a - 1) = -3$

(10.3)   *Use the quadratic formula to solve each equation.*

15. $x^2 - 2x - 4 = 0$ **16.** $-m^2 + 3m + 5 = 0$

17. $3k^2 + 2k + 3 = 0$ **18.** $5p^2 = p + 1$

19. $2p^2 - 3 = 4p$ **20.** $-4a^2 + 7 = 2a$

21. $\dfrac{c^2}{4} = 2 - \dfrac{3}{4}c$ **22.** $\dfrac{3}{2}r^2 = \dfrac{1}{2} - r$

Use the most suitable of the following methods to solve each quadratic equation: factoring, the square root property, completing the square, or the quadratic formula.

23. $(2t - 1)(t + 5) = 0$

24. $(2p + 1)^2 = 4$

25. $(k + 2)(k - 1) = 3$

26. $6t^2 + 7t - 3 = 0$

27. $2x^2 + 3x + 2 = x^2 - 2x$

28. $x^2 + 2x + 1 = 3$

29. $m^2 - 4m + 5 = 0$

30. $k^2 - 9k + 10 = 0$

31. $(3x + 5)(3x + 5) = 7$

32. $\dfrac{1}{2}r^2 = \dfrac{7}{2} - r$

33. $x^2 + 4x - 1 = 0$

34. $7x^2 - 8 = 5x^2 + 16$

Chapter 10 Review Exercises

(10.4) *Sketch the graph of each equation. Identify each vertex.*

35. $y = 3x^2$

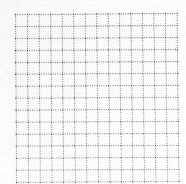

36. $y = -3x^2$

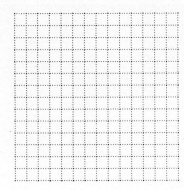

37. $y = -x^2 + 3$

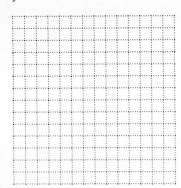

38. $y = x^2 - 1$

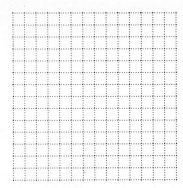

39. $y = -(x - 1)^2 + 4$

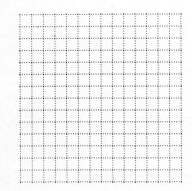

40. $y = (x + 2)^2 - 3$

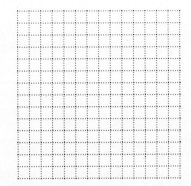

name

CHAPTER 10   TEST

Solve by completing the square.

1. $x^2 = 3x - 2$

2. $2x^2 - 5x = 0$

Solve by using the square root property.

3. $x^2 = 5$

4. $(k - 3)^2 = 2$

5. $(3r - 2)^2 = 72$

Solve by the quadratic formula.

6. $m^2 = 3m + 10$

7. $2k^2 + 5k = 3$

8. $3z^2 + 2 = 7z$

9. $4n^2 + 8n + 5 = 0$

10. $y^2 - \frac{5}{3}y + \frac{1}{3} = 0$

Solve by the best method.

11. $y^2 - 2y = 1$

12. $(2x - 1)^2 = 18$

13. _____ 13. $(x - 5)(3x + 2) = 0$

14. _____ 14. $(x - 5)(3x - 2) = 4$

15. _____ 15. $(x - 5)^2 = 8$

16. _____ 16. $x^2 = 6x - 2$

17. _____ 17. $p^2 + 16 = 8p$

18. _____ 18. $2x^2 = 3x + 5$

Sketch the graph of each equation.

19. 19. $y = -(x + 3)^2$

20. 20. $y = (x - 4)^2 + 1$

FINAL EXAMINATION

Perform each operation, wherever possible.

1. $\dfrac{-4 \cdot 3 + 6}{2 - 4 \cdot 1}$

2. $-9 - (-8)(2) + 6 - (3 + 5)$

3. $8 \div 4 \cdot 3 - 1$

4. $\dfrac{4 \cdot 2 - 4 \cdot 7 - 3(2 + 5)}{3 \cdot 7 - 4 \cdot 5 + (-1)1}$

Identify all numbers from the list $-7, \frac{0}{8}, \sqrt{7}, \frac{3}{8}, \frac{12}{4}, \sqrt{-6}, -\sqrt{4}$ *that belong to each of the following categories.*

5. Whole numbers

6. Integers

7. Rational numbers

8. Irrational numbers

9. Real numbers

10. Not real numbers

Solve each equation.

11. $6x - 5 = 7$

12. $3k - 9k - 8k + 6 - 3 = 31$

13. $2(m - 1) - 6(3 - m) + 5 = 1$

14. $9k - 3(k - 6) + 4 + 5k = 0$

15. Solve the formula $I = prt$ for t.

16. Solve the formula $A = \dfrac{1}{2}bh$ for b.

Final Examination

Solve each inequality. Graph the solution.

17. $-8m < 16$

18. $-9p + 2(8 - p) - 6 \geq 3p - 14 - 4$

Simplify each of the following. Leave answers in exponential form, with positive exponents.

19. $\dfrac{6^{12} \cdot 6^8}{6^9 \cdot 6^{10}}$

20. $\dfrac{(5^2)^3 (5^3)^4}{5^{12} 5^8}$

21. 7^{-2}

Perform the indicated operations.

22. $(5x^5 - 9x^4 + 8x^2) - (-3x^5 + 8x^4 - 9x^2)$

23. $(6r + 1)^2$

24. $\dfrac{3x^3 + 10x^2 - 7x + 4}{x + 4}$

Factor each of the following as completely as possible.

25. $9x - x^2$

26. $x^2 - 6x - 16$

27. $2a^2 - 5a - 3$

28. $25m^2 - 20m + 4$

Perform the following operations. Write all answers in lowest terms.

29. $\dfrac{2}{a - 3} \div \dfrac{5}{2a - 6}$

30. $\dfrac{1}{k} - \dfrac{2}{k - 1}$

31. $\dfrac{2}{a^2 - 4} + \dfrac{3}{a^2 - 4a + 4}$

32. $\dfrac{2 + \dfrac{1}{x}}{3 - \dfrac{1}{x}}$

33. Tom can pick a box of oranges in 8 minutes, Sam in 6. How long would it take them if they worked together?

33. _____

Graph each of the following.

34. $2x + 3y = 6$

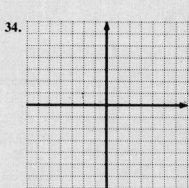

34.

35. $5x - 3y = 15$

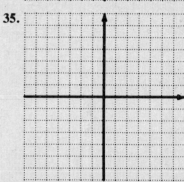

35.

36. $x = -4$

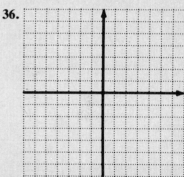

36.

37. $2x - 5y < 10$

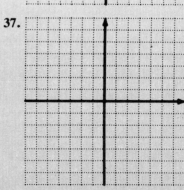

37.

38. Write an equation for the line through $(-1, 4)$ and $(5, 2)$. Give the slope of the line.

38. _____

Final Examination

Solve each system of equations.

39. $2x + y = -4$
    $-3x + 2y = 13$

40. $3x - 5y = 8$
    $-6x + 10y = 16$

41. The sum of two numbers is 82. One number is 5 less than twice the other. Find the two numbers.

Simplify each of the following as much as possible.

42. $\sqrt{100}$

43. $-\sqrt{841}$

44. $\dfrac{5\sqrt{3}}{\sqrt{2}}$

45. $3\sqrt{5} - 2\sqrt{20}$

46. $-2\sqrt{10} + 3\sqrt{40}$

Solve each quadratic equation.

47. $2x^2 - 5x + 3 = 0$

48. $m^2 - 2m - 1 = 0$

49. $2a^2 - 2a = 1$

50. Graph the parabola $y = x^2 - 4$.

Quadratic Equations

Appendices

APPENDIX A SETS

▶ A **set** is a collection of things. The objects in a set are called the **elements** of the set. A set is represented by listing its elements between **set braces,** { }. The order in which the elements of a set are listed is unimportant.

EXAMPLE 1 Represent the following sets by listing the elements.

(a) The set of states in the United States that border on the Pacific Ocean = {California, Oregon, Washington, Hawaii, Alaska}.

(b) The set of all counting numbers less than 6 = {1, 2, 3, 4, 5}.

(c) The set of all coins currently issued in the United States = {penny, nickel, dime, quarter, half-dollar, dollar}. ◀

Work Problem 1 at the side.

▶ Capital letters are used to name sets. To state that 5 is an element of

$$S = \{1, 2, 3, 4, 5\},$$

write $5 \in S$. The statement $6 \notin S$ means that 6 is not an element of S.

Work Problem 2 at the side.

A set with no elements is called the **empty set,** or the **null set.** The symbols $\emptyset$ or { } are used for the empty set. If we let A be the set of all cats that fly, then A is the empty set.

$$A = \emptyset \quad \text{or} \quad A = \{\ \}$$

Do not make the common error of writing the empty set as $\{\emptyset\}$.

In any discussion of sets, there is some set that includes all the elements under consideration. This set is called the **universal set** for that situation. For example, if the discussion is about presidents of the United States, then the set of all presidents of the United States is the universal set. The universal set is denoted U.

▶ In Example 1, there are five elements in the set in part (a), five in part (b), and six in part (c). If the number of elements in a set is either 0 or a counting number, then the set is **finite.** On the other hand, the set of natural numbers, for example, is an **infinite set,** because there is no final number. We can list the elements of the set of natural numbers as

$$N = \{1, 2, 3, 4, \ldots\},$$

where the three dots indicate that the set continues indefinitely. Not all infinite sets can be listed in this way. For example, there is no way to list the elements in the set of all real numbers between 1 and 2.

Objectives

▶ List the elements of a set.
▶ Learn the vocabulary and symbols used to discuss sets.
▶ Decide whether a set is finite or infinite.
▶ Decide whether a given set is a subset of another set.
▶ Find the complement of a set.
▶ Find the union and the intersection of two sets.

1. Represent the following sets by listing the elements.

   (a) The set of whole numbers between 2.5 and 4.8

   (b) The set of all the days of the week

2. Decide whether each statement is true or false for the set $T = \{m, n, p, q\}$.

   (a) $m \in T$

   (b) $n \in T$

   (c) $k \notin T$

   (d) $h \notin T$

ANSWERS
1. (a) {3, 4} (b) {Sunday, Monday, Tuesday, Wednesday, Thursday, Friday, Saturday}
2. (a) true (b) true (c) true (d) true

Appendix A Sets   523

3. List the elements of each set, if possible. Decide whether each set is finite or infinite.

   (a) The set of integers between -2 and 2

   (b) The set of all even integers

   (c) The set of all real numbers between 0 and 1

4. Let $P = \{5, 10, 15\}$, $Q = \{5\}$, $R = \{10, 15\}$, and $S = \{15, 10, 5\}$. Use $=$, $\subset$, or $\not\subset$ to make each statement true.

   (a) $Q \quad\quad P$

   (b) $R \quad\quad P$

   (c) $S \quad\quad P$

   (d) $P \quad\quad Q$

5. Find all subsets of $\{2, 4, 10\}$.

ANSWERS

3. (a) $\{-1, 0, 1\}$; finite
   (b) $\{\ldots, -2, 0, 2, 4, \ldots\}$; infinite
   (c) cannot be listed; infinite
4. (a) $\subset$ (b) $\subset$ (c) $=$ (d) $\not\subset$
5. $\{2, 4, 10\}$, $\{2, 4\}$, $\{2, 10\}$, $\{4, 10\}$, $\{2\}$, $\{4\}$, $\{10\}$, $\emptyset$

EXAMPLE 2 List the elements of each set, if possible. Decide whether each set is finite or infinite.

(a) The set of all integers

One way to list the elements is $\{\ldots, -2, -1, 0, 1, 2, \ldots\}$. The set is infinite.

(b) The set of all natural numbers between 0 and 5

$\{1, 2, 3, 4\}$ The set is finite.

(c) The set of all rational numbers.

This is an infinite set whose elements cannot be listed. ◄

Work Problem 3 at the side.

Two sets are **equal** if they have exactly the same elements. Thus, the set of natural numbers and the set of positive integers are equal sets. Also, the sets

$$\{1, 2, 4, 7\} \quad \text{and} \quad \{4, 2, 7, 1\}$$

are equal. The order of the elements does not make a difference.

▶ If all the elements of a set A are also elements of some new set B, then we say A is a **subset** of B, written $A \subset B$. We use the symbol $A \not\subset B$ to mean that A is not a subset of B.

EXAMPLE 3 Let $A = \{1, 2, 3, 4\}$, $B = \{1, 4\}$, and $C = \{1\}$. Then $B \subset A$, $C \subset A$, and $C \subset B$, but $A \not\subset B$, $A \not\subset C$, and $B \not\subset C$. ◄

Work Problem 4 at the side.

The set $M = \{a, b\}$ has four subsets: $\{a, b\}$, $\{a\}$, $\{b\}$, and $\emptyset$. The empty set is a subset of any set. How many subsets does $N = \{a, b, c\}$ have? There is one subset with 3 elements: $\{a, b, c\}$. There are three subsets with 2 elements:

$$\{a, b\}, \quad \{a, c\}, \quad \text{and} \quad \{b, c\}.$$

There are three subsets with 1 element:

$$\{a\}, \quad \{b\}, \quad \text{and} \quad \{c\}.$$

There is one subset with 0 elements: $\emptyset$. Thus, set N has eight subsets.

In general,

> a set with n elements has 2^n subsets.

Work Problem 5 at the side.

To illustrate the relationships between sets, **Venn diagrams** are often used. A rectangle represents the universal set, U. The sets under discussion are represented by regions within the rectangle. The Venn diagram in Figure 1 shows that $B \subset A$.

▶ For every set A, there is a set A', the **complement** of A, that contains all the elements of U that are not in A. The shaded region in the Venn diagram in Figure 2 represents A'.

Appendices

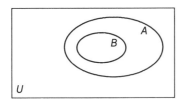

Figure 1   Figure 2

EXAMPLE 4   Given $U = \{a, b, c, d, e, f, g\}$, $A = \{a, b, c\}$, $B = \{a, d, f, g\}$, and $C = \{d, e\}$, find A', B', and C'.

(a) $A' = \{d, e, f, g\}$

(b) $B' = \{b, c, e\}$

(c) $C' = \{a, b, c, f, g\}$ ◀

Work Problem 6 at the side.

▶ The **union** of two sets A and B, written $A \cup B$, is the set of all elements of A together with all elements of B. Thus, for the sets in Example 4,

$$A \cup B = \{a, b, c, d, f, g\}$$

and

$$A \cup C = \{a, b, c, d, e\}.$$

In Figure 3 the shaded region is the union of sets A and B.

EXAMPLE 5   If $M = \{2, 5, 7\}$ and $N = \{1, 2, 3, 4, 5\}$, then

$$M \cup N = \{1, 2, 3, 4, 5, 7\}. \blacktriangleleft$$

Work Problem 7 at the side.

The **intersection** of two sets A and B, written $A \cap B$, is the set of all elements that belong to both A and B. For example, if

$$A = \{\text{Jose, Ellen, Marge, Kevin}\}$$

and

$$B = \{\text{Jose, Patrick, Ellen, Sue}\},$$

then

$$A \cap B = \{\text{Jose, Ellen}\}.$$

The shaded region in Figure 4 represents the intersection of the two sets A and B.

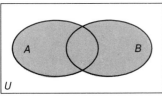   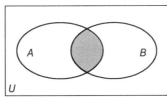

Figure 3   Figure 4

6. Given $U = \{1, 2, 3, 4, 5, 6\}$, $A = \{2, 4, 6\}$, $B = \{1, 2, 3, 4\}$, and $C = \{1, 2, 5, 6\}$, find A', B', and C'.

7. Given $A = \{2, 4, 6\}$ and $B = \{1, 2, 3, 4\}$, find $A \cup B$.

ANSWERS

6. $A' = \{1, 3, 5\}$, $B' = \{5, 6\}$, $C' = \{3, 4\}$

7. $\{1, 2, 3, 4, 6\}$

Appendix A   Sets

8. Let $S = \{a, b, c, d, e, f\}$, $T = \{a, c, k\}$, and $W = \{d, g, h\}$. Find the following.

   (a) $S \cap T$

   (b) $S \cap W$

   (c) $T \cap W$

9. Use the sets in Example 7 to find the following.

   (a) $A \cap C$

   (b) $A \cup C$

   (c) C'

EXAMPLE 6 Suppose that $P = \{3, 9, 27\}$, $Q = \{2, 3, 10, 18, 27, 28\}$, and $R = \{2, 10, 28\}$.

(a) $P \cap Q = \{3, 27\}$

(b) $Q \cap R = \{2, 10, 28\} = R$

(c) $P \cap R = \emptyset$ ◀

Sets like P and R in Example 6 that have no elements in common are called **disjoint sets.** The Venn diagram in Figure 5 shows a pair of disjoint sets.

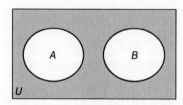

Figure 5

Work Problem 8 at the side.

EXAMPLE 7 Let $U = \{2, 5, 7, 10, 14, 20\}$, $A = \{2, 10, 14, 20\}$, $B = \{5, 7\}$, and $C = \{2, 5, 7\}$. Find each of the following.

(a) $A \cup B = \{2, 5, 7, 10, 14, 20\}$

(b) $A \cap B = \emptyset$

(c) $B \cup C = \{2, 5, 7\} = C$

(d) $B \cap C = \{5, 7\} = B$

(e) $A' = \{5, 7\}$ ◀

Work Problem 9 at the side.

ANSWERS
8. (a) {a, c} (b) {d} (c) ∅
9. (a) {2} (b) {2, 5, 7, 10, 14, 20}
   (c) {10, 14, 20}

name   date   hour

APPENDIX A   EXERCISES

List the elements of each of the following sets. See Examples 1 and 2.

1. The set of all natural numbers less than 8

2. The set of all integers between 4 and 10

3. The set of seasons

4. The set of months of the year

5. The set of women presidents of the United States

6. The set of all living humans more than 200 years old

7. The set of letters of the alphabet between K and M

8. The set of letters of the alphabet between D and H

9. The set of positive even numbers

10. The set of multiples of 5

11. Which of the sets described in Exercises 1–10 are infinite sets?

12. Which of the sets described in Exercises 1–10 are finite sets?

Tell whether each of the following is true or false.

13. $5 \in \{1, 2, 5, 8\}$

14. $6 \in \{1, 2, 3, 4, 5\}$

15. $2 \in \{1, 3, 5, 7, 9\}$

16. $1 \in \{6, 2, 5, 1\}$

17. $7 \notin \{2, 4, 6, 8\}$

18. $7 \notin \{1, 3, 5, 7\}$

19. $\{2, 4, 9, 12, 13\} = \{13, 12, 9, 4, 2\}$

20. $\{7, 11, 4\} = \{7, 11, 4, 0\}$

Let $A = \{1, 3, 4, 5, 7, 8\}$
$B = \{2, 4, 6, 8\}$
$C = \{1, 3, 5, 7\}$
$D = \{1, 2, 3\}$
$E = \{3, 7\}$
$U = \{1, 2, 3, 4, 5, 6, 7, 8, 9, 10\}$.

Tell whether each of the following is true or false. See Examples 3, 5, 6, and 7.

21. $A \subset U$

22. $D \subset A$

23. $\emptyset \subset A$

24. $\{1, 2\} \subset D$

25. $C \subset A$

26. $A \subset C$

27. $D \subset B$

28. $E \subset C$

Appendix A   Exercises   527

29. $D \not\subseteq E$

30. $E \not\subseteq A$

31. There are exactly 4 subsets of E.

32. There are exactly 8 subsets of D.

33. There are exactly 12 subsets of C.

34. There are exactly 16 subsets of B.

35. $\{4, 6, 8, 12\} \cap \{6, 8, 14, 17\} = \{6, 8\}$

36. $\{2, 5, 9\} \cap \{1, 2, 3, 4, 5\} = \{2, 5\}$

37. $\{3, 1, 0\} \cap \{0, 2, 4\} = \{0\}$

38. $\{4, 2, 1\} \cap \{1, 2, 3, 4\} = \{1, 2, 3\}$

39. $\{3, 9, 12\} \cap \emptyset = \{3, 9, 12\}$

40. $\{3, 9, 12\} \cup \emptyset = \emptyset$

41. $\{4, 9, 11, 7, 3\} \cup \{1, 2, 3, 4, 5\}$
$= \{1, 2, 3, 4, 5, 7, 9, 11\}$

42. $\{1, 2, 3\} \cup \{1, 2, 3\} = \{1, 2, 3\}$

43. $\{3, 5, 7, 9\} \cup \{4, 6, 8\} = \emptyset$

44. $\{5, 10, 15, 20\} \cup \{5, 15, 30\} = \{5, 15\}$

Let $U = \{a, b, c, d, e, f, g, h\}$
$A = \{a, b, c, d, e, f\}$
$B = \{a, c, e\}$
$C = \{a, f\}$
$D = \{d\}$

List the elements in the following sets. See Examples 4–7.

45. A'

46. B'

47. C'

48. D'

49. $A \cap B$

50. $B \cap A$

51. $A \cap D$

52. $B \cap D$

53. $B \cap C$

54. $A \cup B$

55. $B \cup D$

56. $B \cup C$

57. $C \cup B$

58. $C \cup D$

59. $A \cap \emptyset$

60. $B \cup \emptyset$

61. Name every pair of disjoint sets among A–D above.

APPENDIX B
COMPLEX SOLUTIONS OF EQUATIONS

In Section 10.3, when we used the quadratic formula to solve an equation, we sometimes found that an equation had no real number solution. For example, the solution

$$x = \frac{-3 + \sqrt{-5}}{2}$$

does not represent a real number because of $\sqrt{-5}$. For every quadratic equation to have a solution, a larger set of numbers is needed.

▶ To find this larger set of numbers, a new number i is defined such that

$$i^2 = -1.$$

This means that $i = \sqrt{-1}$. Further, we now can write numbers like $\sqrt{-5}$, $\sqrt{-4}$, and $\sqrt{-8}$ as multiples of i:

$$\sqrt{-5} = \sqrt{-1 \cdot 5} = \sqrt{-1} \cdot \sqrt{5} = i\sqrt{5},$$
$$\sqrt{-4} = \sqrt{-1} \cdot \sqrt{4} = i\sqrt{4} = i \cdot 2 = 2i,$$
$$\sqrt{-8} = i\sqrt{8} = i \cdot 2 \cdot \sqrt{2} = 2i\sqrt{2}.$$

(Write $i\sqrt{5}$ rather than $\sqrt{5}i$ to avoid confusion with $\sqrt{5i}$.)

Work Problem 1 at the side.

▶ Numbers that are multiples of i are called **imaginary numbers.** The *complex numbers* include all real numbers and all imaginary numbers.

> A complex number is a number of the form $a + bi$, where a and b are real numbers.

For example, the real number 2 is a complex number since it can be written as $2 + 0i$. Also, the imaginary number $3i = 0 + 3i$ is a complex number. Other complex numbers are formed by adding (or subtracting) real and imaginary numbers. Some examples are

$$3 - 2i, \quad 1 + i\sqrt{2}, \quad \text{and} \quad -5 + 4i.$$

A complex number written in the form $a + bi$ (or $a + ib$) is said to be in **standard form.**

EXAMPLE 1 Write the following complex numbers in standard form. Decide whether each real number is *real, imaginary,* or *neither.*

(a) $-\frac{3}{4}$

In standard form $-\frac{3}{4} = -\frac{3}{4} + 0i$. It is a real number.

(b) $5i$

This number is imaginary and is written $0 + 5i$ in standard form.

Objectives

▶ Write imaginary numbers as multiples of i.

▶ Write a complex number in standard form

▶ Add and subtract complex numbers.

▶ Multiply complex numbers.

▶ Write quotients in standard form.

1. Write as multiples of i.

  (a) $\sqrt{-9}$

  (b) $\sqrt{-7}$

  (c) $\sqrt{-12}$

ANSWERS
1. (a) $3i$ (b) $i\sqrt{7}$ (c) $2i\sqrt{3}$

2. Write each number in standard form. Decide whether each number is real or imaginary.

   (a) $-i\sqrt{3}$

   (b) $\sqrt{7}$

   (c) $6 + 2i$

3. What values of x and y will make each pair of complex numbers equal?

   (a) $2 + xi$ and $y - 4i$

   (b) $-x - 5i$ and $3 + 7yi$

4. Add each pair of complex numbers.

   (a) $5 - 2i$ and $1 + 3i$

   (b) $2 + 5i$ and $6 - 8i$

(c) $1 + i\sqrt{10}$

This complex number is neither real nor imginary. It is already in standard form. ◀

Work Problem 2 at the side.

The number a is the **real part** of the complex number $a + bi$, while b is the **imaginary part.** Two complex numbers are **equal** if their real parts are equal and their imaginary parts are equal. For real numbers a, b, c, and d,

$$a + bi = c + di \quad \text{if} \quad a = c \quad \text{and} \quad b = d.$$

EXAMPLE 2   What values of x and y will make $2 + xi = y + 3i$?

From the definition above,

$$2 + xi = y + 3i$$

if $2 = y$ and $x = 3$. ◀

Work Problem 3 at the side.

▶ To add two complex numbers, add the real parts and add the imaginary parts. For real numbers a, b, c, and d,

$$(a + bi) + (c + di) = (a + c) + (b + d)i.$$

EXAMPLE 3   Add $2 - 6i$ and $7 + 4i$.

Use the rule given above.

$$(2 - 6i) + (7 + 4i) = (2 + 7) + (-6 + 4)i$$
$$= 9 + (-2)i$$
$$= 9 - 2i \blacktriangleleft$$

Work Problem 4 at the side.

The negative of a complex number $a + bi$ is $-a - bi$. To subtract complex numbers, as with real numbers, add the negative. For real numbers a, b, c, and d,

$$(a + bi) - (c + di) = (a + bi) + (-c - di).$$

EXAMPLE 4   Find the negative of $4 - 7i$.

The negative of $4 - 7i$ is $-4 + 7i$. Note that

$$(4 - 7i) + (-4 + 7i) = 0 + 0i = 0. \blacktriangleleft$$

The sum of a complex number and its negative is 0, just as with the sum of a real number and its negative.

EXAMPLE 5   Subtract.

(a) $(2 + 6i) - (-4 + i) = (2 + 6i) + (4 - i)$
$$= (2 + 4) + (6 - 1)i = 6 + 5i$$

ANSWERS

2. (a) $0 - i\sqrt{3}$; imaginary
   (b) $\sqrt{7} + 0i$; real   (c) $6 + 2i$; neither
3. (a) $x = -4, y = 2$
   (b) $x = -3, y = -\dfrac{5}{7}$
4. (a) $6 + i$ (b) $8 - 3i$

(b) $(-1 + i) - (2 - 7i) = (-1 + i) + (-2 + 7i)$
$= (-1 - 2) + (1 + 7)i$
$= -3 + 8i$ ◀

Work Problem 5 at the side.

▶ Multiplication of complex numbers is handled in the same way as multiplication of polynomials.

EXAMPLE 6 Find the following products.

(a) $3i(2 - 5i) = 6i - 15i^2$ Distributive property
$= 6i - 15(-1)$ Recall that $i^2 = -1$ by definition
$= 6i + 15$
$= 15 + 6i$

The last step gives the result in standard form.

(b) $(4 - 3i)(2 + 5i) = 4(2) + 4(5i) + (-3i)(2) + (-3i)(5i)$
$= 8 + 20i - 6i - 15i^2$ Use FOIL
$= 8 + 14i - 15(-1)$
$= 8 + 14i + 15$
$= 23 + 14i$

(c) $(4 + 3i)(4 - 3i) = 16 - 12i + 12i - 9i^2$
$= 16 - 9(-1)$
$= 16 + 9$
$= 25$ ◀

Work Problem 6 at the side.

▶ Given the quotient of two complex numbers such as

$$\frac{2 - 5i}{4 + 3i},$$

we want to express it in standard form. This means that the denominator must be a real number. In Example 6(c), we saw that the product

$$(4 + 3i)(4 - 3i)$$

was 25, a real number. Because of this, multiply numerator and denominator of the given quotient by $4 - 3i$, as follows.

$$\frac{2 - 5i}{4 + 3i} \cdot \frac{4 - 3i}{4 - 3i} = \frac{8 - 6i - 20i + 15i^2}{16 - 9i^2}$$
$$= \frac{8 - 26i + 15(-1)}{16 - 9(-1)}$$
$$= \frac{8 - 26i - 15}{16 + 9}$$
$$= \frac{-7 - 26i}{25}$$
$$= \frac{-7}{25} - \frac{26}{25}i$$

5. (a) Find the negative of $8 - 3i$.

(b) Subtract $8 - 3i$ from $-2 + i$.

(c) Find $(6 + 5i) - (4 + 2i)$.

6. Find the following products.

(a) $-i(1 + 2i)$

(b) $4i(-5 + i)$

(c) $(1 + 2i)(-3 + 5i)$

(d) $(8 - 3i)(8 + 3i)$

ANSWERS
5. (a) $-8 + 3i$ (b) $-10 + 4i$
(c) $2 + 3i$
6. (a) $2 - i$ (b) $-4 - 20i$
(c) $-13 - i$ (d) 73

Appendix B   Complex Solutions of Equations

7. Write the quotients in standard form.

(a) $\dfrac{2+3i}{1+i}$

(b) $\dfrac{15-3i}{2i}$

8. Solve each equation

(a) $x^2 - 2x + 3 = 0$

(b) $2p^2 + 6p + 5 = 0$

(c) $r^2 + 41 = 10r$

ANSWERS

7. (a) $\dfrac{5}{2} + \dfrac{1}{2}i$ (b) $-\dfrac{3}{2} - \dfrac{15}{2}i$

8. (a) $1 + i\sqrt{2}, 1 - i\sqrt{2}$

(b) $\dfrac{-3+i}{2}, \dfrac{-3-i}{2}$

(c) $5 + 4i, 5 - 4i$

The last step gives the result in standard form. Recall that this is the same method used to rationalize certain radical expressions in Chapter 9. The complex numbers $4 + 3i$ and $4 - 3i$ are **conjugates**.

EXAMPLE 7 Write the following quotients in standard form.

(a) $\dfrac{-4+i}{2-i}$

Multiply by the conjugate, $2 + i$.

$$\dfrac{-4+i}{2-i} \cdot \dfrac{2+i}{2+i} = \dfrac{-8 - 4i + 2i + i^2}{4 - i^2}$$

$$= \dfrac{-8 - 2i - 1}{4 - (-1)}$$

$$= \dfrac{-9 - 2i}{5}$$

$$= -\dfrac{9}{5} - \dfrac{2}{5}i$$

(b) $\dfrac{3+i}{-i}$

Here, the conjugate of $0 - i$ is $0 + i$, or i.

$$\dfrac{3+i}{-i} \cdot \dfrac{i}{i} = \dfrac{3i + i^2}{-i^2} = \dfrac{-1 + 3i}{-(-1)} = -1 + 3i$$ ◀

Work Problem 7 at the side.

The next example shows how to find complex solutions of an equation.

EXAMPLE 8 Solve $2p^2 = 4p - 5$.

Write the equation as $2p^2 - 4p + 5 = 0$. Then $a = 2$, $b = -4$, and $c = 5$. The solutions are

$$p = \dfrac{-(-4) \pm \sqrt{(-4)^2 - 4(2)(5)}}{2(2)}$$

$$= \dfrac{4 \pm \sqrt{16 - 40}}{4} = \dfrac{4 \pm \sqrt{-24}}{4}.$$

Since $\sqrt{-24} = i\sqrt{24} = i \cdot \sqrt{4} \cdot \sqrt{6} = i \cdot 2 \cdot \sqrt{6} = 2i\sqrt{6}$,

$$p = \dfrac{4 \pm 2i\sqrt{6}}{4}$$

$$= \dfrac{2(2 \pm i\sqrt{6})}{4} = \dfrac{2 \pm i\sqrt{6}}{2}.$$

The solutions are the complex numbers

$$\dfrac{2 + i\sqrt{6}}{2} \quad \text{and} \quad \dfrac{2 - i\sqrt{6}}{2}.$$ ◀

Work Problem 8 at the side.

APPENDIX B EXERCISES

Write the following imaginary numbers as the product of i and a real number.

1. $\sqrt{-9}$

2. $\sqrt{-18}$

3. $\sqrt{-20}$

4. $\sqrt{-27}$

5. $\sqrt{-36}$

6. $\sqrt{-50}$

Decide whether the following complex numbers are real, imaginary, or neither. See Example 1.

7. 14

8. $2 + 5i$

9. $-3i$

10. $i\sqrt{2}$

11. $-\sqrt{7}$

12. $\sqrt{5} + 7i$

Find values of x and y that make each pair of complex numbers equal. See Example 2.

13. $2x + 5i$ and $1 - yi$

14. $-x - 2i$ and $3 + 3yi$

15. $2 - 4xi$ and $3y + 2i$

16. $-6x + i$ and $12 - 4yi$

17. $-7x + 2yi$ and $-8i$

18. $15x - 12yi$ and $30 + 4i$

Add or subtract as indicated. See Examples 3–5.

19. $(2 + 8i) + (3 - 5i)$

20. $(4 - 5i) + (7 - 2i)$

21. $(16 + 5i) + (2 - 7i)$

22. $(6 + 3i) + (-1 - 4i)$

23. $(8 - 3i) - (2 + 6i)$

24. $(1 + i) - (3 - 2i)$

25. $(6 + 7i) - (2i + 5)$

26. $(8 - 6i) + (3i + 2)$

Find each product. See Example 6.

27. $(2 + 9i)(5i)$

28. $(6 + i)(-3i)$

29. $-i(2 - 7i)$

30. $2i(4 - 5i)$

31. $(3 + 2i)(4 - i)$

32. $(9 - 2i)(3 + i)$

Appendix B Exercises 533

33. $(5 - 4i)(3 - 2i)$

34. $(10 + 6i)(8 - 4i)$

35. $(3 + 6i)(3 - 6i)$

36. $(11 - 2i)(11 + 2i)$

Write each quotient in standard form. See Example 7.

37. $\dfrac{-2 + i}{1 - i}$

38. $\dfrac{6 + i}{2 + i}$

39. $\dfrac{3 - 4i}{2 + 2i}$

40. $\dfrac{7 - i}{3 - 2i}$

41. $\dfrac{5 + 2i}{1 + 3i}$

42. $\dfrac{4 + 6i}{2 - 3i}$

Solve each quadratic equation by using the quadratic formula. See Example 8.

43. $m^2 - 2m + 2 = 0$

44. $b^2 + b + 3 = 0$

45. $2r^2 + 3r + 5 = 0$

46. $3q^2 = 2q - 3$

47. $2r^2 + 5 = 4r$

48. $4q^2 + 2q + 3 = 0$

SYMBOLS

$+$	Plus sign, addition
$-$	Minus sign, subtraction
$\times$	Multiplication
$a(b)$, $(a)b$, or $(a)(b)$	Multiplication
$a \cdot b$	Multiplication
ab	Multiplication
$\div$	Division
a/b	Division
$=$	Equals
$\neq$	Is not equal to
$\approx$	Is approximately equal to
$<$	Is less than
$\leq$	Is less than or equal to
$>$	Is greater than
$\geq$	Is greater than or equal to
$\lvert x \rvert$	Absolute value of x
$\{a, b, c\}$	The set containing the elements a, b, and c
$\{x \lvert P\}$	The set of all x satisfying property P
$\emptyset$	Empty set, or null set
(x, y)	Ordered pair
x^2	x squared; $x \cdot x$
x^3	x cubed; $x \cdot x \cdot x$
x^n	x to the power n; x to the nth
x^{-n}	x to the negative n; $x^{-n} = 1/x^n$ if $x \neq 0$
x^0	x to the power 0; $x^0 = 1$ if $x \neq 0$
$P(x)$	Polynomial having the variable x
$\sqrt{a}$	Positive square root of a; $\sqrt{a} \cdot \sqrt{a} = a$
$\sqrt[n]{a}$	Positive nth root of a
i	Imaginary number: $i^2 = -1$
$a + bi$	Complex number

TABLE 1   SELECTED POWERS OF NUMBERS

n	n^2	n^3	n^4	n^5	n^6
2	4	8	16	32	64
3	9	27	81	243	729
4	16	64	256	1024	4096
5	25	125	625	3125	
6	36	216	1296		
7	49	343	2401		
8	64	512	4096		
9	81	729	6561		
10	100	1000	10,000		

TABLE 2 POWERS AND ROOTS

n	n^2	n^3	$\sqrt{n}$	$\sqrt[3]{n}$	$\sqrt{10n}$	n	n^2	n^3	$\sqrt{n}$	$\sqrt[3]{n}$	$\sqrt{10n}$
1	1	1	1.000	1.000	3.162	51	2601	132,651	7.141	3.708	22.583
2	4	8	1.414	1.260	4.472	52	2704	140,608	7.211	3.733	22.804
3	9	27	1.732	1.442	5.477	53	2809	148,877	7.280	3.756	23.022
4	16	64	2.000	1.587	6.325	54	2916	157,464	7.348	3.780	23.238
5	25	125	2.236	1.710	7.071	55	3025	166,375	7.416	3.803	23.452
6	36	216	2.449	1.817	7.746	56	3136	175,616	7.483	3.826	23.664
7	49	343	2.646	1.913	8.367	57	3249	185,193	7.550	3.849	23.875
8	64	512	2.828	2.000	8.944	58	3364	195,112	7.616	3.871	24.083
9	81	729	3.000	2.080	9.487	59	3481	205,379	7.681	3.893	24.290
10	100	1,000	3.162	2.154	10.000	60	3600	216,000	7.746	3.915	24.495
11	121	1,331	3.317	2.224	10.488	61	3721	226,981	7.810	3.936	24.698
12	144	1,728	3.464	2.289	10.954	62	3844	238,328	7.874	3.958	24.900
13	169	2,197	3.606	2.351	11.402	63	3969	250,047	7.937	3.979	25.100
14	196	2,744	3.742	2.410	11.832	64	4096	262,144	8.000	4.000	25.298
15	225	3,375	3.873	2.466	12.247	65	4225	274,625	8.062	4.021	25.495
16	256	4,096	4.000	2.520	12.649	66	4356	287,496	8.124	4.041	25.690
17	289	4,913	4.123	2.571	13.038	67	4489	300,763	8.185	4.062	25.884
18	324	5,832	4.243	2.621	13.416	68	4624	314,432	8.246	4.082	26.077
19	361	6,859	4.359	2.688	13.784	69	4761	328,509	8.307	4.102	26.268
20	400	8,000	4.472	2.714	14.142	70	4900	343,000	8.367	4.121	26.458
21	441	9,261	4.583	2.759	14.491	71	5041	357,911	8.426	4.141	26.646
22	484	10,648	4.690	2.802	14.832	72	5184	373,248	8.485	4.160	26.833
23	529	12,167	4.796	2.844	15.166	73	5329	389,017	8.544	4.179	27.019
24	576	13,824	4.899	2.884	15.492	74	5476	405,224	8.602	4.198	27.203
25	625	15,625	5.000	2.924	15.811	75	5625	421,875	8.660	4.217	27.386
26	676	17,576	5.099	2.962	16.125	76	5776	438,976	8.718	4.236	27.568
27	729	19,683	5.196	3.000	16.432	77	5929	456,533	8.775	4.254	27.749
28	784	21,952	5.292	3.037	16.733	78	6084	474,552	8.832	4.273	27.928
29	841	24,389	5.385	3.072	17.029	79	6241	493,039	8.888	4.291	28.107
30	900	27,000	5.477	3.107	17.321	80	6400	512,000	8.944	4.309	28.284
31	961	29,791	5.568	3.141	17.607	81	6561	531,441	9.000	4.327	28.460
32	1024	32,768	5.657	3.175	17.889	82	6724	551,368	9.055	4.344	28.636
33	1089	35,937	5.745	3.208	18.166	83	6889	571,787	9.110	4.362	28.810
34	1156	39,304	5.831	3.240	18.439	84	7056	592,704	9.165	4.380	28.983
35	1225	42,875	5.916	3.271	18.708	85	7225	614,125	9.220	4.397	29.155
36	1296	46,656	6.000	3.302	18.974	86	7396	636,056	9.274	4.414	29.326
37	1369	50,653	6.083	3.332	19.235	87	7569	658,503	9.327	4.431	29.496
38	1444	54,872	6.164	3.362	19.494	88	7744	681,472	9.381	4.448	29.665
39	1521	59,319	6.245	3.391	19.748	89	7921	704,969	9.434	4.465	29.833
40	1600	64,000	6.325	3.420	20.000	90	8100	729,000	9.487	4.481	30.000
41	1681	68,921	6.403	3.448	20.248	91	8281	753,571	9.539	4.498	30.166
42	1764	74,088	6.481	3.476	20.494	92	8464	778,688	9.592	4.514	30.332
43	1849	79,507	6.557	3.503	20.736	93	8649	804,357	9.644	4.531	30.496
44	1936	85,184	6.633	3.530	20.976	94	8836	830,584	9.695	4.547	30.659
45	2025	91,125	6.708	3.557	21.213	95	9025	857,375	9.747	4.563	30.822
46	2116	97,336	6.782	3.583	21.148	96	9216	884,736	9.798	4.579	30.984
47	2209	103,823	6.856	3.609	21.679	97	9409	912,673	9.849	4.595	31.145
48	2304	110,592	6.928	3.634	21.909	98	9604	941,192	9.899	4.610	31.305
49	2401	117,649	7.000	3.659	22.136	99	9801	970,299	9.950	4.626	31.464
50	2500	125,000	7.071	3.684	22.361	100	10,000	1,000,000	10.000	4.642	31.623

Answers to Selected Exercises

The solutions to selected odd-numbered exercises are given in the section beginning on page 561.

Diagnostic Pretest (page ix)

1. $\frac{2}{3}$ **2.** $\frac{61}{24}$ or $2\frac{13}{24}$ **3.** $\frac{9}{10}$ **4.** 27
5. 7 **6.** 5 **7.** -14 **8.** -11 **9.** 25
10. 8 **11.** 11 **12.** 15 **13.** 3
14. $13\frac{1}{2}$ meters **15.** $z > 5$ **16.** $y \geq -9$
17. 625; base is 5, exponent is 4 **18.** $-8x^{15}$
19. $\frac{y^{15}}{3^6 \cdot 4^3}$ **20.** $5x^3 + 6x^2 - 10$
21. $6p^2 + 8p - 30$ **22.** $4x^2 - 8x + 5$
23. $2 \cdot 5^2$ **24.** $2^6 \cdot 5$ **25.** $(m + 7)(m + 2)$
26. $(4a + 5b)(3a - 4b)$ **27.** $(11z - 2)^2$
28. 2, 3 **29.** 9, 11, or $-1, 1$ **30.** $\frac{x}{x + 1}$
31. $\frac{x^2}{(x + 1)(x - 1)}$
32. -2 **33.**

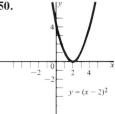

34.

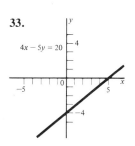

35.

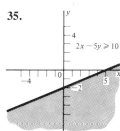

36. $-\frac{5}{6}$ **37.** $(3, -7)$ **38.** same line
39. -32 **40.** 6 **41.** $6\sqrt{2}$ **42.** $2\sqrt[3]{5}$
43. $19\sqrt{3}$ **44.** $15\sqrt{6}$ **45.** $2\sqrt{7}$
46. $\frac{-2(3 + \sqrt{2})}{7}$ **47.** 1 **48.** $1 + \sqrt{2}$, $1 - \sqrt{2}$ **49.** $\frac{-5 + \sqrt{41}}{2}, \frac{-5 - \sqrt{41}}{2}$

50.

[graph of $y = (x - 2)^2$]

CHAPTER 1

Section 1.1 (page 7)

1. $\frac{1}{2}$ **3.** $\frac{5}{6}$ **5.** $\frac{8}{9}$ **7.** $\frac{2}{3}$ **9.** $\frac{2}{3}$ **11.** $\frac{5}{6}$
13. $\frac{9}{20}$ **15.** $\frac{3}{25}$ **17.** $\frac{6}{5}$ **19.** $\frac{3}{10}$ **21.** $\frac{1}{9}$
23. 8 **25.** $\frac{21}{20}$ or $1\frac{1}{20}$ **27.** $\frac{78}{23}$ or $3\frac{9}{23}$
29. $\frac{2211}{70}$ or $31\frac{41}{70}$ **31.** $\frac{1}{2}$ **33.** $\frac{4}{5}$ **35.** $\frac{10}{9}$
37. $\frac{19}{22}$ **39.** $\frac{1}{15}$ **41.** $\frac{8}{15}$ **43.** $\frac{27}{8}$ or $3\frac{3}{8}$
45. $\frac{49}{6}$ or $8\frac{1}{6}$ **47.** $\frac{13}{12}$ or $1\frac{1}{12}$ **49.** $\frac{49}{30}$
51. $\frac{13}{28}$ **53.** $\frac{5}{24}$ **55.** $\frac{17}{24}$ of the debt
57. $14\frac{7}{16}$ tons **59.** $8\frac{23}{24}$ hours **61.** 16 dresses

Section 1.2 (page 15)

1. $80 + 6$ **3.** $600 + 90 + 4$
5. $5000 + 200 + 30 + 7$
7. $30 + 6 + .8 + .01$ **9.** $.5 + .06 + .007$
11. $\frac{4}{5}$ **13.** $\frac{18}{25}$ **15.** $\frac{43}{200}$ **17.** $\frac{5}{8}$
19. 887.859 **21.** 19.57 **23.** 270.58
25. 31.48 **27.** 14.218 **29.** 43.01
31. 80.0904 **33.** 1655.9 **35.** 8.21
37. 6.14 **39.** 16.76 **41.** .375
43. .438 (rounded) **45.** .938 (rounded)
47. .833 (rounded) **49.** .733 (rounded)
51. .632 (rounded) **53.** 75% **55.** 140%

57. .3%   **59.** 98.3%   **61.** .38   **63.** 1.74
65. .11   **67.** .001   **69.** 42   **71.** 300
73. 65%   **75.** 120%   **77.** 2%   **79.** 3.84
81. 7.50   **83.** 76.60%   **85.** $2400
87. 35%   **89.** 15%   **91.** $559   **93.** 6%
95. 11,844 square feet   **97.** $150.79

Section 1.3 (page 21)

1. <   **3.** <   **5.** >   **7.** <   **9.** <
11. >   **13.** ≤   **15.** ≥   **17.** ≤   **19.** ≤
21. < and ≤   **23.** ≥ and >   **25.** ≤ and ≥
27. ≥ and >   **29.** > and ≥   **31.** < and ≤
33. < and ≤   **35.** < and ≤   **37.** $7 = 5 + 2$
39. $3 < \frac{50}{5}$ **41.** $12 \neq 5$ **43.** $0 \geq 0$
45. true   **47.** true   **49.** true   **51.** true
53. true   **55.** false   **57.** false   **59.** false
61. false   **63.** $14 > 6$ **65.** $3 \leq 15$
67. $8 < 9$ **69.** $6 \geq 0$ **71.** $15 \leq 18$
73. $.439 \leq .481$

Section 1.4 (page 27)

1. 36   **3.** 64   **5.** 289   **7.** 125   **9.** 1296
11. 32   **13.** 729   **15.** $\frac{1}{4}$ **17.** $\frac{8}{125}$
19. $\frac{64}{125}$ **21.** .475   **23.** 3.112   **25.** 16
27. $\frac{49}{30}$ **29.** 11   **31.** 87.34   **33.** 65
35. 2   **37.** 3   **39.** 26   **41.** true
43. true   **45.** true   **47.** false   **49.** false
51. false   **53.** false   **55.** $10 - (7 - 3) = 6$
57. no parentheses needed, or $(3 \cdot 5) + 7 = 22$
59. $3 \cdot (5 - 4) = 3$ **61.** $3 \cdot (5 - 2) \cdot 4 = 36$
63. no parentheses needed, or
$(360 \div 18) \div 4 = 5$ **65.** $(6 + 5) \cdot 3^2 = 99$
67. $(8 - 2^2) \cdot 2 = 8$ **69.** no parentheses needed,
or $\frac{1}{2} + \left(\frac{5}{3} \cdot \frac{9}{7}\right) - \frac{3}{2} = \frac{8}{7}$ **71.** $(5 \cdot 52) - 4$
73. $5 + (6 - 1) + (3 - 2)$

Section 1.5 (page 33)

1. (a) 12   (b) 24   **3.** (a) 15   (b) 75
5. (a) $\frac{7}{3}$ (b) $\frac{31}{3}$ **7.** (a) $\frac{4}{3}$ (b) $\frac{16}{3}$ **9.** (a) $\frac{2}{3}$
(b) $\frac{4}{3}$ **11.** (a) 30   (b) 690   **13.** (a) 19.377
(b) 96.885   **15.** (a) .8195   (b) 16.9115
17. (a) 43   (b) 28   **19.** (a) 24   (b) 33
21. (a) 6   (b) $1\frac{4}{5}$ or $\frac{9}{5}$ **23.** (a) $\frac{19}{6}$ (b) $\frac{8}{3}$
25. (a) $\frac{185}{3}$ (b) $\frac{565}{6}$ **27.** (a) 2   (b) $\frac{17}{7}$
29. (a) $\frac{5}{6}$ (b) $\frac{16}{27}$ **31.** (a) 28   (b) 55
33. (a) $\frac{10}{3}$ (b) $\frac{13}{3}$ **35.** (a) $\frac{8}{7}$ (b) $\frac{28}{17}$

37. (a) 14.736   (b) 8.841   **39.** $8x$
41. $5 \cdot x$ or $5x$ **43.** $4 + x$ **45.** $x - 9$
47. $6 + \frac{2}{3}x$ **49.** solution   **51.** not a
solution   **53.** solution   **55.** solution
57. solution   **59.** solution   **61.** not a
solution   **63.** $x + 8 = 12$; 4
65. $16 - \frac{3}{4}x = 10$; 8   **67.** $2x + 5 = 13$; 4
69. $3x = 2 + 2x$; 2   **71.** $\frac{20}{5x} = 2$; 2

Chapter 1 Review Exercises (page 37)

1. $\frac{1}{2}$ **2.** $\frac{1}{3}$ **3.** $\frac{5}{9}$ **4.** $\frac{3}{5}$ **5.** $\frac{1}{3}$ **6.** $\frac{5}{6}$
7. $\frac{6}{7}$ **8.** $\frac{4}{5}$ **9.** $\frac{5}{24}$ **10.** $\frac{7}{50}$ **11.** $\frac{5}{14}$
12. $\frac{4}{15}$ **13.** $\frac{3}{2}$ **14.** $\frac{5}{14}$ **15.** $\frac{3}{4}$ **16.** $\frac{1}{5}$
17. $\frac{4}{15}$ **18.** $\frac{2}{3}$ **19.** $\frac{6}{7}$ **20.** $\frac{1}{2}$ **21.** $\frac{7}{8}$
22. $\frac{9}{20}$ **23.** $\frac{13}{18}$ **24.** $13\frac{1}{6}$ **25.** $472\frac{1}{6}$
26. $103\frac{25}{36}$ **27.** $\frac{7}{12}$ of the room
28. $6\frac{11}{12}$ feet   **29.** 40 awards   **30.** $31\frac{1}{4}$ cubic
yards   **31.** $400 + 70 + 9$
32. $8000 + 400 + 3$ **33.** $10 + 5 + .4$
34. $2 + .8 + .09 + .001$ **35.** $\frac{2}{5}$ **36.** $\frac{27}{50}$
37. $\frac{17}{20}$ **38.** $\frac{101}{200}$ **39.** $\frac{469}{200}$ **40.** $\frac{21879}{5000}$
41. .875   **42.** .25   **43.** .688   **44.** .167
45. .853   **46.** .652   **47.** 287.954
48. 31.3   **49.** 1.675   **50.** 7.9454
51. 219.78   **52.** 1001.698   **53.** 5.22
54. 7.88   **55.** 96%   **56.** 142%   **57.** 50%
58. 513.6%   **59.** 3600   **60.** 223
61. 218.24   **62.** 15.22%   **63.** 124%
64. $1495.20   **65.** $2385   **66.** $520
67. 6%   **68.** ≤, <   **69.** ≤, ≥   **70.** ≥, >
71. >, ≥   **72.** >, ≥   **73.** >, ≥
74. <, ≤   **75.** <, ≤   **76.** $9 \geq 4$
77. $3 \leq 16$ **78.** $10 > 9$ **79.** $25 \geq 23$
80. 64   **81.** 25   **82.** 32   **83.** 216
84. $\frac{1}{16}$ **85.** $\frac{25}{64}$ **86.** $\frac{196}{9}$ or $21\frac{7}{9}$
87. $\frac{343}{27}$ or $12\frac{19}{27}$ **88.** false   **89.** true
90. false   **91.** true   **92.** true   **93.** true
94. true   **95.** true   **96.** false   **97.** 6
98. 17   **99.** $\frac{1}{4}$ **100.** 9   **101.** 47
102. $\frac{2}{3}$ **103.** 60   **104.** 18   **105.** $\frac{137}{3}$
106. $12x$ **107.** $x + 9$ **108.** $4 - x$
109. $3x - 6$ **110.** $x(1 + x)$ or $x(x + 1)$

111. $\dfrac{x+4}{8}$ **112.** $(x+28) - \dfrac{3}{10}x$
113. $\dfrac{10}{3}x - (x+16)$

Chapter 1 Test (page 41)

1. $\dfrac{3}{8}$ **2.** $\dfrac{7}{11}$ **3.** $\dfrac{61}{40}$ **4.** $\dfrac{203}{120}$
5. $\dfrac{111}{8}$ or $13\dfrac{7}{8}$ **6.** $\dfrac{2}{3}$ **7.** $\dfrac{18}{19}$
8. $58\dfrac{4}{9}$ acres **9.** $\dfrac{5}{8}$ **10.** .5625 **11.** 21.77
12. 24.7 **13.** 15.8256 **14.** 11.56
15. 19% **16.** .762 **17.** 13.6 **18.** 653.2
19. $9.60 **20.** 8 **21.** 2 **22.** 23
23. 29 **24.** 30 **25.** 2 **26.** $4x$
27. $2x + 11$ **28.** $\dfrac{9}{x-8}$ **29.** solution
30. not a solution

CHAPTER 2

Section 2.1 (page 49)

1. -8 **3.** 9 **5.** $\dfrac{2}{3}$ **7.** -15 **9.** -8
11. -3 **13.** -14 **15.** $-\dfrac{4}{15}$
17. $|-2|$ or 2 **19.** $|-8|$ or 8
21. $-|-3|$ or -3 **23.** true **25.** true
27. false **29.** true **31.** true **33.** false
35. true **37.** true **39.** true **41.** false
43. true **45.** true
47. ⟵•••••••⟶ (−2, −1, 2, 6)
49. ⟵•••••⟶ (−5, −3, −2, 0, 4)
51. ⟵•••••⟶ ($-3\dfrac{4}{5}$, $-1\dfrac{5}{8}$, $\dfrac{1}{4}$, $2\dfrac{1}{2}$)

Section 2.2 (page 55)

1. 2 **3.** -2 **5.** -8 **7.** -11
9. -12 **11.** 4.18 **13.** 12 **15.** 5
17. 2 **19.** -9 **21.** 13 **23.** -11
25. $\dfrac{1}{2}$ **27.** $-\dfrac{19}{24}$ **29.** $-\dfrac{3}{4}$ **31.** $-.5$
33. -7.7 **35.** -8 **37.** 0 **39.** -20
41. 1.7125 **43.** true **45.** false **47.** false
49. true **51.** false **53.** true **55.** true
57. false **59.** true **61.** -2 **63.** -3
65. -2 **67.** -2 **69.** 2
71. $4 + (-7) + (-3) = -6$
73. $-3 + [15 + (-1)] = 11$
75. $[-8 + (-15)] + (-3) = -26$
77. 2000 feet **79.** $-$8$ **81.** $-26°$
83. $34 **85.** 26.833 inches **87.** 18 **89.** -3

Section 2.3 (page 61)

1. -3 **3.** -4 **5.** -8 **7.** -14 **9.** 9
11. 17 **13.** -4 **15.** 4 **17.** 1 **19.** $\dfrac{3}{4}$
21. $-\dfrac{11}{8}$ **23.** $\dfrac{15}{8}$ or $1\dfrac{7}{8}$ **25.** 11.6
27. -9.9 **29.** -1.72448 **31.** 3.6863
33. 10 **35.** -5 **37.** 11 **39.** -10
41. -18 **43.** $\dfrac{37}{12}$ **45.** -16 **47.** -12
49. -5.90617 **51.** -41.4616
53. $12 - (-6) = 18$
55. $-25 - (-4) = -21$
57. $\dfrac{4}{27} - \left(-\dfrac{5}{24}\right) = \dfrac{77}{216}$
59. $-7.3 - (-8.4) = 1.1$
61. $8 - (-2) = 10$
63. $-11 + (-4 - 2) = -17$
65. $[-12 + (-3)] - 4 = -19$ **67.** $-15°$
69. 14,776 feet **71.** $-172.3°$ **73.** $105,000

Section 2.4 (page 69)

1. 12 **3.** -12 **5.** 5 **7.** 44 **9.** 120
11. -48 **13.** -30 **15.** 0 **17.** $-\dfrac{8}{9}$
19. $-\dfrac{1}{2}$ **21.** 7.77 **23.** -21.147
25. -14 **27.** -3 **29.** -36 **31.** 12
33. 5 **35.** 12 **37.** 18 **39.** -10
41. 45 **43.** 12 **45.** 16 **47.** -43
49. -128 **51.** $\dfrac{34}{9}$ **53.** -64 **55.** -74
57. 28 **59.** -2 **61.** 0 **63.** -3 **65.** -2
67. -1 **69.** 1 **71.** $4(-7) + (-9) = -37$
73. $-4 - 2(-8 \cdot 2) = 28$ **75.** $-2(3) - 3 = -9$

Section 2.5 (page 75)

1. $\dfrac{1}{9}$ **3.** $-\dfrac{1}{4}$ **5.** $\dfrac{3}{2}$ **7.** $-\dfrac{10}{9}$
9. no inverse **11.** 1.150 **13.** -2
15. -3 **17.** -6 **19.** -5 **21.** 2
23. 15 **25.** 36 **27.** 0 **29.** $\dfrac{2}{3}$ **31.** 2.1
33. -5 **35.** -4 **37.** -10 **39.** 5
41. -4 **43.** -60 **45.** -6 **47.** 2
49. 50.58 **51.** 4 **53.** 3 **55.** -3
57. 8 **59.** -1 **61.** meaningless **63.** 3
65. 2 **67.** 8.363 **69.** -8 **71.** -6
73. 0 **75.** 8 **77.** $1, -1, 2, -2, 3, -3, 4, -4,$
$6, -6, 9, -9, 12, -12, 18, -18, 36, -36$
79. $1, -1, 5, -5, 25, -25$ **81.** $1, -1, 2, -2, 4,$
$-4, 5, -5, 8, -8, 10, -10, 20, -20, 40, -40$
83. $1, -1, 17, -17$ **85.** $1, -1, 29, -29$
87. $4x = -32; -8$ **89.** $\dfrac{x}{4} = -2; -8$
91. $\dfrac{x}{-3} = -4; 12$ **93.** $\dfrac{x}{-1} = 2; -2$
95. $\dfrac{x^2}{3} = 12; 6, -6$

Answers to Selected Exercises **539**

Section 2.6 (page 83)

1. commutative property 3. associative property 5. both the associative property and the commutative property 7. commutative property 9. commutative property 11. inverse property 13. identity property 15. inverse property 17. identity property 19. distributive property 21. $k + 9$ 23. m
25. $3r + 3m$ 27. 1 29. 0
31. $(-5) + 5 = 0$ 33. $-3r - 6$ 35. 9
37. $k - 1$ 39. $4z + (2r + 3k)$
41. $5m + 10$ 43. $-4r - 8$ 45. $-8k + 16$
47. $-\frac{2}{3}a - 6$ 49. $\frac{3}{4}r + 2$ 51. $-16 + 2k$
53. $10r + 12m$ 55. $-12x + 16y$
57. $5(8 + 9) = 5(17) = 85$
59. $7.12(2.3 + 7.7) = 7.12(10) = 71.2$
61. $9(p + q)$ 63. $5(7z + 8w)$
65. $-3k - 5$ 67. $-4y + 8$ 69. $4 - p$
71. $1 + 15r$ 73. not commutative
75. commutative 77. no

Section 2.7 (page 89)

1. 15 3. -22 5. 35 7. -9 9. 1
11. -1 13. like 15. unlike 17. like
19. like 21. like 23. $27m$ 25. $7k + 15$
27. $m - 1$ 29. $2x + 6$ 31. $14 - 7m$
33. $-\frac{5}{4}x - \frac{28}{3}$ 35. $11.56x + 7.2$ 37. $9y^2$
39. $5p^2 - 14p^3$ 41. $-2.844q^2 - 5.32q - 18.7$
43. $30t + 66$ 45. $-3n - 15$
47. $-10 + 9t$ 49. $5a - 7$ 51. $4r + 15$
53. $12k - 5$ 55. $-5 - 8x$ 57. $-2k - 3$
59. $-.148y - 20.7872$
61. $(x + 2) - 2x = 2 - x$
63. $9x - (2x - 3x) = 10x$
65. $(4 - 2x) - 9(5x + 4) = -47x - 32$

Chapter 2 Review Exercises (page 93)

1. -9 2. -5 3. -8 4. $-\frac{7}{8}$
5. $\left|-\frac{3}{2}\right|$ 6. $|-7|$ 7. $-|-9|$ 8. $-|-7|$
9. true 10. false 11. false 12. true
13. true 14. false 15. false 16. true
17. true 18. false 19. true 20. false
21. [number line]
22. [number line]
23. [number line]
24. [number line]
25. -6 26. -3 27. -15 28. $\frac{23}{40}$
29. $\frac{13}{36}$ 30. -14.2 31. -1.33 32. -1
33. -13 34. -16 35. -18 36. -15
37. -12 38. $-\$2$ 39. $-7°$
40. $\$21,000$ 41. $2°$ 42. -2 43. 9
44. 16 45. 18 46. -17 47. -39
48. $\frac{17}{12}$ 49. $\frac{1}{2}$ 50. 2.6 51. -25.1
52. 0 53. -13 54. -14 55. -27
56. -21 57. 22 58. 33 59. -85
60. -108 61. $\frac{8}{7}$ 62. $\frac{2}{5}$ 63. -28.25
64. 26.32 65. -16 66. -4 67. -20
68. -60 69. -12 70. 65 71. 45
72. -38 73. -13 74. -27 75. 34
76. -2 77. -378 78. -73 79. 5
80. -40 81. $-\frac{4}{3}$ 82. -11.2 83. 6
84. 10 85. -1 86. -3 87. 2
88. 2 89. 12 90. -27 91. identity property 92. inverse property
93. commutative property 94. associative property 95. inverse property 96. identity property 97. F, H 98. B 99. C, D
100. A, G 101. E 102. $11m$
103. $16p^2$ 104. $16p^2 + 2p$ 105. $-4k + 12$
106. $29 - 2m$ 107. $-5k - 1$

Chapter 2 Test (page 97)

1. [number line] 2. [number line]
3. $-.742$ 4. $-|-8|$ 5. $11 - 2x$
6. $\frac{9}{x - 8}$ 7. -7 8. $\frac{61}{20}$ or $3\frac{1}{20}$ 9. 0
10. -1 11. -14 12. 3 13. 4
14. meaningless 15. -9 16. -4
17. -2 18. -116 19. 6 20. A, E
21. H 22. C, I 23. B, G 24. D, F
25. $4x + 2$ 26. $7m - 1$

Cumulative Review Exercises (page 99)

1. $\frac{3}{8}$ 2. $\frac{3}{5}$ 3. $\frac{3}{4}$ 4. $\frac{13}{8}$ or $1\frac{5}{8}$
5. $\frac{31}{20}$ or $1\frac{11}{20}$ 6. $\frac{551}{40}$ or $13\frac{31}{40}$ 7. 6
8. $\frac{6}{5}$ 9. $\frac{2}{5}$ 10. $20\frac{23}{24}$ pounds
11. 35 yards 12. $4\frac{1}{6}$ cups 13. $8\frac{5}{8}$ meters
14. $\frac{3}{8}$ 15. $.8125$ 16. 34.03 17. 27.31
18. 30.51 19. 13.3 20. 56.3
21. 16.318 22. 45% 23. $.392$ 24. 21.6
25. 174 26. 37.5% 27. 162.5%
28. 25% 29. $\$1187.65$ 30. $\$1261.88$
31. 68% 32. $\$3750$ 33. false 34. false
35. true 36. true 37. false 38. false
39. 7 40. 1 41. 13 42. 15
43. -40 44. -504 45. -12

46. meaningless **47.** -6 **48.** -7 **49.** 28 **50.** 14 **51.** 1 **52.** -6 **53.** meaningless **54.** meaningless **55.** -64 **56.** $\frac{73}{18}$ **57.** -11 **58.** -15 **59.** -8.23 **60.** $-|-7|$ **61.** -3 **62.** -10 **63.** 24 **64.** -162 **65.** -8 **66.** -134 **67.** $-\frac{29}{6}$ **68.** $\frac{2}{3}$ **69.** 3 **70.** -2 **71.** -1 **72.** $\frac{1}{2}x + 3 = 2; -2$ **73.** $\frac{x+6}{x^2} = 1; 3, -2$ **74.** $-x + 7 = 4; 3$ **75.** $4x^2 = 0; 0$ **76.** commutative property **77.** identity property **78.** distributive property **79.** commutative property **80.** inverse property **81.** inverse property

CHAPTER 3

Section 3.1 (page 107)

1. 10 **3.** -2 **5.** 10 **7.** -8 **9.** 4 **11.** -5 **13.** -11 **15.** -6 **17.** $\frac{1}{2}$ **19.** -5 **21.** $\frac{7}{6}$ **23.** -5 **25.** 7 **27.** 8 **29.** -10 **31.** 13.3 **33.** 6 **35.** -2 **37.** 17 **39.** 18 **41.** 26 **43.** $3x = 17 + 2x; 17$ **45.** $5x + 3x = 7x + 9; 9$ **47.** $6(2x + 5) = 13x - 8; 38$

Section 3.2 (page 113)

1. 5 **3.** 25 **5.** -8 **7.** -7 **9.** $-\frac{8}{3}$ **11.** -6 **13.** 0 **15.** -6 **17.** 4 **19.** 4 **21.** 8 **23.** 20 **25.** -3 **27.** 4 **29.** 7 **31.** 32 **33.** 49 **35.** 9 **37.** $\frac{8}{3}$ **39.** -80 **41.** $\frac{15}{2}$ **43.** $\frac{49}{2}$ **45.** 3 **47.** -6.1 **49.** .7 **51.** -1.1 **53.** -5.7 **55.** $4x = 6; \frac{3}{2}$ **57.** $\frac{x}{4} = 62; \$248$ **59.** $\frac{2x}{1.74} = -8.38; -7.2906$

Section 3.3 (page 119)

1. 2 **3.** 24 **5.** 9 **7.** $-\frac{5}{3}$ **9.** -1 **11.** $\frac{9}{11}$ **13.** -12.5 **15.** 2 **17.** -1.5 **19.** $-\frac{10}{3}$ **21.** -3 **23.** 8 **25.** -2 **27.** 6 **29.** $-\frac{5}{2}$ **31.** 0 **33.** 0 **35.** $\frac{10}{7}$ **37.** -4 **39.** 3 **41.** 1.4 **43.** -5.2 **45.** $x - 6$ or $-6 + x$ **47.** $x + 12$ or $12 + x$ **49.** $x - 5$ **51.** $x - 9$ **53.** $9x$ **55.** $3x$ **57.** $\frac{x}{6}$ **59.** $\frac{x}{-4}$ **61.** $8(x + 3)$ **63.** $3\left(\frac{x}{2}\right)$

Section 3.4 (page 127)

1. 8 **3.** 6 **5.** 30 **7.** -1 **9.** $\frac{10}{7}$ **11.** 19 inches **13.** 48 **15.** 36 **17.** 613 gallons **19.** 10, 11 **21.** 55, 57 **23.** 15, 16, 17 **25.** 11, 13, 15 **27.** Bob is 7 and Kevin is 21. **29.** 3 feet by 9 feet **31.** $650

Section 3.5 (page 135)

1. 128 **3.** 36 **5.** 4 **7.** 8 **9.** 14 **11.** 10 **13.** 1.5 **15.** 254.34 **17.** 2 **19.** 3 **21.** 7 **23.** 10 meters **25.** 37.68 feet **27.** 24 meters **29.** 5 inches **31.** 3 meters **33.** 36 square feet **35.** $L = \frac{A}{W}$ **37.** $t = \frac{d}{r}$ **39.** $H = \frac{V}{LW}$ **41.** $t = \frac{I}{pr}$ **43.** $a^2 = c^2 - b^2$ **45.** $b = \frac{2A}{h}$ **47.** $r^2 = \frac{V}{\pi h}$ **49.** $W = \frac{P - 2L}{2}$ **51.** $t = \frac{P - A}{-Ar}$ or $t = \frac{A - P}{Ar}$ **53.** $v = \frac{d - gt^2}{t}$ **55.** $l = \frac{r}{2A}$ **57.** $b = \frac{2A}{h} - B$ or $b = \frac{2A - Bh}{h}$

Section 3.6 (page 143)

1. $\frac{3}{2}$ **3.** $\frac{36}{55}$ **5.** $\frac{2}{5}$ **7.** $\frac{5}{8}$ **9.** $\frac{1}{10}$ **11.** $\frac{8}{5}$ **13.** $\frac{1}{6}$ **15.** $\frac{4}{15}$ **17.** true **19.** true **21.** false **23.** false **25.** true **27.** true **29.** true **31.** false **33.** 49 **35.** 27 **37.** 26 **39.** $\frac{33}{8}$ **41.** $\frac{17}{6}$ **43.** $\frac{35}{12}$ **45.** $-\frac{3}{2}$ **47.** $\frac{25}{19}$ **49.** $\frac{87}{4}$ **51.** 9 tanks **53.** 50 minutes **55.** $5.50 **57.** $75 **59.** 15 sacks **61.** 19.2 yards **63.** 20 ounces

Section 3.7 (page 151)

1. 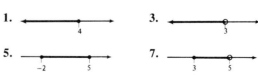 **3.** **5.** **7.** **9.** $a < 2$ **11.** $z \geq 1$ **13.** $x < 6$ **15.** $k \leq -9$ **17.** $n \leq -11$ **19.** $y < 5$

21. $k \geq 44$

23. $k > -21$

25. $6 \leq p \leq 13$

27. $5 < y < 12$

29. $4x + 8 < 3x + 5$; $x < -3$
31. $2x - x \leq 7$; 7 meters
33. $4w - 3w \geq 15$; smallest width is 15, smallest length is 60

Section 3.8 (page 157)

1. $x < 9$

3. $r \geq -3$

5. $k \geq -6$

7. $k \geq -5$

9. $q > -2$

11. $z < 1$

13. $w \geq 3$

15. $k \leq 0$

17. $x < -11$

19. $p < -.98$

21. $-1 \leq x \leq 6$

23. $-26 \leq z \leq 6$

25. The shortest side cannot be longer than 11 feet, and the remaining side cannot be longer than 22 feet. **27.** 30 meters **29.** \$1100 **31.** any number greater than or equal to -16

Chapter 3 Review Exercises (page 159)

1. 6 **2.** -12 **3.** 7 **4.** $\frac{2}{3}$ **5.** 11
6. 17 **7.** 5 **8.** -4 **9.** 5 **10.** -12
11. $\frac{64}{5}$ **12.** -6 **13.** $\frac{13}{4}$ **14.** 2 **15.** 1
16. 6 **17.** 7 **18.** 19, 21 **19.** 61, 96
20. 12 miles **21.** $-3, -2, -1$ **22.** 11
23. 84 **24.** 2 **25.** 4.19 (rounded)
26. $W = \frac{A}{L}$ **27.** $r = \frac{I}{pt}$ **28.** $m = \frac{x^2 - a}{2}$
29. $h = \frac{2A}{b + B}$ **30.** 5 meters **31.** 360 square centimeters **32.** $\frac{5}{3}$ **33.** $\frac{6}{7}$ **34.** $\frac{3}{4}$

35. $\frac{1}{18}$ **36.** true **37.** false **38.** $\frac{7}{2}$ or $3\frac{1}{2}$
39. $\frac{36}{5}$ **40.** 7 **41.** $-\frac{2}{7}$ **42.** 8 quarts
43. 63 hours **44.** 75 yards **45.** 375 kilometers
46.  **47.**

48. $y \geq -3$

49. $y > 8$

50. $k \geq 53$

51. $z \leq -37$

52. $-8 \leq x \leq -2$

53. $7 < y < 9$

54. $k \geq -3$

55. $y > -2$

56. $z < 4$

57. $m \leq -6$

58. $p < -5$

59. $y \leq \frac{27}{10}$

60. $-2 \leq m \leq \frac{3}{2}$

61. $\frac{4}{3} < m \leq 5$

62. 50 meters **63.** 9 centimeters **64.** 6
65. 12 **66.** 90 **67.** 47 **68.** 70 feet
69. 44 meters **70.** 26 inches **71.** 26 inches
72. 71.6° F **73.** \$1536.30 **74.** 12 pounds
75. 125 miles **76.** 35 **77.** 37 small cars
78. 15 **79.** $\frac{2}{3}$ cup **80.** 250 kilometers
81. \$13.98 **82.** \$14.63 **83.** $\frac{279}{8}$ or $34\frac{7}{8}$

Chapter 3 Test (page 167)

1. 4 **2.** 5 **3.** -5 **4.** 0 **5.** 10
6. $\frac{8}{3}$ **7.** $-\frac{11}{2}$ **8.** -3 **9.** -5 **10.** \$100
11. 15, 17 **12.** $p = \frac{I}{rt}$ **13.** $n = \frac{m - bp}{a}$
14. $h = \frac{2A}{b + B}$ **15.** 14 meters **16.** true
17. 1 **18.** $-\frac{22}{3}$ **19.** \$17.60

20. $x \leq 4$ **21.** $m > 7$

22. $k \leq 1$ **23.** $r \leq 4$

24. $-2 < k \leq \frac{14}{3}$

CHAPTER 4

Section 4.1 (page 175)

1. Base is 5; exponent is 12. **3.** Base is $3m$; exponent is 4. **5.** Base is 125; exponent is 3. **7.** Base is $-2x$; exponent is 2. **9.** Base is m; exponent is 2. **11.** Base is r; exponent is 5.
13. 3^5 **15.** $(-2)^5$ **17.** p^5 **19.** $\frac{1}{(-2)^3}$
21. $\frac{1}{2^5}$ **23.** $(-2z)^3$ **25.** 90 **27.** 80
29. 36 **31.** 2 **33.** 2 **35.** $\frac{1}{27}$ **37.** $\frac{1}{25}$
39. $\frac{1}{9}$ **41.** $\frac{1}{36}$ **43.** $\frac{1}{7}$ **45.** 32 **47.** 2
49. $\frac{27}{8}$ **51.** $\frac{5}{6}$ **53.** $\frac{5}{64}$ **55.** 1.041
57. .017 **59.** 4^5 **61.** 9^8 **63.** $\frac{1}{k^3}$
65. $\frac{1}{4^2}$ **67.** $(-3)^5$ **69.** $(-z)^9$ **71.** 4^5
73. $\frac{1}{4^2}$ **75.** $\frac{1}{8^6}$ **77.** $\frac{1}{q^6}$ **79.** d^3 **81.** y^7
83. $\frac{1}{r^2}$ **85.** $\frac{z^7}{6^3 y}$ **87.** $15y^2$; $56y^4$
89. $8q^4$; $15q^8$ **91.** $14a^3$; $54a^9$
93. $-2r^{-4}$; $-48r^{-8}$

Section 4.2 (page 181)

1. 6^6 **3.** $\frac{1}{9^6}$ **5.** 3^{10} **7.** y^5 **9.** $\frac{1}{k^4}$
11. a^3 **13.** $\frac{1}{4^9}$ **15.** $\frac{1}{5^9}$ **17.** m^5 **19.** $\frac{1}{a}$
21. $\frac{1}{r}$ **23.** a^7 **25.** $5^3 m^3$ **27.** $3^4 m^4 n^4$
29. $3^2 x^{10}$ or $(-3)^2 x^{10}$ **31.** $5^3 p^6 q^3$
33. $\frac{3^2}{x^{10}}$ **35.** $\frac{9^2}{y^{10}}$ **37.** $\frac{a^3}{5^3}$ **39.** $\frac{3^5 m^5 n^5}{2^5}$
41. $\frac{bc}{a}$ **43.** $\frac{5^2}{m^2}$ **45.** $\frac{1}{x^{10}}$ **47.** $\frac{1}{b^7}$
49. $\frac{5^3}{3^3 x^2}$ **51.** $\frac{a^{12}}{2b^5}$ **53.** $\frac{2^2 \cdot 3^3}{y^5 z^3}$ **55.** $\frac{9z^2}{4^2 \cdot 5^2 \cdot x^3}$

Section 4.3 (page 185)

1. 6.835×10^9 **3.** 8.36×10^{12} **5.** 2.5×10^4
7. 1.01×10^{-2} **9.** 1.2×10^{-5}
11. 8.34×10^{-1} **13.** 8,100,000,000
15. 9,132,000 **17.** .00032 **19.** 800,000
21. .000004 **23.** 420 **25.** 1440
27. 3,000,000 **29.** .4 **31.** 1300 **33.** .000008
35. 4×10^{-4}; 8×10^{-4} **37.** 3.68×10^{15}
39. 1000; .06102 **41.** 35,000

Section 4.4 (page 191)

1. $8m^5$ **3.** $-r^5$ **5.** cannot be simplified
7. x^5 **9.** $-p^7$ **11.** $6y^2$ **13.** 0
15. $9y^4 + 7y^2$ **17.** $14z^5 - 9z^3 + 8z^2$
19. $2p^7 - 8p^6 + 5p^4 - 9p$
21. $1.171q^2 + 1.401q - .252$
23. already simplified; degree 4; binomial
25. already simplified; degree 9; trinomial
27. already simplified; degree 8; trinomial
29. x^5; degree 5; monomial
31. already simplified; degree 0; monomial
33. always **35.** never **37.** never
39. $5m^2 + 3m$ **41.** $4x^4 - 4x^2$
43. $-n^5 - 12n^3 - 2$ **45.** $12m^3 + m^2 + 12m - 14$
47. $15m^2 - 3m + 4$ **49.** $8b^2 + 2b + 7$
51. $-r^2 - 2r$ **53.** $5m^2 - 14m$
55. $-6s^2 + 5s + 1$ **57.** $2s^2 + 4s$
59. $4x^3 + 2x^2 + 5x$ **61.** $-11y^4 + 8y^2 + 3y$
63. $a^4 - a^2 + 1$ **65.** $5m^2 + 8m - 10$
67. $3.386m^2 - 12.273m + 4.953$
69. $(4 + x^2) + (-9x + 2) > 8$
71. $(5 + x^2) + (3 - 2x) \neq 5$

Section 4.5 (page 199)

1. $-32x^7$ **3.** $15y^{11}$ **5.** $30a^9$ **7.** $6m^2 + 4m$
9. $-6p^4 + 12p^3$ **11.** $-16z^2 - 24z^3 - 24z^4$
13. $6y + 4y^2 + 10y^5$ **15.** $32z^5 + 20xz^4 - 12x^2z^3$
17. $x^2 - 25$ **19.** $3r^2 - 13r + 4$
21. $6p^2 - p - 5$ **23.** $16m^2 + 24m + 9$
25. $6b^2 + 46b - 16$ **27.** $16b^2 + 2ab - 3a^2$
29. $-8h^2 + 6hk - k^2$ **31.** $12x^3 + 26x^2 + 10x + 1$
33. $81a^3 + 27a^2 + 11a + 2$
35. $20m^4 - m^3 - 8m^2 - 17m - 15$
37. $6x^6 - 3x^5 - 4x^4 + 4x^3 - 5x^2 + 8x - 3$
39. $5x^4 - 13x^3 + 20x^2 + 7x + 5$
41. $x^2 + 14x + 49$ **43.** $a^2 - 8a + 16$
45. $4p^2 - 20p + 25$ **47.** $25k^2 + 80k + 64$
49. $m^3 - 15m^2 + 75m - 125$
51. $8a^3 + 12a^2 + 6a + 1$
53. $k^4 + 4k^3 + 6k^2 + 4k + 1$
55. $27r^3 - 54r^2 s + 36rs^2 - 8s^3$

Section 4.6 (page 205)

1. $r^2 + 2r - 3$ **3.** $x^2 - 10x + 21$
5. $6x^2 + x - 2$ **7.** $6z^2 - 13z - 15$
9. $2a^2 + 9a + 4$ **11.** $8r^2 + 2r - 3$
13. $6a^2 + 8a - 8$ **15.** $20 + 9x - 20x^2$

17. $-12 + 5r + 2r^2$ 19. $15 + a - 2a^2$
21. $p^2 + 4pq + 3q^2$ 23. $10y^2 - 3yz - z^2$
25. $8y^2 + 31yz - 45z^2$ 27. $-8r^2 + 2rs + 45s^2$
29. $4.0896y^2 - .5118y - 15.834$
31. $m^2 + 4m + 4$ 33. $x^2 - 4xy + 4y^2$
35. $4z^2 - 10zx + \frac{25}{4}x^2$ 37. $25p^2 + 20pq + 4q^2$
39. $16a^2 - 10ab + \frac{25}{16}b^2$
41. $.7225r^2 + .391rs + .0529s^2$ 43. $p^2 - 4$
45. $4b^2 - 25$ 47. $36a^2 - p^2$ 49. $4m^2 - \frac{25}{9}$
51. $49y^4 - 100z^2$ 53. $.2304q^2 - .1369r^2$
55. $(x + 3)^2 = 5$ 57. $(3 + x)(x - 4) > 7$

Section 4.7 (page 209)

1. $2x$ 3. $2a^2$ 5. $\frac{9k^3}{m}$ 7. $30m^3 - 10m$
9. $5m^4 - 8m + 4m^2$ 11. $4m^4 - 2m^2 + 2m$
13. $m^4 - 2m + 4$ 15. $m - 1 + \frac{5}{2m}$
17. $4x^2 - x + 1$ 19. $9x - 3x^2 + 6x^3$
21. $\frac{12}{x} + 8 + x$ 23. $\frac{1}{3}x + 2 - \frac{1}{3x}$
25. $4k^3 - 6k^2 - k + \frac{7}{2} - \frac{3}{2k}$
27. $-10p^3 + 5p^2 - 3p + \frac{3}{p}$ 29. $\frac{2}{x^3} + \frac{4}{x^2} + \frac{5}{2x}$
31. $4y^3 - 2 + \frac{3}{y}$ 33. $\frac{12}{x} - \frac{6}{x^2} + \frac{14}{x^3} - \frac{10}{x^4}$
35. $12x^5 + 9x^4 - 12x^3 + 6x^2$
37. $-63y^4 - 21y^3 - 35y^2 + 14y$

Section 4.8 (page 215)

1. $x + 2$ 3. $2y - 5$ 5. $p - 4 + \frac{4}{p + 6}$
7. $r - 5$ 9. $6m - 1$ 11. $a - 7 + \frac{37}{2a + 3}$
13. $x + 2 + \frac{1}{2x + 1}$ 15. $2k + 5 + \frac{10}{7k - 8}$
17. $x^2 - x + 2$ 19. $4k^3 - k + 2$ 21. $3y + 1$
23. $3k - 4 + \frac{2}{k^2 - 2}$ 25. $x^2 + 1 + \frac{-6x + 2}{x^2 - 2}$
27. $2p^2 - 5p + 4 + \frac{6}{3p^2 + 1}$
29. $m^3 - 2m^2 + 4 + \frac{-3}{4m^2 - 3}$ 31. $y^2 - y + 1$
33. $a^2 - 1$

Chapter 4 Review Exercises (page 217)

1. 80 2. 2 3. $\frac{1}{16}$ 4. $\frac{1}{6}$ 5. $\frac{1}{49}$ 6. $\frac{64}{25}$
7. 32 8. $\frac{3}{4}$ 9. $-\frac{8}{9}$ 10. 5^{11} 11. $\frac{1}{9^2}$

12. $(-4)^8$ 13. 15^5 14. $\frac{1}{5^{11}}$ 15. 6^2
16. x^2 17. $\frac{1}{p^{12}}$ 18. r^4 19. 2^8 20. $\frac{1}{9^6}$
21. 5^8 22. $\frac{1}{8^{12}}$ 23. $\frac{1}{m^2}$ 24. y^7 25. r^{13}
26. $(-5)^2 m^6$ 27. $\frac{y^{12}}{2^3}$ 28. $\frac{r^2}{6}$ 29. $\frac{3^5}{p^3}$
30. $\frac{6^3 r^6 s^3}{5^3}$ 31. $\frac{1}{a^3 b^5}$ 32. $2 \cdot 6^2 \cdot r^5$ 33. $\frac{2^3 n^{10}}{3m^{13}}$
34. 6.4×10^4 35. 1.58×10^7
36. 2.6954×10^{10} 37. 4.251×10^{-4}
38. 9.76×10^{-5} 39. 7.84×10^{-1} 40. $12,000$
41. $689,000,000$ 42. $.0004253$ 43. $.877$
44. $90,000,000$ 45. 800 46. $.0003$
47. $4,000,000$ 48. $.025$ 49. $.01$ 50. $22m^2$; degree 2   51. $p^3 - p^2 + 4p + 2$; degree 3   52. $4r^4$; degree 4   53. $12a^5 - 9a^4 + 8a^3 + 2a^2 - a + 3$; degree 5   54. $-8x^5 + 9x^3 - 7x$; degree 5
55. $-5z^3 - 6z^2 + 8z + 7$; degree 3   56. $a^3 + 4a^2$
57. $2r^3 - 3r^2 + 3r$ 58. $y^2 - 10y + 9$
59. $-13k^4 - 15k^2 + 18k - 6$ 60. $5m^3 - 6m^2 - 3$
61. $-y^2 - 4y + 4$ 62. $10p^2 - 3p - 5$
63. $7r^4 - 4r^3 + 1$ 64. $-11 + p$
65. $7y^3 - y^2 + 9$ 66. $10x^2 - 55x$
67. $-6p^5 + 15p^4$ 68. $-22y^4 + 4y^3 + 18y^2$
69. $-8m^7 + 10m^6 - 6m^5$ 70. $m^2 - 7m - 18$
71. $6k^2 - 9k - 6$ 72. $144a^2 - 1$
73. $49 - 28k + 4k^2$ 74. $a^3 - 2a^2 - 7a + 2$
75. $6r^3 + 8r^2 - 17r + 6$ 76. $5p^3 + 8p^2 - 7p - 6$
77. $r^3 + 6r^2 + 12r + 8$ 78. $6k^2 - 7k - 3$
79. $10r^2 + 21r - 10$ 80. $2a^2 + 5ab - 3b^2$
81. $12k^2 - 32kq - 35q^2$ 82. $a^2 + 8a + 16$
83. $9p^2 - 12p + 4$ 84. $4r^2 + 20rs + 25s^2$
85. $64z^2 - 48zy + 9y^2$ 86. $36m^2 - 25$
87. $4z^2 - 49$ 88. $25a^2 - 36b^2$
89. $81y^2 - 64z^2$ 90. $\frac{-5y^2}{3}$ 91. $2x^2 y$
92. $y^3 - 2y + 3$ 93. $p - 3 + \frac{5}{2p}$
94. $-2m^2 n + mn^2 + \frac{6n^3}{5}$ 95. $\frac{5}{2} - \frac{4}{5xy} + \frac{3x}{2y^2}$
96. $2r + 7$ 97. $4m + 3 + \frac{5}{3m - 5}$
98. $y^2 + 5y + 1$ 99. $2a^2 + 3a - 1 + \frac{6}{5a - 3}$
100. $k^3 + k^2 + 4k - 2 + \frac{-6}{2k + 1}$
101. $m^2 + 2m - \frac{1}{2} + \frac{-6m + \frac{9}{2}}{2m^2 - 3}$

Chapter 4 Test (page 221)

1. $\frac{1}{81}$ 2. -1 3. $\frac{1}{5}$ 4. 8^5 5. p^{24}
6. $\frac{1}{a^{10} b^4}$ 7. 2.45×10^8 8. 3.79×10^{-4}

9. .0048   10. 400   11. $-x^2 + 6x$; degree 2; binomial   12. $2m^4 + 11m^3 - 8m^2$; degree 4; trinomial   13. already simplified; degree 3; none of these   14. $x^5 - x^2 - 2x + 12$
15. $3y^2 - 2y - 2$
16. $6m^5 + 12m^4 - 18m^3 + 42m^2$
17. $r^2 - 3r - 10$ 18. $6t^2 - tw - 12w^2$
19. $25r^2 - 30rs + 9s^2$ 20. $36p^2 - 64q^2$
21. $2x^3 + x^2 - 16x + 15$ 22. $3y^2 - 5y + 2$
23. $2r^2 + 5r - 3 + \dfrac{8}{5r}$ 24. $3y - 6 + \dfrac{7}{4y + 3}$
25. $3x^2 + 4x + 2$

Cumulative Review Exercises   (page 223)

1. $7p - 14$ 2. $-2k + 4$ 3. $2k - 11$
4. $12m - 23$ 5. 7   6. 1   7. -4 8. -7
9. -1 10. $-\dfrac{3}{5}$ 11. 2   12. 6   13. $\dfrac{19}{3}$
14. -5 15. -13 16. 26   17. $x = 3y - 2$
18. $x = \dfrac{4 - y}{2}$ 19. $x = \dfrac{10 + a}{2}$
20. $x = \dfrac{3}{a} - b$ or $x = \dfrac{3 - ab}{a}$, $a \neq 0$
21. $z \leq 2$ 22. $k < -3$

23. $r \leq 1$ 24. $m > 1$

25. $-7 < x < 5$ 26. $1 \leq y < 5$

27. $-2 \leq x \leq 1$ 28. $\dfrac{5}{2} < x < 4$

29. 12   30. 8   31. $49.50   32. $98.45
33. 30 centimeters   34. 16 inches   35. $\dfrac{1}{2}$
36. -27 37. $\dfrac{16}{9}$ 38. $\dfrac{81}{64}$ 39. $\dfrac{1}{49}$ 40. 1
41. 81   42. 256   43. $\dfrac{1}{4^9}$ 44. $\dfrac{1}{2^4 x^7}$ 45. $\dfrac{1}{p^2}$
46. $\dfrac{1}{m^6}$ 47. $5x^3 + 2x^2 - 5x + 7$
48. $-2x^4 - 4x^2 - 1$ 49. $7x^2 - 15x + 2$
50. $-4k^2 + 2k + 3$ 51. $72x^6 y^7$
52. $6m^7 - 15m^5 + 3m^3$ 53. $2y^2 - 7y - 4$
54. $12z^2 - 2wz - 4w^2$ 55. $3y^3 + 8y^2 + 12y - 5$
56. $r^4 - r^2 - 4r - 4$ 57. $9p^2 + 12p + 4$
58. $16a^2 - 8ab + b^2$ 59. $25k^2 - 16$
60. $4p^2 - 9q^2$ 61. $a^2 - 3a + 5$
62. $4x^3 + 6x^2 - 3x + 10$
63. $6p^2 + 7p + 1 + \dfrac{6}{2p - 2}$
64. $3z^2 - z + 4 + \dfrac{16}{5z - 2}$

CHAPTER 5

Section 5.1   (page 233)

1. 1, 2, 7, 14   3. 1, 3, 9, 27   5. 1, 3, 5, 9, 15, 45
7. 1, 2, 3, 4, 5, 6, 10, 12, 15, 20, 30, 60
9. 1, 2, 4, 5, 10, 20, 25, 50, 100   11. 1, 29
13. $2^3 \cdot 3 \cdot 5$ 15. $2^2 \cdot 3^2 \cdot 5$ 17. $5^2 \cdot 11$
19. $5^2 \cdot 19$ 21. 12   23. $10p^2$ 25. $6m^2 n$
27. 2   29. x 31. $3m^2$ 33. $2z^4$ 35. $2mn^4$
37. $7x^3 y^2$ 39. $12(x + 2)$ 41. $9a(a - 2)$
43. $5y^5(13y^4 - 7)$ 45. $11z^2 - 100$
47. $19y^2 p^2(y + 2p)$ 49. $13y^3(y^3 + 2y^2 - 3)$
51. $9qp^3(5q^3 p^2 - 4p^3 + 9q)$
53. $5z^3 a^3(25z^2 - 12za^2 + 17a)$
55. $11y^3(3y^5 - 4y^9 + 7 + y)$ 57. $(p + 4)(p + 3)$
59. $(a - 2)(a + 5)$ 61. $(z + 2)(7z - a)$
63. $(m + 3p)(5m - 2p)$ 65. $(a^2 + b^2)(3a + 2b)$
67. $(1 - a)(1 - b)$

Section 5.2   (page 239)

1. $x + 3$ 3. $r + 8$ 5. $t - 12$ 7. $x - 8$
9. $m + 6$ 11. $p - 1$ 13. $x - 5y$
15. $(y + 1)(y + 8)$ 17. $(b + 3)(b + 5)$
19. $(m + 5)(m - 4)$ 21. $(n - 2)(n + 6)$
23. $(r - 6)(r + 5)$ 25. prime
27. $(a + 9)(a - 11)$ 29. $(x - 4)(x - 5)$
31. $(z - 7)(z - 7)$ 33. $(r - 6)(r + 7)$
35. $(p - 6)(p + 11)$ 37. $(x + 2m)(x - 3m)$
39. $(z + 5x)(z - 3x)$ 41. $(a + 7y)(a - 8y)$
43. $(m - 3n)(m + n)$ 45. $(c - 4d)(c - d)$
47. $3y^3(y - 1)(y - 5)$ 49. $h^5(h + 2)(h - 7)$
51. $2x^4(x + 3)(x - 7)$ 53. $2y(y - 5)(y + 1)$
55. $mn(m - 3n)(m + n)$
57. $k^5(k + 3m)(k - 5m)$ 59. $x^7(x - 3w)(x + 8w)$
61. The factor $3x + 12$ can be factored further as $3(x + 4)$.   63. $a^2 + 15a + 54$

Section 5.3   (page 245)

1. $x - 1$ 3. $b - 3$ 5. $4y - 3$ 7. $5x + 4$
9. $m + 10$ 11. $3a - 4b$ 13. $k + 3m$
15. $2x^2 - 5x - 3$; $x - 3$ 17. $6m^2 + 7m - 20$; $2m + 5$ 19. $(2x + 1)(x + 3)$
21. $(a + 1)(3a + 7)$ 23. $(4r - 3)(r + 1)$
25. $(5m + 2)(3m - 1)$ 27. $(2m - 3)(4m + 1)$
29. $(a - 2)(5a + 3)$ 31. $(3r - 5)(r + 2)$
33. $(4y + 1)(y + 17)$ 35. $(19x + 2)(2x + 1)$
37. $(2x + 3)(5x - 2)$ 39. $(2w + 5)(3w + 2)$
41. $(2q + 3)(3q + 7)$ 43. $(5m - 4)(2m - 3)$
45. $(2k + 3)(4k - 5)$ 47. $(5m - 8)(2m + 3)$
49. $3(2x - 3)(4x - 1)$ 51. $q(5m + 2)(8m - 3)$
53. $2m(m + 5)(m - 4)$ 55. $2a^2(4a - 1)(3a + 2)$
57. $4z^3(8z + 3)(z - 1)$ 59. $(4p - 3q)(3p + 4q)$
61. $(5a + 3b)(5a + 2b)$ 63. $(3a - 5b)(2a + b)$
65. $m^4 n(2m + n)(3m + 2n)$
67. $3zy(3z + 7y)(2z - 5y)$

Section 5.4   (page 253)

1. $(x + 4)(x - 4)$ 3. $(p + 2)(p - 2)$

Answers to Selected Exercises   545

5. $(3m + 1)(3m - 1)$ **7.** $\left(5m + \frac{4}{7}\right)\left(5m - \frac{4}{7}\right)$
9. $4(3t + 2)(3t - 2)$ **11.** $(5a + 4r)(5a - 4r)$
13. prime   **15.** $(p^2 + 6)(p^2 - 6)$
17. $(a^2 + 1)(a + 1)(a - 1)$
19. $(m^2 + 9)(m + 3)(m - 3)$
21. $(4k^2 + 1)(2k + 1)(2k - 1)$ **23.** $(p + 1)^2$
25. $(y - 4)^2$ **27.** $(m - 10)^2$ **29.** $\left(r + \frac{1}{3}\right)^2$
31. not a perfect square   **33.** $(4a - 5b)^2$
35. $25(2m + 1)^2$ **37.** $(7x + 2y)^2$
39. $(2c + 3d)^2$ **41.** $(5h - 2y)^2$
43. $(y + 1)(y^2 - y + 1)$
45. $(2a + 1)(4a^2 - 2a + 1)$
47. $(3x - 5)(9x^2 + 15x + 25)$
49. $(2p + q)(4p^2 - 2pq + q^2)$
51. $(3a - 4b)(9a^2 + 12ab + 16b^2)$
53. $(4x + 5y)(16x^2 - 20xy + 25y^2)$
55. $(5m - 2p)(25m^2 + 10mp + 4p^2)$
57. $(4y - 11w)(16y^2 + 44wy + 121w^2)$
59. $(m^2 - 2)(m^4 + 2m^2 + 4)$
61. $(5z + 4r^2)(25z^2 - 20r^2z + 16r^4)$
63. $(3r^3 + 5s)(9r^6 - 15r^3s + 25s^2)$

Supplementary Factoring Exercises (page 255)

1. $8m^3(4m^6 + 2m^2 + 3)$ **3.** $7k(2k + 5)(k - 2)$
5. $(z + 5)(6z + 1)$ **7.** $(7z + 4y)(7z - 4y)$
9. $(2p + 3)(4p^2 - 6p + 9)$ **11.** $(5y - 6z)(2y + z)$
13. $(m + 5)(m - 3)$ **15.** $8z(4z - 1)(z + 2)$
17. $(z - 6)^2$ **19.** $(y + 2k)(y - 6k)$
21. $6(y - 2)(y + 1)$ **23.** $(p - 6)(p - 11)$
25. prime   **27.** $(z - 5a)(z + 2a)$
29. $(2k - 3)^2$ **31.** $(4r + 3m)^2$
33. $3k(k - 5)(k + 1)$ **35.** $4(2k - 3)^2$
37. $6y^4(3y + 4)(2y - 5)$ **39.** $5z(z - 7)(z - 2)$
41. $6(3m + 2z)(3m - 2z)$ **43.** $2(3a - 1)(a + 2)$
45. $(3r - 2s)(9r^2 + 6rs + 4s^2)$
47. $5m^2(5m - 3n)(5m - 13n)$ **49.** $(m - 2)^2$
51. $6k^2p(4k^2 + 10kp + 25p^2)$
53. $(4p + 3q)(3p - 2q)$
55. $4(4p + 5m)(4p - 5m)$
57. $(10a + 9y)(10a - 9y)$ **59.** $(a + 4)^2$

Section 5.5 (page 261)

1. $2, -4$ **3.** $-\frac{5}{3}, \frac{1}{2}$ **5.** $-\frac{1}{5}, \frac{1}{2}$ **7.** $-2, -3$
9. $-7, 4$ **11.** $-8, 3$ **13.** $3, -1$
15. $-1, -2$ **17.** -4 **19.** $\frac{1}{3}, -2$
21. $\frac{5}{2}, -2$ **23.** $-\frac{4}{3}, \frac{1}{2}$ **25.** $\frac{1}{3}, -\frac{5}{2}$
27. $-4, \frac{5}{2}$ **29.** $\frac{5}{3}, -4$ **31.** $\frac{2}{5}, -\frac{1}{3}$
33. $-\frac{5}{4}, \frac{5}{4}$ **35.** $5, 2$ **37.** $\frac{3}{2}, -3$
39. $-12, \frac{11}{2}$ **41.** $3, -1$ **43.** $\frac{1}{4}, 4$ **45.** $\frac{5}{2}, \frac{1}{3}, 5$
47. $-\frac{7}{2}, 3, -1$ **49.** $0, -\frac{7}{3}, \frac{7}{3}$ **51.** $0, 4, -2$
53. $0, -5, 4$

Section 5.6 (page 267)

1. length: 11 centimeters; width: 6 centimeters
3. 2 feet by 5 feet   **5.** 4 inches by 8 inches
7. height: 5 centimeters; base: 10 centimeters
9. area of base: 16 square meters; height: 6 meters
11. 19 feet and 18 feet   **13.** length: 6 feet; width: 5 feet   **15.** $4, 5$ or $-1, 0$ **17.** $12, 14, 16$ or $-2, 0, 2$ **19.** $-6, -5$ or $5, 6$ **21.** -2
23. 12 centimeters   **25.** 8 feet   **27.** 256 feet
29. 10 seconds   **31.** 48 feet   **33.** 48 feet

Chapter 5 Review Exercises (page 271)

1. $2^2 \cdot 3$ **2.** $2^4 \cdot 3$ **3.** $2 \cdot 5 \cdot 11$ **4.** $2^2 \cdot 3^2 \cdot 5$
5. $2^3 \cdot 3 \cdot 5^2$ **6.** 29 **7.** $6(p + 2)$
8. $20(2r^2 + 1)$ **9.** $25m^3(m - 2)$
10. $16z(2 + 3z)$ **11.** $6(1 - 3r^5 + 2r^3)$
12. $50y^3(2y^3 - 1 + 6y)$
13. $5m^2n^4(3m - 4n + 10mn^2)$
14. $8y^4r^2(4r - 6y + 3y^3r^3)$ **15.** $(2p + 3)(3p + 2)$
16. $(4y + 3)(y + 2)$ **17.** $(3m + 4p)(5m - 4p)$
18. $(2r + 3q)(6r - 5q)$
19. $6xyz(2xz^2 + 2y - 5x^2yz^3)$
20. $8abc(3b^2c - 7ac^2 + 9ab)$ **21.** $(m + 2)(m + 3)$
22. $(y + 8)(y + 5)$ **23.** $(r - 9)(r + 3)$
24. $(p + 6)(p - 5)$ **25.** $(z - 11)(z + 4)$
26. $(k + 9)(k - 7)$ **27.** $(z - 10x)(z - x)$
28. $(r - 12s)(r + 8s)$ **29.** $(p + 12q)(p - 10q)$
30. $4p(p + 2)(p - 5)$ **31.** $3z^2(z - 2)(z - 8)$
32. $(m - 6n)(m + 3n)$ **33.** $(y - 5z)(y - 3z)$
34. $p^5(p - 2q)(p + q)$ **35.** $3r^3(r + 3s)(r - 5s)$
36. $2a^3(a - 6)(a + 2)$ **37.** $5p^4(p - 7)(p - 2)$
38. $6m^7(m - 9)(m - 5)$ **39.** $(2k - 1)(k - 2)$
40. $(3z - 1)(z + 4)$ **41.** $(3r + 2)(2r - 3)$
42. $(5p + 1)(2p - 1)$ **43.** $(8y - 7)(y + 3)$
44. $(3k + 5)(k + 2)$ **45.** $3m(m - 5)(2m + 3)$
46. $4k^3(2k - 1)(3k - 1)$ **47.** $(7m - 2n)(m + 3n)$
48. $rs(2r + s)(5r + 6s)$ **49.** $z(2z - 1)(z + 5)$
50. $p^2(3p + 4)(p - 2)$ **51.** $(m + 5)(m - 5)$
52. $(5p + 11)(5p - 11)$ **53.** $(10a + 3)(10a - 3)$
54. $(7y + 5z)(7y - 5z)$ **55.** $36(2p + q)(2p - q)$
56. $(y^2 + 25)(y + 5)(y - 5)$ **57.** prime
58. $(z - 4)^2$ **59.** $(3r - 7)^2$ **60.** $(4m + 5n)^2$
61. prime   **62.** $6x(3x - 2)^2$
63. $(p - 3)(p^2 + 3p + 9)$
64. $(y + 2)(y^2 - 2y + 4)$
65. $(2m + 3p)(4m^2 - 6mp + 9p^2)$
66. $(5r - 6s)(25r^2 + 30rs + 36s^2)$
67. $(x + y + 4)(x^2 + 2xy + y^2 - 4x - 4y + 16)$
68. $4ab$ **69.** $\frac{1}{3}, -2$ **70.** $\frac{2}{5}, -4$ **71.** $-\frac{5}{2}, \frac{7}{3}$
72. $-1, -3$ **73.** $2, -5$ **74.** $4, 1$ **75.** $5, 3$
76. $5, 6$ **77.** $\frac{3}{2}, -\frac{1}{4}$ **78.** $5, -\frac{4}{3}$ **79.** $\frac{2}{5}$
80. $\frac{7}{10}, -\frac{7}{10}$ **81.** $6, -1$ **82.** 6

83. $\frac{1}{3}$, -1, -2 **84.** $-\frac{3}{2}$, 3, 1   **85.** 0, 3, -3
86. 2   **87.** width: 4 meters; length: 10 meters
88. width: 2 meters; length: 6 meters   **89.** width: 8 centimeters; length: 10 centimeters   **90.** base: 6 meters; height: 4 meters   **91.** length: 6 meters; width: 4 meters   **92.** length: 6 meters; height: 5 meters   **93.** 4, 6 or -2, 0   **94.** 6, 8 or -8, -6 **95.** width: 3 meters; length: 7 meters
96. width: 4 meters; length: 9 meters   **97.** 1
98. 2 or 3   **99.** -1 or 3   **100.** 5, 6, 7 or -5, -4, -3 **101.** -4, -2 or 8, 10
102. 16, 18 or -6, -4
103. 15 meters, 36 meters, 39 meters
104. 25 miles   **105.** 34 miles   **106.** 8 feet
107. 112 feet   **108.** 192 feet   **109.** 256 feet
110. after 8 seconds   **111.** 2 inches
112. 2 centimeters

Chapter 5 Test (page 277)

1. $8m(2m - 3)$ **2.** $6y(x + 2y)$
3. $14p(2q + 1 + 4p)$ **4.** $3mn(m + 3 + 2n)$
5. prime   **6.** $(x + 5)(x + 6)$
7. $(p + 7)(p - 1)$ **8.** $(2y + 3)(y - 5)$
9. $(2m - 1)(2m + 3)$ **10.** $(3x - 2)(x + 5)$
11. $(2z + 1)(5z + 1)$ **12.** $(5a + 1)(2a - 5)$
13. $(4r + 5)(3r + 1)$ **14.** prime
15. $(a + 5b)(a - 2b)$ **16.** $(3r - 2s)(2r + s)$
17. $(x + 5)(x - 5)$ **18.** $(5m + 7)(5m - 7)$
19. $(2p + 3)^2$ **20.** $(5z - 1)^2$ **21.** $4p(p + 2)^2$
22. $5m^2(2m + 1)(m + 5)$ **23.** -1, -2
24. $\frac{1}{3}$, -2 **25.** $\frac{5}{2}$, -4 **26.** 2, -2
27. 0, 4, -4 **28.** 3 inches by 5 inches
29. -1, 0   **30.** 5 meters, 12 meters, and 13 meters

CHAPTER 6

Section 6.1 (page 283)

1. 0   **3.** 4   **5.** -5 **7.** 3, 5   **9.** -1, $\frac{3}{2}$
11. never meaningless   **13.** (a) 1   (b) $\frac{14}{9}$
15. (a) 2   (b) $-\frac{14}{3}$ **17.** (a) $\frac{256}{15}$
(b) meaningless   **19.** (a) -5 (b) $\frac{5}{23}$
21. (a) -5 (b) $\frac{15}{2}$ **23.** (a) meaningless   (b) $\frac{1}{10}$
25. $2k$ **27.** $\frac{-4y^3}{3}$ **29.** $\frac{4m}{3p}$ **31.** $2r - 3$
33. $4m^2 - 3$ **35.** $\frac{8y + 5}{6}$ **37.** $\frac{3}{4}$ **39.** $\frac{x-1}{x+1}$
41. $\frac{m}{2}$ **43.** $4r + 2s$ or $2(2r + s)$ **45.** $\frac{m-2}{m+3}$

47. $\frac{x + 4}{x + 1}$ **49.** -1 **51.** $-(x + 1)$ or $-x - 1$
53. -1

Section 6.2 (page 289)

1. $\frac{3m}{4}$ **3.** $\frac{3}{32}$ **5.** $2a^4$ **7.** $\frac{1}{4}$ **9.** $\frac{1}{6}$
11. $\frac{2}{r^5}$ **13.** $\frac{6}{a+b}$ **15.** 2   **17.** $\frac{2}{9}$ **19.** $\frac{3}{10}$
21. $\frac{2r}{3}$ **23.** $(y + 4)(y - 3)$
25. $\frac{18}{(m-1)(m+2)}$ **27.** $-\frac{7}{8}$ **29.** -1
31. $\frac{a+1}{a-2}$ **33.** $\frac{k+2}{k+3}$ **35.** $\frac{z+4}{z-4}$
37. $\frac{4k-1}{3k-2}$ **39.** $\frac{m-3}{2m-3}$ **41.** $\frac{m+4p}{m+p}$
43. $\frac{10}{x+10}$

Section 6.3 (page 293)

1. 60   **3.** 120   **5.** 1800   **7.** x^5 **9.** $30p$
11. $150m$ **13.** $180y^4$ **15.** $300r^5$ **17.** $15a^5b^3$
19. $12p(p - 2)$ **21.** $32r^2(r - 2)$ **23.** $18(r - 2)$
25. $12p(p + 5)$ **27.** $p - q$ or $q - p$
29. $a(a + 6)(a - 3)$ **31.** $(k + 7)(k - 5)(k + 8)$
33. $(2y - 1)(y + 4)(y - 3)$
35. $(3r - 5)(2r + 3)(r - 1)$ **37.** $\frac{42}{66}$ **39.** $\frac{54}{6r}$
41. $-\frac{88}{8m}$ **43.** $\frac{24y^2}{70y^3}$ **45.** $\frac{60m^2k^3}{32k^4}$
47. $\frac{57z}{6z - 18}$ **49.** $\frac{-4a}{18a - 36}$
51. $\frac{6(k + 1)}{k(k - 4)(k + 1)}$

Section 6.4 (page 299)

1. $\frac{7}{p}$ **3.** $-\frac{3}{k}$ **5.** 1   **7.** $\frac{-2q}{3}$ **9.** m
11. $q - 2$ **13.** $\frac{2m + 3}{6}$ **15.** $\frac{28 - 5z}{20}$
17. $\frac{4y - 3}{3y}$ **19.** $\frac{m}{6}$ **21.** $\frac{k+2}{2}$ **23.** $\frac{5m+8}{6}$
25. $\frac{5m + 11}{4m}$ **27.** $\frac{6 - 2y}{y^2}$ **29.** $\frac{9p - 8}{2p^2}$
31. $\frac{-y + 3x}{x^2y}$ **33.** $\frac{4(x+6)}{(x-2)(x+2)}$
35. $\frac{x}{2(x+y)}$ **37.** $\frac{m}{3(m+3)(m-3)}$
39. $\frac{3}{(m+1)(m-1)(m+2)}$ **41.** $\frac{43m+1}{5m(m-2)}$
43. $\frac{9r+2}{r(r+2)(r-1)}$ **45.** $\frac{4k^2 - 10k + 5}{(k+4)(k-2)(k-1)}$
47. $\frac{11y^2 - y - 11}{(2y-1)(y+3)(3y+2)}$

Answers to Selected Exercises **547**

Section 6.5 (page 305)

1. $\dfrac{31}{50}$ 3. $\dfrac{5}{18}$ 5. $\dfrac{mp}{40}$ 7. $\dfrac{x(x+1)}{4(x-3)}$ 9. $\dfrac{2}{y}$

11. $\dfrac{8}{x}$ 13. $\dfrac{x^2+1}{4+xy}$ 15. $\dfrac{5(p+3)}{5+p}$

17. $\dfrac{40-12p}{85p}$ 19. $\dfrac{5y-2x}{3+4xy}$ 21. $\dfrac{a-2}{2a}$

23. $\dfrac{z-5}{4}$ 25. $\dfrac{-m}{m+2}$ 27. $\dfrac{3m(m-3)}{(m-1)(m-8)}$

29. $\dfrac{1}{3}$ 31. $\dfrac{19}{15}r$

Section 6.6 (page 311)

1. $\dfrac{2}{5}$ 3. 24   5. 2   7. -7 9. 1   11. 2

13. 5   15. -8 17. 8   19. $\dfrac{2}{7}$ 21. 6

23. 1   25. 4   27. 0   29. 3   31. 3

33. 3   35. no solution   37. 4, 6

39. no solution   41. $\dfrac{1}{2}, -6$ 43. 3   45. $\dfrac{4}{3}, 3$

47. $-\dfrac{6}{7}, 3$ 49. $q = \dfrac{kp}{3p-k}$ 51. $b = \dfrac{5c}{2ac+1}$

53. $r = \dfrac{E-RI}{I}$ 55. $b = \dfrac{2A-hB}{h}$

Supplementary Exercises on Rational Expressions (page 315)

1. $\dfrac{8}{m}$ 3. $\dfrac{3}{4x^2(x+3)}$ 5. $\dfrac{r+4}{r+1}$ 7. -43

9. $\dfrac{17}{3(y-1)}$ 11. $2, \dfrac{1}{5}$ 13. $\dfrac{-19}{18z}$

15. $\dfrac{3m+5}{(m+3)(m+2)(m+1)}$

Section 6.7 (page 321)

1. 9   3. $\dfrac{12}{17}$ 5. 6, 9   7. $\dfrac{2}{3}$ or -3

9. $\dfrac{2}{3}$ or 1   11. father's wage = $120, son's wage = $48   13. $32,000   15. 2 miles per hour

17. 150 miles each way   19. $\dfrac{37}{2}$ or $18\dfrac{1}{2}$ miles per hour   21. $\dfrac{6}{5}$ or $1\dfrac{1}{5}$ hours   23. $\dfrac{84}{19}$ or $4\dfrac{8}{19}$ hours

25. $\dfrac{28}{5}$ or $5\dfrac{3}{5}$ days   27. $\dfrac{100}{11}$ or $9\dfrac{1}{11}$ minutes

29. 3 hours   31. $\dfrac{63}{2}$ 33. 50   35. $\dfrac{5}{2}$

37. 69.08 centimeters   39. 100 amp

Chapter 6 Review Exercises (page 325)

1. 0   2. 5   3. 4, -2 4. $\dfrac{1}{2}, -3$

5. (a) $\dfrac{9}{5}$ (b) 1   6. (a) $\dfrac{18}{5}$ (b) $\dfrac{2}{3}$ 7. (a) $\dfrac{24}{7}$

(b) 8   8. (a) meaningless   (b) $\dfrac{3}{2}$ 9. $\dfrac{2z}{3y^2}$

10. $\dfrac{3x-4}{2}$ 11. -1 12. $\dfrac{2k+5y}{4k-y}$ 13. $36y$

14. $\dfrac{16}{9}$ 15. $\dfrac{40}{3p^2}$ 16. $\dfrac{1}{z^5}$ 17. $\dfrac{5}{4}$ 18. $\dfrac{1}{9}$

19. $\dfrac{7}{2}$ 20. $\dfrac{k-2}{2(3k-5)}$ 21. $\dfrac{a-2}{a-1}$ 22. $\dfrac{z-1}{z-3}$

23. $\dfrac{(2p-1)(2p+3)}{(p+3)^2}$ 24. $\dfrac{8r+1}{r-7}$ 25. 90   26. $48y^3$

27. $y(y+2)(y+4)$ 28. $(z-2)(z+3)(z+1)$

29. $\dfrac{20}{45}$ 30. $\dfrac{60}{5m}$ 31. $\dfrac{9m}{24m^3}$ 32. $\dfrac{96}{8y-32}$

33. $\dfrac{-10k}{15k+75}$ 34. $\dfrac{12y(y-4)}{(y-2)(y+1)(y-4)}$

35. $\dfrac{13}{m}$ 36. $\dfrac{1}{r}$ 37. $\dfrac{12+p}{3p}$ 38. $\dfrac{35-2k}{5k}$

39. $\dfrac{6+9m}{4}$ 40. $\dfrac{16-3r}{2r^2}$

41. $\dfrac{14p+q}{(2p-q)(2p+q)}$ 42. $\dfrac{-y-5}{(y+1)(y-1)}$

43. $\dfrac{16-3p}{5(p+2)(p-2)}$ 44. $\dfrac{z+7}{(z+1)(z-1)^2}$

45. $\dfrac{5m-12}{m(m+5)(m-2)}$ 46. $\dfrac{8p-30}{p(p-2)(p-3)}$

47. $\dfrac{1}{q^3 r^3}$ 48. $\dfrac{3(m-2)}{m+2}$ 49. $\dfrac{8(5k-1)}{4k+3}$

50. 4   51. $\dfrac{(n-m)^2}{mn}$ 52. $\dfrac{zx+1}{zx-1}$

53. $\dfrac{1-a-b}{1+a+b}$ 54. $\dfrac{(p+q)(2+3p-3q)}{(p-q)(5p+5q-4)}$

55. $\dfrac{10}{3}$ 56. -4 57. -6 58. $-2, 3$

59. no solution   60. 3   61. $r = \dfrac{3zk}{5k-z}$

62. $b = \dfrac{-4ac}{2c-a}$ or $\dfrac{4ac}{a-2c}$ 63. $y = \dfrac{xz}{3-2x}$

64. $m = \dfrac{p^2 q}{3p^2 - 4}$ 65. 12   66. $\dfrac{15}{11}$ 67. 1

68. $\dfrac{6}{5}$ or $\dfrac{1}{2}$ 69. small car, $120; large car, $180

70. 3 miles per hour   71. 150 kilometers per hour

72. $\dfrac{40}{13}$ or $3\dfrac{1}{13}$ hours   73. 6 days   74. 2 days

75. $\dfrac{36}{5}$ 76. 12   77. $\dfrac{45}{2}$ 78. $\dfrac{40}{3}$

Chapter 6 Test (page 329)

1. 3, 1   2. $\dfrac{5s(s-1)}{2}$ 3. $\dfrac{4p-3q}{p+q}$ 4. x

5. $\dfrac{40}{27}$ 6. $\dfrac{3m-2}{3m+2}$ 7. $\dfrac{a+3}{a+4}$ 8. $\dfrac{77r}{49r^2}$

9. $\dfrac{15}{24m-48}$ 10. $-\dfrac{1}{x}$ 11. $\dfrac{-13}{6(a+1)}$

12. $\dfrac{5k+2}{(k+2)(2k-1)(k+1)}$

13. $\dfrac{4m+20}{(2m-5)(m+2)(2m+1)}$ 14. $\dfrac{3(1-2k)}{k(1+3k)}$

15. $\dfrac{-2p - 7}{2p + 9}$ **16.** 6 **17.** 1 **18.** $\dfrac{1}{4}$ or $\dfrac{1}{2}$
19. $\dfrac{20}{9}$ hours **20.** $\dfrac{56}{3}$

91. 1, −2 **92.** $\dfrac{2}{5}$, 3 **93.** $\dfrac{6}{5}$ hours **94.** $\dfrac{18}{5}$ hours
95. 3 days **96.** 4 hours **97.** $\dfrac{5}{2}$ **98.** 60

Cumulative Review Exercises (page 331)

1. $8x(x - 3)$ **2.** $11q^3(5 - 8q)$ **3.** $9q^5(r + 2qp)$
4. prime **5.** prime **6.** $6y^3z^2(9yz + 8y^2 - z^2)$
7. $(x + 5)(x + 1)$ **8.** $(m + 2)(m + 7)$
9. $(p + 1)(p - 5)$ **10.** $(r + 5)(r - 8)$
11. $(p + 2q)(p - 9q)$ **12.** $(b + 2d)(b - 11d)$
13. $(2a - 1)(a + 4)$ **14.** $(m + 2)(4m - 7)$
15. $(2m + 3)(5m + 2)$ **16.** $(5x + 3y)(3x - 2y)$
17. $(4t + 3v)(2t + v)$ **18.** $(9r + s)(2r - 3s)$
19. $(3x + 1)^2$ **20.** $(2p - 3)^2$ **21.** $(5m - 3p)^2$
22. $(4t + 7z)^2$ **23.** $(5r + 9t)(5r - 9t)$
24. $(10p - 7q)(10p + 7q)$
25. $(10y - 13p^2)(10y + 13p^2)$ **26.** prime
27. $16(3m + 2p)(3m - 2p)$
28. $25(2p + 5q)(2p - 5q)$ **29.** $m(2a - 1)(3a + 2)$
30. $2pq(3p + 1)(p + 1)$
31. $(6z + 5)(36z^2 - 30z + 25)$
32. $(4r - 3)(16r^2 + 12r + 9)$
33. $(2p - 5q)(4p^2 + 10pq + 25q^2)$
34. $(3m + 4p)(9m^2 - 12mp + 16p^2)$
35. $125(2a - m)(4a^2 + 2am + m^2)$
36. $(5y + 6z)(25y^2 - 30yz + 36z^2)$ **37.** $\dfrac{3}{2}$, −2
38. −3, 2 **39.** −3, 5 **40.** −7, 11
41. $\dfrac{1}{2}$, $-\dfrac{2}{3}$ **42.** $-\dfrac{3}{4}$, −5 **43.** 0, 8
44. −2, 2 **45.** 2, −1, 3 **46.** 5, $-\dfrac{1}{2}$, $\dfrac{5}{3}$
47. 0, −4, 4 **48.** 0, −5, 5 **49.** −2 or −1
50. −1 or 10 **51.** 5 and 3 **52.** 6 meters
53. 4 centimeters **54.** 3 centimeters **55.** $\dfrac{5}{3}$
56. $-\dfrac{5}{2}$ **57.** 2, −5 **58.** $\dfrac{5}{2}$, −4 **59.** none
60. none **61.** $\dfrac{m}{4}$ **62.** $\dfrac{1}{4p^2}$ **63.** $\dfrac{y + 2}{6}$
64. $\dfrac{5}{r + 1}$ **65.** $\dfrac{1}{xy^6}$ **66.** $\dfrac{s^9}{r^{11}}$ **67.** $\dfrac{7}{m}$ **68.** $\dfrac{4}{q}$
69. $\dfrac{4z - 3}{3z}$ **70.** $\dfrac{3r + 28}{7r}$ **71.** $\dfrac{7}{2m}$ **72.** $\dfrac{51}{8y}$
73. $\dfrac{2 + m}{m + 1}$ **74.** $\dfrac{p - 7}{p - 5}$ **75.** $\dfrac{11}{3(a - 2)}$
76. $\dfrac{9}{2(m + 3)}$ **77.** $\dfrac{26}{15(y + 3)}$ **78.** $\dfrac{7}{15(q - 4)}$
79. $\dfrac{3a - 3}{a(a - 2)(a + 1)}$ **80.** $\dfrac{-k - 5}{k(k + 1)(k - 1)}$
81. $\dfrac{1}{3(m - 1)}$ **82.** $\dfrac{7(2z + 1)}{24}$ **83.** $\dfrac{x + 3}{x + 1}$
84. $\dfrac{5a - 2}{4(5a - 1)}$ **85.** $\dfrac{5}{x + 1}$ **86.** $\dfrac{z^4}{3y^5}$
87. $\dfrac{2x - 5}{3x + 4}$ **88.** $\dfrac{195}{29}$ **89.** $\dfrac{21}{2}$ **90.** −5

CHAPTER 7

Section 7.1 (page 339)

1. yes **3.** yes **5.** no **7.** yes **9.** yes
11. no **13.** (2, 11) **15.** (8, 29)
17. (−3, −4) **19.** (3, 14) **21.** (1, 8)
23. (2, 0) **25.** (−4, 24) **27.** (4, −8)
29. (2, 1), (0, −5), (−3, −14) **31.** (4, −22), (0, −2), (−4, 18) **33.** (1, 7), (0, 4), (−2, −2)
35. (0, 4), (3, 0), (−3, 8) **37.** (9, 0), (0, −4), (18, 4) **39.** (7.265, 3.51), (−2.48, −2.987), (4.655, 1.77) **41.** (2.9, 1.533), (8.725, 9.3), (.625, −1.5) **43.** (2, 3), (2, 8), (2, 0)
45. (4, −8), (0, −8), (−4, −8) **47.** (9, −4), (2, −4), (0, −4)

Section 7.2 (page 345)

1. (2, 5) **3.** (−5, 5) **5.** (7, 3)
7.–17.

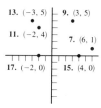

19. I **21.** II **23.** III **25.** II **27.** IV
29. (0, 3), (6, 0), (4, 1), $\left(-1, \dfrac{7}{2}\right)$ **31.** (0, −4), (3, 0), (6, 4), $\left(1\dfrac{1}{2}, -2\right)$

33. (5, −2), (0, −2), (−3, −2), (−2, −2) **35.** (a)

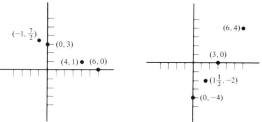

(b) yes

Section 7.3 (page 353)

1. (0, 5), (5, 0), (2, 3)   **3.** (0, 4), (−4, 0), (−2, 2)

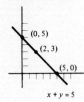
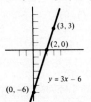

5. (0, −6), (2, 0), (3, 3)

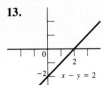

7. (3, 0), (0, 2)   **9.** (3, 0), $\left(0, -\dfrac{9}{5}\right)$
11. (0, 0), (0, 0)

13. 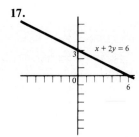   **15.**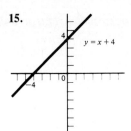

17. 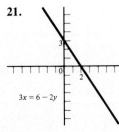   **19.**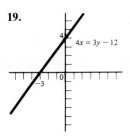

21. 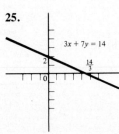   **23.**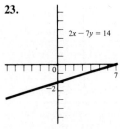

25. 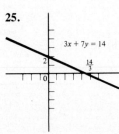   **27.**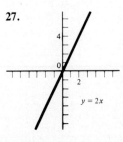

29. 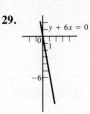   **31.**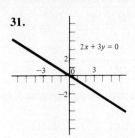

33. 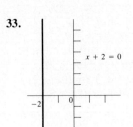   **35.**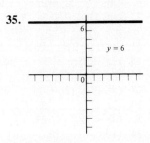

37. (a) about $10,500   **(b)** about $5700
(c) about $6500

Section 7.4 (page 363)

1. $-\dfrac{1}{2}$ **3.** $\dfrac{2}{7}$ **5.** 0   **7.** $\dfrac{7}{6}$ **9.** $\dfrac{9}{2}$

11. $-\dfrac{5}{8}$ **13.** 0   **15.** $\dfrac{1}{3}$ **17.** $-\dfrac{7}{2}$ **19.** 0

21. −1.020   **23.** 5   **25.** 1   **27.** 9   **29.** $\dfrac{3}{2}$

31. $\dfrac{2}{3}$ **33.** $-\dfrac{9}{7}$ **35.** 4; 4; parallel   **37.** 1; 1; parallel   **39.** 1; −1; perpendicular   **41.** $\dfrac{3}{2}$; $-\dfrac{2}{3}$; perpendicular   **43.** $\dfrac{4}{3}$; $\dfrac{4}{3}$; parallel   **45.** $\dfrac{8}{9}$; $-\dfrac{4}{3}$; neither   **47.** −6; $\dfrac{1}{6}$; perpendicular   **49.** 0; 0; parallel

Section 7.5 (page 369)

1. $y = 3x + 5$ **3.** $y = -x - 6$
5. $y = \dfrac{5}{3}x + \dfrac{1}{2}$ **7.** $y = 8x$

9.    **11.**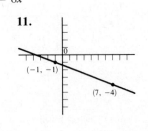

13. 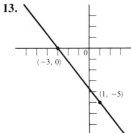 **15.**

17. 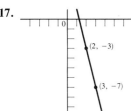 **19.**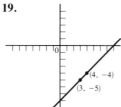

21. 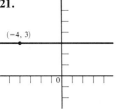 **23.**

25. $3x - y = -1$ **27.** $4x + y = 3$
29. $4x - 5y = 28$ **31.** $5x + 9y = -58$
33. $x - y = 3$ **35.** $5x + 7y = -54$
37. $3x + 2y = -31$ **39.** $x - 2y = 4$
41. $7x + 8y = -35$ **43.** $3x + 9y = 22$
45. $y = 9x + 50$ **47.** $y = 3.5x + 70.5$
49. (a) $(1, 24), (5, 48)$ **(b)** $y = 6x + 18$

Section 7.6 (page 377)

1. 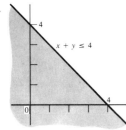 **3.**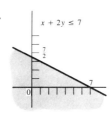

5. 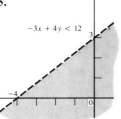 **7.**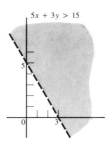

9.  **11.**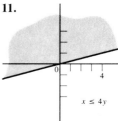

13. 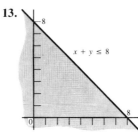 **15.**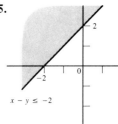

17. 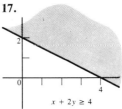 **19.**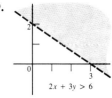

21. 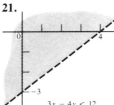 **23.**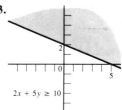

25.  **27.**

29.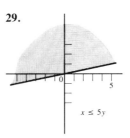

Answers to Selected Exercises 551

31. (a)

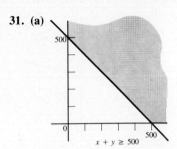

(b) (400, 200), (500, 100), (200, 500) are some of the many possibilities.

Chapter 7 Review Exercises (page 381)

1. $(-1, -5), (0, -2), \left(\frac{7}{3}, 5\right)$

2. $(0, 3), \left(\frac{9}{4}, 0\right), \left(-2, \frac{17}{3}\right)$

3. (0, 0), (8, 4), (−6, −3)
4. (−4, −3), (−4, 0), (−4, 5)   **5.** yes   **6.** no
7. yes   **8.** yes
9.–14.

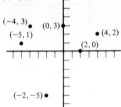

15. II   **16.** III   **17.** none   **18.** none
19. $\left(\frac{5}{2}, 0\right), (0, -5)$ **20.** $\left(\frac{7}{2}, 0\right), (0, 7)$
21. $\left(\frac{8}{3}, 0\right), (0, -8)$

22. 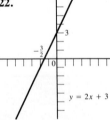   **23.**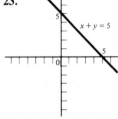

24.    **25.**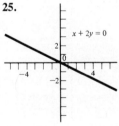

26. 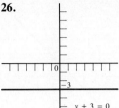   **27.**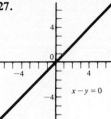

28. $\frac{2}{3}$ **29.** $\frac{1}{3}$ **30.** 2   **31.** undefined   **32.** 3
33. $\frac{2}{3}$ **34.** $-\frac{4}{9}$ **35.** 0   **36.** 0
37. undefined   **38.** $-\frac{6}{7}$ **39.** $\frac{11}{3}$ **40.** $-\frac{8}{3}$
41. parallel   **42.** perpendicular
43. perpendicular   **44.** neither   **45.** $3x - y = 2$
46. $4x + 4y = 3$ **47.** $2x - 3y = -15$
48. $x + 4y = -5$ **49.** $3x + y = 30$
50. $x - y = 7$ **51.** $2x - 3y = -14$
52. $3x + 4y = -1$ **53.** $x + 4y = 6$
54. $4x + 7y = -23$ **55.** $x = 5$ **56.** $y = 6$
57. $x = \frac{1}{2}$

58. 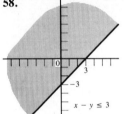   **59.**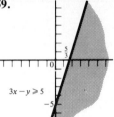

60. 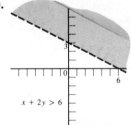   **61.**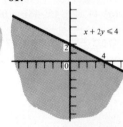

62.    **63.**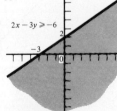

64.    **65.**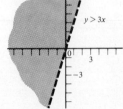

66.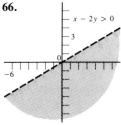

19. 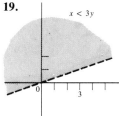   **20.**

Chapter 7 Test (page 385)

1. (0, −6), (−2, −16), (4, 14)
2. (0, −6), (10, 0), (5, −3)   **3.** (0, 0), (6, 2), $\left(8, \frac{8}{3}\right)$, (−12, −4)   **4.** (5, 2), (4, 2), (0, 2), (−3, 2)
5. (4, 0), (0, 4)   **6.** (3, 0), (0, 6)

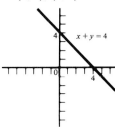   

7. (6, 0), $\left(0, -\frac{9}{2}\right)$ **8.** (0, 0), (0, 0)

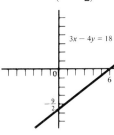   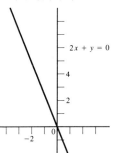

9. (−5, 0), no y-intercept   **10.** no x-intercept; (0, 2)

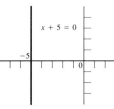

11. $-\frac{3}{7}$ **12.** $-\frac{3}{4}$ **13.** $4x + y = 1$
14. $3x - y = 1$ **15.** $3x + 4y = 8$
16. $9x - y = -12$
17.   **18.**

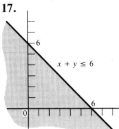   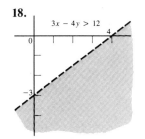

CHAPTER 8

Section 8.1 (page 393)

1. yes   **3.** no   **5.** no   **7.** yes   **9.** no
11. yes   **13.** (4, 2)   **15.** (0, 4)   **17.** (2, 2)
19. (1, 3)   **21.** (3, −2)   **23.** (2, −1)
25. (−4, 1)   **27.** (−4, 3)   **29.** (5, 0)
31. (0, −2)   **33.** (1, 2)   **35.** no solution
37. no solution   **39.** same line   **41.** no solution

Section 8.2 (page 401)

1. (1, −2)   **3.** (2, 0)   **5.** (6, 2)   **7.** (−2, 6)
9. $\left(\frac{1}{2}, 2\right)$ **11.** $\left(\frac{3}{2}, -2\right)$ **13.** (2, −3)
15. (4, 4)   **17.** (2, −5)   **19.** (3, 0)
21. (4, 9)   **23.** (−4, 0)   **25.** (4, −3)
27. (−9, −11)   **29.** (6, 3)   **31.** (6, −5)
33. (−6, 0)   **35.** $\left(\frac{1}{2}, 3\right)$ **37.** $\left(\frac{3}{8}, \frac{5}{6}\right)$
39. (11, 15)   **41.** $\left(\frac{22}{9}, \frac{8}{9}\right)$ **43.** (2, −3)
45. (−7, −1)

Section 8.3 (page 407)

1. no solution   **3.** no solution   **5.** same line
7. no solution   **9.** no solution   **11.** same line
13. (0, 0)   **15.** (3, 2)   **17.** no solution
19. same line   **21.** same line

Section 8.4 (page 413)

1. (2, 4)   **3.** (6, 4)   **5.** (8, −1)   **7.** (2, −4)
9. (5, 1)   **11.** no solution   **13.** (4, 1)
15. (2, −3)   **17.** same line   **19.** (5, −3)
21. (2, 8)   **23.** (4, −6)   **25.** (3, 2)
27. (7, 0)   **29.** same line   **31.** (6, 5)
33. (0, 3)   **35.** (18, −12)   **37.** (3, 2)
39. (3, −1)

Section 8.5 (page 421)

1. 43, 9   **3.** 72, 24   **5.** 55°, 35°
7. width: 10 inches; length: 20 inches   **9.** 22 tens and 63 twenties   **11.** 74 at 8¢ and 96 at 10¢
13. 80 fives and 44 tens   **15.** $7500 at 10% and $2500 at 14%   **17.** 4 liters of 90% and 16 liters of 75%   **19.** 4 liters of 25% and 8 liters of 55%

Answers to Selected Exercises   553

21. 30 pounds of $6 per pound coffee and 60 pounds of $3 per pound coffee   23. 30 barrels of $40 per barrel olives and 20 barrels of $60 per barrel olives   25. boat: 10 miles per hour; current: 2 miles per hour   27. plane: 470 miles per hour; wind: 70 miles per hour   29. John: $3\frac{1}{4}$ miles per hour; Harriet: $2\frac{3}{4}$ miles per hour   31. 3 boys and 4 girls

Section 8.6 (page 427)

1.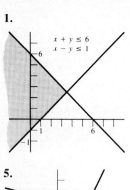
3.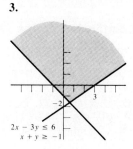
5.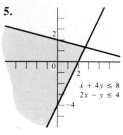
7.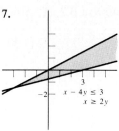
9.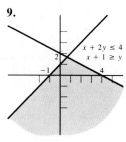
11.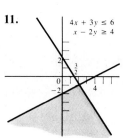
13.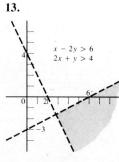
15.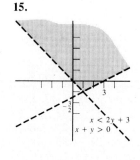
17.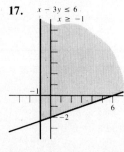

Chapter 8 Review Exercises (page 429)

1. no   2. yes   3. no   4. yes   5. yes
6. yes   7. (2, 1)   8. (−2, 2)   9. (−1, 1)
10. same line   11. (−1, −2)   12. no solution
13. (2, 4)   14. (5, 2)   15. (3, −2)
16. (1, −1)   17. (4, −1)   18. (3, 2)
19. (−1, 2)   20. (3, −2)   21. same line
22. no solution   23. same line   24. same line
25. (4, 2)   26. (−5, −3)   27. (5, 3)
28. (−1, −1)   29. (4, −1)   30. (4, −3)
31. (9, 2)   32. $\left(\frac{10}{7}, -\frac{9}{7}\right)$ 33. (−6, 1)
34. (2, 1)   35. (3, 1)   36. $\left(\frac{4}{7}, \frac{158}{7}\right)$
37. 28, 14   38. 5, 12   39. 8 meters by 12 meters
40. 5 twenties and 15 tens   41. $7000 at 12% and $11,000 at 16%   42. 12.5 pounds of $1.30 candy and 37.5 pounds of $.90 candy   43. 40 liters of 40% and 20 liters of 70%   44. plane: 250 miles per hour; wind: 20 miles per hour

45.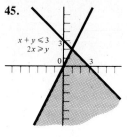
46.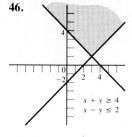
47.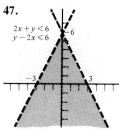
48.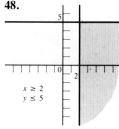
49.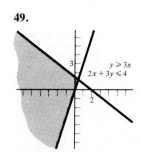
50.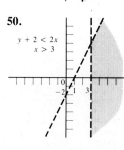

Chapter 8 Test (page 433)

1. (4, −3)   2. (6, −5)   3. (4, 1)   4. (6, 5)
5. (8, −2)   6. (2, −5)   7. no solution
8. (2, 2)   9. $\left(-\frac{11}{14}, -\frac{8}{7}\right)$ 10. (3, −5)
11. (−6, 8)   12. same line   13. (−6, 4)
14. 12, 27   15. 2 at $2.50 and 4 at $3.75

16. 75 liters of 40% and 25 liters of 60%
17. 45 miles per hour and 60 miles per hour
18.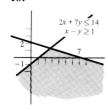
19.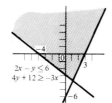
20.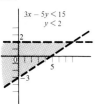

Cumulative Review Exercises (page 437)

1. $(0, 4), (7, 0), \left(3, \frac{16}{7}\right), \left(-\frac{7}{4}, 5\right)$
2. $(0, 6), (-3, 0), (-2, 2), \left(-\frac{3}{2}, 3\right)$
3. $(0, 0), (6, 3), (-10, -5), \left(3, \frac{3}{2}\right)$
4. $(0, -3), (2, -3), (-4, -3), \left(\frac{8}{3}, -3\right)$

5.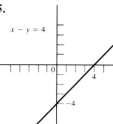
6.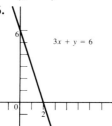
7.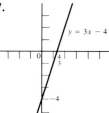
8.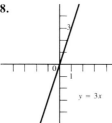
9.
10.

11. $-\frac{4}{3}$ 12. -6 13. 4 14. $-\frac{1}{2}$
15. parallel 16. perpendicular 17. neither
18. perpendicular 19. $2x + y = 3$
20. $4x - 3y = -3$ 21. $10x + 4y = -1$
22. $9x + 24y = -4$ 23. $3x - y = 11$
24. $2x - 3y = 14$

25.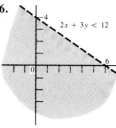
26.

27.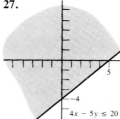
28.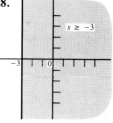

29.
30.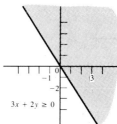

31. no 32. $(-1, 6)$ 33. $(3, -4)$
34. $(2, -1)$ 35. no solution 36. same line
37. $\left(\frac{1}{2}, 1\right)$ 38. $15, 9$ 39. 9 meters
40. 9 fives and 8 tens 41. $30,000 at 10% and $45,000 at 8%

42.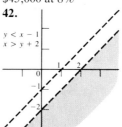
43.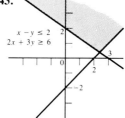

44.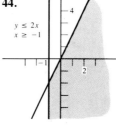

Answers to Selected Exercises 555

CHAPTER 9

Section 9.1 (page 445)

1. $3, -3$ 3. $11, -11$ 5. $\frac{20}{9}, -\frac{20}{9}$
7. $25, -25$ 9. $39, -39$ 11. $63, -63$ 13. 2
15. 5 17. -8 19. 13 21. 30 23. -41
25. $\frac{6}{7}$ 27. $-\frac{10}{13}$ 29. not a real number
31. rational; 4   33. irrational; 3.873
35. irrational; 6.856   37. rational; -11
39. irrational; 10.488   41. rational; 20
43. irrational; -14.142 45. irrational; 23.875
47. $c = 10$ 49. $b = 15$ 51. $c = \sqrt{164}$
53. $a = \sqrt{95}$ 55. 10 57. 5 59. -2
61. -2 63. 1 65. not a real number
67. -3 69. -2 71. 2.280 73. 1.092
75. 4.036

Section 9.2 (page 451)

1. 4 3. 6 5. 21 7. $\sqrt{21}$ 9. $3\sqrt{3}$
11. $2\sqrt{7}$ 13. $3\sqrt{2}$ 15. $4\sqrt{3}$ 17. $5\sqrt{5}$
19. $10\sqrt{7}$ 21. $30\sqrt{3}$ 23. $4\sqrt{5}$ 25. 36
27. 60 29. $7\sqrt{3}$ 31. $15\sqrt{3}$ 33. $20\sqrt{3}$
35. $10\sqrt{10}$ 37. $\frac{10}{3}$ 39. $\frac{6}{7}$ 41. $\frac{\sqrt{5}}{4}$
43. $\frac{\sqrt{30}}{7}$ 45. $\frac{2}{5}$ 47. $\frac{4}{25}$ 49. 5 51. 4
53. $3\sqrt{5}$ 55. $5\sqrt{10}$ 57. y 59. $\sqrt{xz}$
61. x 63. x^2 65. xy^2 67. $x\sqrt{x}$ 69. $\frac{4}{x}$
71. $\frac{\sqrt{11}}{r^2}$ 73. $2\sqrt[3]{5}$ 75. $3\sqrt[3]{2}$ 77. $4\sqrt[3]{2}$
79. $2\sqrt[4]{5}$ 81. $\frac{2}{3}$ 83. $\frac{10}{3}$

Section 9.3 (page 457)

1. $7\sqrt{3}$ 3. $-5\sqrt{7}$ 5. $2\sqrt{6}$ 7. $3\sqrt{17}$
9. $4\sqrt{7}$ 11. $7\sqrt{5}$ 13. $3\sqrt{3}$ 15. $20\sqrt{2}$
17. $-13\sqrt{2}$ 19. $19\sqrt{7}$ 21. $-12\sqrt{5}$
23. $-16\sqrt{2} + 8\sqrt{3}$ 25. $20\sqrt{2} + 6\sqrt{3} + 15\sqrt{5}$
27. $2\sqrt{2}$ 29. $3\sqrt{3} - 2\sqrt{5}$ 31. $3\sqrt{21}$
33. $5\sqrt{3}$ 35. $-\sqrt[3]{2}$ 37. $24\sqrt[3]{3}$
39. $10\sqrt[4]{2} + 4\sqrt[4]{8}$ 41. $6\sqrt{x}$ 43. $13\sqrt{a}$
45. $15x\sqrt{3}$ 47. $2x\sqrt{2}$ 49. $6\sqrt[3]{p^2}$ 51. $14m$

Section 9.4 (page 463)

1. $\frac{6\sqrt{5}}{5}$ 3. $\sqrt{5}$ 5. $\frac{3\sqrt{7}}{7}$ 7. $\frac{8\sqrt{15}}{5}$
9. $\frac{\sqrt{30}}{2}$ 11. $\frac{8\sqrt{3}}{9}$ 13. $\frac{3\sqrt{2}}{10}$ 15. $\sqrt{2}$
17. $\frac{9\sqrt{2}}{8}$ 19. 2 21. $\sqrt{2}$ 23. $\frac{2\sqrt{30}}{3}$
25. $\frac{\sqrt{2}}{2}$ 27. $\frac{\sqrt{70}}{7}$ 29. $\frac{3\sqrt{5}}{5}$ 31. $\frac{\sqrt{21}}{3}$
33. $\frac{\sqrt{15}}{10}$ 35. $\frac{3\sqrt{14}}{4}$ 37. $\frac{\sqrt{3}}{5}$ 39. $\frac{5\sqrt{3}}{12}$
41. $\frac{\sqrt{6p}}{p}$ 43. $\frac{p\sqrt{2pm}}{m}$ 45. $\frac{x\sqrt{y}}{2y}$ 47. $\frac{3a\sqrt{5r}}{5}$
49. $\frac{\sqrt[3]{4}}{2}$ 51. $\frac{\sqrt[3]{2}}{4}$ 53. $\frac{\sqrt[3]{121}}{11}$ 55. $\frac{\sqrt[3]{50}}{5}$
57. $\frac{\sqrt[3]{6}}{2}$ 59. $\frac{\sqrt[3]{42}}{6}$

Section 9.5 (page 469)

1. $27\sqrt{5}$ 3. $21\sqrt{2}$ 5. -4 7. $\sqrt{15} + \sqrt{35}$
9. $2\sqrt{10} + 10$ 11. $-4\sqrt{7}$ 13. $21 - \sqrt{6}$
15. $87 + 9\sqrt{21}$ 17. $34 + 24\sqrt{2}$
19. $37 - 12\sqrt{7}$ 21. 7 23. -4 25. 1
27. $2\sqrt{3} - 2 + 3\sqrt{2} - \sqrt{6}$ 29. $2 + 2\sqrt{2}$
31. $\sqrt{30} + \sqrt{15} + 6\sqrt{5} + 3\sqrt{10}$ 33. $\sqrt{7} - 2$
35. $\frac{\sqrt{3} + 5}{4}$ 37. $\frac{6 - \sqrt{10}}{2}$ 39. $\frac{3 - \sqrt{2}}{7}$
41. $-5(2 - \sqrt{5})$ or $-10 + 5\sqrt{5}$
43. $-2 - \sqrt{11}$ 45. $-\sqrt{2} + 2$
47. $\frac{-\sqrt{5} - 5}{4}$ 49. $3 - \sqrt{3}$ 51. $\frac{-3 + 5\sqrt{3}}{11}$
53. $\frac{\sqrt{6} + \sqrt{2} + 3\sqrt{3} + 3}{2}$
55. $\frac{-6\sqrt{2} + 12 + \sqrt{10} - 2\sqrt{5}}{2}$

Section 9.6 (page 477)

1. 4 3. 1 5. 7 7. -7 9. -5
11. no solution   13. 49 15. no solution
17. 9 19. 16 21. 6 23. 7
25. no solution   27. 6 29. 12 31. 5
33. $0, -1$ 35. 5 37. 3 39. 9 41. 12
43. 21 45. 8 47. (a) 90 miles per hour
(b) 120 miles per hour   (c) 60 miles per hour
(d) 60 miles per hour

Chapter 9 Review Exercises (page 481)

1. $7, -7$ 2. $9, -9$ 3. $14, -14$ 4. $11, -11$
5. $15, -15$ 6. $27, -27$ 7. $49, -49$
8. $87, -87$ 9. 4 10. -6 11. 20
12. -52 13. not a real number   14. -65
15. 3 16. -4 17. -2 18. 2 19. -2
20. not a real number   21. 2 22. -3
23. -2 24. 2 25. irrational; 3.873
26. rational; 8   27. rational; -13
28. irrational; -13.038 29. $\sqrt{10}$ 30. 6
31. $5\sqrt{3}$ 32. 8 33. $3\sqrt{3}$ 34. $4\sqrt{3}$
35. $4\sqrt{10}$ 36. $8\sqrt{5}$ 37. 70 38. 18
39. $16\sqrt{6}$ 40. $25\sqrt{10}$ 41. $\frac{3}{2}$ 42. $\frac{11}{20}$
43. $\frac{\sqrt{3}}{5}$ 44. $\frac{\sqrt{10}}{13}$ 45. $\frac{\sqrt{5}}{6}$ 46. $\frac{2}{15}$

47. $3\sqrt{5}$ **48.** 8 **49.** 15 **50.** p **51.** $\sqrt{km}$
52. y **53.** r^9 **54.** x^5y^8 **55.** $x^4\sqrt{x}$ **56.** $\dfrac{6}{p}$
57. $\dfrac{\sqrt{13}}{k^2}$ **58.** $\dfrac{10}{y^5}$ **59.** $3\sqrt[3]{2}$ **60.** $5\sqrt[3]{2}$
61. $\dfrac{\sqrt[3]{5}}{2}$ **62.** $\dfrac{\sqrt[3]{6}}{5}$ **63.** $2\sqrt[3]{x^2}$ **64.** $3x\sqrt[3]{x}$
65. $2\sqrt{7}$ **66.** $8\sqrt{2}$ **67.** $21\sqrt{3}$ **68.** $12\sqrt{3}$
69. 0 **70.** $3\sqrt{7}$ **71.** $2\sqrt{3} - 3\sqrt{10}$
72. $2\sqrt{2}$ **73.** $6\sqrt{30}$ **74.** $5\sqrt{15} - \sqrt{3}$
75. $8 + 6\sqrt[4]{4}$ **76.** $5\sqrt[3]{2}$ **77.** $11\sqrt{x}$ **78.** 0
79. $-m\sqrt{5}$ **80.** $11k^2\sqrt{2n}$ **81.** $\dfrac{10\sqrt{3}}{3}$
82. $\dfrac{5\sqrt{2}}{2}$ **83.** $\dfrac{3\sqrt{10}}{5}$ **84.** $2\sqrt{2}$ **85.** $\sqrt{6}$
86. $\dfrac{\sqrt{30}}{15}$ **87.** $\dfrac{\sqrt{15}}{5}$ **88.** $\sqrt{10}$ **89.** $\dfrac{\sqrt{30}}{15}$
90. $\dfrac{\sqrt{15}}{5}$ **91.** $\dfrac{\sqrt[3]{9}}{3}$ **92.** $\dfrac{\sqrt[3]{10}}{2}$ **93.** $\dfrac{\sqrt{7x}}{x}$
94. $\dfrac{3m\sqrt{2mp}}{2p}$ **95.** $\dfrac{r\sqrt{5rt}}{5w}$ **96.** $14\sqrt{6}$
97. $-\sqrt{15} + 9$ **98.** $3\sqrt{6} + 12$
99. $22 - 16\sqrt{3}$ **100.** $179 + 20\sqrt{7}$ **101.** -2
102. -13 **103.** $2\sqrt{21} + 12\sqrt{2} - \sqrt{14} - 4\sqrt{3}$
104. $-2 + \sqrt{5}$ **105.** $\dfrac{-2(\sqrt{2} + 3)}{7}$
106. $\dfrac{-2 + 6\sqrt{2}}{17}$
107. $\dfrac{-\sqrt{3}(1 - \sqrt{3})}{2}$ or $\dfrac{-\sqrt{3} + 3}{2}$
108. $\dfrac{-\sqrt{10} + 3\sqrt{5} + \sqrt{2} - 3}{7}$
109. $\dfrac{2\sqrt{3} + 3\sqrt{2} + 2 + \sqrt{6}}{2}$ **110.** 25
111. no solution **112.** 99 **113.** 8 **114.** 121
115. no solution **116.** 1 **117.** 2 **118.** 6
119. -2 **120.** -2 **121.** 4

Chapter 9 Test (page 485)

1. 10 **2.** 8.775 **3.** 13.784 **4.** -3 **5.** 5
6. $\dfrac{8}{13}$ **7.** $5\sqrt{2}$ **8.** $\dfrac{2}{3}$ **9.** $4\sqrt{6}$ **10.** $2\sqrt[3]{4}$
11. $6\sqrt{3}$ **12.** $-\sqrt{5}$ **13.** cannot be simplified
14. $9\sqrt{7}$ **15.** $-7\sqrt{3x}$ **16.** $2y\sqrt[3]{4x^2}$ **17.** 31
18. $6\sqrt{2} + 2 - 3\sqrt{14} - \sqrt{7}$ **19.** $4 - 2\sqrt{3}$
20. $11 + 2\sqrt{30}$ **21.** $\dfrac{4\sqrt{3}}{3}$ **22.** $\sqrt{3}$
23. $\dfrac{\sqrt[3]{15}}{3}$ **24.** $\dfrac{-12 - 3\sqrt{3}}{13}$
25. $\dfrac{3\sqrt{2} + \sqrt{14} + 3 + \sqrt{7}}{2}$ **26.** no solution
27. 23 **28.** 4 **29.** 1, 25 **30.** 1

CHAPTER 10

Section 10.1 (page 489)

1. $5, -5$ **3.** $8, -8$ **5.** $\sqrt{13}, -\sqrt{13}$
7. $\sqrt{2}, -\sqrt{2}$ **9.** $2\sqrt{6}, -2\sqrt{6}$ **11.** $\dfrac{13}{3}, -\dfrac{13}{3}$
13. $\dfrac{3}{4}, -\dfrac{3}{4}$ **15.** $1.1, -1.1$ **17.** $1.6, -1.6$
19. $8.8, -8.8$ **21.** $6, -2$
23. $-4 - \sqrt{10}, -4 + \sqrt{10}$
25. $1 + 4\sqrt{2}, 1 - 4\sqrt{2}$ **27.** $2, -1$
29. $-\dfrac{2}{3}, -\dfrac{8}{3}$ **31.** $\dfrac{5 + \sqrt{30}}{2}, \dfrac{5 - \sqrt{30}}{2}$
33. $\dfrac{1 + 3\sqrt{2}}{3}, \dfrac{1 - 3\sqrt{2}}{3}$ **35.** $\dfrac{5 + 7\sqrt{2}}{2}, \dfrac{5 - 7\sqrt{2}}{2}$
37. $\dfrac{-4 + 2\sqrt{2}}{3}, \dfrac{-4 - 2\sqrt{2}}{3}$ **39.** $4.60, -.80$
41. $-.15, -3.09$ **43.** $1.58, -1.09$
45. about $\dfrac{1}{2}$ second

Section 10.2 (page 495)

1. 1 **3.** 81 **5.** $\dfrac{9}{4}$ **7.** $\dfrac{25}{4}$ **9.** $-1, -3$
11. $-2 + \sqrt{3}, -2 - \sqrt{3}$ **13.** 4
15. $1 + \sqrt{3}, 1 - \sqrt{3}$ **17.** $\dfrac{-5 + \sqrt{37}}{2}, \dfrac{-5 - \sqrt{37}}{2}$
19. $\dfrac{-3}{2}, \dfrac{1}{2}$
21. $\dfrac{2 + \sqrt{14}}{2}, \dfrac{2 - \sqrt{14}}{2}$ **23.** $5 + \sqrt{17}, 5 - \sqrt{17}$
25. $\dfrac{3 + 2\sqrt{6}}{3}, \dfrac{3 - 2\sqrt{6}}{3}$
27. $-2 + \sqrt{3}, -2 - \sqrt{3}$ **29.** $-3, 3$

Section 10.3 (page 501)

1. $a = 3, b = 4, c = -8$ **3.** $a = 2, b = -3, c = 2$ **5.** $a = 3, b = -8, c = 0$ **7.** $a = 1, b = 1, c = -12$ **9.** $a = 9, b = 9, c = -26$
11. $-1 + \sqrt{3}, -1 - \sqrt{3}$ **13.** -2
15. $1, -13$ **17.** $\dfrac{-6 + \sqrt{26}}{2}, \dfrac{-6 - \sqrt{26}}{2}$
19. $\dfrac{1}{5}, -1$ **21.** $-1, \dfrac{5}{2}$ **23.** -3 **25.** $\dfrac{3}{2}$
27. $-1, 0$ **29.** $\dfrac{5}{3}, 0$ **31.** $-2\sqrt{5}, 2\sqrt{5}$
33. $\dfrac{4}{3}, -\dfrac{4}{3}$ **35.** $0, 2$ **37.** no real number solution **39.** no real number solution
41. $\dfrac{-1 + \sqrt{73}}{6}, \dfrac{-1 - \sqrt{73}}{6}$ **43.** $\dfrac{2 + \sqrt{22}}{6}, \dfrac{2 - \sqrt{22}}{6}$
45. $1 + \sqrt{2}, 1 - \sqrt{2}$
47. $\dfrac{1 + \sqrt{3}}{3}, \dfrac{1 - \sqrt{3}}{3}$ **49.** no real number solution

Answers to Selected Exercises

51. no real number solution **53.** 1, 3; both times are reasonable but the actual time is 1 second

Supplementary Exercises on Quadratic Equations (page 505)

1. $\dfrac{-3+\sqrt{5}}{2}, \dfrac{-3-\sqrt{5}}{2}$ **3.** 13, −13

5. $\dfrac{7}{9}, -\dfrac{7}{9}$ **7.** 1.1, −1.1   **9.** −2, −1

11. $\dfrac{-3+\sqrt{17}}{2}, \dfrac{-3-\sqrt{17}}{2}$ **13.** $\dfrac{1+\sqrt{10}}{2}, \dfrac{1-\sqrt{10}}{2}$

15. $\dfrac{-2+\sqrt{5}}{3}, \dfrac{-2-\sqrt{5}}{3}$ **17.** 5, −17

19. $\dfrac{1+4\sqrt{2}}{7}, \dfrac{1-4\sqrt{2}}{7}$ **21.** no real number solution   **23.** $1, -\dfrac{1}{2}$ **25.** no real number solution   **27.** $\dfrac{3}{2}, -\dfrac{5}{4}$

29. $\dfrac{-5+\sqrt{13}}{6}, \dfrac{-5-\sqrt{13}}{6}$ **31.** $\dfrac{-3+\sqrt{17}}{4}, \dfrac{-3-\sqrt{17}}{4}$ **33.** $\dfrac{2}{5}, 4$ **35.** $\dfrac{2+\sqrt{7}}{3}, \dfrac{2-\sqrt{7}}{3}$

37. $\dfrac{1}{4}, 1$ **39.** $\dfrac{-5+\sqrt{41}}{8}, \dfrac{-5-\sqrt{41}}{8}$

41. $-\dfrac{6}{5}, -\dfrac{8}{3}$ **43.** $\dfrac{-5+\sqrt{5}}{2}, \dfrac{-5-\sqrt{5}}{2}$

45. $\dfrac{3+\sqrt{21}}{2}, \dfrac{3-\sqrt{21}}{2}$ **47.** $2, -\dfrac{2}{3}$ **49.** $\dfrac{3}{5}, -4$

51. $\dfrac{5+\sqrt{17}}{4}, \dfrac{5-\sqrt{17}}{4}$ **53.** $-\dfrac{2}{3}, \dfrac{2}{5}$

Section 10.4 (page 511)

1. (0, 0)   **3.** (−1, 0)

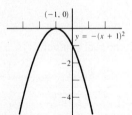

5. (−1, 0)   **7.** (0, 1)

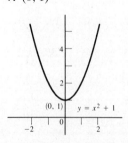

9. (0, 2)   **11.** (0, 2)

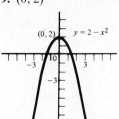 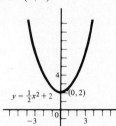

13. (−1, 2)   **15.** (4, −1)

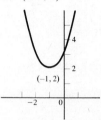 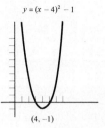

17. (−2, 1)

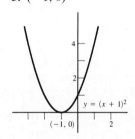

Chapter 10 Review Exercises (page 513)

1. 7, −7   **2.** $\sqrt{15}, -\sqrt{15}$ **3.** $4\sqrt{3}, -4\sqrt{3}$
4. 1, −5   **5.** $3+\sqrt{7}, 3-\sqrt{7}$
6. $\dfrac{-1+\sqrt{11}}{2}, \dfrac{-1-\sqrt{11}}{2}$ **7.** $\dfrac{-2+2\sqrt{3}}{3}, \dfrac{-2-2\sqrt{3}}{3}$ **8.** $\dfrac{-1+\sqrt{3}}{2}, \dfrac{-1-\sqrt{3}}{2}$
9. −1, −5   **10.** $-2+\sqrt{11}, -2-\sqrt{11}$
11. $-1+\sqrt{6}, -1-\sqrt{6}$ **12.** $\dfrac{-4+\sqrt{22}}{2}, \dfrac{-4-\sqrt{22}}{2}$ **13.** $1, -\dfrac{2}{5}$ **14.** no real number solution   **15.** $1+\sqrt{5}, 1-\sqrt{5}$ **16.** $\dfrac{3+\sqrt{29}}{2}, \dfrac{3-\sqrt{29}}{2}$ **17.** no real number solution
18. $\dfrac{1+\sqrt{21}}{10}, \dfrac{1-\sqrt{21}}{10}$ **19.** $\dfrac{2+\sqrt{10}}{2}, \dfrac{2-\sqrt{10}}{2}$
20. $\dfrac{-1+\sqrt{29}}{4}, \dfrac{-1-\sqrt{29}}{4}$

21. $\dfrac{-3 + \sqrt{41}}{2}, \dfrac{-3 - \sqrt{41}}{2}$ **22.** $\dfrac{1}{3}, -1$

23. $\dfrac{1}{2}, -5$ **24.** $\dfrac{1}{2}, -\dfrac{3}{2}$

25. $\dfrac{-1 - \sqrt{21}}{2}, \dfrac{-1 + \sqrt{21}}{2}$ **26.** $\dfrac{1}{3}, -\dfrac{3}{2}$

27. $\dfrac{-5 + \sqrt{17}}{2}, \dfrac{-5 - \sqrt{17}}{2}$

28. $-1 + \sqrt{3}, -1 - \sqrt{3}$ **29.** no real number solution **30.** $\dfrac{9 + \sqrt{41}}{2}, \dfrac{9 - \sqrt{41}}{2}$

31. $\dfrac{-5 + \sqrt{7}}{3}, \dfrac{-5 - \sqrt{7}}{3}$

32. $-1 + 2\sqrt{2}, -1 - 2\sqrt{2}$
33. $-2 + \sqrt{5}, -2 - \sqrt{5}$ **34.** $2\sqrt{3}, -2\sqrt{3}$
35. $(0, 0)$ **36.** $(0, 0)$

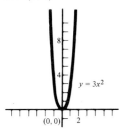

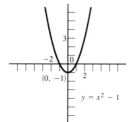

37. $(0, 3)$ **38.** $(0, -1)$

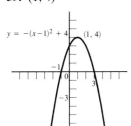

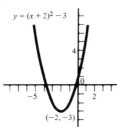

39. $(1, 4)$ **40.** $(-2, -3)$

Chapter 10 Test (page 517)

1. $1, 2$ **2.** $0, \dfrac{5}{2}$ **3.** $\sqrt{5}, -\sqrt{5}$

4. $3 + \sqrt{2}, 3 - \sqrt{2}$ **5.** $\dfrac{2 + 6\sqrt{2}}{3}, \dfrac{2 - 6\sqrt{2}}{3}$

6. $5, -2$ **7.** $\dfrac{1}{2}, -3$ **8.** $2, \dfrac{1}{3}$ **9.** no real number solution **10.** $\dfrac{5 + \sqrt{13}}{6}, \dfrac{5 - \sqrt{13}}{6}$

11. $1 + \sqrt{2}, 1 - \sqrt{2}$ **12.** $\dfrac{1 + 3\sqrt{2}}{2}, \dfrac{1 - 3\sqrt{2}}{2}$

13. $5, -\dfrac{2}{3}$ **14.** $\dfrac{17 + \sqrt{217}}{6}, \dfrac{17 - \sqrt{217}}{6}$

15. $5 + 2\sqrt{2}, 5 - 2\sqrt{2}$ **16.** $3 + \sqrt{7}, 3 - \sqrt{7}$

17. 4 **18.** $\dfrac{5}{2}, -1$

19. **20.**

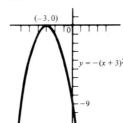

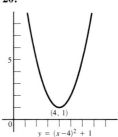

Final Examination (page 519)

1. 3 **2.** 5 **3.** 5 **4.** not a real number

5. $\dfrac{0}{8}, \dfrac{12}{4}$ **6.** $-7, \dfrac{0}{8}, \dfrac{12}{4}, -\sqrt{4}$

7. $-7, \dfrac{0}{8}, \dfrac{3}{8}, \dfrac{12}{4}, -\sqrt{4}$ **8.** $\sqrt{7}$

9. all except $\sqrt{-6}$ **10.** $\sqrt{-6}$ **11.** 2

12. -2 **13.** 2 **14.** -2 **15.** $t = \dfrac{I}{pr}$

16. $b = \dfrac{2A}{h}$

17. **18.**

19. 6 **20.** $\dfrac{1}{5^2}$ **21.** $\dfrac{1}{7^2}$

22. $8x^5 - 17x^4 + 17x^2$ **23.** $36r^2 + 12r + 1$
24. $3x^2 - 2x + 1$ **25.** $x(9 - x)$
26. $(x - 8)(x + 2)$ **27.** $(2a + 1)(a - 3)$
28. $(5m - 2)^2$ **29.** $\dfrac{4}{5}$ **30.** $\dfrac{-k - 1}{k(k - 1)}$

31. $\dfrac{5a + 2}{(a + 2)(a - 2)(a - 2)}$ **32.** $\dfrac{2x + 1}{3x - 1}$

33. $\dfrac{24}{7}$ minutes or $3\dfrac{3}{7}$ minutes

34. **35.**

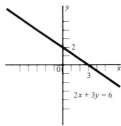

36. 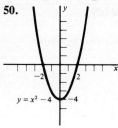 **37.**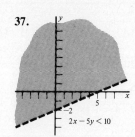

38. $x + 3y = 11$; slope is $-\frac{1}{3}$ **39.** $(-3, 2)$
40. no solution   **41.** 29 and 53   **42.** 10
43. -29 **44.** $\frac{5\sqrt{6}}{2}$ **45.** $-\sqrt{5}$ **46.** $4\sqrt{10}$
47. $\frac{3}{2}, 1$ **48.** $1 + \sqrt{2}, 1 - \sqrt{2}$
49. $\frac{1 + \sqrt{3}}{2}, \frac{1 - \sqrt{3}}{2}$
50. (graph of $y = x^2 - 4$)

APPENDICES

Appendix A   (page 527)

1. {1, 2, 3, 4, 5, 6, 7}
3. {winter, spring, summer, fall}   **5.** ∅   **7.** {L}
9. {2, 4, 6, 8, …}   **11.** The sets in Exercises 9 and 10, since each contains an unlimited number of elements   **13.** true   **15.** false   **17.** true
19. true   **21.** true   **23.** true   **25.** true
27. false   **29.** true   **31.** true   **33.** false
35. true   **37.** true   **39.** false   **41.** true
43. false   **45.** {g, h}   **47.** {b, c, d, e, g, h}
49. {a, c, e}   **51.** {d}   **53.** {a}   **55.** {a, c, d, e}
57. {a, c, e, f}   **59.** ∅   **61.** B and D; C and D

Appendix B   (page 533)

1. $3i$ **3.** $2i\sqrt{5}$ **5.** $6i$ **7.** real
9. imaginary   **11.** real   **13.** $x = \frac{1}{2}$; $y = -5$
15. $x = -\frac{1}{2}$; $y = \frac{2}{3}$ **17.** $x = 0$; $y = -4$
19. $5 + 3i$ **21.** $18 - 2i$ **23.** $6 - 9i$
25. $1 + 5i$ **27.** $-45 + 10i$ **29.** $-7 - 2i$
31. $14 + 5i$ **33.** $7 - 22i$ **35.** 45
37. $-\frac{3}{2} - \frac{1}{2}i$ **39.** $-\frac{1}{4} - \frac{7}{4}i$ **41.** $\frac{11}{10} - \frac{13}{10}i$
43. $1 + i, 1 - i$ **45.** $\frac{-3 + i\sqrt{31}}{4}, \frac{-3 - i\sqrt{31}}{4}$
47. $\frac{2 + i\sqrt{6}}{2}, \frac{2 - i\sqrt{6}}{2}$

Solutions to Selected Exercises

For the answers to all odd-numbered section exercises, all chapter review exercises, all cumulative review exercises, and all chapter tests, see the section beginning on page 537.

If you would like to see more solutions, you may order the *Student's Solutions Manual for Introductory Algebra* from your college bookstore. It contains solutions to the odd-numbered exercises that do not appear in this section, plus solutions to all chapter review exercises, cumulative review exercises, and chapter tests.

CHAPTER 1

Section 1.1 (page 7)

1. $\dfrac{7}{14} = \dfrac{7 \div 7}{14 \div 7} = \dfrac{1}{2}$

5. $\dfrac{16}{18} = \dfrac{16 \div 2}{18 \div 2} = \dfrac{8}{9}$

9. $\dfrac{72}{108} = \dfrac{72 \div 36}{108 \div 36} = \dfrac{2}{3}$

13. $\dfrac{3}{4} \cdot \dfrac{3}{5} = \dfrac{3 \cdot 3}{4 \cdot 5} = \dfrac{9}{20}$

17. $\dfrac{9}{4} \cdot \dfrac{8}{15} = \dfrac{9 \cdot 8}{4 \cdot 15} = \dfrac{72}{60} = \dfrac{6}{5}$

21. $\dfrac{5}{12} \div \dfrac{15}{4} = \dfrac{5}{12} \cdot \dfrac{4}{15} = \dfrac{1}{9}$

25. $3\dfrac{15}{16} \div \dfrac{15}{4} = \dfrac{63}{16} \div \dfrac{15}{4} = \dfrac{63}{16} \cdot \dfrac{4}{15} = \dfrac{21}{20}$ or $1\dfrac{1}{20}$

29. $6\dfrac{7}{10} \cdot 4\dfrac{5}{7} = \dfrac{67}{10} \cdot \dfrac{33}{7} = \dfrac{2211}{70}$ or $31\dfrac{41}{70}$

33. $\dfrac{1}{10} + \dfrac{7}{10} = \dfrac{1+7}{10} = \dfrac{8}{10} = \dfrac{4}{5}$

37. $\dfrac{8}{11} + \dfrac{3}{22} = \dfrac{8 \cdot 2}{11 \cdot 2} + \dfrac{3}{22} = \dfrac{16}{22} + \dfrac{3}{22}$
$= \dfrac{16+3}{22} = \dfrac{19}{22}$

41. $\dfrac{5}{6} - \dfrac{3}{10} = \dfrac{5 \cdot 5}{6 \cdot 5} - \dfrac{3 \cdot 3}{10 \cdot 3} = \dfrac{25}{30} - \dfrac{9}{30} = \dfrac{25-9}{30}$
$= \dfrac{16}{30} = \dfrac{8}{15}$

45. $4\dfrac{1}{2} + 3\dfrac{2}{3} = \dfrac{9}{2} + \dfrac{11}{3} = \dfrac{9 \cdot 3}{2 \cdot 3} + \dfrac{11 \cdot 2}{3 \cdot 2} = \dfrac{27}{6} + \dfrac{22}{6}$
$= \dfrac{27+22}{6} = \dfrac{49}{6}$ or $8\dfrac{1}{6}$

49. $\dfrac{2}{5} + \dfrac{1}{3} + \dfrac{9}{10} = \dfrac{2 \cdot 6}{5 \cdot 6} + \dfrac{1 \cdot 10}{3 \cdot 10} + \dfrac{9 \cdot 3}{10 \cdot 3}$
$= \dfrac{12}{30} + \dfrac{10}{30} + \dfrac{27}{30} = \dfrac{12+10+27}{30} = \dfrac{49}{30}$

53. $\dfrac{3}{4} + \dfrac{1}{8} - \dfrac{2}{3} = \dfrac{3 \cdot 6}{4 \cdot 6} + \dfrac{1 \cdot 3}{8 \cdot 3} - \dfrac{2 \cdot 8}{3 \cdot 8}$
$= \dfrac{18}{24} + \dfrac{3}{24} - \dfrac{16}{24} = \dfrac{18+3-16}{24} = \dfrac{5}{24}$

57. $3\dfrac{1}{4} + 2\dfrac{3}{8} + 7\dfrac{1}{2} + 1\dfrac{5}{16} = 14\dfrac{7}{16}$ tons

61. $36 \div 2\dfrac{1}{4} = 36 \div \dfrac{9}{4} = \dfrac{36}{1} \cdot \dfrac{4}{9} = \dfrac{36 \cdot 4}{1 \cdot 9} = \dfrac{144}{9}$
$= 16$ dresses

Section 1.2 (page 15)

1. $80 + 6$

5. $5000 + 200 + 30 + 7$

9. $.5 + .06 + .007$

13. $.72 = \dfrac{72}{100} = \dfrac{18}{25}$

17. $.625 = \dfrac{625}{1000} = \dfrac{5}{8}$

21. $\begin{array}{r} 27.96 \\ -8.39 \\ \hline 19.57 \end{array}$

25. $\begin{array}{r} 9.71 \\ 4.80 \\ 3.60 \\ 5.20 \\ +\ 8.17 \\ \hline 31.48 \end{array}$

29. $18.7 \times 2.3 = 43.01$

33. $\begin{array}{r} 571 \\ \times\ 2.9 \\ \hline 1655.9 \end{array}$

37. $\dfrac{14.9202}{2.43}$ becomes $\dfrac{1492.02}{243}$

or $243\overline{)1492.02}$ with quotient 6.14
$\begin{array}{r} 1458 \\ \hline 340 \\ 243 \\ \hline 972 \\ 972 \\ \hline 0 \end{array}$

41. $8\overline{)3.000}$ ← Attach zeros, quotient $.375$
$\begin{array}{r} 24 \\ \hline 60 \\ 56 \\ \hline 40 \\ 40 \\ \hline 0 \end{array}$

45. $16\overline{)15.0000}$ ← Attach zeros

$$\begin{array}{r}.9375\\\underline{14\ 4}\\60\\\underline{48}\\120\\\underline{112}\\80\\\underline{80}\\0\end{array}$$

Round answer to .938.

49. $15\overline{)11.0000}$ ← Attach zeros

$$\begin{array}{r}.7333\\\underline{10\ 5}\\50\\\underline{45}\\50\\\underline{45}\\50\\\underline{45}\\5\end{array}$$

Round answer to .733.

53. $.75 = 75(.01) = 75(1\%) = 75\%$
57. $.003 = .3(.01) = .3(1\%) = .3\%$
61. $38\% = 38(1\%) = 38(.01) = .38$
65. $11\% = 11(1\%) = 11(.01) = .11$
69. 12% of $350 = .12(350) = 42$
73. Divide the percentage by the amount:
$$\frac{1300}{2000} = .65 = 65\%.$$
77. Divide the percentage by the amount:
$$\frac{64}{3200} = .02 = 2\%.$$
81. 12.741% of 58.902 is $.12741 \times 58.902$
$= 7.50$ (rounded)
85. The family spends 90% of its income, so it saves 10%. Find 10% of $2000.
$10\% \times \$2000 = .10 \times \$2000 = \$200$
The family saves $200 per month, or, in a year,
$12 \times \$200 = \$2400.$
89. The discount was
$$\frac{\$37.50}{\$250} = .15 \text{ or } 15\%.$$
93. Divide:
$$\frac{\$4620}{\$77,000} = .06 \text{ or } 6\%.$$
97. Multiply:
$7.15\% \times \$2109$
$= .0715 \times \$2109$
$= \$150.79$ (rounded).

Section 1.3 (page 21)

1. Since 6 is smaller than 9, insert $<$ in the blank.
5. 25 is greater than 12, so insert $>$.
9. $\frac{3}{4}$ is less than 1, so insert $<$.
13. 12 is less than 17, so insert $\leq$.
17. 8 is less than 28, so insert $\leq$.
21. 6 is less than 9, so both $<$ and $\leq$ may be used.
25. 5 equals 5, so $\leq$ and $\geq$ may be used.
29. 16 is greater than 10, so $>$ and $\geq$ may be used.
33. Write .61 as .610 to see that .609 is less than .610. Both $<$ and $\leq$ may be used.
37. $7 = 5 + 2$
41. Use $\neq$ to get $12 \neq 5$.
45. Since $8 + 2 = 10$, the statement is true.
49. 0 is less than 15, so the statement is true.
53. 25 is greater than 19, so the statement is true.
57. Since $6 = 5 + 1$, the statement is false.
61. Since 8 is greater than 0, the statement is false.
65. Point the symbol toward the smaller number: $3 \leq 15$.
69. Point the symbol toward the smaller number: $6 \geq 0$.
73. Point the symbol toward the smaller number: $.439 \leq .481$.

Section 1.4 (page 27)

1. $6^2 = 6 \cdot 6 = 36$
5. $17^2 = 17 \cdot 17 = 289$
9. $6^4 = 6 \cdot 6 \cdot 6 \cdot 6 = 1296$
13. $3^6 = 3 \cdot 3 \cdot 3 \cdot 3 \cdot 3 \cdot 3 = 729$
17. $\left(\frac{2}{5}\right)^3 = \frac{2}{5} \cdot \frac{2}{5} \cdot \frac{2}{5} = \frac{8}{125}$
21. $(.83)^4 = (.83) \cdot (.83) \cdot (.83) \cdot (.83)$
$= .475$ (rounded)
25. $9 \cdot 3 - 11 = 27 - 11$ Multiply first
$\qquad\qquad\quad = 16$ Subtract
29. $13 \cdot 2 - 15 \cdot 1 = 26 - 15$ Multiply first
$\qquad\qquad\qquad\quad = 11$ Subtract
33. $5[8 + (2 + 3)] = 5[8 + 5]$ Work in parentheses
$\qquad\qquad\quad = 5[13]$ Work in brackets
$\qquad\qquad\quad = 65$ Multiply
37. $\dfrac{2(5 + 1) - 3(1 + 1)}{5(8 - 6) - 4 \cdot 2}$
$= \dfrac{2(6) - 3(2)}{5(2) - 4 \cdot 2}$ Work in parentheses
$= \dfrac{12 - 6}{10 - 8}$ Multiply
$= \dfrac{6}{2}$
$= 3$
41. On the left,
$2 \cdot 20 - 8 \cdot 5 = 40 - 40 = 0,$
and $0 \geq 0$ is true.
45. $3[5(2) - 3] > 20$
$\ \ 3[10 - 3] > 20$
$\qquad 3[7] > 20$
$\qquad\ \ 21 > 20$ True

49. $\dfrac{9(7-1) - 8 \cdot 2}{4(6-1)}$

$= \dfrac{9(6) - 8 \cdot 2}{4(5)}$ Subtract inside parentheses on top and bottom

$= \dfrac{54 - 16}{20}$ Multiply on top and bottom

$= \dfrac{38}{20}$ Subtract on top

$= 1\dfrac{9}{10}$ Divide

The inequality is now

$1\dfrac{9}{10} > 2,$

which is false.

53. $21.92 \le 7.43^2 - 5.77^2$
$21.92 \le 55.2049 - 33.2929$
$21.92 \le 21.912$ False

57. The statement $3 \cdot 5 + 7 = 22$ is true as it stands, since

$3 \cdot 5 + 7 = 15 + 7 = 22.$

Use parentheses as follows (if desired):

$(3 \cdot 5) + 7 = 22.$

61. The statement $3 \cdot 5 - 2 \cdot 4 = 36$ will be true if parentheses are inserted around $5 - 2$.

$3 \cdot (5 - 2) \cdot 4 = 3(3) \cdot 4$ Subtract inside parentheses

$= 9 \cdot 4$ Multiply 3 by 3 first

$= 36$ Multiply

36 is what we wanted.

65. The statement $6 + 5 \cdot 3^2 = 99$ will be true if parentheses are inserted around $6 + 5$.

$(6 + 5) \cdot 3^2 = 11 \cdot 3^2$ Add inside parentheses

$= 11 \cdot 3 \cdot 3$ Use the exponent

$= 33 \cdot 3$ Multiply 11 by 3

$= 99$ Multiply

99 is what we wanted.

69. No parentheses are needed, or

$\dfrac{1}{2} + \left(\dfrac{5}{3} \cdot \dfrac{9}{7}\right) - \dfrac{3}{2} = \dfrac{8}{7}.$

73. No parentheses are needed, or

$5 + (6 - 1) + (3 - 2).$

Section 1.5 (page 33)

1. (a) Let $x = 3$; then $x + 9 = 3 + 9 = 12$.
(b) Let $x = 15$; then $x + 9 = 15 + 9 = 24$.

5. (a) Let $x = 3$; then

$\dfrac{2}{3}x + \dfrac{1}{3} = \dfrac{2}{3}(3) + \dfrac{1}{3} = \dfrac{6}{3} + \dfrac{1}{3} = \dfrac{7}{3}.$

(b) Let $x = 15$; then

$\dfrac{2}{3}x + \dfrac{1}{3} = \dfrac{2}{3}(15) + \dfrac{1}{3} = \dfrac{30}{3} + \dfrac{1}{3} = \dfrac{31}{3}.$

9. (a) $\dfrac{3x - 5}{2x} = \dfrac{3 \cdot 3 - 5}{2 \cdot 3}$ Replace x with 3

$= \dfrac{9 - 5}{6}$ Multiply on the top and bottom where indicated

$= \dfrac{4}{6}$ Subtract on top

$= \dfrac{2}{3}$ Lowest terms

(b) $\dfrac{3x - 5}{2x} = \dfrac{3 \cdot 15 - 5}{2 \cdot 15}$ Let $x = 15$

$= \dfrac{45 - 5}{30}$ Multiply

$= \dfrac{40}{30}$ Subtract

$= \dfrac{4}{3}$ Lowest terms

13. (a) $6.459x = 6.459 \cdot 3$ Let $x = 3$
$= 19.377$ Multiply

(b) $6.459x = 6.459 \cdot 15$ Let $x = 15$
$= 96.885$ Multiply

17. (a) $8x + 3y + 5 = 8 \cdot 4 + 3 \cdot 2 + 5$

Let $x = 4$ and $y = 2$

$= 32 + 6 + 5$ Multiply first
$= 43$ Add

(b) $8x + 3y + 5 = 8 \cdot 1 + 3 \cdot 5 + 5$

Let $x = 1$ and $y = 5$

$= 8 + 15 + 5$ Multiply
$= 28$ Add

21. (a) $x + \dfrac{4}{y} = 4 + \dfrac{4}{2}$ Let $x = 4$ and $y = 2$

$= 4 + 2$ Divide 4 by 2 first
$= 6$ Add

(b) $x + \dfrac{4}{y} = 1 + \dfrac{4}{5}$ Let $x = 1$ and $y = 5$

$= 1\dfrac{4}{5}$ or $\dfrac{9}{5}$ Add

25. (a) Let $x = 4$ and $y = 2$; then

$5\left(\dfrac{4}{3}x + \dfrac{7}{2}y\right) = 5\left(\dfrac{4}{3} \cdot 4 + \dfrac{7}{2} \cdot 2\right)$

$= 5\left(\dfrac{16}{3} + 7\right)$

$= 5\left(\dfrac{16}{3}\right) + 5(7)$ Distributive property

$= \dfrac{80}{3} + \dfrac{105}{3}$ $35 = \dfrac{105}{3}$

$= \dfrac{80 + 105}{3} = \dfrac{185}{3}.$

(b) Let $x = 1$ and $y = 5$; then

$5\left(\dfrac{4}{3} \cdot 1 + \dfrac{7}{2} \cdot 5\right) = 5\left(\dfrac{4}{3} + \dfrac{35}{2}\right)$

$= 5\left(\dfrac{8}{6} + \dfrac{105}{6}\right)$

$= 5\left(\dfrac{113}{6}\right) = \dfrac{565}{6}.$

29. (a) $\dfrac{2x + 4y - 6}{5y + 2} = \dfrac{2 \cdot 4 + 4 \cdot 2 - 6}{5 \cdot 2 + 2}$ Let $x = 4$ and $y = 2$

$= \dfrac{8 + 8 - 6}{10 + 2}$

Multiply where indicated on the top and bottom

$= \dfrac{16 - 6}{12}$

Add on the top, then the bottom

$= \dfrac{10}{12}$ Subtract

$= \dfrac{5}{6}$ Lowest terms

(b) $\dfrac{2x + 4y - 6}{5y + 2} = \dfrac{2 \cdot 1 + 4 \cdot 5 - 6}{5 \cdot 5 + 2}$ Let $x = 1$ and $y = 5$

$= \dfrac{2 + 20 - 6}{25 + 2}$ Multiply

$= \dfrac{16}{27}$ Simplify

33. (a) $\dfrac{x^2 + y^2}{x + y} = \dfrac{4^2 + 2^2}{4 + 2}$ Let $x = 4$ and $y = 2$

$= \dfrac{16 + 4}{4 + 2}$ Use exponents

$= \dfrac{20}{6}$ Add

$= \dfrac{10}{3}$ Lowest terms

(b) $\dfrac{x^2 + y^2}{x + y} = \dfrac{1^2 + 5^2}{1 + 5}$ Let $x = 1$ and $y = 5$

$= \dfrac{1 + 25}{1 + 5}$ Use exponents

$= \dfrac{26}{6}$ Add

$= \dfrac{13}{3}$ Lowest terms

37. (a) $.841x^2 + .32y^2 = .841(4)^2 + .32(2)^2$

Let $x = 4$ and $y = 2$

$= .841(16) + .32(4)$

Use exponents

$= 13.456 + 1.28$ Multiply

$= 14.736$ Add

(b) $.841x^2 + .32y^2 = .841(1)^2 + .32(5)^2$

Let $x = 1$ and $y = 5$

$= .841(1) + .32(25)$

Use exponents

$= .841 + 8$ Multiply

$= 8.841$ Add

41. Five times a number

$\underbrace{5 \cdot x}$ or $5x$

45. Nine subtracted from a number

$x - 9$

49. $p - 5 = 12$

$17 - 5 = 12$ Let $p = 17$

$12 = 12$ True

17 is a solution.

53. $2y + 3(y - 2) = 14$

$2 \cdot 4 + 3(4 - 2) = 14$ Let $y = 4$

$2 \cdot 4 + 3 \cdot 2 = 14$

Work in parentheses

$8 + 6 = 14$ Multiply

$14 = 14$ True

4 is a solution.

57. $3r^2 - 2 = 46$

$3 \cdot 4^2 - 2 = 46$ Let $r = 4$

$3 \cdot 16 - 2 = 46$ Use exponent

$48 - 2 = 46$ Multiply

$46 = 46$ True

4 is a solution.

61. $9.54x + 3.811 = 0.4273x + 16.57718$

$9.54(1.4) + 3.811 = 0.4273(1.4) + 16.57718$

Let $x = 1.4$

$13.356 + 3.811 = .59822 + 16.57718$ Multiply

$17.167 = 17.1754$ Add

Results are different, so 1.4 is not a solution.

65. Sixteen minus three-fourths of a number is ten.

$16 \quad - \quad \dfrac{3}{4}x \quad = \quad 10$

Since $16 - \dfrac{3}{4}x = 16 - \dfrac{3}{4}(8) = 16 - 6 = 10$, and since 8 is on the list, 8 is the solution.

69. "Three times" indicates multiplication by 3; "more than" indicates addition; "twice" indicates multiplication by 2. So "three times a number is equal to two more than twice the number" translates to $3x = 2 + 2x$.

Since $3 \cdot 2 = 2 + 2 \cdot 2$

or $6 = 2 + 4$

$= 6,$

and since 2 is on the list, 2 is the solution.

CHAPTER 2

Section 2.1   (page 49)

1. Since the sum $-8 + 8 = 0$, the additive inverse of 8 is -8.

5. Since $-\dfrac{2}{3} + \dfrac{2}{3} = 0$, the additive inverse of $-\dfrac{2}{3}$ is $\dfrac{2}{3}$.

9. The value of $|-8|$ is 8; the additive inverse of this number is -8.

13. Since -14 is to the left of -9 on the number line, -14 is the smaller of the two numbers.

17. Since $|-2| = 2$, $|-2|$ is smaller than 5.

21. $-|-2| = -(2) = -2$, and $-|-3| = -(3) = -3$, so $-|-3|$, or -3, is smaller.

25. The statement "-3 is greater than or equal to -7" is true, since -3 is greater than -7.

29. Since $-(-4)$ is the same as 4 and since $-8 \le 4$ is true, $-8 \le -(-4)$ is true.
33. As a number line shows, -3.8 is to the left of -3.2, so the statement "$-3.2 < -3.8$" is false.
37. Since $-\left(-\dfrac{2}{3}\right) = \dfrac{2}{3}$ and since $-\dfrac{1}{2} < \dfrac{2}{3}$, the original statement is true.
41. Since $-|8| = -8$ and $|-9| = 9$, and since $-8 > 9$ is false, the statement $-|8| > |-9|$ is false.
45. Since $|-8| = 8$ and $|-2| = 2$, and since $8 > 2$, the given statement is true.
49. Since $-|-2| = -2$, and $|-4| = 4$, place dots at $-5, -3, -2, 0,$ and 4. See the graph in the answer section.

Section 2.2   (page 55)

1. Find the difference in the absolute values of both numbers:
$5 - 3 = 2$.
Since the larger number in absolute value is 5, the answer is positive. Therefore, the answer is 2.
5. Since numbers with the same sign are being added, add the absolute values of both numbers:
$6 + 2 = 8$.
Since both numbers are negative, the sign of the answer is negative:
$-6 + (-2) = -8$.
9. Since you are adding numbers with the same sign, simply add the absolute values of both numbers:
$3 + 9 = 12$.
Since both numbers are negative, the sign of the answer is negative:
$-3 + (-9) = -12$.
13. Do the work inside brackets first:
$[13 + (-5)] = 8$.
Now, $4 + [13 + (-5)] = 4 + 8 = 12$.
17. $-2 + [5 + (-1)] = -2 + 4 = 2$
21. $[9 + (-2)] + 6 = 7 + 6 = 13$
25. $-\dfrac{1}{6} + \dfrac{2}{3} = -\dfrac{1}{6} + \dfrac{4}{6}$ Write each fraction with a common denominator
$= \dfrac{3}{6}$ Add the numerators: $-1 + 4 = 3$
$= \dfrac{1}{2}$ Lowest terms
29. $2\dfrac{1}{2} + \left(-3\dfrac{1}{4}\right) = \dfrac{5}{2} + \left(\dfrac{-13}{4}\right)$ Change each fraction to an improper fraction
$= \dfrac{10}{4} + \left(\dfrac{-13}{4}\right)$ Write each fraction with a common denominator
$= -\dfrac{3}{4}$ Add the numerators: $10 + (-13) = -3$

33. Do the work inside the brackets first:
$[3.2 + (-4.8)] = -1.6$.
So $-6.1 + [3.2 + (-4.8)] = -6.1 + (-1.6)$
$= -7.7$.
37. Do the work inside the brackets first:
$[-4 + (-3)] = -7$ and $[8 + (-1)] = 7$.
So $[-4 + (-3)] + [8 + (-1)] = -7 + 7 = 0$.
41. $(-9.648 + 11.237) + [(-4.9123 + 1.8769)$
$+ 3.1589]$
$= 1.589 + (-3.0354 + 3.1589)$
$= 1.589 + .1235$
$= 1.7125$
45. $-8 + 12 = 4$ and $8 + (-12) = -4$, so the statement is false.
49. On the left,
$-\dfrac{3}{2} + \dfrac{5}{8} = -\dfrac{12}{8} + \dfrac{5}{8} = -\dfrac{7}{8}$;
on the right,
$\dfrac{5}{8} + \left(-\dfrac{3}{2}\right) = \dfrac{5}{8} + \left(-\dfrac{12}{8}\right) = -\dfrac{7}{8}$,
so the statement is true.
53. $|12 - 3| = |9| = 9$ and $12 - 3 = 9$, so the statement is true.
57. $-7 + [-5 + (-3)] = -7 + (-8) = -15$ and $[(-7) + (-5)] + 3 = (-12) + 3 = -9$, so the statement is false.
61. Try all numbers in the list. The only number that works is -2, which is the solution.
65. From the list, the only number that works is -2.
69. Only 2 works, so 2 is the solution.
73. $-3 + [15 + (-1)] = -3 + 14 = 11$
77. $6000 + (-4000) = 2000$ feet
81. $-14° + (-12°) = -26°$
85. $3.589 + 3.589 + 9.089 + 9.089 + 9.089$
$+ (-7.612) = 26.833$ inches
89. Since $-6 + (-3) = -9$, the number is -3.

Section 2.3   (page 61)

1. $3 - 6 = 3 + (-6) = -3$
5. $-6 - 2 = -6 + (-2) = -8$
9. $6 - (-3) = 6 + (3) = 9$
13. $-6 - (-2) = -6 + (2) = -4$
17. $-2 - (5 - 8) = -2 - (-3) = -2 + (3) = 1$
21. $-\dfrac{3}{4} - \dfrac{5}{8} = -\dfrac{3}{4} + \left(-\dfrac{5}{8}\right)$
$= -\dfrac{6}{8} + \left(-\dfrac{5}{8}\right)$
$= -\dfrac{11}{8}$
25. $3.4 - (-8.2) = 3.4 + (8.2) = 11.6$
29. $-4.1128 - (7.418 - 9.80632)$
$= -4.1128 - (-2.38832)$
$= -4.1128 + 2.38832$
$= -1.72448$
33. $(4 - 6) + 12 = (-2) + 12 = 10$
37. $6 - (-8 + 3) = 6 - (-5) = 6 + (5) = 11$

Solutions to Selected Exercises

41. $(-5 - 6) - (9 - 2)$
$= [-5 + (-6)] - [9 + (-2)]$
$= -11 - (7)$
$= -11 + (-7)$
$= -18$

45. $-9 - [(3 - 2) - (-4 - 2)]$
$= -9 - [1 - (-6)]$
$= -9 - (7) = -16$

49. $-9.1237 + [(-4.8099 - 3.2516) + 11.27903]$
$= -9.1237 + (-8.0615 + 11.27903)$
$= -9.1237 + 3.21753$
$= -5.90617$

53. "Subtract -6 from 12" can be read "from 12 subtract -6," which becomes
$12 - (-6) = 12 + (6) = 18.$

57. Subtract the smaller number, $-\frac{5}{24}$, from the larger number, $\frac{4}{27}$.
$\frac{4}{27} - \left(-\frac{5}{24}\right) = \frac{32}{216} + \frac{45}{216} = \frac{77}{216}$

61. $8 - (-2) = 10$

65. $[-12 + (-3)] - 4 = -15 - 4 = -19$

69. $14,494 - (-282) = 14,494 + 282 = 14,776$ feet

73. $\$76,000 - (-\$29,000) = \$105,000$

Section 2.4 (page 69)

1. $(-3)(-4) = 3 \cdot 4 = 12$
5. $(-1)(-5) = 1 \cdot 5 = 5$
9. $(-10)(-12) = 10 \cdot 12 = 120$
13. $(-6)(5) = -(6 \cdot 5) = -30$
17. $\left(-\frac{7}{3}\right)\left(\frac{8}{21}\right) = \frac{-7 \cdot 8}{3 \cdot 21} = \frac{-56}{63} = -\frac{8}{9}$
21. $(-3.7)(-2.1) = 7.77$
25. $6 - 4 \cdot 5 = 6 - 20$ Order of operations: multiply
$= 6 + (-20)$ Subtract
$= -14$
29. $9(6 - 10) = 9[6 + (-10)]$
$= 9[-4]$
$= -36$
33. $(4 - 9)(2 - 3) = (-5)(-1)$
$= 5$
37. $(-4 - 3)(-2) + 4 = (-7)(-2) + 4$
$= 14 + 4$
$= 18$
41. $(-8 - 2)(-4) - (-5) = (-10)(-4) - (-5)$
$= 40 + 5$
$= 45$
45. $|2|(-4) + |6| \cdot |-4| = 2(-4) + 6 \cdot 4$
$= -8 + 24 = 16$
49. Let $x = -2$, $y = 3$, and $a = -4$ to get
$(5x - 2y)(-2a)$
$= [5(-2) - 2(3)][-2(-4)]$
$= [-10 - 6][8]$
$= (-16)(8)$
$= -128.$

53. Let $x = -2$, $y = 3$, and $a = -4$ to get
$(6 - x)(5 + y)(3 + a)$
$= [6 - (-2)][5 + 3][3 + (-4)]$
$= 8(8)(-1)$
$= -64.$

57. Let $x = -2$ and $y = 3$ to get
$4y^2 - 2x^2 = 4(3)^2 - 2(-2)^2$
$= 4(9) - 2(4)$
$= 36 - 8$
$= 28.$

61. Try each number in the domain to find that 0 is the only solution.

65. Try each number in the domain; the only solution is -2.

69. Try each number in the domain; the only solution is 1.

73. $-4 - 2(-8 \cdot 2) = -4 - 2(-16)$
$= -4 + 32$
$= 28$

Section 2.5 (page 75)

1. Since $9 \cdot \frac{1}{9} = 1$, the multiplicative inverse is $\frac{1}{9}$.

5. Since $\frac{2}{3} \cdot \frac{3}{2} = 1$, the multiplicative inverse is $\frac{3}{2}$.

9. The reciprocal of 0 is $\frac{1}{0}$, but $\frac{1}{0}$ is meaningless, so 0 does not have a multiplicative inverse.

13. $\frac{-10}{5} = -10 \cdot \frac{1}{5} = -2$

17. $\frac{18}{-3} = 18 \cdot \left(-\frac{1}{3}\right) = -6$

21. $\frac{-12}{-6} = -12 \cdot \left(-\frac{1}{6}\right) = 2$

25. $\frac{-180}{-5} = -180\left(-\frac{1}{5}\right) = 36$

29. $-\frac{1}{2} \div \left(-\frac{3}{4}\right) = -\frac{1}{2} \cdot \left(-\frac{4}{3}\right)$ Multiply by the reciprocal of $-\frac{3}{4}$
$= \frac{2}{3}$ Multiply

33. $\frac{4}{-.8}$ indicates the division problem $4 \div (-.8)$. So $.8\overline{)4}$ becomes $8\overline{)40}$ by moving the decimal one place to the right in both the divisor and the dividend. The answer is therefore -5.

37. $\frac{50}{2 - 7} = \frac{50}{-5} = 50 \cdot \left(-\frac{1}{5}\right) = -10$

41. $\frac{-40}{8 - (-2)} = \frac{-40}{8 + 2} = \frac{-40}{10} = -4$

45. $\frac{-15 - 3}{3} = \frac{-18}{3} = -18 \cdot \left(\frac{1}{3}\right) = -6$

49. $\frac{-6.42 - (-3.891)}{-.05} = \frac{-6.42 + 3.891}{-.05} = \frac{-2.529}{-.05}$
$= 50.58$

53. $\frac{-15(2)}{-7 - 3} = \frac{-30}{-10} = 3$

57. $\frac{-5(2) + 3(-2)}{-3 - (-1)} = \frac{-10 + (-6)}{-3 + 1} = \frac{-16}{-2} = 8$

61. $\dfrac{4(-2) - 5(-3)}{2[-1 + (-3)] - (-8)} = \dfrac{-8 + 15}{2[-4] + 8}$
$= \dfrac{7}{-8 + 8}$
$= \dfrac{7}{0}$

The denominator is 0, so this quotient is meaningless (division by zero is meaningless).

65. $\dfrac{3^2 + 5^2}{4^2 + 1^2} = \dfrac{9 + 25}{16 + 1} = \dfrac{34}{17} = 2$

69. Try all numbers in the domain to find the solution, -8.

73. Try all numbers in the domain to find the solution, 0.

77. The integers that divide (with remainder 0) into 36 are
$\quad$ 1, -1, 2, -2, 3, -3, 4, -4, 6, -6, 9,
$\quad$ -9, 12, -12, 18, -18, 36, -36.

81. The integers that divide (with remainder 0) into 40 are
$\quad$ 1, -1, 2, -2, 4, -4, 5, -5, 8, -8,
$\quad$ 10, -10, 20, -20, 40, -40.

85. The numbers that divide (with remainder 0) into 29 are 1, -1, 29, -29.

89. $\dfrac{x}{4} = -2$; -8

93. $\dfrac{x}{-1} = 2$; -2

Section 2.6 (page 83)

1. The numbers are being added in a different order, indicating the commutative property.
5. Both the parentheses and the order are being changed, showing both the associative and commutative properties.
9. The numbers are being added in a different order, indicating the commutative property.
13. The sum of a number and 0 is the number, indicating the identity property.
17. The product of a number and 1 is the number, indicating the identity property.
21. $9 + k = k + 9$ $\quad$ Add in opposite order
25. $3(r + m) = 3 \cdot r + 3 \cdot m$ $\quad$ Multiply both terms in parentheses by 3
$\quad\quad\quad\quad = 3r + 3m$
29. The sum of a number and its inverse is 0.
33. Multiply -3 times r and times 2, to get $-3r - 6$.
37. $(k + 5) + (-6) = k + [5 + (-6)]$
$\quad$ Keep the numbers in the same order, but shift parentheses
$\quad\quad\quad\quad = k - 1$ $\quad$ Simplify
41. $5(m + 2) = 5 \cdot m + 5 \cdot 2 = 5m + 10$
45. $-8(k - 2) = -8k - (-8)(2) = -8k + 16$
49. $\left(r + \dfrac{8}{3}\right)\dfrac{3}{4} = r \cdot \dfrac{3}{4} + \dfrac{8}{3} \cdot \dfrac{3}{4}$
$\quad\quad\quad\quad = \dfrac{3}{4}r + 2$
53. $2(5r + 6m) = 2(5r) + 2(6m) = 10r + 12m$
57. $5 \cdot 8 + 5 \cdot 9 = 5(8 + 9) = 5(17) = 85$

61. $9p + 9q = 9(p + q)$
65. $-(3k + 5) = -1(3k + 5) = -1(3k) + (-1)5$
$\quad\quad = -3k - 5$
69. $-(-4 + p) = -1(-4) + (-1)(p) = 4 - p$
73. It makes a difference which activity you do first, so they are not commutative.
77. $25 - (6 - 2) = 25 - 4 = 21$
$\quad$ and $(25 - 6) - 2 = 19 - 2 = 17$,
$\quad$ so subtraction is not commutative.

Section 2.7 (page 89)

1. $15\underbrace{y}$
$\quad\quad\uparrow$
$\quad$ Numerical coefficient

5. $35\underbrace{a^4b^2}$
$\quad\quad\uparrow$
$\quad$ Numerical coefficient

9. $y^2 = \underbrace{1} \cdot y^2$
$\quad\quad\uparrow$
$\quad$ Numerical coefficient

13. $6m$ and $-14m$ are like terms since the variables are the same.
17. $25y$, $-14y$, and $8y$ are like terms since the variables are the same.
21. p, $-5p$, and $12p$ are like terms since the variables are the same.
25. $2k + 9 + 5k + 6$
$\quad = (2 + 5)k + (9 + 6) = 7k + 15$
29. $-2x + 3 + 4x - 17 + 20$
$\quad = -2x + 4x + 3 - 17 + 20$
$\quad = (-2 + 4)x + (3 - 17 + 20)$
$\quad = 2x + 6$
33. $-\dfrac{10}{3} + x + \dfrac{1}{4}x - 6 - \dfrac{5}{2}x$
$\quad = \left(-\dfrac{10}{3} - 6\right) + \left(1 + \dfrac{1}{4} - \dfrac{5}{2}\right)x$
$\quad = -\dfrac{28}{3} - \dfrac{5}{4}x$
37. $6y^2 + 11y^2 - 8y^2 = (6 + 11 - 8)y^2 = 9y^2$
41. $-7.913q^2 + 2.804q - 11.723 + 5.069q^2 - 8.124q - 6.977$
$\quad = (-7.913 + 5.069)q^2 + (2.804 - 8.124)q + (-11.723 - 6.977)$
$\quad = -2.844q^2 - 5.32q - 18.7$
45. $-3(n + 5) = -3n + (-3)5 = -3n - 15$
49. $2a + 3(a - 2) - 1 = 2a + 3a - 6 - 1 = 5a - 7$
53. $8(2k - 1) - (4k - 3)$
$\quad = 16k - 8 - 4k + 3$
$\quad = 12k - 5$
57. $-2(-3k + 2) - (5k - 6) - 3k - 5$
$\quad = 6k - 4 - 5k + 6 - 3k - 5$
$\quad = (6 - 5 - 3)k + (-4 + 6 - 5)$
$\quad = -2k - 3$
61. Think of this as "the sum of a number and 2 minus two times a number."

$\quad$ The sum of a $\quad\quad\quad$ two times
$\quad$ $\underbrace{\text{number and 2}}$ $\:$ minus $\:$ $\underbrace{\text{a number}}$
$\quad\quad\quad\downarrow\quad\quad\quad\quad\quad\downarrow\quad\quad\quad\downarrow$
$\quad\quad(x + 2)\quad\quad\quad\quad-\quad\quad\quad 2x$
$\quad\quad(x + 2) - 2x = 2 - x$ $\quad$ Simplify

Solutions to Selected Exercises

65. Think of this as "the difference of 4 and twice the number, minus 9 times the sum of five times a number and 4."

$$\underbrace{\text{The difference of 4 and twice the number}}_{(4-2x)} \underbrace{\text{minus}}_{-} \underbrace{\text{9 times}}_{9\cdot} \underbrace{\text{the sum of five times a number and 4}}_{(5x+4)}$$

$$4 - 2x - 9(5x + 4) = 4 - 2x - 45x - 36$$
$$= -47x - 32$$

CHAPTER 3

Section 3.1 (page 107)

Each of these answers should be checked by substituting in the original equation.

1. $x - 3 = 7$
    $x - 3 + 3 = 7 + 3$ Add 3 to both sides
    $x - 10$

5. $3r = 2r + 10$
    $3r - 2r = 2r + 10 - 2r$ Subtract $2r$
    $r = 10$

9. $2p + 6 = 10 + p$
    $2p + 6 - p = 10 + p - p$ Subtract p
    $p + 6 = 10$ Simplify
    $p + 6 - 6 = 10 - 6$ Subtract 6
    $p = 4$

13. $x - 5 = 2x + 6$
    $x - 5 - x = 2x + 6 - x$ Subtract x
    $-5 = x + 6$
    $-5 - 6 = x + 6 - 6$ Subtract 6
    $-11 = x$

17. $2p = p + \dfrac{1}{2}$
    $2p - p = p + \dfrac{1}{2} - p$ Subtract p
    $p = \dfrac{1}{2}$

21. $\dfrac{11}{4}r - \dfrac{1}{2} = \dfrac{7}{4}r + \dfrac{2}{3}$
    $-\dfrac{7}{4}r + \dfrac{11}{4}r - \dfrac{1}{2} = -\dfrac{7}{4}r + \dfrac{7}{4}r + \dfrac{2}{3}$
    Subtract $\dfrac{7}{4}r$
    $\dfrac{4}{4}r - \dfrac{1}{2} = \dfrac{2}{3}$ Simplify
    $r - \dfrac{1}{2} + \dfrac{1}{2} = \dfrac{2}{3} + \dfrac{1}{2}$ Add $\dfrac{1}{2}$ to both sides
    $r = \dfrac{4}{6} + \dfrac{3}{6}$ Write each fraction with a common denominator
    $r = \dfrac{7}{6}$ Simplify

25. $4x + 3 + 2x - 5x = 2 + 8$
    $x + 3 = 10$ Simplify each side
    $x + 3 - 3 = 10 - 3$ Subtract 3 on both sides
    $x = 7$

29. $11z + 2 + 4z - 3z = 5z - 8 + 6z$
    $12z + 2 = 11z - 8$ Simplify
    $-11z + 12z + 2 = -11z + 11z - 8$
    Subtract $11z$
    $z + 2 = -8$ Simplify
    $-2 + z + 2 = -2 + (-8)$ Subtract 2
    $z = -10$

33. $(5y + 6) - (3 + 4y) = 9$
    $5y + 6 - 3 - 4y = 9$ Distributive property
    $y + 3 = 9$ Simplify
    $-3 + y + 3 = -3 + 9$ Subtract 3
    $y = 6$

37. $-6(2a + 1) + 1(13a - 7) = 4$
    $-6(2a) - 6(1) + 1(13a) + 1(-7) = 4$
    Distributive property
    $-12a - 6 + 13a - 7 = 4$
    $a - 13 = 4$ Simplify left side
    $a - 13 + 13 = 4 + 13$ Add 13 to both sides
    $a = 17$

41. $-2(8p + 7) - 3(4 - 7p) = 2(3 + 2p) - 6$
    $-2(8p) - 2(7) - 3(4) - 3(-7p) = 2(3) + 2(2p) - 6$
    Distributive property
    $-16p - 14 - 12 + 21p = 6 + 4p - 6$
    $5p - 26 = 4p$ Simplify each side
    $5p - 26 - 4p = 4p - 4p$ Subtract $4p$ on each side
    $p - 26 = 0$
    $p - 26 + 26 = 0 + 26$ Add 26 to each side
    $p = 26$

45. "If five times a number is added to three times the number, the result is the sum of seven times the number and 9" becomes
    $5 \cdot x + 3 \cdot x = 7 \cdot x + 9$
    $5x + 3x = 7x + 9$
    $8x = 7x + 9$ Simplify the left side
    $-7x + 8x = -7x + 7x + 9$ Subtract $7x$ on each side
    $x = 9.$

Section 3.2 (page 113)

Each of these answers should be checked by substituting in the original equation.

1. $5x = 25$
    $\dfrac{5x}{5} = \dfrac{25}{5}$ Divide both sides by 5
    $1x = 5$ Simplify
    $x = 5$

5. $3a = -24$
$\dfrac{3a}{3} = \dfrac{-24}{3}$ Divide by 3
$1a = -8$ Simplify
$a = -8$

9. $-6x = 16$
$\dfrac{-6x}{-6} = \dfrac{16}{-6}$ Divide both sides by -6
$x = -\dfrac{8}{3}$

13. $5r = 0$
$\dfrac{5r}{5} = \dfrac{0}{5}$ Divide both sides by 5
$1r = 0$ 0 times any number is 0
$r = 0$

17. $-n = -4$
$(-1)\cdot(-n) = (-1)\cdot(-4)$ Multiply by -1
$n = 4$

21. $5m + 6m - 2m = 72$
$9m = 72$ Simplify the left side
$\dfrac{9m}{9} = \dfrac{72}{9}$ Divide each side by 9
$m = 8$

25. $3r - 5r = 6$
$-2r = 6$ Simplify the left side
$\dfrac{-2r}{-2} = \dfrac{6}{-2}$ Divide each side by -2
$r = -3$

29. $-7y + 8y - 9y = -56$
$-8y = -56$ Simplify
$y = 7$ Divide each side by -8

33. $\dfrac{x}{7} = 7$
$\dfrac{1}{7}x = 7$ Definition of division
$7\cdot\dfrac{1}{7}x = 7\cdot 7$ Multiply each side by 7
$1x = 49$ Simplify
$x = 49$

37. $\dfrac{15}{2}z = 20$
$\dfrac{2}{15}\cdot\dfrac{15}{2}z = \dfrac{2}{15}\cdot 20$ Multiply each side by $\dfrac{2}{15}$
$1z = \dfrac{8}{3}$
$z = \dfrac{8}{3}$

41. $\dfrac{2}{3}k = 5$
$\dfrac{3}{2}\left(\dfrac{2}{3}k\right) = \dfrac{3}{2}\cdot 5$ Multiply each side by $\dfrac{3}{2}$
$1k = \dfrac{15}{2}$
$k = \dfrac{15}{2}$

45. $1.7p = 5.1$
$\dfrac{1.7p}{1.7} = \dfrac{5.1}{1.7}$ Divide each side by 1.7
$p = 3$

49. $8.974z = 6.2818$
$\dfrac{8.974z}{8.974} = \dfrac{6.2818}{8.974}$ Divide each side by 8.974
$z = .7$

53. $.9123p = -5.2011$
$\dfrac{.9123p}{.9123} = \dfrac{-5.2011}{.9123}$ Divide each side by .9123
$p = -5.7$

57. Let $x =$ the amount to be divided. Then
$\dfrac{x}{4} = 62$
$4\left(\dfrac{x}{4}\right) = 4(62)$ Multiply by 4
$x = 248.$
The amount divided was $248.

Section 3.3 (page 119)

Each of these answers should be checked by substituting in the original equation.

1. $4h + 8 = 16$
$4h + 8 - 8 = 16 - 8$ Subtract 8 on each side
$4h = 8$
$\dfrac{4h}{4} = \dfrac{8}{4}$ Divide each side by 4
$h = 2$

5. $12p + 18 = 14p$
$-12p + 12p + 18 = -12p + 14p$ Subtract 12p
$18 = 2p$
$\dfrac{18}{2} = \dfrac{2p}{2}$ Divide by 2
$9 = p$

9. $2(2r - 1) = -3(r + 3)$
$4r - 2 = -3r - 9$ Distributive property
$7r - 2 = -9$ Add 3r to each side
$7r = -7$ Add 2 to each side
$\dfrac{7r}{7} = \dfrac{-7}{7}$ Divide each side by 7
$r = -1$ Simplify

13. $3(5 + 1.4x) = 3x$
$15 + 4.2x = 3x$
Distributive property
$15 + 4.2x - 3x = 3x - 3x$
Subtract 3x from each side
$15 + 1.2x = 0$
$-15 + 15 + 1.2x = -15 + 0$
Subtract 15 from each side
$1.2x = -15$
$\dfrac{1.2x}{1.2} = \dfrac{-15}{1.2}$
Divide each side by 1.2
$x = -\dfrac{15}{1.2}$
$x = -12.5$

Solutions to Selected Exercises 569

17. $.291z + 3.715 = -.874z + 1.9675$
  $.874z + .291z + 3.715 = .874z - .874z + 1.9675$ Add $.874z$
  $1.165z + 3.715 = 1.9675$
  $-3.715 + 1.165z + 3.715 = -3.715 + 1.9675$ Subtract 3.715
  $1.165z = -1.7475$
  $\dfrac{1.165z}{1.165} = \dfrac{-1.7475}{1.165}$ Divide by 1.165
  $z = -1.5$

21. $-5k - 8 = 2(k + 6) + 1$
  $-5k - 8 = 2k + 12 + 1$ Distributive property
  $-5k - 8 = 2k + 13$ Simplify
  $5k - 5k - 8 = 5k + 2k + 13$ Add $5k$
  $-8 = 7k + 13$ Simplify
  $-13 - 8 = -13 + 7k + 13$ Subtract 13
  $-21 = 7k$ Simplify
  $\dfrac{-21}{7} = \dfrac{7k}{7}$ Divide by 7
  $-3 = k$

25. $5(4t + 3) = 6(3t + 2) - 1$
  $20t + 15 = 18t + 12 - 1$ Distributive property
  $20t + 15 = 18t + 11$ Simplify
  $-18t + 20t + 15 = -18t + 18t + 11$ Subtract $18t$
  $2t + 15 = 11$
  $-15 + 2t + 15 = -15 + 11$ Subtract 15
  $2t = -4$ Simplify
  $\dfrac{2t}{2} = \dfrac{-4}{2}$ Divide by 2
  $t = -2$

29. $-2(3s + 9) - 6 = -3(4s + 11) - 6$
  $-6s - 18 - 6 = -12s - 33 - 6$ Distributive property
  $-6s - 24 = -12s - 39$ Simplify
  $12s - 6s - 24 = 12s - 12s - 39$ Add $12s$
  $6s - 24 = -39$ Simplify
  $24 + 6s - 24 = 24 - 39$ Add 24
  $6s = -15$ Simplify
  $\dfrac{6s}{6} = \dfrac{-15}{6}$ Divide by 6
  $s = -\dfrac{5}{2}$

33. $-(4m + 2) - (-3m - 5) = 3$
  $-4m - 2 + 3m + 5 = 3$
  $-m + 3 = 3$ Simplify
  $-3 - m + 3 = -3 + 3$ Subtract 3
  $-m = 0$
  $(-1)(-m) = (-1)0$ Multiply by -1
  $m = 0$

37. $3(4x + 2) - 2(5x - 1) = 0$
  $12x + 6 - 10x + 2 = 0$ Simplify
  $2x + 8 = 0$
  $2x + 8 - 8 = 0 - 8$ Subtract 8
  $2x = -8$
  $\dfrac{2x}{2} = \dfrac{-8}{2}$ Divide by 2
  $x = -4$

41. $1.2(x + 5) = 3(2x - 8) + 23.28$
  $1.2x + 6.0 = 6x - 24 + 23.28$ Simplify
  $1.2x + 6.0 = 6x - .72$
  $-1.2x + 1.2x + 6.0 = -1.2x + 6x - .72$ Subtract $1.2x$
  $6.0 = 4.8x - .72$
  $6.0 + .72 = 4.8x - .72 + .72$ Add $.72$
  $6.72 = 4.8x$
  $\dfrac{6.72}{4.8} = \dfrac{4.8x}{4.8}$ Divide by 4.8
  $\dfrac{6.72}{4.8} = x$
  $1.4 = x$

45.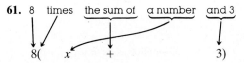
  $x + -6$ or $x - 6$

49. 5 less than a number is the same as a number minus 5.
  $x - 5$

53.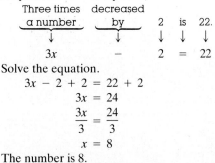
  or $9x$

57. The quotient of a number and 6
  $\dfrac{x}{6}$

61. 8 times the sum of a number and 3
  $8(\quad x \quad + \quad 3)$

Section 3.4   (page 127)

1. Use x to represent the unknown number. Translate the problem into an equation.
  Three times a number   decreased by   2 is 22.
  $3x \qquad - \qquad 2 = 22$
  Solve the equation.
  $3x - 2 + 2 = 22 + 2$
  $3x = 24$
  $\dfrac{3x}{3} = \dfrac{24}{3}$
  $x = 8$
  The number is 8.

5. Use x to represent the unknown number. Translate into an equation.
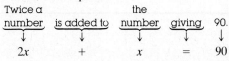

570   Solutions to Selected Exercises

Solve.
$$3x = 90$$
$$\frac{3x}{3} = \frac{90}{3}$$
$$x = 30$$
The number is 30.

9. Use x to represent the unknown number. Translate into an equation.

   Five times a number added to twice the number is 10.
   $$5x + 2x = 10$$

   Solve.
   $$7x = 10$$
   $$\frac{7x}{7} = \frac{10}{7}$$
   $$x = \frac{10}{7}$$
   The number is $\frac{10}{7}$.

13. Let x represent the lowest grade and $x + 42$ represent the highest grade.

    Lowest grade added to highest grade was 138.
    $$x + x + 42 = 138$$

    Solve.
    $$2x + 42 = 138$$
    $$2x + 42 - 42 = 138 - 42$$
    $$2x = 96$$
    $$x = 48$$
    The lowest grade was 48.

17. Let x represent the amount of milk from Bossie. Then $x - 238$ represents the amount of milk Bessie gave.

    The amount Bossie gave was 375 gallons less than twice the amount produced by Bessie.
    $$x = 2(x - 238) - 375$$
    $$x = 2(x - 238) - 375$$
    $$x = 2x - 476 - 375$$
    $$x = 2x - 851$$
    $$x - 2x = 2x - 851 - 2x$$
    $$-x = -851$$
    $$x = 851$$
    Bossie gave 851 gallons and Bessie gave $851 - 238 = 613$ gallons of milk.

21. Let x be the first odd integer and $x + 2$ the next odd integer.

    Twice the larger added to the smaller is 169.
    $$2(x + 2) + x = 169$$

$$2x + 4 + x = 169$$
$$3x + 4 = 169$$
$$3x + 4 - 4 = 169 - 4$$
$$3x = 165$$
$$\frac{3x}{3} = \frac{165}{3}$$
$$x = 55$$
The integers are 55 and $55 + 2 = 57$.

25. Let x, $x + 2$, and $x + 4$ be the three odd integers.

    9 added to the largest equals the sum of the first and second.
    $$9 + x + 4 = x + (x + 2)$$
    $$x + 13 = 2x + 2$$
    $$-x + x + 13 = -x + 2x + 2$$
    $$13 = x + 2$$
    $$11 = x$$
    The integers are 11, 13, and 15.

29. Let x be the width of the table. Then $3x$ is the length. Three feet shorter is $3x - 3$ and 3 feet wider is $x + 3$. A square has length equal to width, so

    Length equals width.
    $$3x - 3 = x + 3$$
    $$-x + 3x - 3 = -x + x + 3$$
    $$2x - 3 = 3$$
    $$3 + 2x - 3 = 3 + 3$$
    $$2x = 6$$
    $$x = 3$$
    The width is 3 feet and the length is $3(3) = 9$ feet.

Section 3.5 (page 135)

1. $P = 4s$
   $= 4(32)$ Substitute 32 for s
   $= 128$

5. $d = rt$
   $8 = 2t$ Let $d = 8$, $r = 2$
   $4 = t$

9. $P = 2L + 2W$
   $40 = 2L + 2(6)$ Let $P = 40$, $W = 6$
   $40 - 12 = 2L + 12 - 12$
   $28 = 2L$
   $14 = L$

13. $C = 2\pi r$
    $9.42 = 2(3.14)r$ Let $C = 9.42$, $\pi = 3.14$
    $9.42 = 6.28r$ Multiply
    $\frac{9.42}{6.28} = \frac{6.28r}{6.28}$ Divide by 6.28
    $1.5 = r$

17. $I = prt$
    $100 = 500(.10)t$ Let $I = 100$, $p = 500$, $r = .10$
    $100 = 50t$
    $2 = t$

Solutions to Selected Exercises

21. $A = \frac{1}{2}(b + B)h$

$42 = \frac{1}{2}(5 + 7)h$ Let $A = 42, b = 5, B = 7$

$2(42) = 2\left(\frac{1}{2}\right)(12)h$ Multiply by 2

$84 = 12h$

$\frac{84}{12} = \frac{12h}{12}$ Divide by 12

$7 = h$

25. $C = 2\pi r$ Formula for the circumference of a circle

$C = 2(3.14)(6)$ Substitute 3.14 for π, 6 for r

$C = 37.68$

The circumference is 37.68 feet.

29. Use the formula for area of a circle:

$A = \pi r^2$

$78.5 = 3.14 r^2$ Let $A = 78.5, \pi = 3.14$

$\frac{78.5}{3.14} = \frac{3.14 r^2}{3.14}$ Divide by 3.14

$25 = r^2$

$5 = r$

The radius is 5 inches.

33. Use the formula for volume of a right pyramid:

$V = \frac{1}{3}Bh$

$144 = \frac{1}{3}(B)(12)$ Let $V = 144, h = 12$

$3(144) = 3\left(\frac{1}{3}\right)(12)B$ Multiply by 3

$432 = 12B$

$\frac{432}{12} = \frac{12B}{12}$ Divide by 12

$36 = B$.

The area of the base is 36 square feet.

37. $d = rt$

$\frac{d}{r} = \frac{rt}{r}$ Divide each side by r

$\frac{d}{r} = t$ or $t = \frac{d}{r}$

41. $I = prt$

$\frac{I}{pr} = \frac{prt}{pr}$ Divide each side by pr

$\frac{I}{pr} = t$ or $t = \frac{I}{pr}$

45. $A = \frac{1}{2}bh$

$2A = 2\left(\frac{1}{2}bh\right)$ Multiply each side by 2

$2A = bh$

$\frac{2A}{h} = \frac{bh}{h}$ Divide each side by h

$\frac{2A}{h} = b$ or $b = \frac{2A}{h}$

49. $P = 2L + 2W$

$P - 2L = 2W$ Subtract $2L$

$\frac{P - 2L}{2} = W$ Divide by 2

53. $d = gt^2 + vt$

$d - gt^2 = vt$ Subtract gt^2 on each side

$\frac{d - gt^2}{t} = v$ Divide each side by t

57. $A = \frac{1}{2}(b + B)h$

$2A = (b + B)h$ Multiply each side by 2

$\frac{2A}{h} = b + B$ Divide each side by h

$\frac{2A}{h} - B = b$ Subtract B on each side

The answer may also be written $b = \frac{2A}{h} - \frac{B}{1} \cdot \frac{h}{h}$

$= \frac{2A - Bh}{h}$.

Section 3.6 (page 143)

1. $\frac{30}{20} = \frac{3}{2}$

5. 5 yards = $5 \times 3 = 15$ feet; ratio, $\frac{6}{15} = \frac{2}{5}$

9. $\frac{12}{120} = \frac{1}{10}$; 1 hour = 60 minutes, so 2 hours = $2 \cdot 60 = 120$ minutes

13. $\frac{20}{120} = \frac{1}{6}$; 1 day = 24 hours, so 5 days = $5 \cdot 24 = 120$ hours

17. $\frac{9}{10} \times \frac{18}{20}$ $\begin{array}{l} 10 \cdot 18 = 180 \\ 9 \cdot 20 = 180 \end{array}$ Same, so proportion is true

21. $\frac{12}{7} = \frac{36}{20}$

$12 \cdot 20 = 240$ and $7 \cdot 36 = 252$

$240 \neq 252$, so proportion is not true.

25. $\frac{19}{30} = \frac{57}{90}$

$19 \cdot 90 = 1710$ and $30 \cdot 57 = 1710$

Since $1710 = 1710$, the proportion is true.

29. $\frac{.612}{1.05} = \frac{1.0404}{1.785}$

$.612(1.785) = 1.09242$ and $1.05(1.0404) = 1.09242$

Since both answers equal 1.09242, the proportion is true.

33. $\frac{z}{56} = \frac{7}{8}$

$8z = 7(56)$

$8z = 392$

$z = \frac{392}{8} = 49$

37. $\frac{25}{100} = \frac{8}{m + 6}$

$25(m + 6) = 8(100)$

$25m + 150 = 800$

$25m = 650$

$m = \frac{650}{25} = 26$

41. $\dfrac{3}{4} = \dfrac{3n-1}{10}$

$3(10) = 4(3n-1)$

$30 = 12n - 4$

$34 = 12n$

$n = \dfrac{34}{12} = \dfrac{17}{6}$

45. $\dfrac{r+1}{r} = \dfrac{1}{3}$

$3(r+1) = 1(r)$

$3r + 3 = r$

$3r = r - 3$

$2r = -3$

$r = -\dfrac{3}{2}$

49. $\dfrac{r+8}{r-9} = \dfrac{7}{3}$

$3(r+8) = 7(r-9)$

$3r + 24 = 7r - 63$

$-4r = -87$

$r = \dfrac{-87}{-4} = \dfrac{87}{4}$

53. Let x represent the number of minutes.

$\dfrac{12}{15} = \dfrac{40}{x}$

$12x = 15(40)$

$12x = 600$

$x = \dfrac{600}{12} = 50$

He needs 50 minutes.

57. Let x represent the required charge.

$\dfrac{30}{x} = \dfrac{50}{125}$

$30(125) = 50x$

$3750 = 50x$

$\dfrac{3750}{50} = x$

$75 = x$

The charge will be $75.

61. Let x represent the number of yards of material.

$\dfrac{12}{x} = \dfrac{5}{8}$

$5x = 12(8)$

$5x = 96$

$x = \dfrac{96}{5} = 19.2$

19.2 yards are required.

Section 3.7 (page 151)

All the graphs for this section can be found in the answer section.

1. Use a dot at 4, and draw an arrow to the left.

5. Use a dot at -2 and one at 5; draw a line between them.

9. $a + 6 < 8$

$-6 + a + 6 < -6 + 8$ Subtract 6

$a < 2$

13. $4x < 3x + 6$

$4x - 3x < 3x - 3x + 6$

Subtract $3x$ on each side

$x < 6$

17. $3n + 5 \le 2n - 6$

$-2n + 3n + 5 \le -2n + 2n - 6$

Subtract $2n$ on each side

$n + 5 \le -6$

$n + 5 - 5 \le -6 - 5$

Subtract 5 on each side

$n \le -11$

21. $-6(k+2) + 3 \ge -7(k-5)$

$-6k - 12 + 3 \ge -7k + 35$

Distributive property

$-6k - 9 \ge -7k + 35$

$-6k + 7k - 9 \ge -7k + 7k + 35$

Add $7k$ to each side

$k - 9 \ge 35$

$k - 9 + 9 \ge 35 + 9$

Add 9 to each side

$k \ge 44$

25. $8 \le p + 2 \le 15$

$8 - 2 \le p + 2 - 2 \le 15 - 2$ Subtract 2 from each part of the inequality

$6 \le p \le 13$

29. Let x represent the number. Then we have the following inequality.

Four times a number $\underbrace{}$ added to $\underbrace{}$ 8; $\underbrace{}$ the result is less than $\underbrace{}$ three times the number $\underbrace{}$ added to $\underbrace{}$ 5.

$4x + 8 < 3x + 5$

Solve the inequality.

$-3x + 4x + 8 < -3x + 3x + 5$

Subtract $3x$ on each side

$x + 8 < 5$

$x + 8 - 8 < 5 - 8$

Subtract 8 on each side

$x < -3$

The number we want is less than -3.

33. Let w represent the width. Then $4w$ is the length. The phrase "at least" means "greater than or equal to," so we have the following.

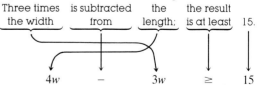

Solve: $w \ge 15$. Since the width must be greater than or equal to 15, the smallest width is 15. The smallest length is $4(15) = 60$.

Section 3.8 (page 157)

All the graphs for this section can be found in the answer section.

1. $3x < 27$
$\dfrac{3x}{3} < \dfrac{27}{3}$ Divide each side by 3
$x < 9$

5. $-2k \leq 12$
$\dfrac{-2k}{-2} \geq \dfrac{12}{-2}$ Divide by -2
⸺ Reverse symbol
$k \geq -6$

9. $4q + 1 - 5 < 8q + 4$
$4q - 4 < 8q + 4$ Simplify
$4 + 4q - 4 < 4 + 8q + 4$ Add 4
$4q < 8 + 8q$ Simplify
$-8q + 4q < -8q + 8 + 8q$ Subtract $8q$
$-4q < 8$ Simplify
$\dfrac{-4q}{-4} > \dfrac{8}{-4}$
⸺ Reverse symbol
$q > -2$

13. $12 - w \leq 4 + 4w - 7$
$12 - w \leq 4w - 3$
$12 \leq 5w - 3$ Add w
$15 \leq 5w$ Add 3
$3 \leq w$ Divide by 5
or $w \geq 3$

17. $2(x - 5) + 3x < 4(x - 6) + 3$
$2x - 10 + 3x < 4x - 24 + 3$
Distributive property
$5x - 10 < 4x - 21$ Simplify
$-4x + 5x - 10 < -4x + 4x - 21$ Subtract $4x$
$x - 10 < -21$
$10 + x - 10 < 10 - 21$ Add 10
$x < -11$

21. $-5 \leq 2x - 3 \leq 9$
$-2 \leq 2x \leq 12$ Add 3
$-1 \leq x \leq 6$ Divide by 2

25. Let x equal the length of the first side. Then the length of the second side is $2x$. Use the formula $P = a + b + c$. Since the perimeter cannot be more than 50,

First side + second side + third side cannot be more than 50.
$x + 2x + 17 \leq 50$

Solve this inequality.
$3x + 17 \leq 50$
$3x \leq 33$ Subtract 17
$x \leq 11$

The shortest side cannot be longer than 11 feet, and the remaining side cannot be longer than $2 \cdot 11 = 22$ feet.

29. Let x represent the earnings needed to qualify.
$\dfrac{900 + 1200 + 1040 + 760 + x}{5} \geq 1000$
$3900 + x \geq 5000$ Add and multiply by 5
$x \geq 1100$ Subtract 3900

The employee must earn at least $1100.

CHAPTER 4

Section 4.1 (page 175)

1. 5^{12} ← Exponent
⸺ Base

5. The negative sign in front of 125 is not part of the base, so the base is 125 and the exponent is 3.

9. Only m is under the exponent, so m is the base and 2 the exponent. The number 3 is a coefficient here.

13. 3 is used as a factor 5 times, so 3 is the base and 5 is the exponent: 3^5.

17. p is used as a factor 5 times, so p is the base and 5 the exponent: p^5.

21. 2 is used as a factor 5 times, giving $\dfrac{1}{2^5}$.

25. $3^2 + 3^4 = 3 \cdot 3 + 3 \cdot 3 \cdot 3 \cdot 3$
$= 9 + 81$
$= 90$

29. $2^2 + 2^5 = 2 \cdot 2 + 2 \cdot 2 \cdot 2 \cdot 2 \cdot 2$
$= 4 + 32$
$= 36$

33. $(-9)^0 + 9^0 = 1 + 1 = 2$
Any nonzero number raised to the zero power equals 1

37. $5^{-2} = \dfrac{1}{5^2} = \dfrac{1}{5 \cdot 5} = \dfrac{1}{25}$

41. $(-6)^{-2} = \dfrac{1}{(-6)^2} = \dfrac{1}{(-6)(-6)} = \dfrac{1}{36}$

45. $\left(\dfrac{1}{2}\right)^{-5} = \left(\dfrac{2}{1}\right)^5 = \dfrac{2^5}{1} = \dfrac{2 \cdot 2 \cdot 2 \cdot 2 \cdot 2}{1} = 32$

49. $\left(\dfrac{2}{3}\right)^{-3} = \left(\dfrac{3}{2}\right)^3 = \dfrac{3^3}{2^3} = \dfrac{3 \cdot 3 \cdot 3}{2 \cdot 2 \cdot 2} = \dfrac{27}{8}$

53. $4^{-2} + 4^{-3} = \dfrac{1}{4^2} + \dfrac{1}{4^3} = \dfrac{1}{4 \cdot 4} + \dfrac{1}{4 \cdot 4 \cdot 4} = \dfrac{1}{16} + \dfrac{1}{64}$
$= \dfrac{4}{64} + \dfrac{1}{64} = \dfrac{5}{64}$

57. $(3.918)^{-3} = \dfrac{1}{(3.918)^3} = .017$

61. $9^5 \cdot 9^3 = 9^{5+3} = 9^8$ Add exponents

65. $4^3 \cdot 4^5 \cdot 4^{-10} = 4^{3+5+(-10)}$ Add exponents
$= 4^{-2} = \dfrac{1}{4^2}$

69. $(-z)^3(-z)^6 = (-z)^{3+6}$ Add exponents
$= (-z)^9$

73. $\dfrac{4^2}{4^4} = 4^{2-4}$ Subtract exponents
$= 4^{-2} = \dfrac{1}{4^2}$

77. $\dfrac{q^{-4}}{q^2} = q^{-4-2} = q^{-6} = \dfrac{1}{q^6}$ Subtract exponents

81. $\dfrac{y^2}{y^{-5}} = y^{2-(-5)} = y^{2+5} = y^7$ Subtract exponents

85. $\dfrac{6^{-1}y^{-2}z^5}{6^2 y^{-1} z^{-2}}$
$= 6^{-1-2}y^{-2-(-1)}z^{5-(-2)}$ Subtract exponents
$= 6^{-3}y^{-2+1}z^{5+2}$
$= 6^{-3} \cdot y^{-1} \cdot z^7$
$= \dfrac{1}{6^3} \cdot \dfrac{1}{y} \cdot z^7$
$= \dfrac{z^7}{6^3 y}$

89. $3q^4 + 5q^4 = (3+5)q^4 = 8q^4$ Distributive property
$3q^4 \cdot 5q^4 = 3 \cdot 5 \cdot q^4 \cdot q^4$ Commutative and associative properties
$= 15q^{4+4}$ Add exponents
$= 15q^8$

93. $6r^{-4} + (-8r^{-4}) = [6 + (-8)]r^{-4}$
Distributive property
$= -2r^{-4}$
$(6r^{-4})(-8r^{-4}) = 6(-8)(r^{-4})(r^{-4})$
Commutative and associative properties
$= -48r^{-4+(-4)}$
Add exponents
$= -48r^{-8}$

Section 4.2 (page 181)

1. $(6^3)^2 = 6^{3 \cdot 2}$ Use the power rule
$= 6^6$

5. $(3^{-5})^{-2} = 3^{-5 \cdot (-2)}$ Power rule
$= 3^{10}$

9. $\dfrac{(k^2)^4}{(k^6)^2} = \dfrac{k^8}{k^{12}}$ Power rule
$= k^{8-12}$ Quotient rule
$= k^{-4}$
$= \dfrac{1}{k^4}$

13. $\dfrac{4^3 \cdot 4^{-5}}{4^7} = \dfrac{4^{3+(-5)}}{4^7}$
$= \dfrac{4^{-2}}{4^7} = 4^{-2-7} = 4^{-9} = \dfrac{1}{4^9}$

17. $\dfrac{m^4 \cdot m^{-5}}{m^{-6}} = \dfrac{m^{4+(-5)}}{m^{-6}}$
$= \dfrac{m^{-1}}{m^{-6}} = m^{-1-(-6)} = m^5$

21. $\dfrac{r^5 \cdot r^{-8}}{r^{-6} \cdot r^4} = \dfrac{r^{5+(-8)}}{r^{-6+4}}$
$= \dfrac{r^{-3}}{r^{-2}} = r^{-3-(-2)} = r^{-3+2} = r^{-1} = \dfrac{1}{r}$

25. $(5m)^3 = (5^1 m^1)^3$
$= 5^{1 \cdot 3} m^{1 \cdot 3}$ Use the power rule; multiply the exponents
$= 5^3 m^3$

29. $(-3x^5)^2 = [(-3)^1 x^5]^2$
$= (-3)^{1 \cdot 2} x^{5 \cdot 2} = (-3)^2 x^{10} = 3^2 x^{10}$

33. $(3x^{-5})^2 = (3^1 x^{-5})^2$
$= 3^{1 \cdot 2} x^{-5 \cdot 2}$
$= 3^2 x^{-10}$
$= 3^2 \cdot \dfrac{1}{x^{10}} = \dfrac{3^2}{x^{10}}$

37. $\left(\dfrac{a}{5}\right)^3 = \left(\dfrac{a^1}{5^1}\right)^3$
$= \dfrac{a^{1 \cdot 3}}{5^{1 \cdot 3}}$ Use the power rule
$= \dfrac{a^3}{5^3}$

41. $\left(\dfrac{a}{bc}\right)^{-1} = \dfrac{a^{-1}}{(bc)^{-1}} = \dfrac{bc}{a}$

45. $\dfrac{(x^3)^2}{x^9 x^7} = \dfrac{x^6}{x^{16}} = \dfrac{1}{x^{10}}$

49. $\dfrac{(3x^2)^{-2}(5x^{-1})^3}{3x^{-5}} = \dfrac{3^{-2}x^{-4} \cdot 5^3 x^{-3}}{3^1 x^{-5}}$
$= \dfrac{3^{-2} \cdot 5^3 \cdot x^{-4+(-3)}}{3^1 x^{-5}}$
$= \dfrac{3^{-2} \cdot 5^3 \cdot x^{-7}}{3^1 x^{-5}}$
$= 3^{-2-1} \cdot 5^3 \cdot x^{-7-(-5)}$
$= 3^{-3} \cdot 5^3 \cdot x^{-2}$
$= \dfrac{1}{3^3} \cdot 5^3 \cdot \dfrac{1}{x^2} = \dfrac{5^3}{3^3 x^2}$

53. $\dfrac{(2y^{-1}z^2)^2 (3y^{-2}z^{-3})^3}{(y^3 z^2)^{-1}} = \dfrac{2^2 y^{-2} z^4 \cdot 3^3 y^{-6} z^{-9}}{y^{-3} z^{-2}}$
$= \dfrac{2^2 \cdot 3^3 \cdot y^{-2+(-6)} z^{4+(-9)}}{y^{-3} z^{-2}}$
$= \dfrac{2^2 \cdot 3^3 y^{-8} z^{-5}}{y^{-3} z^{-2}}$
$= 2^2 \cdot 3^3 y^{-8-(-3)} z^{-5-(-2)}$
$= 2^2 \cdot 3^3 y^{-5} z^{-3}$
$= 2^2 \cdot 3^3 \cdot \dfrac{1}{y^5} \cdot \dfrac{1}{z^3}$
$= \dfrac{2^2 \cdot 3^3}{y^5 z^3}$

Section 4.3 (page 185)

1. In 6,835,000,000, count from the right of the first nonzero digit (the 6), all the way to the decimal point (9 places). Using 9 as the exponent of 10, we have
6.835×10^9.

5. 25,000   4 places to right
2.5×10^4

9. .000012   5 places to left
1.2×10^{-5}

Solutions to Selected Exercises   575

13. 8.1×10^9
Because of the nine, move the decimal point 9 places to the right. Fill in with zeros as necessary.
8,100,000,000   9 places

17. 3.2×10^{-4}
Move the decimal point 4 places to the left, since -4 is a negative number.
.00032

21. $(4 \times 10^{-1}) \times (1 \times 10^{-5})$
$= (4 \times 1) \times (10^{-1} \times 10^{-5})$
$= 4 \times 10^{-6}$
$= 4. \times 10^{-6}$
$= .000004$
Move the decimal point six places to the left.

25. $(1.2 \times 10^2) \times (5 \times 10^{-3}) \times (2.4 \times 10^3)$
$= (1.2 \times 5 \times 2.4) \times (10^2 \times 10^{-3} \times 10^3)$
$= 14.4 \times 10^2$
$= 1440$
Move the decimal point two places to the right.

29. $\dfrac{8 \times 10^{-3}}{2 \times 10^{-2}} = 4 \times 10^{-3-(-2)}$
$= 4 \times 10^{-3+2}$
$= 4 \times 10^{-1}$
$= .4$

33. $\dfrac{7.2 \times 10^{-3} \times 1.6 \times 10^5}{4 \times 10^{-2} \times 3.6 \times 10^9} = \dfrac{7.2 \times 1.6}{4 \times 3.6} \times \dfrac{10^{-3} \times 10^5}{10^{-2} \times 10^9}$
$= .8 \times \dfrac{10^2}{10^7}$
$= .8 \times 10^{-5}$
$= .000008$

37. To write 3,680,000,000,000,000 in scientific notation, move the decimal point 15 places to the left, getting 3.68×10^{15}.

41. To write 3.5×10^4 without exponents, add zeros and move the decimal point 4 places to the right, getting 35,000.

Section 4.4 (page 191)

1. $3m^5 + 5m^5 = (3 + 5)m^5$
$= 8m^5$

5. $2m^5 - 5m^2$
Since the exponents are different, the terms are unlike terms, so the expression cannot be simplified.

9. $-4p^7 + 8p^7 - 5p^7 = (-4 + 8 - 5)p^7$
$= -1p^7$
$= -p^7$

13. $-5p^5 + 8p^5 - 2p^5 - 1p^5 = (-5 + 8 - 2 - 1)p^5$
$= 0 \cdot p^5$
$= 0$ Zero times any number is zero

17. $4z^5 - 9z^3 + 8z^2 + 10z^5$
$= (4z^5 + 10z^5) - 9z^3 + 8z^2$
$= 14z^5 - 9z^3 + 8z^2$

21. $-.823q^2 + 1.725q - .374 + 1.994q^2 - .324q + .122$
$= (-.823q^2 + 1.994q^2) + (1.725q - .324q) + (-.374 + .122)$ Like terms together
$= 1.171q^2 + 1.401q - .252$

25. The polynomial is already simplified; the degree is the highest exponent, 9; there are three terms, so it is a trinomial.

29. $\dfrac{3}{5}x^5 + \dfrac{2}{5}x^5 = \dfrac{5}{5}x^5 = 1x^5 = x^5$
The degree is 5; it is a monomial (one term).

33. A binomial is a special kind of polynomial, so the statement is always true.

37. A binomial must have just two terms and a trinomial has three, so it is never true.

41. Subtract. $12x^4 - x^2$
                $8x^4 + 3x^2$
Change all signs in the second row; then add.
$12x^4 - 1x^2$
$-8x^4 - 3x^2$
$4x^4 - 4x^2$

45. $9m^3 - 5m^2 + 4m - 8$
$3m^3 + 6m^2 + 8m - 6$
$12m^3 + m^2 + 12m - 14$ Add in columns

49. $5b^2 + 6b + 2$
$3b^2 - 4b + 5$
$8b^2 + 2b + 7$ Add in columns

53. $(8m^2 - 7m) - (3m^2 + 7m)$
$= 8m^2 - 7m - 3m^2 - 7m$
$= 5m^2 - 14m$

57. $(8s - 3s^2) + (-4s + 5s^2) = 8s - 3s^2 - 4s + 5s^2$
$= 2s^2 + 4s$

61. $(7y^4 + 3y^2 + 2y) - (18y^4 - 5y^2 - y)$
$= 7y^4 + 3y^2 + 2y - 18y^4 + 5y^2 + y$
$= -11y^4 + 8y^2 + 3y$

65. $[(8m^2 + 4m - 7) - 1(2m^2 - 5m + 2)] - (m^2 + m + 1)$
$= 8m^2 + 4m - 7 - 2m^2 + 5m - 2 - m^2 - m - 1$
$= 8m^2 - 2m^2 - 1m^2 + 4m + 5m - 1m - 7 - 2 - 1$
$= 5m^2 + 8m - 10$

69. When $(4 + x^2)$ is added to $(-9x + 2)$ the result is larger than 8.
$(4 + x^2) + (-9x + 2) > 8$

Section 4.5 (page 199)

1. $(-4x^5)(8x^2) = (-4)(8)(x^5)(x^2) = -32x^7$

5. $(15a^4)(2a^5) = (15 \cdot 2)a^{4+5} = 30a^9$

9. $3p(-2p^3 + 4p^2) = (3p)(-2p^3) + (3p)(4p^2)$
$= -6p^4 + 12p^3$

13. $2y(3 + 2y + 5y^4)$
$= (2y)(3) + (2y)(2y) + (2y)(5y^4)$
$= 6y + 4y^2 + 10y^5$

17. $x + 5$
$x - 5$
$\overline{-5x - 25}$
$x^2 + 5x$
$\overline{x^2 - 25}$

21. $6p + 5$
$p - 1$
$\overline{-6p - 5}$
$6p^2 + 5p$
$\overline{6p^2 - p - 5}$

25.
$$\begin{array}{r} b + 8 \\ 6b - 2 \\ \hline -2b - 16 \\ 6b^2 + 48b \\ \hline 6b^2 + 46b - 16 \end{array}$$

29.
$$\begin{array}{r} -4h + k \\ 2h - k \\ \hline 4hk - k^2 \\ -8h^2 + 2hk \\ \hline -8h^2 + 6hk - k^2 \end{array}$$

33.
$$\begin{array}{r} 9a^2 + a + 1 \\ 9a + 2 \\ \hline 18a^2 + 2a + 2 \\ 81a^3 + 9a^2 + 9a \\ \hline 81a^3 + 27a^2 + 11a + 2 \end{array}$$

37.
$$\begin{array}{r} 3x^5 \quad - 2x^3 + x^2 - 2x + 3 \\ 2x - 1 \\ \hline -3x^5 \quad + 2x^3 - 1x^2 + 2x - 3 \\ 6x^6 \quad - 4x^4 + 2x^3 - 4x^2 + 6x \\ \hline 6x^6 - 3x^5 - 4x^4 + 4x^3 - 5x^2 + 8x - 3 \end{array}$$

41. $(x + 7)^2 = (x + 7)(x + 7)$
Multiply.
$$\begin{array}{r} x + 7 \\ x + 7 \\ \hline 7x + 49 \\ x^2 + 7x \\ \hline x^2 + 14x + 49 \end{array}$$

45. $(2p - 5)^2 = (2p - 5)(2p - 5)$
$$\begin{array}{r} 2p - 5 \\ 2p - 5 \\ \hline -10p + 25 \\ 4p^2 - 10p \\ \hline 4p^2 - 20p + 25 \end{array}$$

49. $(m - 5)^3 = (m - 5)(m - 5)(m - 5)$
First multiply $(m - 5)(m - 5)$.
$$\begin{array}{r} m - 5 \\ m - 5 \\ \hline -5m + 25 \\ m^2 - 5m \\ \hline m^2 - 10m + 25 \end{array}$$
Now multiply this last result by $(m - 5)$.
$$\begin{array}{r} m^2 - 10m + 25 \\ m - 5 \\ \hline -5m^2 + 50m - 125 \\ m^3 - 10m^2 + 25m \\ \hline m^3 - 15m^2 + 75m - 125 \end{array}$$

53. $(k + 1)^4 = (k + 1)^2(k + 1)^2$
First find $(k + 1)^2 = (k + 1)(k + 1)$.
$$\begin{array}{r} k + 1 \\ k + 1 \\ \hline k + 1 \\ k^2 + k \\ \hline k^2 + 2k + 1 \end{array}$$
Now $(k + 1)^4 = (k + 1)^2(k + 1)^2$
$= (k^2 + 2k + 1)(k^2 + 2k + 1)$.
Finally,
$$\begin{array}{r} k^2 + 2k + 1 \\ k^2 + 2k + 1 \\ \hline 1k^2 + 2k + 1 \\ 2k^3 + 4k^2 + 2k \\ k^4 + 2k^3 + 1k^2 \\ \hline k^4 + 4k^3 + 6k^2 + 4k + 1. \end{array}$$

Section 4.6 (page 205)

1. $(r - 1)(r + 3)$
       F   O   I   L
   $= (r)(r) + (r)(3) + (-1)(r) + (-1)(3)$
   $= r^2 + 3r - 1r - 3$
   $= r^2 + 2r - 3$ Simplify

5. $(2x - 1)(3x + 2) = 2x(3x) + (2x)(2) - 1(3x) - 1(2)$
                           F   O   I   L
   $= 6x^2 + 4x - 3x - 2$
   $= 6x^2 + x - 2$

9. $(a + 4)(2a + 1)$
   $= (a)(2a) + (a)(1) + (4)(2a) + (4)(1)$
   $= 2a^2 + 1a + 8a + 4$
   $= 2a^2 + 9a + 4$

13. $(2a + 4)(3a - 2)$
    $= (2a)(3a) + (2a)(-2) + (4)(3a) + (4)(-2)$
    $= 6a^2 - 4a + 12a - 8$
    $= 6a^2 + 8a - 8$

17. $(-3 + 2r)(4 + r)$
    $= (-3)(4) + (-3)(r) + (2r)(4) + (2r)(r)$
    $= -12 - 3r + 8r + 2r^2$
    $= -12 + 5r + 2r^2$

21. $(p + 3q)(p + q)$
    $= (p)(p) + (p)(q) + (3q)(p) + (3q)(q)$
    $= p^2 + 1pq + 3pq + 3q^2$
    $= p^2 + 4pq + 3q^2$

25. $(8y - 9z)(y + 5z)$
    $= (8y)(y) + (8y)(5z) + (-9z)(y) + (-9z)(5z)$
    $= 8y^2 + 40yz - 9yz - 45z^2$
    $= 8y^2 + 31yz - 45z^2$

29. $(2.13y + 4.06)(1.92y - 3.9)$
    $= (2.13y)(1.92y) + (2.13y)(-3.9)$
    $\quad + 4.06(1.92y) + 4.06(-3.9)$
    $= 4.0896y^2 - 8.307y + 7.7952y - 15.834$
    $= 4.0896y^2 - .5118y - 15.834$

33. $(x - 2y)^2 = (x)^2 - 2(x)(2y) + (2y)^2$
    $= x^2 - 4xy + 2^2y^2$
    $= x^2 - 4xy + 4y^2$

37. $(5p + 2q)^2 = (5p)^2 + 2(5p)(2q) + (2q)^2$
    $= 25p^2 + 20pq + 4q^2$

41. $(.85r + .23s)^2$
    $= (.85r)^2 + 2(.85r)(.23s) + (.23s)^2$
    $= .7225r^2 + .391rs + .0529s^2$

45. $(2b + 5)(2b - 5) = (2b)^2 - (5)^2$
    $= 4b^2 - 25$

49. $\left(2m - \dfrac{5}{3}\right)\left(2m + \dfrac{5}{3}\right) = (2m)^2 - \left(\dfrac{5}{3}\right)^2 = 4m^2 - \dfrac{25}{9}$

53. $(.48q + .37r)(.48q - .37r)$
    $= (.48q)^2 - (.37r)^2$
    $= .2304q^2 - .1369r^2$

57. 3 plus a number   multiplied by   the number less 4   is greater than   7.
    $(3 + x)$ $\cdot$ $(x - 4)$ $>$ 7

Section 4.7 (page 209)

1. $\dfrac{4x^2}{2x} = \dfrac{4}{2} \cdot \dfrac{x^2}{x^1} = 2x^{2-1} = 2x^1 = 2x$

Solutions to Selected Exercises   577

5. $\dfrac{27k^4m^5}{3km^6} = \dfrac{27}{3} \cdot \dfrac{k^4}{k^1} \cdot \dfrac{m^5}{m^6}$

$= 9 \cdot k^{4-1} \cdot m^{5-6} = 9k^3m^{-1} = \dfrac{9k^3}{m}$

9. $\dfrac{10m^5 - 16m^2 + 8m^3}{2m} = \dfrac{10m^5}{2m} - \dfrac{16m^2}{2m} + \dfrac{8m^3}{2m}$

$= 5m^4 - 8m + 4m^2$

13. $\dfrac{2m^5 - 4m^2 + 8m}{2m} = \dfrac{2m^5}{2m} - \dfrac{4m^2}{2m} + \dfrac{8m}{2m}$

$= m^4 - 2m + 4$

17. $\dfrac{12x^4 - 3x^3 + 3x^2}{3x^2} = \dfrac{12x^4}{3x^2} - \dfrac{3x^3}{3x^2} + \dfrac{3x^2}{3x^2}$

$= 4x^2 - x + 1$

21. $\dfrac{36x + 24x^2 + 3x^3}{3x^2} = \dfrac{36x}{3x^2} + \dfrac{24x^2}{3x^2} + \dfrac{3x^3}{3x^2}$

$= \dfrac{12}{x} + 8 + x$

25. $\dfrac{8k^4 - 12k^3 - 2k^2 + 7k - 3}{2k}$

$= \dfrac{8k^4}{2k} - \dfrac{12k^3}{2k} - \dfrac{2k^2}{2k} + \dfrac{7k}{2k} - \dfrac{3}{2k}$

$= 4k^3 - 6k^2 - k + \dfrac{7}{2} - \dfrac{3}{2k}$

29. $\dfrac{8x + 16x^2 + 10x^3}{4x^4} = \dfrac{8x}{4x^4} + \dfrac{16x^2}{4x^4} + \dfrac{10x^3}{4x^4}$

$= \dfrac{2}{x^3} + \dfrac{4}{x^2} + \dfrac{5}{2x}$

33. $\dfrac{120x^{11} - 60x^{10} + 140x^9 - 100x^8}{10x^{12}}$

$= \dfrac{120x^{11}}{10x^{12}} - \dfrac{60x^{10}}{10x^{12}} + \dfrac{140x^9}{10x^{12}} - \dfrac{100x^8}{10x^{12}}$

$= \dfrac{12}{x} - \dfrac{6}{x^2} + \dfrac{14}{x^3} - \dfrac{10}{x^4}$

37. Multiplying the quotient by $-7y^2$ gives the original polynomial.

$-7y^2\left(9y^2 + 3y + 5 - \dfrac{2}{y}\right)$

$= -7y^2(9y^2) - 7y^2(3y) - 7y^2(5) - 7y^2\left(-\dfrac{2}{y}\right)$

$= -63y^4 - 21y^3 - 35y^2 + 14y$

Section 4.8 (page 215)

1. $x - 3 \overline{) x^2 - x - 6}$ with quotient $x + 2$

$\underline{x^2 - 3x}$ ← Subtract by changing signs in the second row and then adding

$2x - 6$

$\underline{2x - 6}$

0

There is no remainder. The quotient is $x + 2$.

5. $p + 6 \overline{) p^2 + 2p - 20}$ with quotient $p - 4$

$\underline{p^2 + 6p}$

$-4p - 20$ ← Change signs and add

$\underline{-4p - 24}$

4

The remainder is 4. The quotient is

$p - 4 + \dfrac{4}{p + 6}$.

9. $2m - 3 \overline{) 12m^2 - 20m + 3}$ with quotient $6m - 1$

$\underline{12m^2 - 18m}$ ← Change signs and add

$-2m + 3$

$\underline{-2m + 3}$

0

There is no remainder. The quotient is $6m - 1$.

13. $2x + 1 \overline{) 2x^2 + 5x + 3}$ with quotient $x + 2$

$\underline{2x^2 + x}$

$4x + 3$ ← Change signs and add

$\underline{4x + 2}$

1

The remainder is 1. The quotient is

$x + 2 + \dfrac{1}{2x + 1}$.

17. $2x + 1 \overline{) 2x^3 - x^2 + 3x + 2}$ with quotient $x^2 - x + 2$

$\underline{2x^3 + x^2}$

$-2x^2 + 3x$

$\underline{-2x^2 - x}$ ← Change signs and add

$4x + 2$

$\underline{4x + 2}$

0

There is no remainder. The quotient is $x^2 - x + 2$.

21. $y^2 + 0y + 1 \overline{) 3y^3 + y^2 + 3y + 1}$ with quotient $3y + 1$

$\underline{3y^3 + 0y^2 + 3y}$ ← Change signs and add

$y^2 \qquad + 1$

$\underline{y^2 \qquad + 1}$

0

There is no remainder. The quotient is $3y + 1$.

25. $x^2 + 0x - 2 \overline{) x^4 + 0x^3 - x^2 - 6x}$ with quotient $x^2 + 1$

$\underline{x^4 + 0x^3 - 2x^2}$ ← Change signs and add

$x^2 - 6x$

$\underline{x^2 + 0x - 2}$

$-6x + 2$

The remainder is $-6x + 2$. The quotient is

$x^2 + 1 + \dfrac{-6x + 2}{x^2 - 2}$.

29. $4m^2 + 0m - 3 \overline{) 4m^5 - 8m^4 - 3m^3 + 22m^2 + 0m - 15}$ with quotient $m^3 - 2m^2 + 4$

$\underline{4m^5 - 0m^4 - 3m^3}$ ← Change signs and add

$-8m^4 \qquad + 22m^2$

$\underline{-8m^4 + 0m^3 + 6m^2}$

$16m^2 \qquad - 15$

$\underline{16m^2 + 0m - 12}$

-3

The remainder is -3. The quotient is

$m^3 - 2m^2 + 4 + \dfrac{-3}{4m^2 - 3}$.

33. $a^2 + 0a + 1 \overline{) a^4 - 0a^3 + 0a^2 + 0a - 1}$ with quotient $a^2 - 1$

$\underline{a^4 + 0a^3 + a^2}$ ← Change signs and add

$-a^2 + 0a - 1$

$\underline{-a^2 + 0a - 1}$

0

There is no remainder. The quotient is $a^2 - 1$.

CHAPTER 5

Section 5.1 (page 233)

1. Write all the multiplication problems with integer factors having 14 as an answer.
$$1 \cdot 14 = 14$$
$$2 \cdot 7 = 14$$
List all the numbers used in the multiplication problems, in increasing order: 1, 2, 7, 14. These are the positive integer factors of 14.

5. The multiplication problems with 45 as an answer are
$$1 \cdot 45 = 45$$
$$3 \cdot 15 = 45$$
$$5 \cdot 9 = 45.$$
The positive factors of 45 are 1, 3, 5, 9, 15, 45.

9. The multiplication problems with 100 as an answer are
$$1 \cdot 100 = 100$$
$$2 \cdot 50 = 100$$
$$4 \cdot 25 = 100$$
$$5 \cdot 20 = 100$$
$$10 \cdot 10 = 100$$
The positive factors of 100 are 1, 2, 4, 5, 10, 20, 25, 50, and 100.

Hint for Exercises 13–20: A quick and simple way of finding prime factors of a number is to use "short division." For example, find the prime factors of 20 by first dividing 20 by the smallest possible prime, 2.

$$2 \underline{|20}$$
$$10$$

Now use 2 again, and divide it into 10.

$$2 \underline{|20}$$
$$2 \underline{|10}$$
$$5$$

The process is finished when we obtain a prime as an answer (the 5). The 2, 2, and 5 are the prime factors of 20, so the prime factored form for 20 is $2 \cdot 2 \cdot 5$ or $2^2 \cdot 5$.

13. Use the "short division" method explained above.
$$2\underline{|120}$$
$$2\underline{|60}$$
$$2\underline{|30}$$
$$3\underline{|15}$$
$$5$$
Therefore
$$120 = 2 \cdot 2 \cdot 2 \cdot 3 \cdot 5$$
$$= 2^3 \cdot 3 \cdot 5$$

17. Use the "short division" explained above.
$$5\underline{|275}$$
$$5\underline{|55}$$
$$11 \quad 275 = 5^2 \cdot 11$$

21. In factored form,
$12y = 2 \cdot 2 \cdot 3 \cdot y$ and $24 = 2 \cdot 2 \cdot 2 \cdot 3$.
To get the greatest common factor, take each prime the least number of times it appears in factored form. The least number of times the 2 appears is twice, and the 3 appears one time. The y is not in 24 so it will not be in the greatest common factor. Thus,
$$\text{greatest common factor} = 2 \cdot 2 \cdot 3 = 12.$$

25. In factored form,
$$18m^2n^2 = 2 \cdot 3 \cdot 3 \cdot m^2 \cdot n^2$$
$$36m^4n^5 = 2 \cdot 2 \cdot 3 \cdot 3 \cdot m^4 \cdot n^5$$
$$12m^3n = 2 \cdot 2 \cdot 3 \cdot m^3n.$$
The greatest common numerical factor is $2 \cdot 3 = 6$. We use the lowest exponent of the common variable factors to get m^2n. Therefore, the greatest common factor is $6m^2n$.

29. Since $3x^2 = 3x \cdot x$, place x in the blank.

33. Since $-8z^9 = -4z^5 \cdot (2z^4)$, place $2z^4$ in the blank.

37. Since $14x^4y^3 = 2xy(7x^3y^2)$, place $7x^3y^2$ in the blank.

41. The greatest common factor for $9a^2$ and $18a$ is $9a$.
$$9a^2 - 18a = 9a \cdot a - 9a \cdot 2 = 9a(a - 2)$$

45. The greatest common factor for $11z^2$ and 100 is 1. Therefore, we cannot factor a greatest common factor from $11z^2 - 100$.

49. The greatest common factor of $13y^6 + 26y^5 - 39y^3$ is $13y^3$. Thus,
$$13y^6 + 26y^5 - 39y^3$$
$$= 13y^3(y^3) + 13y^3(2y^2) - 13y^3(3)$$
$$= 13y^3(y^3 + 2y^2 - 3).$$

53. The greatest common factor is $5z^3a^3$. Thus,
$$125z^5a^3 - 60z^4a^5 + 85z^3a^4$$
$$= 5z^3a^3(25z^2) - 5z^3a^3(12za^2)$$
$$ + 5z^3a^3(17a)$$
$$= 5z^3a^3(25z^2 - 12za^2 + 17a).$$

57. $p^2 + 4p + 3p + 12 = p(p + 4) + 3(p + 4)$
$$= (p + 4)(p + 3)$$

61. $7z^2 + 14z - az - 2a = 7z(z + 2) - a(z + 2)$
$$= (z + 2)(7z - a)$$

65. $3a^3 + 3ab^2 + 2a^2b + 2b^3$
$$= 3a(a^2 + b^2) + 2b(a^2 + b^2)$$
$$= (a^2 + b^2)(3a + 2b)$$

Section 5.2 (page 239)

1. To get the last term of the trinomial, we use $7 \cdot 3 = 21$ and notice that $7 + 3 = 10$, which matches the coefficient of the middle term, $10x$. So
$$x^2 + 10x + 21 = (x + 7)(x + 3).$$

5. To get the last term of the trinomial, we use $-2 \cdot (-12) = 24$ and notice that $-2 + (-12) = -14$, which matches the coefficient of the middle term, $-14t$. So
$$t^2 - 14t + 24 = (t - 2)(t - 12).$$

9. To get the last term of the trinomial, we use $-4 \cdot 6 = -24$ and notice that $-4 + 6 = 2$, which matches the coefficient of the middle term, $2m$. So
$$m^2 + 2m - 24 = (m - 4)(m + 6).$$

13. Since $(-2y) \cdot (-5y) = 10y^2$ and since $-2y + (-5y) = -7y$, the missing expression is $x - 5y$, and $x^2 - 7xy + 10y^2$
$$= (x - 2y)(x - 5y).$$

17. Write all pairs of factors whose product is 15.
$$1 \cdot 15 = 15 \qquad -1(-15) = 15$$
$$3 \cdot 5 = 15 \qquad -3(-5) = 15$$
The pair 3 and 5 has a sum of 8, so
$$b^2 + 8b + 15 = (b + 3)(b + 5).$$

Solutions to Selected Exercises 579

21. The factors of -12 are
$$-3 \cdot 4 = -12 \quad 3(-4) = -12$$
$$-2 \cdot 6 = -12 \quad 2(-6) = -12$$
$$-1 \cdot 12 = -12 \quad 1(-12) = -12$$
Since $-2 + 6 = 4$, and 4 is the coefficient of n,
$$n^2 + 4n - 12 = (n - 2)(n + 6).$$

25. List the pairs of factors of 12.
$$3 \cdot 4 = 12 \quad -3(-4) = 12$$
$$2 \cdot 6 = 12 \quad -2(-6) = 12$$
$$1 \cdot 12 = 12 \quad -1(-12) = 12$$
None of these pairs of numbers has a sum of 11, so $h^2 + 11h + 12$ is prime.

29. The only pair of numbers having a product of 20 and a sum of -9 is -4 and -5, so
$$x^2 - 9x + 20 = (x - 4)(x - 5).$$

33. The pair of numbers having a product of -42 and a sum of 1 is -6 and 7, so
$$r^2 + r - 42 = (r - 6)(r + 7).$$

37. The last term can be factored as
$$(-2m)(3m) = -6m^2,$$
$$(2m)(-3m) = -6m^2,$$
$$(-m)(6m) = 6m^2,$$
or $(m)(-6m) = -6m^2.$
Since $2m + (-3m) = -m$,
$$x^2 - mx - 6m^2 = (x + 2m)(x - 3m).$$

41. The last term, $-56y^2$, can be factored as
$$(-7y)(8y) \quad (7y)(-8y)$$
$$(-4y)(14y) \quad (4y)(-14y)$$
$$(-2y)(28y) \quad (2y)(-28y)$$
$$(-y)(56y) \quad (y)(-56y).$$
Only $7y + (-8y) = -y$, so
$$a^2 - ay - 56y^2 = (a + 7y)(a - 8y).$$

45. The terms whose product is $4d^2$ and whose sum is $-5d$ are $-4d$ and $-d$, so
$$c^2 - 5cd + 4d^2 = (c - 4d)(c - d).$$

49. The greatest common factor is h^5. Thus,
$$h^7 - 5h^6 - 14h^5 = h^5(h^2 - 5h - 14).$$
Now factor $h^2 - 5h - 14$. The pair of numbers with a product of -14 and a sum of -5 is 2 and -7, so
$$h^2 - 5h - 14 = (h + 2)(h - 7).$$
Putting this result together with the common factor gives
$$h^7 - 5h^6 - 14h^5 = h^5(h + 2)(h - 7).$$

53. The greatest common factor is $2y$.
$$2y^3 - 8y^2 - 10y = 2y(y^2 - 4y - 5).$$
Factor $y^2 - 4y - 5$ as $(y - 5)(y + 1)$. Putting these results together gives
$$2y^3 - 8y^2 - 10y = 2y(y - 5)(y + 1).$$

57. The greatest common factor is k^5, so
$$k^7 - 2k^6m - 15k^5m^2 = k^5(k^2 - 2km - 15m^2).$$
The terms whose product is $-15m^2$ and whose sum is $-2m$ are $3m$ and $-5m$, so
$$k^7 - 2k^6m - 15k^5m^2 = k^5(k + 3m)(k - 5m).$$

61. $3x^2 + 9x - 12$ is incorrectly factored as $(x - 1)(3x + 12)$ because the factor $3x + 12$ can be factored further:
$$3x + 12 = 3(x + 4),$$
and $3x^2 + 9x - 12 = 3(x - 1)(x + 4).$

Section 5.3 (page 245)

1. Since $2x \cdot x = 2x^2$ and $1 \cdot -1 = -1$, let the missing factor be $x - 1$. We can see that in
$$(2x + 1)(x - 1),$$
the inner product is $(1)(x) = x$, and the outer product is $(2x)(-1) = -2x$. Since $x + (-2x) = -x$, the middle term of the original trinomial,
$$2x^2 - x - 1 = (2x + 1)(x - 1).$$

5. Since $4y^2 = y \cdot 4y$ and $-15 = 5(-3)$, try $4y - 3$ as the missing factor:
$$(y + 5)(4y - 3).$$
Multiply to see that this product is correct.

9. Since $2m^2 = m \cdot 2m$ and $-10 = -1 \cdot 10$, check to see that the product is $(2m - 1)(m + 10)$.

13. Since $4k^2 = k \cdot 4k$ and $3m^2 = m \cdot 3m$, the product is
$$(4k + m)(k + 3m).$$

17. First, factor out a common factor of m^4.
$$6m^6 + 7m^5 - 20m^4 = m^4(6m^2 + 7m - 20)$$
Now factor within the parentheses to get
$$6m^6 + 7m^5 - 20m^4 = m^4(3m - 4)(2m + 5).$$

21. $3a^2 + 10a + 7 \quad 3 \cdot 7 = 21$
We need two numbers whose product is 21 and whose sum is 10. The numbers are 3 and 7. Write $10a$ as $3a + 7a$.
$$3a^2 + 10a + 7 = 3a^2 + 3a + 7a + 7$$
Factor by grouping.
$$3a^2 + 3a + 7a + 7 = 3a(a + 1) + 7(a + 1)$$
$$= (a + 1)(3a + 7)$$

25. $15m^2 + m - 2 \quad 15 \cdot (-2) = -30$
We need two numbers whose product is -30 and whose sum is 1. The numbers are 6 and -5. Write m as $1m$, or $6m - 5m$.
$$15m^2 + m - 2 = 15m^2 + 6m - 5m - 2$$
$$= 3m(5m + 2) - 1(5m + 2)$$
Factor by grouping
$$= (5m + 2)(3m - 1)$$

29. $5a^2 - 7a - 6 \quad 5 \cdot (-6) = -30$
Two numbers whose product is -30 and whose sum is -7 are -10 and 3. Therefore,
$$5a^2 - 7a - 6 = 5a^2 - 10a + 3a - 6$$
$$= 5a(a - 2) + 3(a - 2)$$
$$= (a - 2)(5a + 3).$$

33. Here we factor by the alternate method. Write the first term $4y^2 = 4y \cdot y$ and the last term $17 = 1 \cdot 17$. So we have
$$(4y + 1)(y + 17).$$
The outer product $= (4y)(17) = 68y$, and inner product $(1)(y) = y$. The sum of the inner and outer products is $68y + 1y = 69y$, which matches the middle term. So
$$4y^2 + 69y + 17 = (4y + 1)(y + 17).$$

37. Write the first term $10x^2 = 2x \cdot 5x$ and the last term $-6 = 3 \cdot -2$. So we have
$$(2x + 3)(5x - 2).$$
The inner product $= (3)(5x) = 15x$, and the outer product $= (2x)(-2) = -4x$. Since $15x + (-4x) = 11x$,
$$10x^2 + 11x - 6 = (2x + 3)(5x - 2).$$

41. $6q^2 + 23q + 21$
    $\uparrow \underline{\qquad} \uparrow \quad 6 \cdot 21 = 126$
    Two numbers whose product is 126 and whose sum is 23 are 9 and 14. So
    $$6q^2 + 23q + 21 = 6q^2 + 9q + 14q + 21$$
    $$= 3q(2q + 3) + 7(2q + 3)$$
    $$= (2q + 3)(3q + 7).$$

45. $8k^2 + 2k - 15$
    $\uparrow \underline{\qquad} \uparrow \quad 8 \cdot (-15) = -120$
    Two numbers whose product is -120 and whose sum is 2 are 12 and -10, so
    $$8k^2 + 2k - 15 = 8k^2 + 12k - 10k - 15$$
    $$= 4k(2k + 3) - 5(2k + 3)$$
    $$= (2k + 3)(4k - 5).$$

49. $24x^2 - 42x + 9$
    First factor out the common factor of 3 to get $3(8x^2 - 14x + 3)$. Use trial and error to factor $8x^2 - 14x + 3$. List some possible factors.
    $(8x - 3)(x - 1) = 8x^2 - 11x + 3$ Wrong
    $(2x - 1)(4x - 3) = 8x^2 - 10x + 3$ Wrong
    $(4x - 1)(2x - 3) = 8x^2 - 14x + 3$ Right
    The final result is
    $$24x^2 - 42x + 9 = 3(4x - 1)(2x - 3).$$

53. Factor out a common factor of $2m$.
    $$2m^3 + 2m^2 - 40m = 2m(m^2 + m - 20)$$
    Factor $m^2 + m - 20$ to get
    $$2m^3 + 2m^2 - 40m = 2m(m + 5)(m - 4).$$

57. Factor out the common factor of $4z^3$.
    $$32z^5 - 20z^4 - 12z^3 = 4z^3(8z^2 - 5z - 3)$$
    Factor $8z^2 - 5z - 3$ to get
    $$32z^5 - 20z^4 - 12z^3 = 4z^3(8z + 3)(z - 1).$$

61. $25a^2 + 25ab + 6b^2$
    $\uparrow \underline{\qquad} \uparrow \quad 25 \cdot 6 = 150$
    Two numbers whose product is 150 and whose sum is 25 are 15 and 10, so
    $$25a^2 + 25ab + 6b^2$$
    $$= 25a^2 + 15ab + 10ab + 6b^2$$
    $$= 5a(5a + 3b) + 2b(5a + 3b)$$
    $$= (5a + 3b)(5a + 2b).$$

65. Factor out the greatest common factor of m^4n to get
    $$6m^6n + 7m^5n^2 + 2m^4n^3 = m^4n(6m^2 + 7mn + 2n^2).$$
    Factor $6m^2 + 7mn + 2n^2$ as $(2m + n)(3m + 2n)$. Putting both results together gives
    $$6m^6n + 7m^5n^2 + 2m^4n^3 = m^4n(2m + n)(3m + 2n).$$

Section 5.4 (page 253)

1. Since $x^2 - 16 = (x^2) - (4)^2$, the expression $x^2 - 16$ is a difference of two squares. Therefore,
$$x^2 - 16 = (x + 4)(x - 4).$$

5. $9m^2 - 1 = (3m)^2 - (1)^2$
   $= (3m + 1)(3m - 1)$

9. The greatest common factor is 4, so
   $$36t^2 - 16 = 4(9t^2 - 4).$$
   Factor $9t^2 - 4$.
   $$9t^2 - 4 = (3t)^2 - (2)^2$$
   $$= (3t + 2)(3t - 2)$$
   Putting the above results together, we have
   $$36t^2 - 16 = 4(3t + 2)(3t - 2).$$

13. $81x^2 + 16$ is the *sum* of two squares, which is prime.

17. $a^4 - 1 = (a^2)^2 - (1)^2$
    $= (a^2 + 1)(a^2 - 1)$
    Factor $a^2 - 1$.
    $a^2 - 1 = (a)^2 - (1)^2$
    $= (a + 1)(a - 1)$
    Putting the above results together, we have
    $$a^4 - 1 = (a^2 + 1)(a + 1)(a - 1).$$

21. $16k^4 - 1 = (4k^2)^2 - 1^2$
    $= (4k^2 + 1)(4k^2 - 1)$
    Now factor $4k^2 - 1$:
    $4k^2 - 1 = (2k)^2 - 1^2$
    $= (2k + 1)(2k - 1).$
    $4k^2 + 1$ is prime. Putting all this together gives
    $$16k^4 - 1 = (4k^2 + 1)(2k + 1)(2k - 1).$$

25. Since $y^2 = (y)^2$ and $16 = (4)^2$, try $(y - 4)^2$. Since $2(-4)(y) = -8y$, which is the middle term,
    $$y^2 - 8y + 16 = (y - 4)^2.$$

29. Since $r^2 = (r)^2$ and $\dfrac{1}{9} = \left(\dfrac{1}{3}\right)^2$, try $\left(r + \dfrac{1}{3}\right)^2$. Since $2(r)\left(\dfrac{1}{3}\right) = \dfrac{2}{3}r$, which is the middle term,
    $$r^2 + \dfrac{2}{3}r + \dfrac{1}{9} = \left(r + \dfrac{1}{3}\right)^2.$$

33. Since $16a^2 = (4a)^2$ and $25b^2 = (5b)^2$, try $(4a - 5b)^2$. Since $2(4a)(-5b) = -40ab$, the middle term,
    $$16a^2 - 40ab + 25b^2 = (4a - 5b)^2.$$

37. Since $49x^2 = (7x)^2$ and $4y^2 = (2y)^2$, try $(7x + 2y)^2$. Since $2(7x)(2y) = 28xy$, the middle term,
    $$49x^2 + 28xy + 4y^2 = (7x + 2y)^2.$$

41. Since $25h^2 = (5h)^2$ and $4y^2 = (2y)^2$, try $(5h - 2y)^2$. Since $2(5h)(-2y) = -20hy$, the middle term,
    $$25h^2 - 20hy + 4y^2 = (5h - 2y)^2.$$

45. $8a^3 + 1 = (2a)^3 + 1^3$
    $= (2a + 1)[(2a)^2 - (2a)(1) + (1)^2]$
    $= (2a + 1)(4a^2 - 2a + 1)$

49. $8p^3 + q^3 = (2p)^3 + q^3$
    $= (2p + q)[(2p)^2 - (2p)(q) + q^2]$
    $= (2p + q)(4p^2 - 2pq + q^2)$

53. $64x^3 + 125y^3$
    $= (4x)^3 + (5y)^3$
    $= (4x + 5y)[(4x)^2 - (4x)(5y) + (5y)^2]$
    $= (4x + 5y)(16x^2 - 20xy + 25y^2)$

57. $64y^3 - 1331w^3$
    $= (4y)^3 - (11w)^3$
    $= (4y - 11w)[(4y)^2 + (4y)(11w) + (11w)^2]$
    $= (4y - 11w)(16y^2 + 44yw + 121w^2)$

61. $125z^3 + 64r^6$
$= (5z)^3 + (4r^2)^3$
$= (5z + 4r^2)[(5z)^2 - (5z)(4r^2) + (4r^2)^2]$
$= (5z + 4r^2)(25z^2 - 20r^2z + 16r^4)$

Supplementary Factoring Exercises (page 255)

1. The greatest common factor is $8m^3$.
$32m^9 + 16m^5 + 24m^3 = 8m^3(4m^6 + 2m^2 + 3)$

5. $6z^2 + 31z + 5$

$6 \cdot 5 = 30$

Find a pair of numbers whose product is 30 and whose sum is 31. The numbers are 30 and 1. Write $31z$ as $30z + 1z$:
$$6z^2 + 31z + 5 = 6z^2 + 30z + 1z + 5.$$
Factor by grouping.
$$6z^2 + 30z + 1z + 5 = 6z(z + 5) + 1(z + 5)$$
$$= (z + 5)(6z + 1)$$
Then $6z^2 + 31z + 5 = (z + 5)(6z + 1)$.

9. Since $8p^3 = (2p)^3$ and $27 = 3^3$, use the pattern for the sum of cubes.
$8p^3 + 27 = (2p)^3 + 3^3$
$= (2p + 3)[(2p)^2 - 6p + 3^2]$
$= (2p + 3)(4p^2 - 6p + 9)$

13. Look for a pair of numbers whose product is -15 and whose sum is 2. The numbers are 5 and -3, so
$$m^2 + 2m - 15 = (m + 5)(m - 3).$$

17. Since $z^2 = (z)^2$ and $36 = 6^2$, try $(z - 6)^2$. Since $2(z)(-6) = -12z$, the middle term,
$$z^2 - 12z + 36 = (z - 6)^2.$$

21. Factoring out the greatest common factor, 6, gives
$$6y^2 - 6y - 12 = 6(y^2 - y - 2).$$
Now factor $y^2 - y - 2$. Look for a pair of numbers whose product is -2 and whose sum is 1. The numbers are -2 and 1, so
$$y^2 - y - 2 = (y - 2)(y + 1).$$
Putting everything together gives
$$6y^2 - 6y - 12 = 6(y - 2)(y + 1).$$

25. The polynomial $k^2 + 9$ cannot be factored, since it is the sum of two squares. Thus $k^2 + 9$ is prime.

29. Since $4k^2 = (2k)^2$ and $9 = 3^2$, try $(2k - 3)^2$. Since $2(2k)(-3) = -12k$, the middle term,
$$4k^2 - 12k + 9 = (2k - 3)^2.$$

33. Factoring out the greatest common factor of $3k$ gives
$$3k^3 - 12k^2 - 15k = 3k(k^2 - 4k - 5).$$
To factor $k^2 - 4k - 5$, look for a pair of numbers whose product is -5 and whose sum is -4. The numbers are -5 and 1, so
$$k^2 - 4k - 5 = (k - 5)(k + 1).$$
Putting everything together gives
$$3k^3 - 12k^2 - 15k = 3k(k - 5)(k + 1).$$

37. Factoring out the greatest common factor of $6y^4$ gives
$$36y^6 - 42y^5 - 120y^4 = 6y^4(6y^2 - 7y - 20).$$
To factor $6y^2 - 7y - 20$, use trial and error. The correct factors are $(3y + 4)(2y - 5)$, so
$$36y^6 - 42y^5 - 120y^4 = 6y^4(3y + 4)(2y - 5).$$

41. Factor out the greatest common factor of 6 to get
$$54m^2 - 24z^2 = 6(9m^2 - 4z^2).$$
Since $9m^2 = (3m)^2$ and $4z^2 = (2z)^2$, use the difference of square pattern to factor $9m^2 - 4z^2$.
$9m^2 - 4z^2 = (3m)^2 - (2z)^2$
$= (3m + 2z)(3m - 2z)$
Now put this all together to get
$$54m^2 - 24z^2 = 6(3m + 2z)(3m - 2z).$$

45. Use the pattern for the difference of two cubes.
$27r^3 - 8s^3$
$= (3r)^3 - (2s)^3$
$= (3r - 2s)[(3r)^2 + (3r)(2s) + (2s)^2]$
$= (3r - 2s)(9r^2 + 6rs + 4s^2)$

49. Since $m^2 = (m)^2$ and $4 = (2)^2$, try $(m - 2)^2$. Since $2(m)(-2) = -4m$, the middle term,
$$m^2 - 4m + 4 = (m - 2)^2.$$

53. $12p^2 + pq - 6q^2$

$12(-6) = -72$

Look for two numbers whose product is -72 and whose sum is 1. The numbers are 9 and -8. Write pq as $9pq - 8pq$. Then
$12p^2 + pq - 6q^2 = 12p^2 + 9pq - 8pq - 6q^2$
$= 3p(4p + 3q) - 2q(4p + 3q)$

Factor by grouping

$= (4p + 3q)(3p - 2q).$

57. $100a^2 - 81y^2 = (10a)^2 - (9y)^2$
$= (10a + 9y)(10a - 9y)$

Section 5.5 (page 257)

1. $(x - 2)(x + 4) = 0$
Set each factor equal to 0.
$x - 2 = 0$ or $x + 4 = 0$
Solve each equation.
$x = 2$ or $x = -4$
The solutions are $x = 2$ or $x = -4$.

5. $(5p + 1)(2p - 1) = 0$
Set each factor equal to 0 and solve each equation.
$5p + 1 = 0$ or $2p - 1 = 0$
$5p = -1$ $\qquad$ $2p = 1$
$p = -\dfrac{1}{5}$ or $p = \dfrac{1}{2}$
Solutions: $-\dfrac{1}{5}, \dfrac{1}{2}$

9. $m^2 + 3m - 28 = 0$
Factor: $(m + 7)(m - 4) = 0$. Set each factor equal to 0 and solve.
$m + 7 = 0$ or $m - 4 = 0$
$m = -7$ or $m = 4$
Solutions: $-7, 4$

13. $x^2 = 3 + 2x$
Get 0 alone on one side.
$x^2 - 2x - 3 = 0$
Factor the left side.
$(x - 3)(x + 1) = 0$
Set each factor equal to 0 and solve.
$x - 3 = 0$ or $x + 1 = 0$
$x = 3$ or $x = -1$
Solutions: $3, -1$

17. $m^2 + 8m + 16 = 0$
Factor the left side.
$$(m + 4)(m + 4) = 0$$
Set $m + 4$ equal to 0 and solve.
$$m + 4 = 0$$
$$m = -4$$
Solution: -4

21. $2k^2 - k - 10 = 0$
Factor: $(2k - 5)(k + 2) = 0$. Set each factor equal to 0 and solve.
$$2k - 5 = 0 \text{ or } k + 2 = 0$$
$$2k = 5$$
$$k = \frac{5}{2} \text{ or } \quad k = -2$$
Solutions: $\frac{5}{2}, -2$

25. $\quad 6a^2 = 5 - 13a$
Get equation equal to 0.
$$6a^2 + 13a - 5 = 0$$
Factor: $(3a - 1)(2a + 5) = 0$. Set factors equal to 0 and solve.
$$3a - 1 = 0 \text{ or } 2a + 5 = 0$$
$$a = \frac{1}{3} \text{ or } \quad a = -\frac{5}{2}$$
Solutions: $\frac{1}{3}, -\frac{5}{2}$

29. $\quad 3a^2 + 7a = 20$
Get equation equal to 0.
$$3a^2 + 7a - 20 = 0$$
Factor: $(3a - 5)(a + 4) = 0$. Set factors equal to 0 and solve.
$$3a - 5 = 0 \text{ or } a + 4 = 0$$
$$a = \frac{5}{3} \text{ or } \quad a = -4$$
Solutions: $\frac{5}{3}, -4$

33. $16r^2 - 25 = 0$
Factor: $16r^2 - 25 = (4r + 5)(4r - 5) = 0$.
Set each factor equal to 0 and solve.
$$4r + 5 = 0 \quad \text{or } 4r - 5 = 0$$
$$4r = -5 \text{ or } \quad 4r = 5$$
$$r = \frac{-5}{4} \text{ or } \quad r = \frac{5}{4}$$
Solutions: $-\frac{5}{4}, \frac{5}{4}$

37. $\qquad b(2b + 3) = 9$
$\qquad 2b^2 + 3b = 9$ Multiply on left side
$\qquad 2b^2 + 3b - 9 = 0$ Get 0 on one side
$\qquad (2b - 3)(b + 3) = 0$ Factor
Set each factor equal to 0.
$$2b - 3 = 0 \text{ or } b + 3 = 0$$
$$2b = 3$$
$$b = \frac{3}{2} \text{ or } \quad b = -3 \quad \text{Solve}$$
Solutions: $\frac{3}{2}, -3$

41. $3r(r + 1) = (2r + 3)(r + 1)$
$\quad 3r^2 + 3r = 2r^2 + 5r + 3$ Simplify
Get 0 alone on one side.
$$r^2 - 2r - 3 = 0$$
Factor the left side.
$$(r - 3)(r + 1) = 0$$
$$r - 3 = 0 \text{ or } r + 1 = 0$$
$$r = 3 \text{ or } \quad r = -1$$
Solutions: $3, -1$

45. $(2r - 5)(3r^2 - 16r + 5) = 0$
Factor the trinomial.
$$(2r - 5)(3r - 1)(r - 5) = 0$$
Set each factor equal to zero and solve.
$$2r - 5 = 0 \text{ or } 3r - 1 = 0 \text{ or } r - 5 = 0$$
$$2r = 5 \qquad 3r = 1$$
$$r = \frac{5}{2} \text{ or } \qquad r = \frac{1}{3} \text{ or } \quad r = 5$$
Solutions: $\frac{5}{2}, \frac{1}{3}, 5$

49. $9y^3 - 49y = 0$
Factor out y.
$$y(9y^2 - 49) = 0$$
Factor the difference of two squares.
$$y(3y + 7)(3y - 7) = 0$$
Set each factor equal to zero and solve.
$$y = 0 \text{ or } 3y + 7 = 0 \quad \text{or } 3y - 7 = 0$$
$$3y = -7 \qquad 3y = 7$$
$$y = 0 \text{ or } \qquad y = -\frac{7}{3} \text{ or } \quad y = \frac{7}{3}$$
Solutions: $0, -\frac{7}{3}, \frac{7}{3}$

53. $a^3 + a^2 - 20a = 0$
Factor: $a(a^2 + a - 20) = 0$
$$a(a + 5)(a - 4) = 0$$
Set factors equal to 0 and solve.
$$a = 0 \text{ or } a + 5 = 0 \quad \text{or } a - 4 = 0$$
$$a = 0 \text{ or } \quad a = -5 \text{ or } \quad a = 4$$
Solutions: $0, -5, 4$

Section 5.6 (page 267)

1. Let x represent the width. Then $x + 5$ represents the length. The area of a rectangle equals length times width. So, with an area of 66 square centimeters,

Area	equals	length	times	width.
↓	↓	↓	↓	↓
66	=	$(x + 5)$	·	x

Solve.
$$66 = x^2 + 5x \qquad \text{Simplify}$$
$$0 = x^2 + 5x - 66 \qquad \text{Get 0 alone on one side}$$
$$0 = (x + 11)(x - 6) \qquad \text{Factor}$$
Set each factor equal to zero and solve.
$$x + 11 = 0 \quad \text{or } x - 6 = 0$$
$$x = -11 \text{ or } \quad x = 6$$
We reject $x = -11$ because a negative width is nonsense. So the width is 6 centimeters and the length is $6 + 5 = 11$ centimeters.

5. Let the width be x. Then the length is $2x$. With an area of 48 square inches for the new rectangle,

$$\underbrace{\text{Length}}_{2x} \underbrace{\text{times}}_{\cdot} \underbrace{\text{the width increased by 2}}_{(x+2)} \underbrace{\text{equals}}_{=} \underbrace{\text{the area.}}_{48}$$

Solve.

$2x^2 + 4x = 48$ Simplify
$2x^2 + 4x - 48 = 0$ Get 0 alone on one side
$2(x^2 + 2x - 24) = 0$
$2(x + 6)(x - 4) = 0$ Factor

Set each variable factor equal to zero and solve.
$x + 6 = 0$ or $x - 4 = 0$
$x = -6$ or $x = 4$

Reject the negative solution because a negative width is nonsense. So the width is 4 inches and the length is $2(4) = 8$ inches.

9. Let x represent the area of the base. Then $x - 10$ represents the height. The volume is $V = \frac{1}{3}Bh$.

$32 = \frac{1}{3}(x)(x - 10)$

Let $V = 32$, $B = x$, $h = x - 10$
$96 = x(x - 10)$ Multiply by 3
$96 = x^2 - 10x$
$0 = x^2 - 10x - 96$ Get 0 on one side
$0 = (x + 6)(x - 16)$ Factor
$x + 6 = 0$ or $x - 16 = 0$
$x = -6$ or $x = 16$

Reject the negative solution; area cannot be negative. So the area of the base is 16 square meters and the height is $16 - 10 = 6$ meters.

13. Let x represent the length of the box. Then $x - 1$ represents the width. The box is 4 feet high and has a volume of 120 cubic feet. Since the formula for the volume is $V = LWH$,

$120 = x(x - 1)(4)$
Let $V = 120$, $L = x$, $W = x - 1$, $H = 4$
$120 = 4x^2 - 4x$ Simplify
$0 = 4x^2 - 4x - 120$ Get 0 on one side
$0 = 4(x - 6)(x + 5)$ Factor
$x - 6 = 0$ or $x + 5 = 0$
    Set each factor equal to 0
$x = 6$ or $x = -5$

Reject the negative solution; the length cannot be negative. So the length is 6 feet and the width is $6 - 1 = 5$ feet.

17. Let x, $x + 2$, and $x + 4$ represent the integers.

$$\underbrace{\text{Four times}}_{4} \cdot \underbrace{\text{the sum of all three}}_{[x+(x+2)+(x+4)]} \underbrace{\text{equals}}_{=} \underbrace{\text{the product of the smaller two.}}_{x(x+2)}$$

$4(3x + 6) = x(x + 2)$
$12x + 24 = x^2 + 2x$ Simplify
$0 = x^2 - 10x - 24$
    Get 0 on one side
$0 = (x + 2)(x - 12)$ Factor

$x + 2 = 0$ or $x - 12 = 0$
    Set each factor equal to 0
$x = -2$ or $x = 12$

Both solutions can be used since we want integers (not measurements). If the first even integer is -2, the second is $-2 + 2 = 0$, and the third is $-2 + 4 = 2$. A second solution is 12, $12 + 2 = 14$, and $12 + 4 = 16$.

21. Let x represent the integer.

$$\underbrace{\text{Three times}}_{3} \cdot \underbrace{\text{the square of an integer}}_{x^2} \underbrace{\text{and}}_{+} \underbrace{\text{twice the integer}}_{2x} \underbrace{\text{is}}_{=} \underbrace{8.}_{8}$$

$3x^2 + 2x = 8$
$3x^2 + 2x - 8 = 0$ Get 0 on one side
$(3x - 4)(x + 2) = 0$ Factor
$3x - 4 = 0$ or $x + 2 = 0$ Set each factor equal to 0
$x = \frac{4}{3}$ or $x = -2$

Reject $\frac{4}{3}$ since it is not an integer. The only solution is -2.

25. Let x represent the length of the shorter leg. Then $2x + 1$ represents the length of the hypotenuse, and $2x - 1$ represents the length of the longer leg. In a right triangle, the square of the hypotenuse equals the sum of the squares of the legs, so
$(2x + 1)^2 = x^2 + (2x - 1)^2$
$4x^2 + 4x + 1 = x^2 + 4x^2 - 4x + 1$
$4x = x^2 - 4x$
    Subtract $4x^2$ and 1 on both sides
$0 = x^2 - 8x$
    Get 0 on one side
$0 = x(x - 8)$ Factor
$x = 0$ or $x - 8 = 0$
$x = 0$ or $x = 8$

Reject 0 because the side of a triangle cannot measure 0. The shorter leg has a length of 8 feet.

29. Let d, the distance, equal 1600 feet in the formula. Also, $g = 32$.

$d = \frac{1}{2}gt^2$

$1600 = \frac{1}{2}(32)t^2$ Let $d = 1600$ and $g = 32$
$3200 = 32t^2$ Multiply by 2
$100 = t^2$ Divide by 32
$0 = t^2 - 100$ Get 0 on one side
$0 = (t + 10)(t - 10)$ Factor
$t = -10$ or $t = 10$

Reject -10; the time is 10 seconds.

33. Let $t = 3$, $v_0 = 64$ and substitute into $h = v_0 t - 16t^2$.

$h = (64)(3) - 16(3)^2$
$= 192 - 16(9)$
$= 192 - 144$
$= 48$

The height after 3 seconds is 48 feet.

CHAPTER 6

Section 6.1 (page 283)

1. The denominator $4x$ will be zero when $x = 0$, so $\dfrac{3}{4x}$ is meaningless at $x = 0$.

5. Set the denominator $x + 5$ equal to 0. Then $x + 5 = 0$, and $x = -5$, so $\dfrac{x^2}{x + 5}$ is meaningless at -5.

9. To find the numbers that make the denominator 0, we must solve
$$2r^2 - r - 3 = 0.$$
$$(2r - 3)(r + 1) = 0 \quad \text{Factor}$$
$$2r - 3 = 0 \quad r + 1 = 0 \quad \text{Set each}$$
$$2r = 3 \quad r = -1 \quad \text{factor equal}$$
$$r = \dfrac{3}{2} \quad \quad \text{to zero and solve}$$

The expression $\dfrac{8r + 2}{2r^2 - r - 3}$ is meaningless at $r = \dfrac{3}{2}$ or $r = -1$.

13. (a) Let $x = 2$.
$$\dfrac{4x - 2}{3x} = \dfrac{4 \cdot 2 - 2}{3 \cdot 2} = \dfrac{8 - 2}{6} = \dfrac{6}{6} = 1$$
(b) Let $x = -3$.
$$\dfrac{4x - 2}{3x} = \dfrac{4(-3) - 2}{3(-3)} = \dfrac{-12 - 2}{-9} = \dfrac{-14}{-9} = \dfrac{14}{9}$$

17. (a) $\dfrac{(-8x)^2}{3x + 9} = \dfrac{(-8 \cdot 2)^2}{3(2) + 9} \quad \text{Let } x = 2$
$$= \dfrac{(-16)^2}{6 + 9}$$
$$= \dfrac{256}{15}$$
(b) $\dfrac{(-8x)^2}{3x + 9} = \dfrac{(-8 \cdot -3)^2}{3(-3) + 9} \quad \text{Substitute } -3 \text{ for } x$
$$= \dfrac{(24)^2}{-9 + 9} \quad \text{Simplify}$$
$$= \dfrac{(24)^2}{0}$$
The expression is meaningless because of the 0 in the denominator.

21. (a) $\dfrac{5x^2}{6 - 3x - x^2} = \dfrac{5 \cdot 2^2}{6 - 3 \cdot 2 - 2^2} \quad \text{Let } x = 2$
$$= \dfrac{5 \cdot 4}{6 - 6 - 4}$$
$$= \dfrac{20}{-4}$$
$$= -5$$
(b) $\dfrac{5x^2}{6 - 3x - x^2} = \dfrac{5 \cdot (-3)^2}{6 - 3 \cdot (-3) - (-3)^2}$
$$\text{Let } x = -3$$
$$= \dfrac{5 \cdot 9}{6 + 9 - 9}$$
$$= \dfrac{45}{6}$$
$$= \dfrac{15}{2}$$

25. $\dfrac{12k^2}{6k} = \dfrac{2 \cdot 2 \cdot 3 \cdot k \cdot k}{2 \cdot 3 \cdot k} \quad \text{Factor}$
$$= \dfrac{(2 \cdot 3 \cdot k)2 \cdot k}{(2 \cdot 3 \cdot k)}$$
$$= 2k \quad \text{Lowest terms}$$

29. $\dfrac{12m^2p}{9mp^2} = \dfrac{12}{9} \cdot \dfrac{m^2}{m} \cdot \dfrac{p}{p^2} = \dfrac{4}{3} \cdot \dfrac{m}{1} \cdot \dfrac{1}{p} = \dfrac{4m}{3p}$

33. $\dfrac{12m^2 - 9}{3} = \dfrac{3(4m^2 - 3)}{3}$
$$= 4m^2 - 3$$

37. $\dfrac{6y + 12}{8y + 16} = \dfrac{6(y + 2)}{8(y + 2)}$
$$= \dfrac{6}{8} = \dfrac{3}{4}$$

41. $\dfrac{5m^2 - 5m}{10m - 10} = \dfrac{5m(m - 1)}{10(m - 1)}$
$$= \dfrac{5m(m - 1)}{2 \cdot 5(m - 1)} \quad \text{Factor}$$
$$= \dfrac{m}{2} \quad \text{Lowest terms}$$

45. $\dfrac{m^2 - 4m + 4}{m^2 + m - 6} = \dfrac{(m - 2)(m - 2)}{(m + 3)(m - 2)} \quad \text{Factor}$
$$= \dfrac{m - 2}{m + 3} \quad \text{Lowest terms}$$

49. $\dfrac{m - 5}{5 - m} = -1$ since the numerator and denominator differ only in sign.

53. $\dfrac{m^2 - 4m}{4m - m^2} = \dfrac{m^2 - 4m}{-1(m^2 - 4m)}$
$4m - m^2$ is the same as $-1(m^2 - 4m)$
$$= \dfrac{1}{-1}$$
$$= -1 \quad \text{Lowest terms}$$

Section 6.2 (page 289)

1. $\dfrac{9m^2}{16} \cdot \dfrac{4}{3m} = \dfrac{9m^2 \cdot 4}{16 \cdot 3m}$
$$= \dfrac{36m^2}{48m} \quad \text{Multiply}$$
$$= \dfrac{3m}{4} \quad \text{Divide numerator and denominator by } 12m$$

5. $\dfrac{8a^4}{12a^3} \cdot \dfrac{9a^5}{3a^2} = \dfrac{8 \cdot 9a^{4 + 5}}{12 \cdot 3a^{3 + 2}}$
$$= \dfrac{72a^9}{36a^5}$$
$$= \dfrac{72}{36} \cdot \dfrac{a^9}{a^5}$$
$$= 2a^4$$

9. $\dfrac{3m^2}{(4m)^3} \div \dfrac{9m^3}{32m^4} = \dfrac{3m^2}{64m^3} \cdot \dfrac{32m^4}{9m^3} \quad \text{Invert and multiply}$
$$= \dfrac{96m^6}{576m^6}$$
$$= \dfrac{1}{6} \quad \text{Divide numerator and denominator by } 96m^6$$

Solutions to Selected Exercises 585

13. $\dfrac{a+b}{2} \cdot \dfrac{12}{(a+b)^2} = \dfrac{12(a+b)}{2(a+b)^2}$ Multiply

$= \dfrac{6}{a+b}$ Lowest terms

17. $\dfrac{2k+8}{6} \div \dfrac{3k+12}{2}$

$= \dfrac{2(k+4)}{6} \div \dfrac{3(k+4)}{2}$ Factor

$= \dfrac{2(k+4)}{6} \cdot \dfrac{2}{3(k+4)}$ Invert and multiply

$= \dfrac{2 \cdot 2(k+4)}{6 \cdot 3(k+4)}$

$= \dfrac{2}{3 \cdot 3} = \dfrac{2}{9}$

21. $\dfrac{3r+12}{8} \cdot \dfrac{16r}{9r+36}$

$= \dfrac{3(r+4)}{8} \cdot \dfrac{16r}{9(r+4)}$ Factor

$= \dfrac{3 \cdot 16r \cdot (r+4)}{8 \cdot 9 \cdot (r+4)}$ Multiply

$= \dfrac{2r}{3}$

25. $\dfrac{6(m+2)}{3(m-1)^2} \div \dfrac{(m+2)^2}{9(m-1)}$

$= \dfrac{6(m+2)}{3(m-1)^2} \cdot \dfrac{9(m-1)}{(m+2)^2}$ Invert and multiply

$= \dfrac{6 \cdot 9 \cdot (m+2)(m-1)}{3 \cdot (m-1)^2(m+2)^2}$

$= \dfrac{2 \cdot 9}{(m-1)(m+2)} = \dfrac{18}{(m-1)(m+2)}$

29. $\dfrac{8-r}{8+r} \div \dfrac{r-8}{r+8}$

$= \dfrac{8-r}{8+r} \cdot \dfrac{r+8}{r-8}$ Invert and multiply

$= \dfrac{(8-r)(r+8)}{(8+r)(r-8)}$

$= \dfrac{8-r}{r-8}$ $r+8$ and $8+r$ are the same

$= -1$ $8-r$ and $r-8$ are negatives of each other

33. $\dfrac{k^2-k-6}{k^2+k-12} \div \dfrac{k^2+2k-3}{k^2+3k-4}$

$= \dfrac{(k-3)(k+2)}{(k+4)(k-3)} \div \dfrac{(k+3)(k-1)}{(k+4)(k-1)}$ Factor

$= \dfrac{(k-3)(k+2)}{(k+4)(k-3)} \cdot \dfrac{(k+4)(k-1)}{(k+3)(k-1)}$ Invert and multiply

$= \dfrac{(k-3)(k+2)(k+4)(k-1)}{(k+4)(k-3)(k+3)(k-1)} = \dfrac{k+2}{k+3}$

37. $\dfrac{2k^2+3k-2}{6k^2-7k+2} \cdot \dfrac{4k^2-5k+1}{k^2+k-2}$

$= \dfrac{(2k-1)(k+2)}{(2k-1)(3k-2)} \cdot \dfrac{(4k-1)(k-1)}{(k+2)(k-1)}$ Factor

$= \dfrac{(2k-1)(k+2)(4k-1)(k-1)}{(2k-1)(3k-2)(k+2)(k-1)}$ Multiply

$= \dfrac{4k-1}{3k-2}$

41. $\dfrac{m^2+2mp-3p^2}{m^2-3mp+2p^2} \div \dfrac{m^2+4mp+3p^2}{m^2+2mp-8p^2}$

$= \dfrac{(m+3p)(m-p)}{(m-2p)(m-p)} \div \dfrac{(m+3p)(m+p)}{(m+4p)(m-2p)}$ Factor

$= \dfrac{(m+3p)(m-p)}{(m-2p)(m-p)} \cdot \dfrac{(m+4p)(m-2p)}{(m+3p)(m+p)}$ Invert

$= \dfrac{m+4p}{m+p}$ Multiply and write in lowest terms

Section 6.3 (page 293)

1. Factor each denominator.
$12 = 2 \cdot 2 \cdot 3$; $10 = 2 \cdot 5$
Take each factor the greatest number of times it appears:
least common denominator $= 2 \cdot 2 \cdot 3 \cdot 5 = 60$.

5. Factor each denominator: $100 = 2^2 \cdot 5^2$, $120 = 2^3 \cdot 3 \cdot 5$, $180 = 2^2 \cdot 3^2 \cdot 5$. Take each factor the greatest number of times it appears: least common denominator $= 2^3 \cdot 3^2 \cdot 5^2 = 1800$.

9. The least common denominator for 5 and 6 is 30. The only variable is p, so the least common denominator is $30p$.

13. Factor each denominator.
$15y^2 = 3 \cdot 5 \cdot y^2$
$36y^4 = 2 \cdot 2 \cdot 3 \cdot 3 \cdot y^4$
Take each factor the greatest number of times it appears, and use the highest exponent on y:
least common denominator $= 2 \cdot 2 \cdot 3 \cdot 3 \cdot 5 \cdot y^4$
$= 180y^4$

17. Factor 15 as $3 \cdot 5$, and then take each factor the greatest number of times it appears. Use the highest exponents on the variables.
Least common denominator $= 3 \cdot 5 \cdot a^5 b^3$
$= 15a^5 b^3$

21. Factor each denominator.
$32r^2 = 2^5 \cdot r^2$
$16r - 32 = 16(r-2) = 2^4(r-2)$
Least common denominator $= 2^5 r^2 (r-2)$
$= 32r^2(r-2)$

25. $12p + 60 = 12(p+5)$
$= 2 \cdot 2 \cdot 3 \cdot (p+5)$;
$p^2 + 5p = p(p+5)$
Least common denominator
$= 2 \cdot 2 \cdot 3 \cdot p \cdot (p+5)$
$= 12p(p+5)$

29. $a^2 + 6a = a(a+6)$;
$a^2 + 3a - 18 = (a+6)(a-3)$
Least common denominator
$= a(a+6)(a-3)$

33. $2y^2 + 7y - 4 = (2y-1)(y+4)$
$2y^2 - 7y + 3 = (2y-1)(y-3)$
Least common denominator $=$
$(2y-1)(y+4)(y-3)$

37. $\dfrac{7}{11} = \dfrac{7(6)}{11(6)}$ Multiply numerator and denominator by 6

$= \dfrac{42}{66}$

41. $\dfrac{-11}{m} = \dfrac{-11(8)}{m(8)}$ Multiply numerator and denominator by 8
$= \dfrac{-88}{8m}$

45. $\dfrac{15m^2}{8k} = \dfrac{15m^2(4k^3)}{8k(4k^3)}$ Multiply by $4k^3$
$= \dfrac{60m^2k^3}{32k^4}$

49. $\dfrac{-2a}{9a-18} = \dfrac{-2a(2)}{(9a-18)(2)}$ Multiply numerator and denominator by 2
$= \dfrac{-4a}{18a-36}$

Section 6.4 (page 299)

1. $\dfrac{2}{p} + \dfrac{5}{p} = \dfrac{2+5}{p} = \dfrac{7}{p}$

5. $\dfrac{y}{y+1} + \dfrac{1}{y+1} = \dfrac{y+1}{y+1} = 1$

9. $\dfrac{m^2}{m+6} + \dfrac{6m}{m+6} = \dfrac{m^2+6m}{m+6}$
$= \dfrac{m(m+6)}{m+6}$ Factor
$= m$ Lowest terms

13. $\dfrac{m}{3} + \dfrac{1}{2}$
Least common denominator $= 6$
$\dfrac{m}{3} + \dfrac{1}{2} = \dfrac{m \cdot 2}{3 \cdot 2} + \dfrac{1 \cdot 3}{2 \cdot 3} = \dfrac{2m}{6} + \dfrac{3}{6} = \dfrac{2m+3}{6}$

17. $\dfrac{4}{3} - \dfrac{1}{y} = \dfrac{4 \cdot y}{3 \cdot y} - \dfrac{1 \cdot 3}{y \cdot 3} = \dfrac{4y}{3y} - \dfrac{3}{3y} = \dfrac{4y-3}{3y}$

21. $\dfrac{4+2k}{5} + \dfrac{2+k}{10} = \dfrac{(4+2k)(2)}{5(2)} + \dfrac{2+k}{10}$
$= \dfrac{8+4k}{10} + \dfrac{2+k}{10}$
$= \dfrac{8+4k+2+k}{10}$
$= \dfrac{5k+10}{10}$
$= \dfrac{5(k+2)}{10}$ Factor
$= \dfrac{k+2}{2}$ Lowest terms

25. $\dfrac{m+2}{m} + \dfrac{m+3}{4m} = \dfrac{(m+2) \cdot 4}{m \cdot 4} + \dfrac{m+3}{4m}$
$= \dfrac{4m+8}{4m} + \dfrac{m+3}{4m}$
$= \dfrac{4m+8+m+3}{4m}$
$= \dfrac{5m+11}{4m}$

29. $\dfrac{9}{2p} - \dfrac{4}{p^2} = \dfrac{9(p)}{2p(p)} - \dfrac{4(2)}{p^2(2)}$
$= \dfrac{9p}{2p^2} - \dfrac{8}{2p^2}$
$= \dfrac{9p-8}{2p^2}$

33. $\dfrac{8}{x-2} - \dfrac{4}{x+2}$
$= \dfrac{8(x+2)}{(x-2)(x+2)} - \dfrac{4(x-2)}{(x+2)(x-2)}$
$= \dfrac{8x+16-4x+8}{(x-2)(x+2)}$
$= \dfrac{4x+24}{(x-2)(x+2)} = \dfrac{4(x+6)}{(x-2)(x+2)}$

37. $\dfrac{1}{m^2-9} + \dfrac{1}{3m+9}$
$= \dfrac{1}{(m+3)(m-3)} + \dfrac{1}{3(m+3)}$ Factor
$= \dfrac{3 \cdot 1}{3(m+3)(m-3)} + \dfrac{1 \cdot (m-3)}{3(m+3)(m-3)}$
$= \dfrac{3}{3(m+3)(m-3)} + \dfrac{m-3}{3(m+3)(m-3)}$
$= \dfrac{3+m-3}{3(m+3)(m-3)}$ Add
$= \dfrac{m}{3(m+3)(m-3)}$

41. $\dfrac{8}{m-2} + \dfrac{3}{5m} + \dfrac{7}{5m(m-2)}$
$= \dfrac{5m \cdot 8}{5m(m-2)} + \dfrac{3(m-2)}{5m(m-2)} + \dfrac{7}{5m(m-2)}$
$= \dfrac{40m}{5m(m-2)} + \dfrac{3m-6}{5m(m-2)} + \dfrac{7}{5m(m-2)}$
$= \dfrac{40m+3m-6+7}{5m(m-2)}$ Add
$= \dfrac{43m+1}{5m(m-2)}$

45. $\dfrac{k-1}{k^2+2k-8} + \dfrac{3k-2}{k^2+3k-4}$
$= \dfrac{k-1}{(k+4)(k-2)} + \dfrac{3k-2}{(k+4)(k-1)}$ Factor
$= \dfrac{(k-1)(k-1)}{(k+4)(k-2)(k-1)} + \dfrac{(3k-2)(k-2)}{(k+4)(k-1)(k-2)}$
$= \dfrac{k^2-2k+1}{(k+4)(k-2)(k-1)} + \dfrac{3k^2-8k+4}{(k+4)(k-2)(k-1)}$
$= \dfrac{k^2-2k+1+3k^2-8k+4}{(k+4)(k-2)(k-1)}$
$= \dfrac{4k^2-10k+5}{(k+4)(k-2)(k-1)}$

Section 6.5 (page 305)

Only one method is shown for each exercise in this section.

1. $\dfrac{\dfrac{5}{8}+\dfrac{2}{3}}{\dfrac{7}{3}-\dfrac{1}{4}} = \dfrac{24\left(\dfrac{5}{8}+\dfrac{2}{3}\right)}{24\left(\dfrac{7}{3}-\dfrac{1}{4}\right)}$
$= \dfrac{24\left(\dfrac{5}{8}\right)+24\left(\dfrac{2}{3}\right)}{24\left(\dfrac{7}{3}\right)-24\left(\dfrac{1}{4}\right)}$ Distributive property
$= \dfrac{15+16}{56-6} = \dfrac{31}{50}$

Solutions to Selected Exercises

5. $\dfrac{\dfrac{m^3p^4}{5m}}{\dfrac{8mp^5}{p^2}} = \dfrac{m^3p^4}{5m} \cdot \dfrac{p^2}{8mp^5}$

$= \dfrac{m^3p^4 \cdot p^2}{5m \cdot 8mp^5}$

$= \dfrac{mp}{40}$

9. $\dfrac{\dfrac{3}{y} + 1}{\dfrac{3+y}{2}} = \dfrac{2y\left(\dfrac{3}{y} + 1\right)}{2y\left(\dfrac{3+y}{2}\right)}$

$= \dfrac{2y\left(\dfrac{3}{y}\right) + 2y(1)}{2y\left(\dfrac{3+y}{2}\right)}$

$= \dfrac{6 + 2y}{y(3+y)}$

$= \dfrac{2(3+y)}{y(3+y)}$ Factor

$= \dfrac{2}{y}$ Lowest terms

13. $\dfrac{x + \dfrac{1}{x}}{\dfrac{4}{x} + y} = \dfrac{x\left(x + \dfrac{1}{x}\right)}{x\left(\dfrac{4}{x} + y\right)}$

$= \dfrac{x^2 + 1}{4 + xy}$

17. $\dfrac{\dfrac{2}{p^2} - \dfrac{3}{5p}}{\dfrac{4}{p} + \dfrac{1}{4p}} = \dfrac{20p^2\left(\dfrac{2}{p^2} - \dfrac{3}{5p}\right)}{20p^2\left(\dfrac{4}{p} + \dfrac{1}{4p}\right)}$

$= \dfrac{40 - 4p(3)}{20p(4) + 5p(1)}$

$= \dfrac{40 - 12p}{80p + 5p} = \dfrac{40 - 12p}{85p}$

21. $\dfrac{\dfrac{1}{4} - \dfrac{1}{a^2}}{\dfrac{1}{2} + \dfrac{1}{a}} = \dfrac{4a^2\left(\dfrac{1}{4} - \dfrac{1}{a^2}\right)}{4a^2\left(\dfrac{1}{2} + \dfrac{1}{a}\right)}$

$= \dfrac{a^2 - 4}{2a^2 + 4a}$

$= \dfrac{(a+2)(a-2)}{2a(a+2)}$ Factor

$= \dfrac{a-2}{2a}$

25. $\dfrac{\dfrac{1}{m+1} - 1}{\dfrac{1}{m+1} + 1} = \dfrac{(m+1)\left(\dfrac{1}{m+1} - 1\right)}{(m+1)\left(\dfrac{1}{m+1} + 1\right)}$

$= \dfrac{1 - (m+1)}{1 + (m+1)}$

$= \dfrac{1 - m - 1}{1 + m + 1} = \dfrac{-m}{m+2}$

29. $1 - \dfrac{1}{1 + \dfrac{1}{1+1}} = 1 - \dfrac{1}{1 + \dfrac{1}{2}}$

$= 1 - \dfrac{1}{\dfrac{2}{2} + \dfrac{1}{2}}$

$= 1 - \dfrac{1}{\dfrac{3}{2}}$

$= 1 - 1 \cdot \dfrac{2}{3}$

$= 1 - \dfrac{2}{3}$

$= \dfrac{3}{3} - \dfrac{2}{3}$

$= \dfrac{1}{3}$

Section 6.6   (page 311)

1. $\dfrac{6}{x} - \dfrac{4}{x} = 5$

$x\left(\dfrac{6}{x} - \dfrac{4}{x}\right) = x \cdot 5$ Multiply both sides by x

$x\left(\dfrac{6}{x}\right) - x\left(\dfrac{4}{x}\right) = 5x$ Distributive property

$6 - 4 = 5x$

$2 = 5x$

$\dfrac{2}{5} = x$ Divide by 5

Check $\dfrac{2}{5}$ in the original equation.

$\dfrac{6}{x} - \dfrac{4}{x} = 5$

$\dfrac{6}{\frac{2}{5}} - \dfrac{4}{\frac{2}{5}} = 5$ Let $x = \dfrac{2}{5}$

$6 \cdot \dfrac{5}{2} - 4 \cdot \dfrac{5}{2} = 5$

$15 - 10 = 5$

$5 = 5$ True

The solution is $\dfrac{2}{5}$.

5. $\dfrac{9}{m} = 5 - \dfrac{1}{m}$

$m\left(\dfrac{9}{m}\right) = m\left(5 - \dfrac{1}{m}\right)$ Multiply by m

$9 = 5m - 1$ Distributive property

$10 = 5m$ Add 1

$2 = m$ Divide by 5

Check 2 in the original equation.

9. $\dfrac{x+1}{2} = \dfrac{x+2}{3}$

$6\left(\dfrac{x+1}{2}\right) = 6\left(\dfrac{x+2}{3}\right)$

Multiply each term by the common denominator, 6

588   Solutions to Selected Exercises

13.
$$3(x + 1) = 2(x + 2)$$
$$3x + 3 = 2x + 4$$
$$x + 3 = 4 \quad \text{Subtract } 2x$$
$$x = 1 \quad \text{Subtract } 3$$
Check this solution in the original equation.

13.
$$\frac{2p + 8}{9} = \frac{10p + 4}{27}$$
$$27\left(\frac{2p + 8}{9}\right) = 27\left(\frac{10p + 4}{27}\right)$$
$$3(2p + 8) = 10p + 4$$
$$6p + 24 = 10p + 4$$
$$24 = 4p + 4$$
$$20 = 4p$$
$$5 = p$$
Check this solution in the original equation.

17.
$$\frac{a - 4}{4} = \frac{a}{16} + \frac{1}{2}$$
$$16\left(\frac{a - 4}{4}\right) = 16\left(\frac{a}{16} + \frac{1}{2}\right) \quad \text{Multiply both sides by 16}$$
$$4(a - 4) = 16\left(\frac{a}{16}\right) + 16\left(\frac{1}{2}\right) \quad \text{Distributive property}$$
$$4a - 16 = a + 8$$
$$3a - 16 = 8 \quad \text{Subtract } a$$
$$3a = 24 \quad \text{Add } 16$$
$$a = 8 \quad \text{Divide by } 3$$
Check this solution in the original equation.

21.
$$\frac{m - 2}{4} + \frac{m + 1}{3} = \frac{10}{3}$$
$$12\left(\frac{m - 2}{4} + \frac{m + 1}{3}\right) = 12\left(\frac{10}{3}\right) \quad \text{Multiply both sides by 12}$$
$$12\left(\frac{m - 2}{4}\right) + 12\left(\frac{m + 1}{3}\right) = 4(10)$$
$$3(m - 2) + 4(m + 1) = 40$$
$$3m - 6 + 4m + 4 = 40$$
$$7m - 2 = 40$$
$$7m = 42$$
$$m = 6$$
Check this solution in the original equation.

25.
$$\frac{p}{2} - \frac{p - 1}{4} = \frac{5}{4}$$
$$4\left(\frac{p}{2} - \frac{p - 1}{4}\right) = 4\left(\frac{5}{4}\right) \quad \text{Multiply both sides by 4}$$
$$4\left(\frac{p}{2}\right) - 4\left(\frac{p - 1}{4}\right) = 5$$
$$2p - (p - 1) = 5$$
$$2p - p + 1 = 5$$
$$p + 1 = 5$$
$$p = 4 \quad \text{Subtract } 1$$
Check this solution in the original equation.

29.
$$\frac{y - 1}{2} - \frac{y - 3}{4} = 1$$
$$4\left(\frac{y - 1}{2} - \frac{y - 3}{4}\right) = 4(1) \quad \text{Multiply both sides by 4}$$
$$4\left(\frac{y - 1}{2}\right) - 4\left(\frac{y - 3}{4}\right) = 4 \quad \text{Distributive property}$$
$$2(y - 1) - (y - 3) = 4$$
$$2y - 2 - y + 3 = 4$$
$$y + 1 = 4$$
$$y = 3$$
Check this solution in the original equation.

33.
$$\frac{m}{2m + 2} = \frac{2m - 3}{m + 1} - \frac{m}{2m + 2}$$
$$\frac{m}{2(m + 1)} = \frac{2m - 3}{m + 1} - \frac{m}{2(m + 1)} \quad \text{Factor}$$
$$2(m + 1)\left[\frac{m}{2(m + 1)}\right]$$
$$= 2(m + 1)\left[\frac{2m - 3}{m + 1} - \frac{m}{2(m + 1)}\right]$$
Multiply both sides by $2(m + 1)$
$$m = 2(m + 1)\left(\frac{2m - 3}{m + 1}\right) - 2(m + 1)\left(\frac{m}{2(m + 1)}\right)$$
Distributive property
$$m = 2(2m - 3) - m$$
$$m = 4m - 6 - m \quad \text{Distributive property}$$
$$-2m = -6 \quad \text{Subtract } 3m$$
$$m = 3 \quad \text{Divide by } -2$$
Check this solution in the original equation.

37.
$$\frac{2}{y} = \frac{y}{5y - 12}$$
$$y(5y - 12)\left(\frac{2}{y}\right) = y(5y - 12)\left(\frac{y}{5y - 12}\right)$$
Multiply both sides by $y(5y - 12)$
$$(5y - 12)(2) = y^2$$
$$10y - 24 = y^2 \quad \text{Distributive property}$$
$$0 = y^2 - 10y + 24$$
This quadratic equation can be solved by factoring.
$$0 = (y - 6)(y - 4) \quad \text{Factor}$$
$$y - 6 = 0 \text{ or } y - 4 = 0$$
$$y = 6 \text{ or } \quad y = 4$$
Check these solutions in the original equation.

41.
$$\frac{3y}{y^2 + 5y + 6} = \frac{5y}{y^2 + 2y - 3} - \frac{2}{y^2 + y - 2}$$
Factor
$$\frac{3y}{(y + 2)(y + 3)} = \frac{5y}{(y + 3)(y - 1)} - \frac{2}{(y + 2)(y - 1)}$$
$$3y(y - 1) = 5y(y + 2) - 2(y + 3)$$
Multiply by $(y + 2)(y + 3)(y - 1)$
$$3y^2 - 3y = 5y^2 + 10y - 2y - 6$$
$$3y^2 - 3y = 5y^2 + 8y - 6$$
$$0 = 2y^2 + 11y - 6$$
Factor: $0 = (2y - 1)(y + 6)$. Set each factor equal to 0.
$$2y - 1 = 0 \text{ or } y + 6 = 0$$
Solve: $2y = 1$
$$y = \frac{1}{2} \text{ or } \quad y = -6.$$
Check each solution in the original equation.

Solutions to Selected Exercises

45.
$$\frac{6}{r} + \frac{1}{r-2} = 3$$
$$r(r-2)\left(\frac{6}{r} + \frac{1}{r-2}\right) = r(r-2)(3)$$
Multiply both sides by $r(r-2)$
$$r(r-2)\left(\frac{6}{r}\right) + r(r-2)\left(\frac{1}{r-2}\right) = 3r(r-2)$$
$$6(r-2) + r = 3r(r-2)$$
$$6r - 12 + r = 3r^2 - 6r$$
Distributive property
$$7r - 12 = 3r^2 - 6r$$
$$0 = 3r^2 - 13r + 12$$
$$0 = (3r-4)(r-3)$$
Factor
$$3r - 4 = 0 \text{ or } r - 3 = 0$$
$$3r = 4$$
$$r = \frac{4}{3} \text{ or } r = 3$$
Check each solution in the original equation.

49. Solve $\dfrac{3}{k} = \dfrac{1}{p} + \dfrac{1}{q}$ for q.
$$\frac{3}{k} = \frac{1}{p} + \frac{1}{q}$$
$$kpq\left(\frac{3}{k}\right) = kpq\left(\frac{1}{p} + \frac{1}{q}\right)$$
Multiply both sides by kpq
$$kpq\left(\frac{3}{k}\right) = kpq\left(\frac{1}{p}\right) + kpq\left(\frac{1}{q}\right)$$
Distributive property
$$3pq = kq + kp$$
$$3pq - kq = kp \quad \text{Get all terms with } q$$
on one side
$$q(3p - k) = kp \quad \text{Factor}$$
$$q = \frac{kp}{3p - k}$$
Divide by $3p - k$

53. Solve $I = \dfrac{E}{R + r}$ for r.
$$(R + r)I = (R + r)\left(\frac{E}{R + r}\right) \quad \text{Multiply both sides by } R + r$$
$$RI + rI = E$$
$$rI = E - RI \quad \text{Get all terms with } r \text{ on one side}$$
$$r = \frac{E - RI}{I}$$

Supplementary Exercises on Rational Expressions (page 315)

1. $\dfrac{6}{m} + \dfrac{2}{m} = \dfrac{6+2}{m} = \dfrac{8}{m}$

5. $\dfrac{2r^2 - 3r - 9}{2r^2 - r - 6} \cdot \dfrac{r^2 + 2r - 8}{r^2 - 2r - 3}$
$= \dfrac{(2r+3)(r-3)}{(2r+3)(r-2)} \cdot \dfrac{(r+4)(r-2)}{(r-3)(r+1)}$ Factor
$= \dfrac{r+4}{r+1}$ Multiply; lowest terms

9. $\dfrac{5}{y-1} + \dfrac{2}{3y-3} = \dfrac{5}{y-1} + \dfrac{2}{3(y-1)}$ Factor
$= \dfrac{3 \cdot 5}{3(y-1)} + \dfrac{2}{3(y-1)}$
$= \dfrac{15 + 2}{3(y-1)} = \dfrac{17}{3(y-1)}$

13. $\dfrac{4}{9z} - \dfrac{3}{2z} = \dfrac{2 \cdot 4}{2 \cdot 9z} - \dfrac{9 \cdot 3}{9 \cdot 2z}$
$= \dfrac{8}{18z} - \dfrac{27}{18z}$
$= \dfrac{8 - 27}{18z}$
$= \dfrac{-19}{18z}$

Section 6.7 (page 321)

1. Let x represent the number. Then

one-half of a number is 3 more than one-sixth of the same number.

$$\frac{1}{2} \cdot x = 3 + \frac{1}{6} \cdot x$$
Solve.
$$6\left(\frac{1}{2}x\right) = 6(3) + 6\left(\frac{1}{6}x\right) \quad \text{Multiply each term by the common denominator, 6}$$
$$3x = 18 + x$$
$$2x = 18$$
$$x = 9$$
The number is 9.

5. Let the smaller number be x. Then the larger number is $x + 3$.

The smaller number added to two-thirds the larger

$$x + \frac{2}{3} \cdot (x + 3)$$

equals four-fifths the sum of the original numbers.

$$= \frac{4}{5} \cdot (x + x + 3)$$

Solve.
$$x + \frac{2}{3}(x + 3) = \frac{4}{5}(2x + 3)$$
$$15(x) + 15\left[\frac{2}{3}(x + 3)\right] = 15\left[\frac{4}{5}(2x + 3)\right]$$
$$15x + 10(x + 3) = 12(2x + 3)$$
$$15x + 10x + 30 = 24x + 36$$
$$25x + 30 = 24x + 36$$
$$x = 6$$
The smaller number is 6, and the larger number is $6 + 3 = 9$.

9. Let x represent the number. Then

three times a number → $3 \cdot x$
added to → $+$
twice → $2 \cdot$
its reciprocal → $\dfrac{1}{x}$
is → $=$
5. → 5

Solve.
$$3x + \frac{2}{x} = 5$$
$$x(3x) + x\left(\frac{2}{x}\right) = 5x \quad \text{Multiply each term by the common denominator, } x$$
$$3x^2 + 2 = 5x$$
$$3x^2 - 5x + 2 = 0 \quad \text{Get 0 alone on one side}$$
$$(3x - 2)(x - 1) = 0 \quad \text{Factor}$$
$$3x - 2 = 0 \text{ or } x - 1 = 0 \quad \text{Set each factor equal to 0}$$
$$3x = 2$$
$$x = \frac{2}{3} \text{ or } \quad x = 1$$

The solutions are $\dfrac{2}{3}$ or 1.

13. Let x represent the salary of an experienced professor. Then $\dfrac{3}{4}$ of x, or $\dfrac{3}{4}x$, represents the new instructor's salary, which gives the following equation.

Professor's salary plus instructor's salary equals total salary paid.

$$x + \frac{3}{4}x = 56{,}000$$

Solve.
$$4(x) + 4\left(\frac{3}{4}x\right) = 4(56{,}000) \quad \text{Multiply by 4}$$
$$4x + 3x = 224{,}000$$
$$7x = 224{,}000$$
$$x = 32{,}000$$

The professor's salary is $32,000.

17. Let x represent the distance each way (the distance to the destination is the same as the distance from the destination). We fill in the chart as follows, realizing that the time column is filled in by using the formula $t = \dfrac{d}{r}$.

	d	r	t
To	x	60	$\dfrac{x}{60}$
From	x	50	$\dfrac{x}{50}$

Using the numbers in the time column in the chart above, we have the following equation.

"From" time is $\dfrac{1}{2}$ hour more than "To" time.

$$\frac{x}{50} = \frac{1}{2} + \frac{x}{60}$$

Solve.
$$300\left(\frac{x}{50}\right) = 300\left(\frac{1}{2}\right) + 300\left(\frac{x}{60}\right)$$
$$6x = 150 + 5x$$
$$x = 150$$

She traveled 150 miles each way.

21. Let x represent the number of hours it takes for Paul and Marco to tune the Toyota together. Since Paul can tune the car in 2 hours, he can complete $\dfrac{1}{2}$ of the tuneup in 1 hour. Also, since Marco can tune the car in 3 hours, he can complete $\dfrac{1}{3}$ of the job in 1 hour. The amount of work Paul can do in 1 hour plus the amount of work Marco can do in 1 hour must equal the amount of work they can do together in 1 hour, or $\dfrac{1}{x}$ of the job. So,

$$\frac{1}{2} + \frac{1}{3} = \frac{1}{x}.$$

Solve.
$$6x\left(\frac{1}{2}\right) + 6x\left(\frac{1}{3}\right) = 6x\left(\frac{1}{x}\right)$$
$$3x + 2x = 6$$
$$5x = 6$$
$$x = \frac{6}{5}$$

Working together, it takes Paul and Marco $\dfrac{6}{5} = 1\dfrac{1}{5}$ hrs. = 1 hour 12 minutes to tune the Toyota.

25. Let x represent the number of days it takes Sue to complete the job. Then she completes $\dfrac{1}{x}$ of the job in 1 day. Since Dennis can do the job in 4 days, then he does $\dfrac{1}{4}$ of the job in 1 day. Finally, if it takes $2\dfrac{1}{3} = \dfrac{7}{3}$ days to complete the job working together, then they can complete $\dfrac{1}{\frac{7}{3}} = \dfrac{3}{7}$ of the job in 1 day.

So we have
$$\frac{1}{x} + \frac{1}{4} = \frac{3}{7}.$$

Solve.
$$28x\left(\frac{1}{x}\right) + 28x\left(\frac{1}{4}\right) = 28x\left(\frac{3}{7}\right)$$
$$28 + 7x = 12x$$
$$28 = 5x$$
$$\frac{28}{5} = x$$

It takes Sue $\dfrac{28}{5} = 5\dfrac{3}{5}$ days to complete the job.

Solutions to Selected Exercises

29. Let x represent the amount of time required for an experienced employee. Then $\frac{1}{x}$ of the job is done in 1 hour. The new employee then requires $2x$ hours, and so completes $\frac{1}{2x}$ of the job in 1 hour working alone. They can complete $\frac{1}{2}$ of the job in 1 hour working together. So we have
$$\frac{1}{x} + \frac{1}{2x} = \frac{1}{2}.$$
Solve.
$$2x\left(\frac{1}{x}\right) + 2x\left(\frac{1}{2x}\right) = 2x\left(\frac{1}{2}\right)$$
$$2 + 1 = x$$
$$3 = x$$
It will take the experienced employee 3 hours working alone.

33. Since m varies directly as p, there is a number k such that
$$m = kp.$$
Find k from the given fact that $m = 20$ when $p = 2$.
$$20 = k \cdot 2 \quad \text{Let } m = 20, p = 2$$
$$k = 10$$
Let $k = 10$ in $m = kp$ to get
$$m = 10p.$$
If $p = 5$, then
$$m = 10 \cdot 5 \quad \text{Let } p = 5$$
$$m = 50$$

37. If the circumference varies directly as the radius, and if c represents the circumference and r the radius, then there is a number k such that
$$c = kr.$$
Find k by letting $r = 7$ and $c = 43.96$.
$$43.96 = k \cdot 7$$
$$6.28 = k$$
Let $k = 6.28$ to get
$$c = 6.28r.$$
If $r = 11$, then
$$c = 6.28(11) = 69.08 \text{ centimeters.}$$

CHAPTER 7

Section 7.1 (page 339)

1. The equation is $x + y = 9$, and we want to know if $(2, 7)$ is a solution. Replace x with 2 and y with 7 to get $2 + 7 = 9$. This is true, so $(2, 7)$ is a solution.

5. To see whether $(1, 2)$ is a solution of $4x - 3y = 6$, replace x with 1 and y with 2 to get $4(1) - 3(2) = 6$, a false statement. The ordered pair is not a solution.

9. To see whether $(-6, 8)$ is a solution of $x = -6$, replace x with -6 (there is no replacement for y), to get $-6 = -6$, a true statement. The ordered pair is a solution.

13. Let $x = 2$; then $y = 3x + 5 = 3(2) + 5 = 6 + 5 = 11$. The ordered pair is $(2, 11)$.

17. Let $x = -3$; then $y = 3(-3) + 5 = -9 + 5 = -4$. The ordered pair is $(-3, -4)$.

21. Let $y = 8$; then $8 = 3x + 5$. Add -5 to both sides to get $3 = 3x$. Divide both sides by 3 to get $x = 1$. The ordered pair is $(1, 8)$.

25. Let $y = 24$; then $24 = -4x + 8$. Add -8 to both sides to get $16 = -4x$. Divide both sides by -4 to get $x = -4$. The ordered pair is $(-4, 24)$.

29. Let $x = 2$; then $y = 3(2) - 5 = 6 - 5 = 1$. Let $x = 0$; then $y = 3(0) - 5 = 0 - 5 = -5$. Let $x = -3$; then $y = 3(-3) - 5 = -9 - 5 = -14$. The ordered pairs are $(2, 1), (0, -5)$, and $(-3, -14)$.

33. Let $m = 1$; then $-3(1) + n = 4$ or $-3 + n = 4$. Add 3 to each side to get $n = 7$. Let $m = 0$; then $-3(0) + n = 4$ or $0 + n = 4$ or $n = 4$. Let $m = -2$; then $-3(-2) + n = 4$ or $6 + n = 4$. Subtract 6 from each side to get $n = -2$. The three ordered pairs are $(1, 7), (0, 4)$, and $(-2, -2)$.

37. Let $y = 0$; then $4x - 9(0) = 36$ or $4x - 0 = 36$ or $4x = 36$. Divide both sides by 4 to get $x = 9$. Let $x = 0$; then $4(0) - 9y = 36$ or $0 - 9y = 36$ or $-9y = 36$. Divide both sides by -9 to get $y = -4$. Let $y = 4$; then $4x - 9(4) = 36$ or $4x - 36 = 36$. Add 36 to both sides to get $4x = 72$. Divide both sides by 4 to get $x = 18$. The ordered pairs are $(9, 0), (0, -4)$, and $(18, 4)$.

41. Let $x = 2.9$; then $4(2.9) - 3y = 7$ or $11.6 - 3y = 7$. Subtract 11.6 from both sides to get $-3y = -4.6$. Divide both sides by -3 to get $y = 1.533$. Let $y = 9.3$; then $4x - 3(9.3) = 7$ or $4x - 27.9 = 7$. Add 27.9 to both sides to get $4x = 34.9$. Divide both sides by 4 to get $x = 8.725$. Let $y = -1.5$; then $4x - 3(-1.5) = 7$ or $4x + 4.5 = 7$. Subtract 4.5 from both sides to get $4x = 2.5$. Divide both sides by 4 to get $x = .625$. The ordered pairs are $(2.9, 1.533), (8.725, 9.3)$, and $(.625, -1.5)$.

45. The equation $y = -8$ says y is always -8. The ordered pairs are $(4, -8), (0, -8)$, and $(-4, -8)$.

Section 7.2 (page 345)

1. $(2, 5)$

5. $(7, 3)$

9.–17. See the graph in the answer section.

21. The x-coordinate is negative and the y-coordinate is positive, so the point is in Quadrant II.

25. The x-coordinate is negative and the y-coordinate is positive, so the point is in Quadrant II.

29. The ordered pairs are $(0, 3), (6, 0), (4, 1)$, $\left(-1, \frac{7}{2}\right)$; see the graph in the answer section.

33. The ordered pairs are $(5, -2), (0, -2)$, $(-3, -2), (-2, -2)$; see the graph in the answer section.

Section 7.3 (page 353)

All the graphs for these exercises may be found in the answer section.

1. $(0, 5), (5, 0), (2, 3)$
5. $(0, -6), (2, 0), (3, 3)$
9. To find the x-intercept, let $y = 0$. Then $3x - 5(0) = 9; 3x - 0 = 9; 3x = 9$. Divide both sides by 3 to get $x = 3$, giving $(3, 0)$ as the x-intercept. To find the y-intercept, let $x = 0$. Then $3(0) - 5y = 9; 0 - 5y = 9; -5y = 9$. Divide on both sides by -5 to get $y = -\frac{9}{5}$, giving $\left(0, -\frac{9}{5}\right)$ as the y-intercept.
13. To graph $x - y = 2$, find the intercepts. If $x = 0$, then $0 - y = 2$ or $y = -2$, which gives the point $(0, -2)$. For $y = 0$, then $x - 0 = 2$ or $x = 2$, which gives $(2, 0)$. A third point can be found as a check. Let $x = 4$; then $4 - y = 2$, or $-y = -2$, or $y = 2$, which gives the point $(4, 2)$. Plot these three points and draw a line through them.
17. To graph $x + 2y = 6$, find the intercepts. If $x = 0$, then $0 + 2y = 6, 2y = 6$, or $y = 3$. If $y = 0$, then $x + 2(0) = 6, x + 0 = 6$, or $x = 6$. The two points are $(0, 3)$ and $(6, 0)$. Find one more point as a check. Let $x = 3$; then $3 + 2y = 6, 2y = 3$, or $y = \frac{3}{2}$, giving the point $\left(3, \frac{3}{2}\right)$. Plot these three points and draw a line through them.
21. To graph $3x = 6 - 2y$, find the intercepts. Let $x = 0$; then $3(0) = 6 - 2y$, or $0 = 6 - 2y$ or $6 = 2y$ or $y = 3$. Let $y = 0$; then $3x = 6 - 2(0)$, $3x = 6 - 0, 3x = 6, x = 2$. The points are $(0, 3)$ and $(2, 0)$. Find one more point as a check. Let $x = 4$; then $3(4) = 6 - 2y, 12 = 6 - 2y$, $6 = -2y$, or $y = -3$, giving $(4, -3)$. Plot these three points and draw a line through them.
25. To graph $3x + 7y = 14$, find the intercepts. If $x = 0$, then $3(0) + 7y = 14, 0 + 7y = 14$, $7y = 14$, or $y = 2$. If $y = 0$, then $3x + 7(0) = 14$, $3x + 0 = 14, 3x = 14, x = \frac{14}{3}$ or $4\frac{2}{3}$. The points are $(0, 2)$ and $\left(4\frac{2}{3}, 0\right)$. Find one more point as a check. Let $x = 2$; then $3(2) + 7y = 14, 6 + 7y = 14, 7y = 8, y = \frac{8}{7}$ or $1\frac{1}{7}$. The point is $\left(2, 1\frac{1}{7}\right)$. Plot these three points and draw a line through them.
29. To graph $y + 6x = 0$, let $x = 0$ to get $y + 6(0) = 0, y + 0 = 0$, or $y = 0$. This gives the point $(0, 0)$. If we let $y = 0$, we will get the same point. Choose two other values for x and solve for y. If we let $x = \frac{1}{2}$, we get $y + 6\left(\frac{1}{2}\right) = 0, y + 3 = 0$, or $y = -3$. If $x = -\frac{1}{2}$, then $y + 6\left(-\frac{1}{2}\right) = 0$, $y - 3 = 0$, or $y = 3$. The points are $\left(\frac{1}{2}, -3\right)$ and $\left(-\frac{1}{2}, 3\right)$. Plot these three points and draw a line through them.
33. To graph $x + 2 = 0$, find three points on the line. The equation can be written as $x = -2$, which says that y can be anything as long as x is -2. Plot the points $(-2, -3), (-2, 0)$, and $(-2, 3)$. The graph is a vertical line through these points.
37. (a) Look at where the graph marked "Private" crosses the vertical line labeled '86. Read across the vertical axis to get an estimated cost of about $10.5 or $10,500.
    (b) The graph labeled "Public" crosses the line for 1987 at about $5.7, so the estimated cost is $5,700.
    (c) The graph representing private schools crosses the line for 1987 at about $12.7. The graph for public schools crosses the 1987 line at about $6.2. Since $12.7 - 6.2 = 6.5$, the difference in cost is about $6,500.

Section 7.4 (page 363)

1. Locate two points on the graph. The graph goes through $(-5, 2)$ and $(5, -3)$. Use the slope formula:
$$m = \frac{-3 - 2}{5 - (-5)} = \frac{-5}{10} = -\frac{1}{2}.$$
5. Locate two points on the graph. The graph goes through $(-4, 3)$ and $(2, 3)$. Use the slope formula:
$$m = \frac{3 - 3}{2 - (-4)} = \frac{0}{6} = 0.$$
9. Let $(x_1, y_1) = (-1, 2)$ and $(x_2, y_2) = (-3, -7)$ in the slope formula. Then $m = \frac{-7 - 2}{-3 - (-1)}$
$= \frac{-9}{-2} = \frac{9}{2}.$
13. Let $(x_1, y_1) = (-1, 6)$ and $(x_2, y_2) = (4, 6)$ in the slope formula. Then $m = \frac{6 - 6}{4 - (-1)} = \frac{0}{5} = 0.$
17. Let $(x_1, y_1) = \left(\frac{3}{7}, -\frac{1}{4}\right)$ and $(x_2, y_2) = \left(\frac{5}{7}, -\frac{5}{4}\right)$.
$$\text{Slope} = m = \frac{-\frac{5}{4} - \left(-\frac{1}{4}\right)}{\frac{5}{7} - \frac{3}{7}}$$
$$= \frac{-\frac{5}{4} + \frac{1}{4}}{\frac{2}{7}}$$
$$= \frac{-1}{\frac{2}{7}} = -1 \cdot \frac{7}{2} = -\frac{7}{2}.$$

Solutions to Selected Exercises

21. Let $(x_1, y_1) = (0.03, 1.57)$ and $(x_2, y_2) = (3.54, -2.01)$.
$$\text{Slope} = m = \frac{-2.01 - 1.57}{3.54 - 0.03}$$
$$= \frac{-3.58}{3.51} = -1.020$$

25. The slope is the coefficient of x in $y = x + 1 = 1x + 1$, so the slope is 1.

29. Solve for y: $3x - 2y = 5$
$$-2y = -3x + 5$$
$$y = \frac{3}{2}x - \frac{5}{2}.$$
The slope is the coefficient of x, which is $\frac{3}{2}$.

33. Solve for y: $9x + 7y = 5$
$$7y = -9x + 5$$
$$y = -\frac{9}{7}x + \frac{5}{7}.$$
The slope is the coefficient of x, which is $-\frac{9}{7}$.

37. Solve each equation for y:
$y - x = 3$ gives $y = x + 3$;
$y - x = 5$ gives $y = x + 5$.
Both lines have a slope of 1. Since the slopes are equal, the lines are parallel.

41. Solve each equation for y:
$$3x - 2y = 4 \quad \text{gives} \quad y = \frac{3}{2}x - 2;$$
$$2x + 3y = 1 \quad \text{gives} \quad y = -\frac{2}{3}x + \frac{1}{3}.$$
The slopes, $\frac{3}{2}$ and $-\frac{2}{3}$, have a product of -1, so the lines are perpendicular.

45. The slope of $8x - 9y = 2$ is $\frac{8}{9}$, and the slope of $8x + 6y = 1$ is $-\frac{4}{3}$. These lines are neither parallel nor perpendicular.

49. Both of the lines $y = 2$ and $y - 4 = 8$ or $y = 12$ have a slope of 0, so they are parallel.

Section 7.5 (page 369)

1. Substitute $m = 3$ and $b = 5$ into the slope-intercept form,
$$y = mx + b$$
$$y = 3x + 5$$

5. Substitute $m = \frac{5}{3}$ and $b = \frac{1}{2}$ into the slope-intercept form.
$$y = mx + b$$
$$y = \frac{5}{3}x + \frac{1}{2}$$

9. Starting at $(2, 5)$, the slope $\frac{1}{2}$ says to go 1 up and 2 over to find a second point on the line. See the graph in the answer section.

13. Starting at $(-3, 0)$, the slope $-\frac{5}{4}$ says to go 4 over and 5 down to find a second point on the line. See the graph in the answer section.

17. Starting at $(2, -3)$, the slope of -4, or $-\frac{4}{1}$, says to go 1 over and 4 down to find a second point on the line. See the graph in the answer section.

21. A line with a slope of 0 is horizontal, so draw a horizontal line through $(-4, 3)$. See the graph in the answer section.

25. Substitute the values $m = 3$, $x_1 = 1$, and $y_1 = 4$ into the point-slope form.
$$y - y_1 = m(x - x_1)$$
$$y - 4 = 3(x - 1)$$
$$y - 4 = 3x - 3 \quad \text{Distributive property}$$
$$y = 3x + 1 \quad \text{Add 4 to each side}$$
$$-3x + y = 1 \quad \text{Subtract } 3x \text{ from each side}$$
$$3x - y = -1 \quad \text{Multiply each side by } -1$$

29. Substitute the values $m = \frac{4}{5}$, $x_1 = 2$, and $y_1 = -4$ into the point-slope form.
$$y - y_1 = m(x - x_1)$$
$$y - (-4) = \frac{4}{5}(x - 2)$$
$$y + 4 = \frac{4}{5}(x - 2)$$
$$5y + 20 = 4(x - 2) \quad \text{Multiply each side by 5}$$
$$5y + 20 = 4x - 8 \quad \text{Distributive property}$$
$$5y = 4x - 28 \quad \text{Subtract 28 from each side}$$
$$-4x + 5y = -28 \quad \text{Subtract } 4x \text{ from each side}$$
$$4x - 5y = 28 \quad \text{Multiply each side by } -1$$

33. First find the slope of the line by using the definition of slope.
$$m = \frac{4 - 5}{7 - 8}$$
$$= \frac{-1}{-1}$$
$$= 1$$
Now use either point and the point-slope form. Using $(7, 4)$ gives the following result.
$$y - y_1 = m(x - x_1)$$
$$y - 4 = 1(x - 7)$$
$$y - 4 = x - 7 \quad \text{Distributive property}$$
$$y = x - 3 \quad \text{Add 4 to each side}$$
$$-x + y = -3 \quad \text{Subtract } x \text{ from each side}$$
$$x - y = 3 \quad \text{Multiply each side by } -1$$

37. First, find the slope of the line, using the definition of slope.
$$m = \frac{-5 - (-2)}{-7 - (-9)}$$
$$= \frac{-5 + 2}{-7 + 9} = \frac{-3}{2}$$

Now use either point and the point-slope form.
Using $(-7, -5)$ gives the following results.
$$y - y_1 = m(x - x_1)$$
$$y - (-5) = \frac{-3}{2}[x - (-7)]$$
$$y + 5 = \frac{-3}{2}(x + 7)$$
$$2y + 10 = -3(x + 7) \quad \text{Multiply each side by 2}$$
$$2y + 10 = -3x - 21 \quad \text{Distributive property}$$
$$2y = -3x - 31 \quad \text{Subtract 10 from each side}$$
$$3x + 2y = -31 \quad \text{Add } 3x \text{ to each side}$$

41. First, find the slope of the line, using the definition of slope.
$$m = \frac{-7 - 0}{3 - (-5)} = \frac{-7}{8}$$
Now use either point and the point-slope form. Using $(-5, 0)$ gives
$$y - y_1 = m(x - x_1)$$
$$y - 0 = \frac{-7}{8}[x - (-5)]$$
$$y = \frac{-7}{8}(x + 5)$$
$$8y = -7(x + 5) \quad \text{Multiply each side by 8}$$
$$8y = -7x - 35 \quad \text{Distributive property}$$
$$7x + 8y = -35. \quad \text{Add } 7x \text{ to each side}$$

45. Let $m = 9$ and $b = 50$ in the equation $y = mx + b$. The cost equation is $y = 9x + 50$.

49. (a) Use the ordered pairs $(1, 24)$ and $(5, 48)$.
(b) The slope is
$$m = \frac{48 - 24}{5 - 1} = \frac{24}{4} = 6.$$
Use $(1, 24)$ and $m = 6$ in the point-slope form.
$$y - y_1 = m(x - x_1)$$
$$y - 24 = 6(x - 1)$$
$$y - 24 = 6x - 6$$
$$y = 6x + 18$$

Section 7.6 (page 377)

All the graphs for these exercises may be found in the answer section.

1. Use $(0, 0)$ as a test point.
$$x + y \leq 4 \quad \text{Original inequality}$$
$$0 + 0 \leq 4 \quad \text{Let } x = 0, y = 0$$
$$0 \leq 4 \quad \text{True}$$
Since this last statement is true, shade the side of the graph containing $(0, 0)$.

5. Use $(0, 0)$ as a test point.
$$-3x + 4y < 12$$
$$-3(0) + 4(0) < 12$$
$$0 + 0 < 12$$
$$0 < 12 \quad \text{True}$$
Shade the side of the graph containing $(0, 0)$.

9. Use $(0, 0)$ as a test point.
$$x < 4 \quad \text{Original inequality}$$
$$0 < 4 \quad \text{Use } x = 0; \text{ true}$$
Since this statement is true, shade the side of the graph that contains $(0, 0)$.

13. *Step 1*: Graph $x + y = 8$.
If $x = 0$, then $y = 8$, giving $(0, 8)$.
If $y = 0$, then $x = 8$, giving $(8, 0)$.
Graph these points and the line through them. The line is solid because of the "$\leq$" symbol.
Step 2: Use $(0, 0)$ as a test point.
$$x + y \leq 8$$
$$0 + 0 \leq 8$$
$$0 \leq 8 \quad \text{True}$$
Step 3: Since the statement is true, shade the side of the graph containing $(0, 0)$.

17. *Step 1*: Graph $x + 2y = 4$.
If $x = 0$, then $0 + 2y = 4$ or $y = 2$, giving $(0, 2)$.
If $y = 0$, then $x + 2(0) = 4$ or $x = 4$, giving $(4, 0)$.
Graph these points and the line through them. The line is solid because of the "$\geq$" symbol.
Step 2: Use $(0, 0)$ as a test point.
$$x + 2y \geq 4$$
$$0 + 2(0) \geq 4$$
$$0 \geq 4 \quad \text{False}$$
Step 3: Since the statement is false, shade the side that does not contain $(0, 0)$.

21. *Step 1*: Graph $3x - 4y = 12$.
If $x = 0$, then $3(0) - 4y = 12$
$$-4y = 12$$
$$y = -3, \quad \text{giving } (0, -3).$$
If $y = 0$, then $3x - 4(0) = 12$
$$3x = 12$$
$$x = 4, \quad \text{giving } (4, 0).$$
Graph these points and the line through them. The line is dotted because of the "$<$" symbol.
Step 2: Use $(0, 0)$ as a test point.
$$3x - 4y < 12$$
$$3(0) - 4(0) < 12$$
$$0 < 12 \quad \text{True}$$
Step 3: Since the statement is true, shade the side containing $(0, 0)$.

25. *Step 1*: Graph $x = -2$.
This is a vertical line through the x-axis at $x = -2$. The line is dotted because of the "$<$" sign.
Step 2: Use $(0, 0)$ as a test point.
$$x < -2 \quad \text{Original inequality}$$
$$0 < -2 \quad \text{Let } x = 0$$
Step 3: Since the last statement is false, shade the side of the graph that does not contain $(0, 0)$.

29. *Step 1*: Graph $x = 5y$.
The line goes through $(0, 0)$. Let $y = 1$ to get $x = 5 \cdot 1 = 5$. Draw the line through $(0, 0)$ and $(5, 1)$.
Step 2: The line goes through $(0, 0)$, so this point cannot be used as a test point. Choose any point

Solutions to Selected Exercises

not on the line. Let us choose (0, 4). Substitute $x = 0$ and $y = 4$ in the original inequality.
$$x \leq 5y$$
$$0 \leq 5 \cdot 4$$
$$0 \leq 20 \quad \text{True}$$
Step 3: Because the statement is true, shade the side including (0, 4).

CHAPTER 8

Section 8.1 (page 393)

1. Replace x with 2 and y with -5. In the first equation we get $3(2) + (-5) = 1$ or $6 + (-5) = 1$, which is true. In the second equation we get $2(2) + 3(-5) = -11$ or $4 + (-15) = -11$, which is true. Since $(2, -5)$ makes both equations true, it is the solution of the system.

5. Replace x with 2 and y with 0. In the first equation we get $3(2) + 5(0) = 6$ or $6 + 0 = 6$, which is true. In the second equation we get $4(2) + 2(0) = 5$ or $8 + 0 = 5$, which is false. This ordered pair is not a solution.

9. Replace x with 6 and y with -8. In the first equation we get $6 + 2(-8) + 10 = 0$ or $6 + (-16) + 10 = 0$, which is true. In the second equation we get $2(6) - 3(-8) + 30 = 0$ or $12 + 24 + 30 = 0$, which is false. This ordered pair is not a solution.

13. Graph both lines on the same axes. For $x + y = 6$, use the intercepts, (0, 6) and (6, 0). For $x - y = 2$, use the intercepts, (0, -2) and (2, 0). These lines intersect at (4, 2).

17. Graph both lines on the same axes. For $2x + y = 6$ use the intercepts, (3, 0) and (0, 6). For $x - 3y = -4$ use the intercepts, (-4, 0) and $\left(0, \frac{4}{3}\right)$. These lines intersect at (2, 2).

21. Graph both lines on the same axes. For $5x + 4y = 7$ use the points (-1, 3) and (3, -2). For $2x - 3y = 12$ use the intercepts, (6, 0) and (0, -4). These lines intersect at (3, -2).

25. Graph both lines on the same axes. For $3x + 2y = -10$ use the points (0, -5) and (-2, -2). For $x - 2y = -6$ use the intercepts, (-6, 0) and (0, 3). These lines intersect at (-4, 1).

29. Graph both lines on the same axes. For $x - 2y = 5$ use the points (5, 0) and (-1, -3). For $x + \frac{3}{4}y = 5$, use the points (5, 0) and (2, 4). These lines intersect at (5, 0).

33. Graph both lines on the same axes. For $2x + 5y = 12$ use the points (6, 0) and (1, 2). For $x - 2y = -3$ use the points (-3, 0) and (-1, 1). These lines intersect at (1, 2).

37. Graph each line. The two lines are parallel, so there is no solution.

41. Graph each line. The two lines are parallel, so there is no solution.

Section 8.2 (page 401)

1. Add the equations, getting $2x = 2$ or $x = 1$. Replace x with 1 in either equation. If we use $x - y = 3$, we get $1 - y = 3$ or $-y = 2$ or $y = -2$. The solution is (1, -2).

5. Add the equations, getting $3x = 18$ or $x = 6$. Replace x with 6 in either equation. If we use $2x + y = 14$, we get $2(6) + y = 14$ or $12 + y = 14$ or $y = 2$. The solution is (6, 2).

9. Add the equations, getting $4y = 8$ or $y = 2$. Replace y with 2 in either equation. If we use $6x - y = 1$, we get $6x - 2 = 1$ or $6x = 3$ or $x = \frac{1}{2}$. The solution is $\left(\frac{1}{2}, 2\right)$.

13. Multiply both sides of the first equation by 2 and add.
$$\begin{aligned} 4x - 2y &= 14 \\ 3x + 2y &= 0 \\ \hline 7x &= 14 \\ x &= 2 \end{aligned}$$
Let $x = 2$ in the second equation to get
$$3x + 2y = 0$$
$$3(2) + 2y = 0$$
$$6 + 2y = 0$$
$$2y = -6$$
$$y = -3.$$
The solution is (2, -3).

17. Multiply both sides of the first equation by -3 and then add.
$$\begin{aligned} -3x - 12y &= 54 \\ 3x + 5y &= -19 \\ \hline -7y &= 35 \\ y &= -5 \end{aligned}$$
Let $y = -5$ in the first equation to get
$$x + 4y = -18$$
$$x + 4(-5) = -18$$
$$x - 20 = -18$$
$$x = 2.$$
The solution is (2, -5).

21. One way to proceed is to multiply the first equation by 2 to get $6x - 4y = -12$. Then add.
$$\begin{aligned} 6x - 4y &= -12 \\ -5x + 4y &= 16 \\ \hline x &= 4 \end{aligned}$$
Replace x with 4 in either equation to find that $y = 9$, giving (4, 9).

25. One way to proceed is to multiply the first equation by -3 to get $-6x - 3y = -15$. Then add.
$$\begin{aligned} -6x - 3y &= -15 \\ 5x + 3y &= 11 \\ \hline -x &= -4 \\ x &= 4 \end{aligned}$$
Replace x with 4 in either equation to find that $y = -3$, giving (4, -3).

29. One way to proceed is to multiply the top equation by 3 to get $9x + 15y = 99$, and the bottom equation by 5, to get $20x - 15y = 75$. Add these two equations to get $29x = 174$ or $x = 6$. Replace

x with 6 in either of the original equations to find $y = 3$, giving the solution $(6, 3)$.

33. One way to proceed is to multiply the first equation by 7 to get $14x + 21y = -84$, and the second equation by 3 to get $15x - 21y = -90$. Add these equations to get $29x = -174$ or $x = -6$. Replace x with -6 in either of the original equations to find $y = 0$, giving the solution $(-6, 0)$.

37. Multiply the first equation by 3 to get $72x + 36y = 57$, and the second equation by 2 to get $32x - 36y = -18$. Add these equations to get $104x = 39$ or $x = \frac{3}{8}$. Replace x with $\frac{3}{8}$ in one of the original equations to find $y = \frac{5}{6}$, giving the solution $\left(\frac{3}{8}, \frac{5}{6}\right)$.

41. Multiply the first equation by 3 to get $15x - 21y = 18$, and the second equation by -5 to get $-15x + 30y = -10$. Add these equations to get $9y = 8$ or $y = \frac{8}{9}$. Replace y with $\frac{8}{9}$ in either of the original equations to find $x = \frac{22}{9}$, giving the solution $\left(\frac{22}{9}, \frac{8}{9}\right)$.

45. Multiply the top equation by -3 to get $-.39x + 1.56y = 1.17$. Add this to the second equation to get $1.64y = -1.64$ or $y = -1$. Replace y with -1 in the first equation to get $x = -7$, giving the solution $(-7, -1)$.

Section 8.3 (page 407)

1. Multiply the top equation by -1, giving $-x - y = -4$. Then add the two equations, giving $0 = -6$. This false statement shows that the system has no solution.

5. Multiply the top equation by -2, giving $-2x - 6y = -10$. Add the two equations, giving $0 = 0$. This true statement shows that the two equations represent the same line. The system has an infinite number of solutions.

9. Multiply the first equation by -1 to get $-5x = -y - 4$. Add this result to the second equation to get $0 = -8$, a false statement. This shows that the system has no solution.

13. Multiply the top equation by -2, getting $-4x + 6y = 0$; then add to get $11y = 0$ or $y = 0$. Replace y with 0 to see that $x = 0$, giving the solution $(0, 0)$.

17. Multiply the first equation by -2 to get $-8x + 4y = -2$. Add this result to the second equation to get $0 = -1$. This false statement shows that the system has no solution.

21. Multiply the second equation by 2 to get $-4x + y = -3$. Add this result to the first equation to get $0 = 0$. This true statement shows that these equations represent the same line. The system has an infinite number of solutions.

Section 8.4 (page 413)

1. The second equation says that $y = 2x$. Replace y with $2x$ in the first equation to get $x + 2x = 6$ or $3x = 6$ or $x = 2$. Replace x with 2 in the equation $y = 2x$ to get $y = 2(2) = 4$. The solution is $(2, 4)$.

5. Solve the second equation for x to get $x = 2y + 10$. Replace x with $2y + 10$ in the first equation to get $2y + 10 + 5y = 3$, from which $7y + 10 = 3$, $7y = -7$, or $y = -1$. Replace y with -1 in $x = 2y + 10$ to get $x = 2(-1) + 10 = 8$. The solution is $(8, -1)$.

9. Solve either equation for either variable. If we solve the first equation for x, we get $x = 6 - y$. Replace x with $6 - y$ in the second equation to get $6 - y - y = 4$, from which $6 - 2y = 4$, $-2y = -2$, or $y = 1$. Replace y with 1 in $x = 6 - y$ to get $x = 6 - 1 = 5$. The solution is $(5, 1)$.

13. Solve the second equation for y to get
$$y = 2x - 7.$$
Replace y with $2x - 7$ in the first equation.
$$2x + 3y = 11$$
$$2x + 3(2x - 7) = 11$$
$$2x + 6x - 21 = 11$$
$$8x = 32$$
$$x = 4$$
Find y from $y = 2x - 7$.
$$y = 2(4) - 7 = 8 - 7 = 1$$
The solution is $(4, 1)$.

17. To solve the system by the substitution method, solve the second equation for x.
$$3x = 4y + 2$$
$$x = \frac{4}{3}y + \frac{2}{3}$$
Substitute for x in the first equation.
$$6\left(\frac{4}{3}y + \frac{2}{3}\right) - 8y = 4$$
$$6\left(\frac{4}{3}y\right) + 6\left(\frac{2}{3}\right) - 8y = 4$$
$$8y + 4 - 8y = 4$$
$$4 = 4 \quad \text{True}$$
This true statement shows that the equations of the system represent the same line. The system has an infinite number of solutions.

21. Replace y in the first equation with $4x$ to get $x + 4(4x) = 34$ or $x + 16x = 34$, from which $17x = 34$ or $x = 2$. Replace x with 2 in the second equation to get $y = 8$. The solution is $(2, 8)$.

25. Combine terms in each equation to get $x - 6y = -9$ for the first equation and $2x + y = 8$ for the second equation. Solve the first equation for x: $x = 6y - 9$. Replace x with $6y - 9$ in the second equation to get $2(6y - 9) + y = 8$, from which $12y - 18 + y = 8$ or $13y = 26$, so $y = 2$. Replace y with 2 in $x = 6y - 9$ to get $x = 6(2) - 9 = 12 - 9 = 3$. The solution is $(3, 2)$.

29. Combine terms in each equation to get $-2x + y = 12$ and $2x - y = -12$. Add these two equations, getting $0 = 0$, which indicates that the equations represent the same line. The system has an infinite number of solutions.

Solutions to Selected Exercises   597

33. Multiply both sides of the first equation by 3 to clear fractions, getting $3x + y = 3y - 6$. Combine terms to get $3x - 2y = -6$. Multiply the second equation on both sides by 4 to clear fractions, getting $x - 4y = 4x - 4y$. Combine terms to get $-3x = 0$ or $x = 0$. Replace x with 0 in $3x - 2y = -6$ to get $-2y = -6$ or $y = 3$. The solution is $(0, 3)$.

37. Multiply the first equation on both sides by the common denominator, 12, to get $4x - 9y = -6$. Multiply the second equation on both sides by the common denominator, 6, to get $4x + 3y = 18$. Multiply the first result by -1 and add to the second result, getting $12y = 24$ or $y = 2$. Replace y with 2 in the first result to get $4x - 9(2) = -6$ or $4x - 18 = -6$, from which $4x = 12$ or $x = 3$. The solution is $(3, 2)$.

Section 8.5 (page 421)

1. Let x and y represent the two numbers. Then $x + y = 52$ and $x - y = 34$. Add the two equations to get $2x = 86$, from which $x = 43$. From the first equation, $43 + y = 52$ so $y = 9$. The numbers are 43 and 9.

5. Let x represent the larger angle and y the smaller angle.
$$x + y = 90 \quad \text{Sum is 90}$$
$$x - y = 20 \quad \text{Difference is 20}$$
Solve by addition.
$$\begin{aligned} x + y &= 90 \\ x - y &= 20 \\ \hline 2x &= 110 \\ x &= 55 \quad \text{Solve for } x \\ 55 + y &= 90 \quad \text{Replace } x \text{ with 55 in} \\ y &= 35 \quad \text{the first equation} \end{aligned}$$
The angles are 55° and 35°.

9. Let x represent the number of $10 bills. Let y represent the number of $20 bills. Since there are 85 bills altogether,
$$x + y = 85.$$
The monetary value of the $10 bills is $10 times the number of bills, or $10x$. In the same manner, the value of the $20 bills is $20y$. Since the total value of all the bills is $1480,
$$10x + 20y = 1480.$$
So we wish to solve the system
$$x + y = 85$$
$$10x + 20y = 1480.$$
$$\begin{aligned} -10x - 10y &= -850 \quad \text{Multiply the first} \\ 10x + 20y &= 1480 \quad \text{equation by } -10 \\ \hline 10y &= 630 \quad \text{Add the equations} \\ y &= 63 \quad \text{Solve for } y \\ x + 63 &= 85 \quad \text{Substitute } y = 63 \text{ into the} \\ x &= 22 \quad \text{first equation} \end{aligned}$$
There are 22 $10 bills and 63 $20 bills.

13. Let x represent the number of $5 bills and y represent the number of $10 bills. There are 124 bills altogether, so
$$x + y = 124.$$
The monetary value of the $5 bills is $5 times the number of bills, or $5x$. In the same manner, the value of the $10 bills is $10y$. The total value of the money is $840, so
$$5x + 10y = 840.$$
Solve, using the addition method.
$$x + y = 124$$
$$5x + 10y = 840$$
$$\begin{aligned} -5x - 5y &= -620 \quad \text{Multiply the first} \\ 5x + 10y &= 840 \quad \text{equation by } -5; \\ \hline 5y &= 220 \quad \text{add the two equations} \\ y &= 44 \quad \text{Solve for } y \\ x + 44 &= 124 \quad \text{Substitute } y = 44 \text{ in the} \\ x &= 80 \quad \text{first equation to find } x \end{aligned}$$
There are 80 $5 bills and 44 $10 bills.

17. Let x be the number of liters of 90% solution and y the number of liters of 75%. Make a chart.

Liters of solution	Percent	Liters of pure antifreeze
x	90	$.90x$
y	75	$.75y$
20	78	$20(.78) = 15.6$

Write two equations: $x + y = 20$ and $.90x + .75y = 15.6$. Solve this system to get 4 liters of 90% and 16 liters of 75%.

21. Let x be the number of pounds of $6 coffee and y be the number of pounds of $3 coffee. Make a chart.

Pounds	Dollars per pound	Total price
x	6	$6x$
y	3	$3y$
90	4	$90(4) = 360$

Write two equations: $x + y = 90$ and $6x + 3y = 360$. Solve the system to get 30 pounds of $6 coffee and 60 pounds of $3 coffee.

25. Since $d = rt$, the first row of the chart gives
$$3(x + y) = 36,$$
and the second row gives
$$3(x - y) = 24.$$
Use the distributive property to get the system
$$3x + 3y = 36.$$
$$3x - 3y = 24.$$
Add to get
$$\begin{aligned} 3x + 3y &= 36 \\ 3x - 3y &= 24 \\ \hline 6x &= 60 \\ x &= 10. \end{aligned}$$
Use $x = 10$ and $3x + 3y = 36$ to get
$$3x + 3y = 36$$
$$3(10) + 3y = 36$$
$$30 + 3y = 36$$
$$3y = 6$$
$$y = 2.$$
The boat goes 10 miles per hour and the current is 2 miles per hour.

29. Let x represent John's speed and let y represent Harriet's speed. If they walk in the same direction for 60 hours, John overtakes Harriet. Walking 60 hours at x miles per hour, John walks $60x$ miles. Also, Harriet walks $60y$ miles. John walks 30 miles further, so
$$60x = 60y + 30.$$
In 5 hours, John walks $5x$ miles and Harriet walks $5y$ miles. They are 30 miles apart, so
$$5x + 5y = 30.$$
Rewrite this equation as
$$5x = -5y + 30$$
to get the system
$$60x = 60y + 30$$
$$5x = -5y + 30.$$
Multiply both sides of the second equation by 12 and add.
$$60x = 60y + 30$$
$$60x = -60y + 360$$
$$\overline{120x = 390}$$
$$x = \frac{390}{120} = \frac{39}{12} = \frac{13}{4},$$
or $3\frac{1}{4}$ miles per hour for John. Use this result to find $y = \frac{11}{4}$, or $2\frac{3}{4}$ miles per hour for Harriet.

Section 8.6 (page 427)

1. *Step 1*: Graph the line $x + y = 6$.
If $x = 0$, then $y = 6$, giving the ordered pair $(0, 6)$.
If $y = 0$, then $x = 6$, giving the ordered pair $(6, 0)$.
Graph these two points and the solid line through them.
Step 2: Use $(0, 0)$ as a test point.
$$\begin{aligned} x + y &\le 6 \quad \text{Original inequality} \\ 0 + 0 &\le 6 \quad \text{Substitute } x = 0, y = 0 \\ 0 &\le 6 \quad \text{True} \end{aligned}$$
Step 3: Since the inequality is true for $(0, 0)$, shade the side of the graph containing $(0, 0)$.
Go through the same steps above for the second inequality.
Step 1: Graph the line $x - y = 1$.
If $x = 0$, then $-y = 1$ or $y = -1$, giving the ordered pair $(0, -1)$.
If $y = 0$, then $x = 1$, giving the ordered pair $(1, 0)$.
Graph these two points and the solid line through them.
Step 2: Use $(0, 0)$ as a test point.
$$\begin{aligned} x - y &\le 1 \quad \text{Original inequality} \\ 0 - 0 &\le 1 \quad \text{Substitute } x = 0, y = 0 \\ 0 &\le 1 \quad \text{True} \end{aligned}$$
Step 3: Since the inequality is true for $(0, 0)$, shade the side of the graph containing $(0, 0)$.
The solution is where the two shaded regions overlap, or the darkest shaded region and the lines that border this region. See the graph in the answer section.

5. Graph the line $x + 4y = 8$. Make it a solid line. Use $(0, 0)$ as a test point.
$$\begin{aligned} x + 4y &\le 8 \\ 0 + 4 \cdot 0 &\le 8 \\ 0 + 0 &\le 8 \\ 0 &\le 8 \quad \text{True} \end{aligned}$$
Shade the area on the side of the line containing $(0, 0)$.
Graph the line $2x - y = 4$. Make it a solid line. Use $(0, 0)$ for a test point.
$$\begin{aligned} 2x - y &\le 4 \\ 2 \cdot 0 - 0 &\le 4 \\ 0 - 0 &\le 4 \\ 0 &\le 4 \quad \text{True} \end{aligned}$$
Shade the area on the side of this line containing $(0, 0)$.
The solution is the overlapped portion of the shaded areas. See the graph in the answer section.

9. Graph the line $x + 2y = 4$. Make the line solid. Use $(0, 0)$ as a test point. The inequality $x + 2y \le 4$ is true at $(0, 0)$. Shade the area on the side of the line containing $(0, 0)$.
Graph the line $x + 1 = y$. Make the line solid. Use $(0, 0)$ as a test point. The inequality $x + 1 \ge y$ is true at $(0, 0)$. Shade the area on the side of this line containing $(0, 0)$.
The solution is the overlapped portion of the shaded areas. See the graph in the answer section.

13. Graph the line $x - 2y = 6$. Make the line dashed. Use $(0, 0)$ for a test point. The inequality $x - 2y > 6$ is false at $(0, 0)$. Shade the area on the side of the line that doesn't contain $(0, 0)$.
Graph the line $2x + y = 4$. Make the line dashed. Use $(0, 0)$ for a test point. The inequality $2x + y > 4$ is false at $(0, 0)$. Shade the area on the side of this line that doesn't contain $(0, 0)$.
The solution is the overlapped portion of the shaded area. See the graph in the answer section.

17. Graph the line $x - 3y = 6$. Make it a solid line. Use $(0, 0)$ for a test point. The inequality $x - 3y \le 6$ is true at $(0, 0)$. Shade the area on the side of the line containing $(0, 0)$.
Graph the line $x = -1$. Make it a solid line. Use $(0, 0)$ for a test point. The inequality $x \ge -1$ is true at $(0, 0)$. Shade the area on the side of this line that contains $(0, 0)$.
The solution is the overlapped portion of the shaded areas. See the graph in the answer section.

CHAPTER 9

Section 9.1 (page 445)

1. Since $3^2 = 9$ and $(-3)^2 = 9$, the square roots of 9 are 3 and -3.

5. Since $\left(\frac{20}{9}\right)^2 = \frac{400}{81}$ and $\left(-\frac{20}{9}\right)^2 = \frac{400}{81}$, the square roots of $\frac{400}{81}$ are $\frac{20}{9}$ and $-\frac{20}{9}$.

9. Since $39^2 = 1521$ and $(-39)^2 = 1521$, the square roots of 1521 are 39 and -39.
13. The square root is 2. The number -2 is not correct, although $(-2)^2 = 4$, because the radical sign indicates only the nonnegative square root.
17. The answer is -8. The number 8 is not correct, although $8^2 = 64$, because the negative square root is indicated here.
21. $\sqrt{900}$ is the nonnegative square root of 900, which is 30.
25. $\sqrt{\frac{36}{49}} = \frac{6}{7}$ since $\left(\frac{6}{7}\right)^2 = \frac{36}{49}$, and $\frac{6}{7}$ is positive.
29. Not a real number, since there is no real number whose square is -9.
33. There is no rational number whose square is 15, so $\sqrt{15}$ is irrational. From the table, or a calculator $\sqrt{15} = 3.873$.
37. $-\sqrt{121}$ is rational, since it equals the rational number -11.
41. $\sqrt{400}$ is rational, since it equals the rational number 20.
45. Find $\sqrt{570}$ in the table by locating 57 at the left. Then look in the $\sqrt{10n}$ column to find 23.875. The number $\sqrt{570}$ is irrational.
49. Let $c = 17$ and $a = 8$ in $a^2 + b^2 = c^2$.
$$a^2 + b^2 = c^2$$
$$8^2 + b^2 = 17^2$$
$$64 + b^2 = 289$$
$$b^2 = 225 \quad \text{Subtract 64}$$
$$b = 15$$
Use only the positive square root for the length of a side of a triangle.
53. Let $c = 12$ and $b = 7$.
$$a^2 + b^2 = c^2$$
$$a^2 + 7^2 = 12^2$$
$$a^2 + 49 = 144$$
$$a^2 = 95$$
$$a = \sqrt{95}$$
57. $\sqrt[3]{125} = 5$ since $5^3 = 125$.
61. $\sqrt[3]{-8} = -2$ since $(-2)^3 = -8$.
65. There is no real number whose fourth power is -16, so $\sqrt[4]{-16}$ is not a real number.
69. $\sqrt[5]{-32} = -2$ since $(-2)^5 = -32$.
73. Use a calculator with a square root key. Enter 1.42 and push the square root key twice to get 1.092 (rounded).

Section 9.2 (page 451)

1. $\sqrt{8} \cdot \sqrt{2} = \sqrt{8 \cdot 2} = \sqrt{16} = 4$
5. $\sqrt{21} \cdot \sqrt{21} = \sqrt{21^2} = 21$
9. $\sqrt{27} = \sqrt{9 \cdot 3} = \sqrt{9} \cdot \sqrt{3} = 3\sqrt{3}$
13. $\sqrt{18} = \sqrt{9} \cdot \sqrt{2} = 3\sqrt{2}$
17. $\sqrt{125} = \sqrt{25} \cdot \sqrt{5} = 5\sqrt{5}$
21. $10\sqrt{27} = 10(3\sqrt{3}) = 30\sqrt{3}$
25. $\sqrt{27} \cdot \sqrt{48} = 3\sqrt{3} \cdot 4\sqrt{3}$
$= (3 \cdot 4)(\sqrt{3} \cdot \sqrt{3}) = 12(3) = 36;$
alternate solution: $\sqrt{27} \cdot \sqrt{48}$
$= \sqrt{27 \cdot 48} = \sqrt{1296} = 36$

29. $\sqrt{7} \cdot \sqrt{21} = \sqrt{7 \cdot 21} = \sqrt{147} = \sqrt{49 \cdot 3}$
$= \sqrt{49} \cdot \sqrt{3} = 7\sqrt{3}$
33. $\sqrt{80} \cdot \sqrt{15} = \sqrt{16} \cdot \sqrt{5} \cdot \sqrt{15}$
$= 4\sqrt{5} \cdot \sqrt{5} \cdot \sqrt{3} = 4 \cdot \sqrt{25} \cdot \sqrt{3}$
$= 4 \cdot 5 \cdot \sqrt{3} = 20\sqrt{3}$
37. $\sqrt{\frac{100}{9}} = \frac{\sqrt{100}}{\sqrt{9}} = \frac{10}{3}$
41. $\sqrt{\frac{5}{16}} = \frac{\sqrt{5}}{\sqrt{16}} = \frac{\sqrt{5}}{4}$
45. First multiply $\frac{1}{5}$ and $\frac{4}{5}$, getting $\frac{4}{25}$. The square root of $\frac{4}{25}$ is $\frac{2}{5}$.
49. Divide 3 into 75, getting 25. The square root of 25 is 5.
53. $\frac{15\sqrt{10}}{5\sqrt{2}} = \frac{15}{5} \cdot \frac{\sqrt{10}}{\sqrt{2}} = 3 \cdot \sqrt{\frac{10}{2}} = 3\sqrt{5}$
57. $\sqrt{y} \cdot \sqrt{y} = \sqrt{y^2} = y$ Absolute value bars are not necessary, since we assume $y > 0$.
61. $\sqrt{x^2} = x$ Absolute value bars are not necessary, since we assume $x > 0$.
65. $\sqrt{x^2y^4} = xy^2$ Absolute value bars are not necessary, since we assume $x > 0$.
69. $\sqrt{\frac{16}{x^2}} = \frac{\sqrt{16}}{\sqrt{x^2}} = \frac{4}{x}$ Absolute value bars are not necessary, since we assume $x > 0$.
73. $\sqrt[3]{40} = \sqrt[3]{8 \cdot 5} = \sqrt[3]{8} \cdot \sqrt[3]{5} = 2\sqrt[3]{5}$
77. $\sqrt[3]{128} = \sqrt[3]{64 \cdot 2} = \sqrt[3]{64} \cdot \sqrt[3]{2} = 4\sqrt[3]{2}$
81. $\sqrt[3]{\frac{8}{27}} = \frac{\sqrt[3]{8}}{\sqrt[3]{27}} = \frac{2}{3}$

Section 9.3 (page 457)

1. $2\sqrt{3} + 5\sqrt{3} = (2 + 5)\sqrt{3} = 7\sqrt{3}$
5. $\sqrt{6} + \sqrt{6} = 1 \cdot \sqrt{6} + 1 \cdot \sqrt{6} = 2\sqrt{6}$
9. $5\sqrt{7} - \sqrt{7} = 5\sqrt{7} - 1\sqrt{7} = (5 - 1)\sqrt{7} = 4\sqrt{7}$
13. $-\sqrt{12} + \sqrt{75} = -(\sqrt{4} \cdot \sqrt{3}) + \sqrt{25} \cdot \sqrt{3}$
$= -(2 \cdot \sqrt{3}) + 5 \cdot \sqrt{3} = (-2 + 5)\sqrt{3} = 3\sqrt{3}$
17. $-5\sqrt{32} + \sqrt{98} = -5 \cdot \sqrt{16} \cdot \sqrt{2} + \sqrt{49} \cdot \sqrt{2}$
$= -5 \cdot 4 \cdot \sqrt{2} + 7 \cdot \sqrt{2} = -20\sqrt{2} + 7\sqrt{2}$
$= -13\sqrt{2}$
21. $6\sqrt{5} + 3\sqrt{20} - 8\sqrt{45}$
$= 6\sqrt{5} + 3 \cdot \sqrt{4} \cdot \sqrt{5} - 8 \cdot \sqrt{9} \cdot \sqrt{5}$
$= 6\sqrt{5} + 3 \cdot 2\sqrt{5} - 8 \cdot 3 \cdot \sqrt{5}$
$= 6\sqrt{5} + 6\sqrt{5} - 24\sqrt{5} = -12\sqrt{5}$
25. $4\sqrt{50} + 3\sqrt{12} + 5\sqrt{45}$
$= 4\sqrt{25} \cdot \sqrt{2} + 3\sqrt{4} \cdot \sqrt{3} + 5\sqrt{9} \cdot \sqrt{5}$
$= 4 \cdot 5 \cdot \sqrt{2} + 3 \cdot 2 \cdot \sqrt{3} + 5 \cdot 3 \cdot \sqrt{5}$
$= 20\sqrt{2} + 6\sqrt{3} + 15\sqrt{5}$
29. $\frac{3}{5}\sqrt{75} - \frac{2}{3}\sqrt{45} = \frac{3}{5}\sqrt{25} \cdot \sqrt{3} - \frac{2}{3}\sqrt{9} \cdot \sqrt{5}$
$= \frac{3}{5} \cdot 5\sqrt{3} - \frac{2}{3} \cdot 3\sqrt{5}$
$= 3\sqrt{3} - 2\sqrt{5}$
33. $\sqrt{6} \cdot \sqrt{2} + 3\sqrt{3} = \sqrt{12} + 3\sqrt{3}$
$= \sqrt{4} \cdot \sqrt{3} + 3\sqrt{3}$
$= 2\sqrt{3} + 3\sqrt{3}$
$= (2 + 3)\sqrt{3} = 5\sqrt{3}$

37. $3\sqrt[3]{24} + 6\sqrt[3]{81} = 3\sqrt[3]{8} \cdot \sqrt[3]{3} + 6\sqrt[3]{27} \cdot \sqrt[3]{3}$
$= 3 \cdot 2\sqrt[3]{3} + 6 \cdot 3\sqrt[3]{3}$
$= 6\sqrt[3]{3} + 18\sqrt[3]{3}$
$= (6 + 18)\sqrt[3]{3} = 24\sqrt[3]{3}$

41. $\sqrt{9x} + \sqrt{49x} - \sqrt{16x}$
$= \sqrt{9}\sqrt{x} + \sqrt{49}\sqrt{x} - \sqrt{16}\sqrt{x}$
$= 3\sqrt{x} + 7\sqrt{x} - 4\sqrt{x}$
$= (3 + 7 - 4)\sqrt{x} = 6\sqrt{x}$

45. $\sqrt{75x^2} + x\sqrt{300}$
$= \sqrt{25} \cdot \sqrt{x^2} \cdot \sqrt{3} + x \cdot \sqrt{100} \cdot \sqrt{3}$
$= 5x\sqrt{3} + 10x\sqrt{3}$
$= (5x + 10x)\sqrt{3} = 15x\sqrt{3}$

49. $6\sqrt[3]{8p^2} - 2\sqrt[3]{27p^2} = 6\sqrt[3]{8}\sqrt[3]{p^2} - 2\sqrt[3]{27}\sqrt[3]{p^2}$
$= 6 \cdot 2\sqrt[3]{p^2} - 2 \cdot 3\sqrt[3]{p^2}$
$= 12\sqrt[3]{p^2} - 6\sqrt[3]{p^2}$
$= (12 - 6)\sqrt[3]{p^2} = 6\sqrt[3]{p^2}$

Section 9.4 (page 463)

1. Multiply numerator and denominator by $\sqrt{5}$, getting a final answer of $\dfrac{6\sqrt{5}}{5}$.

5. $\dfrac{3}{\sqrt{7}} = \dfrac{3}{\sqrt{7}} \cdot \dfrac{\sqrt{7}}{\sqrt{7}} = \dfrac{3\sqrt{7}}{7}$

9. Multiply numerator and denominator by $\sqrt{3}$ and then write in lowest terms, getting $\dfrac{\sqrt{30}}{2}$.

13. Here it is only necessary to multiply by $\sqrt{2}$ (and not $\sqrt{50}$), giving an answer of $\dfrac{3\sqrt{2}}{10}$.

17. $\dfrac{9}{\sqrt{32}} = \dfrac{9}{\sqrt{32}} \cdot \dfrac{\sqrt{2}}{\sqrt{2}} = \dfrac{9\sqrt{2}}{\sqrt{64}} = \dfrac{9\sqrt{2}}{8}$

21. $\dfrac{\sqrt{10}}{\sqrt{5}} = \dfrac{\sqrt{10}}{\sqrt{5}} \cdot \dfrac{\sqrt{5}}{\sqrt{5}} = \dfrac{\sqrt{50}}{5} = \dfrac{\sqrt{25} \cdot \sqrt{2}}{5}$
$= \dfrac{5\sqrt{2}}{5} = \sqrt{2}$

25. Multiply numerator and denominator, inside the radical, by 2, giving $\sqrt{\dfrac{2}{4}}$, or $\dfrac{\sqrt{2}}{\sqrt{4}}$, or $\dfrac{\sqrt{2}}{2}$.

29. $\sqrt{\dfrac{9}{5}} = \dfrac{\sqrt{9}}{\sqrt{5}} = \dfrac{3}{\sqrt{5}} \cdot \dfrac{\sqrt{5}}{\sqrt{5}} = \dfrac{3\sqrt{5}}{5}$

33. $\sqrt{\dfrac{3}{4}} \cdot \sqrt{\dfrac{1}{5}} = \sqrt{\dfrac{3}{4} \cdot \dfrac{1}{5}} = \sqrt{\dfrac{3}{4 \cdot 5}} = \dfrac{\sqrt{3}}{\sqrt{4 \cdot 5}}$
$= \dfrac{\sqrt{3}}{2\sqrt{5}} \cdot \dfrac{\sqrt{5}}{\sqrt{5}} = \dfrac{\sqrt{15}}{2 \cdot 5} = \dfrac{\sqrt{15}}{10}$

37. $\sqrt{\dfrac{2}{5}} \cdot \sqrt{\dfrac{3}{10}} = \sqrt{\dfrac{2}{5} \cdot \dfrac{3}{10}} = \sqrt{\dfrac{6}{50}} = \dfrac{\sqrt{6}}{\sqrt{50}} \cdot \dfrac{\sqrt{2}}{\sqrt{2}}$
$= \dfrac{\sqrt{12}}{\sqrt{100}} = \dfrac{\sqrt{4} \cdot \sqrt{3}}{10} = \dfrac{2\sqrt{3}}{10} = \dfrac{\sqrt{3}}{5}$

41. $\sqrt{\dfrac{6}{p}} = \dfrac{\sqrt{6}}{\sqrt{p}} \cdot \dfrac{\sqrt{p}}{\sqrt{p}} = \dfrac{\sqrt{6p}}{\sqrt{p^2}} = \dfrac{\sqrt{6p}}{p}$

45. $\sqrt{\dfrac{x^2}{4y}} = \dfrac{\sqrt{x^2}}{\sqrt{4y}} = \dfrac{x}{2\sqrt{y}} \cdot \dfrac{\sqrt{y}}{\sqrt{y}} = \dfrac{x\sqrt{y}}{2y}$

49. $\sqrt[3]{\dfrac{1}{2}} = \dfrac{\sqrt[3]{1}}{\sqrt[3]{2}}$ Quotient rule

$= \dfrac{1 \cdot \sqrt[3]{4}}{\sqrt[3]{2} \cdot \sqrt[3]{4}}$ Get the denominator to the perfect cube 8 by multiplying by $\sqrt[3]{4}$

$= \dfrac{\sqrt[3]{4}}{\sqrt[3]{8}} = \dfrac{\sqrt[3]{4}}{2}$ $\sqrt[3]{8} = 2$ since $2^3 = 8$

53. $\sqrt[3]{\dfrac{1}{11}} = \dfrac{\sqrt[3]{1}}{\sqrt[3]{11}}$ Quotient rule

$= \dfrac{1 \cdot \sqrt[3]{121}}{\sqrt[3]{11} \cdot \sqrt[3]{121}}$ Denominator becomes perfect cube 11^3 by multiplying by $\sqrt[3]{121}$

$= \dfrac{\sqrt[3]{121}}{11}$

57. $\sqrt[3]{\dfrac{3}{4}} = \dfrac{\sqrt[3]{3}}{\sqrt[3]{4}}$ Quotient rule

$= \dfrac{\sqrt[3]{3} \cdot \sqrt[3]{2}}{\sqrt[3]{4} \cdot \sqrt[3]{2}}$ Get the denominator to the perfect cube 8 by multiplying by $\sqrt[3]{2}$

$= \dfrac{\sqrt[3]{6}}{\sqrt[3]{8}}$

$= \dfrac{\sqrt[3]{6}}{2}$ $\sqrt[3]{8} = 2$ since $2^3 = 8$

Section 9.5 (page 469)

1. $3\sqrt{5} + 8\sqrt{45} = 3\sqrt{5} + 8(\sqrt{9 \cdot 5})$
$= 3\sqrt{5} + 8(\sqrt{9} \cdot \sqrt{5}) = 3\sqrt{5} + 8(3\sqrt{5})$
$= 3\sqrt{5} + 24\sqrt{5} = 27\sqrt{5}$

5. $\sqrt{2}(\sqrt{8} - \sqrt{32}) = \sqrt{2 \cdot 8} - \sqrt{2 \cdot 32}$
$= \sqrt{16} - \sqrt{64} = 4 - 8 = -4$

9. $2\sqrt{5}(\sqrt{2} + \sqrt{5}) = 2\sqrt{10} + 2\sqrt{25}$
$= 2\sqrt{10} + 2 \cdot 5 = 2\sqrt{10} + 10$

13. Use FOIL.
$(2\sqrt{6} + 3)(3\sqrt{6} - 5)$
$= (2\sqrt{6})(3\sqrt{6}) - 2\sqrt{6}(5) + 3(3\sqrt{6}) - 3(5)$
$= (2 \cdot 3)(\sqrt{6} \cdot \sqrt{6}) - 10\sqrt{6} + 9\sqrt{6} - 15$
$= 6(6) - \sqrt{6} - 15 = 21 - \sqrt{6}$

17. Use FOIL.
$(3\sqrt{2} + 4)(3\sqrt{2} + 4)$
$= 3\sqrt{2} \cdot 3\sqrt{2} + 3\sqrt{2} \cdot 4 + 4 \cdot 3\sqrt{2} + 4 \cdot 4$
$= 9\sqrt{4} + 12\sqrt{2} + 12\sqrt{2} + 16$
$= 9 \cdot 2 + 24\sqrt{2} + 16$
$= 18 + 24\sqrt{2} + 16$
$= 34 + 24\sqrt{2}$

21. Use the pattern $(a - b)(a + b) = a^2 - b^2$.
$(3 - \sqrt{2})(3 + \sqrt{2})$
$= 3^2 - (\sqrt{2})^2 = 9 - 2 = 7$

25. Use the pattern $(a - b)(a + b) = a^2 - b^2$.
$(\sqrt{6} - \sqrt{5})(\sqrt{6} + \sqrt{5}) = (\sqrt{6})^2 - (\sqrt{5})^2$
$= 6 - 5 = 1$

29. Use FOIL.
$(\sqrt{8} - \sqrt{2})(\sqrt{2} + \sqrt{4})$
$= \sqrt{8} \cdot \sqrt{2} + \sqrt{8} \cdot \sqrt{4} - \sqrt{2} \cdot \sqrt{2}$
$ - \sqrt{2} \cdot \sqrt{4}$
$= \sqrt{16} + \sqrt{32} - \sqrt{4} - \sqrt{8}$
$= 4 + 4\sqrt{2} - 2 - 2\sqrt{2} = 2 + 2\sqrt{2}$

Solutions to Selected Exercises

33. Factor the numerator as $5(\sqrt{7} - 2)$. Then divide numerator and denominator by the common factor, 5, to get $\sqrt{7} - 2$.
37. Factor the numerator as $2(6 - \sqrt{10})$. Then divide numerator and denominator by the common factor, 2, to get $\dfrac{6 - \sqrt{10}}{2}$.
41. Multiply numerator and denominator by $2 - \sqrt{5}$. The numerator becomes $5(2 - \sqrt{5})$, and the denominator becomes $4 - 5$ or -1. The quotient is $-5(2 - \sqrt{5})$ or $-10 + 5\sqrt{5}$.
45. Multiply numerator and denominator by $1 - \sqrt{2}$. The numerator becomes $\sqrt{2}(1 - \sqrt{2}) = \sqrt{2} - 2$. The denominator becomes $1 - 2$ or -1. The quotient is $-\sqrt{2} + 2$.
49. Multiply numerator and denominator by $\sqrt{3} - 1$. The numerator becomes $\sqrt{12}(\sqrt{3} - 1) = \sqrt{36} - \sqrt{12} = 6 - 2\sqrt{3}$. The denominator becomes $3 - 1$, or 2. Factor the numerator to get $2(3 - \sqrt{3})$. Divide numerator and denominator by the common factor, 2, to get $3 - \sqrt{3}$ for the quotient.
53. Multiply numerator and denominator by $\sqrt{3} + 1$. The numerator becomes $(\sqrt{2} + 3)(\sqrt{3} + 1) = \sqrt{6} + \sqrt{2} + 3\sqrt{3} + 3$, using FOIL. The denominator becomes $3 - 1 = 2$. The quotient is $\dfrac{\sqrt{6} + \sqrt{2} + 3\sqrt{3} + 3}{2}$.

Section 9.6 (page 477)

1. Square both sides of $\sqrt{x} = 2$ to get $x = 4$. Check in the original equation; $\sqrt{4} = 2$ is true. The solution is 4.
5. Square both sides of $\sqrt{t - 3} = 2$ to get $t - 3 = 4$ or $t = 7$. Check this solution in the original equation: $\sqrt{7 - 3} = 2$ is true. The solution is 7.
9. Square both sides of $\sqrt{m + 5} = 0$ to get $m + 5 = 0$ or $m = -5$. Check this solution in the original equation: $\sqrt{-5 + 5} = 0$ is true. The solution is -5.
13. To get the radical alone on one side of the equation, add 2 to both sides to get $\sqrt{k} = 7$. Square both sides: $k = 49$. Check this solution in the original equation: $\sqrt{49} - 2 = 5$ or $7 - 2 = 5$, which is true. The solution is 49.
17. Square both sides to get $5t - 9 = 4t$ (be careful on the right side). Solve this equation and check to see that the answer is 9.
21. Square both sides to get $5y - 5 = 4y + 1$. Solve this equation to get $y = 6$. Check this solution in the original equation.
25. Square both sides to get $p^2 = p^2 - 3p - 12$. Subtract p^2 on both sides to get $0 = -3p - 12$. Solve this equation; you should get $p = -4$. Check this solution in the original equation: $-4 = \sqrt{16 + 12 - 12}$ or $-4 = 4$, which is false. There is no solution.

29. Square both sides of the equation. On the left you get $2x + 1$, and on the right you get $(x - 7)^2 = x^2 - 14x + 49$. The new equation is $2x + 1 = x^2 - 14x + 49$. Make the equation equal to 0 by adding $-2x$ and -1 to both sides. This gives $x^2 - 16x + 48 = 0$. Factor this equation as $(x - 12)(x - 4) = 0$. Solve each of these equations: $x - 12 = 0$ gives $x = 12$, and $x - 4 = 0$ gives $x = 4$. Now go back to the original equation: $x = 12$ gives a true statement, but $x = 4$ does not. The only solution is 12.
33. Add 1 to each side to get $\sqrt{x + 1} = x + 1$. Square both sides, getting $x + 1 = x^2 + 2x + 1$. (Remember that $(x + 1)^2 = x^2 + 2x + 1$.) Set one side equal to 0 by adding $-x$ and -1 to both sides: $0 = x^2 + x$. Factor on the right: $0 = x(x + 1)$, from which $x = 0$ or $x = -1$. Check each solution in the original equation to see that both 0 and -1 are solutions.
37. Square both sides, to get $9(x + 13) = x^2 + 18x + 81$. Multiply on the left, and make the left side of the equation 0 by adding $-9x$ and -117 to both sides: $0 = x^2 + 9x - 36$. Factor on the right to get $0 = (x + 12)(x - 3)$, from which $x = -12$ or $x = 3$. Check both solutions in the original equation. Since -12 leads to a false statement and 3 leads to a true statement, the only solution is 3.
41. Add 4 to both sides, getting $\sqrt{3x} = x - 6$. Square both sides, to get $3x = x^2 - 12x + 36$. Add $-3x$ to both sides to get $0 = x^2 - 15x + 36$. Factor on the right, which gives $0 = (x - 3)(x - 12)$, from which $x = 3$ or $x = 12$. Check in the original equation. Since 3 leads to a false statement and 12 to a true statement, the only solution is 12.
45. Write an equation, letting x represent the unknown number: $3\sqrt{2} = \sqrt{x + 10}$. Square both sides to get $9 \cdot 2 = x + 10$ or $18 = x + 10$, from which $x = 8$. Check it in the words of the original problem. The solution is 8.

CHAPTER 10

Section 10.1 (page 489)

1. Take the square root of each side to get 5 and -5.
5. Take the square root of each side to get $\sqrt{13}$ and $-\sqrt{13}$.
9. Simplify $\sqrt{24}$: $\sqrt{24} = \sqrt{4 \cdot 6} = \sqrt{4} \cdot \sqrt{6} = 2\sqrt{6}$. The two solutions are $2\sqrt{6}$ and $-2\sqrt{6}$.
13. By the square root property,
$$k = \sqrt{\dfrac{9}{16}} \quad \text{or} \quad k = -\sqrt{\dfrac{9}{16}}.$$
Since $\sqrt{\dfrac{9}{16}} = \dfrac{3}{4}$, the solutions are $\dfrac{3}{4}$ and $-\dfrac{3}{4}$.

17. By the square root property,
$$k = \sqrt{2.56} \quad \text{or} \quad k = -\sqrt{2.56}.$$
Since $\sqrt{2.56} = 1.6$, the solutions are 1.6 and -1.6.

21. By the square root property,
$$x - 2 = \sqrt{16} \quad \text{or} \quad x - 2 = -\sqrt{16}.$$
Since $\sqrt{16} = 4$,
$$x - 2 = 4 \quad \text{or} \quad x - 2 = -4$$
$$x = 6 \quad \text{or} \quad x = -2.$$

25. By the square root property,
$$x - 1 = \sqrt{32} \quad \text{or} \quad x - 1 = -\sqrt{32}.$$
Then
$$x = 1 + \sqrt{32} \text{ or } x = 1 - \sqrt{32}.$$
Simplify $\sqrt{32}$ as $\sqrt{32} = \sqrt{16 \cdot 2} = \sqrt{16} \cdot \sqrt{2} = 4\sqrt{2}$, so the solutions are $1 + 4\sqrt{2}$ and $1 - 4\sqrt{2}$.

29. Use the square root property to get
$$3z + 5 = \sqrt{9} \quad \text{or} \quad 3z + 5 = -\sqrt{9}$$
$$3z + 5 = 3 \quad \text{or} \quad 3z + 5 = -3$$
$$3z = -2 \quad \text{or} \quad 3z = -8$$
$$z = -\frac{2}{3} \quad \text{or} \quad z = -\frac{8}{3}.$$

33. Use the square root property to get $3p - 1 = \sqrt{18}$ or $3p - 1 = -\sqrt{18}$. We can simplify $\sqrt{18}$ as $3\sqrt{2}$. Add 1 to both sides of each equation to give $3p = 1 + 3\sqrt{2}$ and $3p - 1 - 3\sqrt{2}$. Divide both sides of each equation by 3 to end up with the solutions, $\dfrac{1 + 3\sqrt{2}}{3}$ and $\dfrac{1 - 3\sqrt{2}}{3}$.

37. Use the square root property to get $3m + 4 = \sqrt{8}$ or $3m + 4 = -\sqrt{8}$. Add -4 to both sides of each equation to get $3m = -4 + \sqrt{8}$ or $3m = -4 - \sqrt{8}$. Simplify $\sqrt{8}$ as $\sqrt{8} = \sqrt{4} \cdot \sqrt{2} = 2\sqrt{2}$, giving $3m = -4 + 2\sqrt{2}$ and $3m = -4 - 2\sqrt{2}$. Finally, divide both sides of each equation by 3 to get the solutions, $\dfrac{-4 + 2\sqrt{2}}{3}$ and $\dfrac{-4 - 2\sqrt{2}}{3}$.

41. Use the square root property to get
$$2.11p + 3.42 = \sqrt{9.58} \quad \text{or} \quad 2.11p + 3.42 = -\sqrt{9.58}.$$
Round to the nearest hundredth to get $\sqrt{9.58} = 3.10$. Then $2.11p + 3.42 = 3.10$ or $2.11p + 3.42 = -3.10$. Solve each equation.
$$2.11p = -.32 \quad \text{or} \quad 2.11p = -6.52$$
$$p = -.15 \quad \text{or} \quad p = -3.09$$

45. Since the coin falls 4 feet, let $d = 4$ to get
$$4 = 16t^2.$$
Divide both sides by 16 to get
$$\frac{1}{4} = t^2.$$
Use the square root property to get
$$\frac{1}{2} = t \quad \text{or} \quad -\frac{1}{2} = t.$$
In this problem, time can't be negative, so the solution is $\dfrac{1}{2}$ second.

Section 10.2 (page 495)

1. Take half of 2, which is 1. Square this to get the answer: $1^2 = 1$.

5. Half of 3 is $\dfrac{3}{2}$ and $\left(\dfrac{3}{2}\right)^2 = \dfrac{9}{4}$, so $\dfrac{9}{4}$ should be added.

9. Half of 4 is 2, and $2^2 = 4$, so add 4 to each side.
$$x^2 + 4x + 4 = -3 + 4$$
$$x^2 + 4x + 4 = 1$$
Factor on the left.
$$(x + 2)^2 = 1$$
Now use the square root property to get
$$x + 2 = 1 \quad \text{or} \quad x + 2 = -1$$
$$x = -1 \quad \text{or} \quad x = -3.$$
The solutions are -1 and -3.

13. Half of -8 is -4, and $(-4)^2 = 16$. Add 16 to each side.
$$q^2 - 8q + 16 = -16 + 16$$
Factor on the left.
$$(q - 4)^2 = 0$$
Use the square root property. (Since $\sqrt{0} = 0$ and $-0 = 0$, there is only one equation.)
$$q - 4 = 0$$
$$q = 4$$
The solution is 4.

17. Add 3 on each side to get
$$k^2 + 5k = 3.$$
Half of 5 is $\dfrac{5}{2}$, and $\left(\dfrac{5}{2}\right)^2 = \dfrac{25}{4}$.
Add $\dfrac{25}{4}$ on each side.
$$k^2 + 5k + \frac{25}{4} = 3 + \frac{25}{4}$$
Factor.
$$\left(k + \frac{5}{2}\right)^2 = \frac{37}{4}$$
Use the square root property.
$$k + \frac{5}{2} = \sqrt{\frac{37}{4}} \quad \text{or} \quad k + \frac{5}{2} = -\sqrt{\frac{37}{4}}$$
Simplify the radical and then solve each equation.
$$k + \frac{5}{2} = \frac{\sqrt{37}}{2} \quad \text{or} \quad k + \frac{5}{2} = -\frac{\sqrt{37}}{2}$$
$$k = -\frac{5}{2} + \frac{\sqrt{37}}{2} \quad \text{or} \quad k = -\frac{5}{2} - \frac{\sqrt{37}}{2}$$
Add.
$$k = \frac{-5 + \sqrt{37}}{2} \quad \text{or} \quad k = \frac{-5 - \sqrt{37}}{2}$$

21. Divide through by 2 to get the coefficient of m^2 equal to 1.
$$2m^2 - 4m - 5 = 0$$
$$m^2 - 2m - \frac{5}{2} = 0$$
Get the variable terms alone on one side.
$$m^2 - 2m = \frac{5}{2}$$
Take $\dfrac{1}{2}$ the middle term: $\dfrac{1}{2}(2) = 1$.

Solutions to Selected Exercises

Square 1: $(1)^2 = 1$. Add 1 to both sides.
$$m^2 - 2m + 1 = \frac{5}{2} + 1$$
Factor the left side. Rewrite the right side with a common denominator.
$$(m - 1)^2 = \frac{5}{2} + \frac{2}{2}$$
$$(m - 1)^2 = \frac{7}{2}$$
Take the square root of each side.
$$m - 1 = \sqrt{\frac{7}{2}} \text{ or } m - 1 = -\sqrt{\frac{7}{2}}$$
Since $\sqrt{\frac{7}{2}} = \frac{\sqrt{7}}{\sqrt{2}} = \frac{\sqrt{7} \cdot \sqrt{2}}{\sqrt{2} \cdot \sqrt{2}} = \frac{\sqrt{14}}{2}$,
$$m - 1 = \frac{\sqrt{14}}{2} \text{ or } m - 1 = -\frac{\sqrt{14}}{2}.$$
Add 1 to both sides in each equation.
$$m = 1 + \frac{\sqrt{14}}{2} \text{ or } m = 1 - \frac{\sqrt{14}}{2}$$
Write 1 as $\frac{2}{2}$, then add.
$$m = \frac{2}{2} + \frac{\sqrt{14}}{2} \text{ or } m = \frac{2}{2} - \frac{\sqrt{14}}{2}$$
$$m = \frac{2 + \sqrt{14}}{2} \text{ or } m = \frac{2 - \sqrt{14}}{2}$$
The solutions are $\frac{2 + \sqrt{14}}{2}$ or $\frac{2 - \sqrt{14}}{2}$.

25. Add $-6r$ to both sides to get $3r^2 - 2 - 6r = 3$. Add 2 to both sides, getting $3r^2 - 6r = 5$. Divide by 3 on both sides to get $r^2 - 2r = \frac{5}{3}$. Take half of 2, which is 1, and square it to get 1. Add 1 to both sides, getting $r^2 - 2r + 1 = 1 + \frac{5}{3}$. Factor on the left and add on the right to get $(r - 1)^2 = \frac{3}{3} + \frac{5}{3}$ $= \frac{8}{3}$. Use the square root property on $(r - 1)^2 = \frac{8}{3}$ to get $r - 1 = \sqrt{\frac{8}{3}}$ or $r - 1 = -\sqrt{\frac{8}{3}}$. Simplify $\sqrt{\frac{8}{3}}$, getting $\frac{2\sqrt{2}}{\sqrt{3}} = \frac{2\sqrt{6}}{3}$. Replace $\sqrt{\frac{8}{3}}$ with this result and add 1 to both sides of each equation to get $r = 1 + \frac{2\sqrt{6}}{3}$ and $r = 1 - \frac{2\sqrt{6}}{3}$. To combine terms on the right, first change 1 to $\frac{3}{3}$, getting $r = \frac{3 + 2\sqrt{6}}{3}$ or $r = \frac{3 - 2\sqrt{6}}{3}$.

29. Multiply on the left.
$$r^2 - 4 = 5,$$
or $r^2 = 9$
By the square root property,
$$r = 3 \text{ or } r = -3.$$

Section 10.3 (page 501)

1. The number in front of the x^2 term is 3, so that $a = 3$. The number in front of the x term is 4, so that $b = 4$. The number left over is -8, so that $c = -8$. (Note: One side of the equation was equal to 0 before we started.)

5. Since 0 is on one side, the value of a is 3, while $b = -8$. No constant term is given, so $c = 0$.

9. Multiply the expression on the left and combine terms to get $9x^2 + 9x - 26 = 0$, so $a = 9, b = 9$, and $c = -26$.

13. We have $a = 1, b = 4$, and $c = 4$. Using the quadratic formula, we get
$$y = \frac{-4 \pm \sqrt{4^2 - 4(1)(4)}}{2(1)} = \frac{-4 \pm \sqrt{16 - 16}}{2}$$
$$= \frac{-4 \pm 0}{2} = -2. \text{ The only solution is } -2.$$

17. Here $a = 2, b = 12$, and $c = 5$. In the quadratic formula this gives $w = \frac{-12 \pm \sqrt{12^2 - 4(2)(5)}}{2(2)}$
$$= \frac{-12 \pm \sqrt{144 - 40}}{4} = \frac{-12 \pm \sqrt{104}}{4}.$$
Simplify: $\sqrt{104} = 2\sqrt{26}$, so we have
$\frac{-12 \pm 2\sqrt{26}}{4}$. Factor out 2 in the numerator, and divide numerator and denominator by 2 to write $\frac{2(-6 \pm \sqrt{26})}{4}$ as $\frac{-6 \pm \sqrt{26}}{2}$. The solutions are
$$\frac{-6 + \sqrt{26}}{2} \text{ and } \frac{-6 - \sqrt{26}}{2}.$$

21. Add $-3x$ and -5 to both sides to get $2x^2 - 3x - 5 = 0$. Here $a = 2, b = -3$, and $c = -5$. By the quadratic formula,
$$x = \frac{-(-3) \pm \sqrt{(-3)^2 - 4(2)(-5)}}{2(2)}$$
$$= \frac{3 \pm \sqrt{9 + 40}}{4} = \frac{3 \pm \sqrt{49}}{4} = \frac{3 \pm 7}{4}. \text{ Now we}$$
can find the two answers. First, use the plus sign; x
$= \frac{3 + 7}{4} = \frac{10}{4} = \frac{5}{2}$. Then use the minus sign; x
$= \frac{3 - 7}{4} = \frac{-4}{4} = -1.$

25. In this equation $a = 4, b = -12$, and $c = 9$. Substitute into the quadratic formula.
$$p = \frac{-b \pm \sqrt{b^2 - 4ac}}{2a}$$
$$= \frac{-(-12) \pm \sqrt{(-12)^2 - 4(4)(9)}}{2 \cdot 4}$$
$$= \frac{12 \pm \sqrt{144 - 144}}{8}$$
$$= \frac{12 \pm 0}{8}$$
$$= \frac{12}{8} = \frac{3}{2}$$
The solution is $\frac{3}{2}$.

29. Write $6p^2 = 10p$ as $6p^2 - 10p + 0 = 0$. Now $a = 6$, $b = -10$, and $c = 0$. Substitute these values into the quadratic formula.
$$p = \frac{-(-10) \pm \sqrt{(-10)^2 - 4(6)(0)}}{2(6)}$$
$$= \frac{10 \pm \sqrt{100}}{12} = \frac{10 \pm 10}{12}$$
The solutions are $\frac{10 + 10}{12} = \frac{20}{12} = \frac{5}{3}$ and $\frac{10 - 10}{12} = \frac{0}{12} = 0$.

33. Write $9r^2 - 16 = 0$ as $9r^2 + 0r - 16 = 0$. Now $a = 9$, $b = 0$, and $c = -16$. Substitute these values into the quadratic formula.
$$r = \frac{-0 \pm \sqrt{(0)^2 - 4(9)(-16)}}{2(9)}$$
$$= \frac{\pm\sqrt{576}}{18} = \pm\frac{24}{18}$$
The solutions are $\frac{24}{18} = \frac{4}{3}$ and $\frac{-24}{18} = \frac{-4}{3}$.

37. Here $a = 2$, $b = 1$, and $c = 7$. By the quadratic formula, $x = \frac{-1 \pm \sqrt{1^2 - 4(2)(7)}}{2(2)}$
$$= \frac{-1 \pm \sqrt{1 - 56}}{4} = \frac{-1 \pm \sqrt{-55}}{4}.$$ Since $\sqrt{-55}$ is not a real number, this equation has no real number solution.

41. Multiply both sides by 6 to get $3x^2 = 6 - x$. Add x and -6 to both sides to get $3x^2 + x - 6 = 0$. Then $a = 3$, $b = 1$, and $c = -6$. By the quadratic formula,
$$x = \frac{-1 \pm \sqrt{1^2 - 4(3)(-6)}}{2(3)}$$
$$= \frac{-1 \pm \sqrt{1 + 72}}{6} = \frac{-1 \pm \sqrt{73}}{6}.$$
The solutions are $\frac{-1 + \sqrt{73}}{6}$ and $\frac{-1 - \sqrt{73}}{6}$.

45. Multiply both sides by 2 to get
$$r^2 = 2r + 1$$
or $\quad r^2 - 2r - 1 = 0$.
Let $a = 1$, $b = -2$, and $c = -1$, and use the quadratic formula.
$$r = \frac{-(-2) \pm \sqrt{(-2)^2 - 4(1)(-1)}}{2(1)}$$
$$= \frac{2 \pm \sqrt{4 + 4}}{2} = \frac{2 \pm 2\sqrt{2}}{2} = 1 \pm \sqrt{2}.$$
The solutions are $1 + \sqrt{2}$ and $1 - \sqrt{2}$.

49. Multiply both sides by 2 to get $m^2 = m - 2$. Add $-m$ and 2 to both sides, giving $m^2 - m + 2 = 0$. Use the quadratic formula, with $a = 1$, $b = -1$, and $c = 2$ to get
$$m = \frac{1 \pm \sqrt{(-1)^2 - 4(1)(2)}}{2(1)}$$
$$= \frac{1 \pm \sqrt{1 - 8}}{2} = \frac{1 \pm \sqrt{-7}}{2}.$$
Since $\sqrt{-7}$ is not a real number, there is no real number solution.

53. Rewrite $48 = 64t - 16t^2$ as
$$16t^2 - 64t + 48 = 0.$$
Divide both sides by 16 to get
$$t^2 - 4t + 3 = 0.$$
Then $a = 1$, $b = -4$, and $c = 3$. Use the quadratic formula to find t.
$$t = \frac{-(-4) \pm \sqrt{(-4)^2 - 4(1)(3)}}{2(1)}$$
$$= \frac{4 \pm \sqrt{16 - 12}}{2}$$
$$= \frac{4 \pm \sqrt{4}}{2} = \frac{4 \pm 2}{2}$$
Use the $+$ sign:
$$t = \frac{4 + 2}{2} = \frac{6}{2} = 3.$$
Use the $-$ sign:
$$t = \frac{4 - 2}{2} = \frac{2}{2} = 1.$$
Both times are reasonable, but experiments have shown that the actual time is 1 second.

Supplementary Exercises on Quadratic Equations (page 505)

1. This equation will not factor, so use the quadratic formula, with $a = 1$, $b = 3$, $c = 1$.
$$y = \frac{-b \pm \sqrt{b^2 - 4ac}}{2a}$$
$$= \frac{-3 \pm \sqrt{3^2 - 4(1)(1)}}{2 \cdot 1}$$
$$= \frac{-3 \pm \sqrt{9 - 4}}{2}$$
$$= \frac{-3 \pm \sqrt{5}}{2}$$
The solutions are $\frac{-3 + \sqrt{5}}{2}$ and $\frac{-3 - \sqrt{5}}{2}$.
Check each solution.

5. Use the square root property to get $\frac{7}{9}$ and $-\frac{7}{9}$.
Check each solution.

9. Factor to get
$$(x + 2)(x + 1) = 0$$
$x + 2 = 0 \quad$ or $\quad x + 1 = 0$
$x = -2 \quad$ or $\quad x = -1$.
Check each solution.

13. Use the square root property to get
$2p - 1 = \sqrt{10} \quad$ or $\quad 2p - 1 = -\sqrt{10}$
$2p = 1 + \sqrt{10} \quad$ or $\quad 2p = 1 - \sqrt{10}$
$p = \frac{1 + \sqrt{10}}{2} \quad$ or $\quad p = \frac{1 - \sqrt{10}}{2}$.

17. Use the square root property to get
$y + 6 = \sqrt{121} \quad$ or $\quad y + 6 = -\sqrt{121}$.
Since $\sqrt{121} = 11$,
$y + 6 = 11 \quad$ or $\quad y + 6 = -11$,
$y = 5 \quad$ or $\quad y = -17$.

21. Since $\sqrt{-1}$ is not a real number, the equation has no real number solutions.

25. Rewrite $2p^2 + 1 = p$ as $2p^2 - p + 1 = 0$. Then $a = 2$, $b = -1$, and $c = 1$. Use the quadratic formula.
$$p = \frac{-(-1) \pm \sqrt{(-1)^2 - 4(2)(1)}}{2(2)}$$
$$p = \frac{1 \pm \sqrt{1 - 8}}{4} = \frac{1 \pm \sqrt{-7}}{4}$$
Since $\sqrt{-7}$ is not a real number, this equation has no real number solutions.

29. Let $a = 3$, $b = 5$, and $c = 1$. The quadratic formula gives
$$p = \frac{-5 \pm \sqrt{5^2 - 4(3)(1)}}{2(3)}$$
$$p = \frac{-5 \pm \sqrt{25 - 12}}{6} = \frac{-5 \pm \sqrt{13}}{6}.$$
The solutions are
$$\frac{-5 + \sqrt{13}}{6} \text{ and } \frac{-5 - \sqrt{13}}{6}.$$

33. Write $5k^2 + 8 = 22k$ as
$$5k^2 - 22k + 8 = 0.$$
$(5k - 2)(k - 4) = 0$ Factor
$5k - 2 = 0$ or $k - 4 = 0$ Solve each
$5k = 2$ for k
$k = \frac{2}{5}$ or $k = 4$

The solutions are $\frac{2}{5}$ and 4.

37. Write $4x^2 = 5x - 1$ as
$$4x^2 - 5x + 1 = 0.$$
$(4x - 1)(x - 1) = 0$ Factor
$4x - 1 = 0$ or $x - 1 = 0$ Solve each
$4x = 1$ for x
$x = \frac{1}{4}$ or $x = 1$

The solutions are $\frac{1}{4}$ and 1.

41. Here $a = 15$, $b = 58$, and $c = 48$. Use the quadratic formula.
$$t = \frac{-58 \pm \sqrt{58^2 - 4(15)(48)}}{2(15)}$$
$$= \frac{-58 \pm \sqrt{3364 - 2880}}{30}$$
$$= \frac{-58 \pm \sqrt{484}}{30}$$
$$t = \frac{-58 \pm 22}{30}$$
Use + to get
$$\frac{-58 + 22}{30} = \frac{-36}{30} = -\frac{6}{5}.$$
Use − to get
$$\frac{-58 - 22}{30} = \frac{-80}{30} = \frac{-8}{3}.$$
[This equation also could have been solved by factoring, since $15t^2 + 58t + 48 = (5t + 6)(3t + 8)$.]

45. Here $a = 1$, $b = -3$, and $c = -3$. By the quadratic formula,
$$k = \frac{3 \pm \sqrt{(-3)^2 - 4(1)(-3)}}{2(1)}$$
$$= \frac{3 \pm \sqrt{9 + 12}}{2}$$
$$= \frac{3 \pm \sqrt{21}}{2}.$$
The solutions are
$$\frac{3 + \sqrt{21}}{2} \text{ and } \frac{3 - \sqrt{21}}{2}.$$

49. Rewrite the equation as
$$5k^2 + 17k - 12 = 0.$$
Then $a = 5$, $b = 17$, $c = -12$. By the quadratic formula,
$$k = \frac{-17 \pm \sqrt{17^2 - 4(5)(-12)}}{2(5)}$$
$$= \frac{-17 \pm \sqrt{289 + 240}}{10}$$
$$= \frac{-17 \pm \sqrt{529}}{10}$$
$$= \frac{-17 \pm 23}{10}.$$
Use + to get
$$k = \frac{-17 + 23}{10} = \frac{6}{10} = \frac{3}{5}.$$
Use − to get
$$k = \frac{-17 - 23}{10} = \frac{-40}{10} = -4.$$

53. $k^2 + \frac{4}{15}k = \frac{4}{15}$
$15k^2 + 4k = 4$ Multiply both sides by 15
$15k^2 + 4k - 4 = 0$ Get 0 on one side
$(3k + 2)(5k - 2) = 0$ Factor
$3k + 2 = 0$ or $5k - 2 = 0$
$3k = -2$ $5k = 2$
$k = -\frac{2}{3}$ $k = \frac{2}{5}$

Section 10.4 (page 511)

All the graphs in this section may be found in the answer section.

1. The graph of $y = 2x^2$ is a parabola with vertex at $(0, 0)$. Because of the 2, the graph is narrower than that of $y = x^2$ and opens upward. Some additional points on the graph are $(1, 2)$, $(-1, 2)$, $(2, 8)$, and $(-2, 8)$.

5. The graph of $y = -(x + 1)^2$ is a parabola with vertex at $(-1, 0)$. The coefficient of x^2 is -1, indicating that the graph opens downward and has the same shape as that of $y = x^2$. Additional points on the graph are $(-3, -4)$, $(-2, -1)$, $(0, -1)$, and $(1, -4)$.

9. The graph of $y = 2 - x^2$ is a parabola with vertex at (0, 2). Since the coefficient of x^2 is -1, the graph opens downward and has the same shape as that of $y = x^2$. Additional points on the graph are $(-2, -2), (-1, 1), (1, 1)$, and $(2, -2)$.
13. The graph of $y = (x + 1)^2 + 2$ is a parabola with vertex at $(-1, 2)$. The graph opens upward and has the same shape as that of $y = x^2$. Some additional points on the graph are $(-3, 6)$, $(-2, 3), (0, 3)$, and $(1, 6)$.
17. The graph of $y = 1 - (x + 2)^2$ is a parabola with vertex at $(-2, 1)$. The graph opens downward and has the same shape as that of $y = x^2$. Additional points on the graph are $(-4, -3), (-3, 0)$, $(-1, 0)$, and $(0, -3)$.

APPENDICES

Appendix A (page 527)

1. The natural numbers less than 8 are the counting numbers 1, 2, 3, 4, 5, 6, and 7. Remember to use set braces in the answer.
5. There have been no women presidents of the United States, so this set is empty: ∅.
9. The set of positive even numbers is the infinite set $\{2, 4, 6, 8, \ldots\}$.
13. Since 5 is an element of $\{1, 2, 5, 8\}$, the statement is true.
17. Since 7 is not an element of $\{2, 4, 6, 8\}$, the statement is true.
21. True, since every element of set A is in set U.
25. True, because every element in set C is also in set A.
29. True, since the numbers 1 and 2 belong to set D, but not to set E.
33. False; set C has 4 elements and therefore has $2^4 = 16$ subsets, not 12 subsets.
37. True. The intersection of the given sets is the set $\{0\}$.
41. True. Every element of the first set and every element of the second set are in the union.
45. A', the complement of A, contains the elements of U that are not in A: $A' = \{g, h\}$.
49. The intersection of sets A and B contains only those elements that are in both A and B: a, c, and e.
53. The intersection of sets B and C contains the elements that are in both B and C, so a is the only element in $B \cap C$.
57. The union of C and B contains every element in C and every element in B, so the elements are a, c, e, and f.
61. Disjoint sets have no elements in common, so sets B and D are disjoint, and sets C and D are disjoint.

Appendix B (page 533)

1. $\sqrt{-9} = \sqrt{9 \cdot -1} = \sqrt{9} \cdot \sqrt{-1} = 3i$
5. $\sqrt{-36} = \sqrt{36 \cdot -1} = \sqrt{36} \cdot \sqrt{-1} = 6i$
9. Since $-3i$ is a multiple of i, it is imaginary.
13. $2x + 5i = 1 - yi$
$2x = 1$ and $5i = -yi$
$x = \frac{1}{2}$ $\quad 5 = -y$
$\quad y = -5$
17. $-7x + 2yi = -8i = 0 - 8i$
$-7x = 0$ and $2yi = -8i$
$\quad\quad\quad\quad\quad\quad 2y = -8$
$x = 0$ $\quad\quad\quad y = -4$
21. $(16 + 5i) + (2 - 7i) = (16 + 2) + (5 - 7)i$
$= 18 + (-2)i = 18 - 2i$
25. $(6 + 7i) - (2i + 5) = (6 + 7i) - (5 + 2i)$
$= (6 - 5) + (7 - 2)i = 1 + 5i$
29. $-i(2 - 7i) = -2i + 7i^2 = -2i + 7(-1)$
$= -2i - 7 = -7 - 2i$
33. $(5 - 4i)(3 - 2i) = 5(3) - 5(2i) - (4i)(3)$
$\quad\quad\quad\quad\quad\quad\quad - (4i)(-2i)$
$= 15 - 10i - 12i + 8i^2$
$= 15 - 22i + 8(-1)$
$= 7 - 22i$
39. $\dfrac{3 - 4i}{2 + 2i} \cdot \dfrac{2 - 2i}{2 - 2i}$ Multiply by the conjugate of the denominator
$= \dfrac{6 - 6i - 8i + 8i^2}{4 - 4i^2} = \dfrac{6 - 14i + 8(-1)}{4 - 4(-1)}$
$= \dfrac{-2 - 14i}{8} = \dfrac{-2(1 + 7i)}{8}$
$= \dfrac{-(1 + 7i)}{4} = \dfrac{-1 - 7i}{4} = -\dfrac{1}{4} - \dfrac{7}{4}i$
43. $m^2 - 2m + 2$; $a = 1, b = -2, c = 2$
$m = \dfrac{-(-2) \pm \sqrt{(-2)^2 - 4(1)(2)}}{2(1)}$
$= \dfrac{2 \pm \sqrt{4 - 8}}{2} = \dfrac{2 \pm \sqrt{-4}}{2}$
$= \dfrac{2 \pm 2i}{2} = 1 \pm i$
The solutions are $1 + i$ and $1 - i$.
47. $2r^2 + 5 = 4r$
$2r^2 - 4r + 5 = 0$ Get 0 on one side
$a = 2, b = -4, c = 5$
$r = \dfrac{-(-4) \pm \sqrt{(-4)^2 - 4(2)(5)}}{2(2)}$
$= \dfrac{4 \pm \sqrt{16 - 40}}{4}$
$= \dfrac{4 \pm \sqrt{-24}}{4}$
$= \dfrac{4 \pm 2i\sqrt{6}}{4} = \dfrac{2 \pm i\sqrt{6}}{2}$
The solutions are $\dfrac{2 + i\sqrt{6}}{2}$ and $\dfrac{2 - i\sqrt{6}}{2}$.

Index

Absolute value, 47
Addition
  of fractions, 4
  of like terms, 187
  of polynomials, 189
  of radicals, 455–56
  of rational expressions, 295
  of real numbers, 51
Addition method for linear systems, 397
Addition property of equality, 104
Addition property of inequality, 147
Additive inverse, 46
Algebraic expression, 29
Associative properties, 79
Axes of a coordinate system, 341

Base of an exponential expression, 169
Binomial, 188
  product of binomials, 201
  square of, 203
Brackets, 25

Coefficient, 87, 187
Common denominator, 4
Common factor, 229
Commutative properties, 79
Complement of a set, 524
Completing the square, 491
Complex fraction, 301
Complex numbers, 529
  division of, 531
  multiplication of, 531
Composite number, 228
Conjugates, 467
  of complex numbers, 532
Consecutive integers, 125, 264
Consistent system, 392
Coordinate, 341
Coordinate system, 341, 507
Cross product, 140
Cube root, 444.
    See table on page 536
Cubes, sum and difference of, 251

Decimal numbers, 9
  addition of, 10
  division of, 11
  multiplication of, 10
  and percents, 13
  repeating, 13
  subtraction of, 10
Degree of a polynomial, 188
Denominator, 1
  least common, 4
  rationalizing, 459
Dependent system, 392
Descending powers, 188
Difference, 5
Difference of two cubes, 251
Difference of two squares, 249
Direct variation, 319
Disjoint sets, 526
Distributive property, 81
Division
  of complex numbers, 531
  of decimal numbers, 11
  of fractions, 3
  of polynomials, 211
  of radicals, 448
  of rational expressions, 286
  of real numbers, 71
Domain of a function, 31
Double negative rule, 46

Element of a set, 523
Empty set, 523
Equality, multiplication property of, 109
Equal sets, 524
Equation, 31
  linear, 103
  quadratic, 257, 487
  with radicals, 473
  with rational expressions, 307
  solution of, 31.
    See also System of equations
Expanded form, 9
Exponent, 23, 169
  negative, 171
  power rule, 177
  product rule, 170
  quotient rule, 173
  rules for, 174, 179
  and scientific notation, 183–84
  zero, 171
Exponential expression, 169

Factor, 1, 23, 74, 227
  common, 229
  zero-factor property, 257
Factored form, 227
  of a polynomial, 235
  prime, 228
Factoring by grouping, 231–32
Finite set, 523
First-degree equation. *See* Linear equation
FOIL method, 201
Formulas, 131
  quadratic formula, 498.
    See lists of formulas inside front and back covers
Fourth root, 444
Fractions, 1
  addition of, 4
  complex, 301
  division of, 3
  improper, 1
  and mixed numbers, 1
  multiplication of, 2
  proper, 1
  subtraction of, 6
Fundamental property of rational expressions, 280

Graph
  of a linear equation, 347
  of a linear inequality, 373
  of a quadratic equation, 507
Graphical solution
  of a system of linear equations, 390–92
  of a system of linear inequalities, 425
"Greater than" symbol, 19
Greatest common factor, 229
Grouping, factoring by, 231–32

Horizontal line, 351

Identity element, 80
Identity properties, 80
Imaginary numbers, 529
Improper fraction, 1
Inconsistent system, 392
Inequalities, 147
Infinite set, 523
Inner product, 201
Integers, 44
  consecutive, 125, 264
Intercepts, 349
Intersection, 525
Inverse, additive, 46
Inverse properties, 80
Inverse variation, 320
Irrational numbers, 44, 442

Least common denominator, 4, 271
"Less than" symbol, 19
Like radicals, 455
Like terms, 87
Linear equation, 103
  in two variables, 335
  solving, 115
Linear inequality, 373
Linear system, 389
Lowest terms, 2

Mixed number, 1
Mixture problems, 419
Monomial, 188
Multiplication
  of complex numbers, 531
  of decimal numbers, 111
  of fractions, 2
  of polynomials, 195
  of radicals, 447
  of rational expressions, 285
  of real numbers, 67
Multiplication property
  of equality, 109
  of inequality, 102
Multiplicative inverse (reciprocal), 3

Negative exponent, 171
Negative number, 43
Null set, 523
Number line, 43
Numbers
  complex, 529
  composite, 228
  decimal, 9
  fraction, 1
  imaginary, 529
  integer, 44
  irrational, 44, 442
  mixed, 1
  negative, 43
  positive, 43
  prime, 1

  rational, 44
  real, 45
  signed, 43–44
  whole, 1, 43
Numerator, 1
Numerical coefficient, 87, 187

Opposites, 46
Ordered pair, 335
Order of operations, 24
Origin, 341
Outer product, 201

Pair, ordered, 335
Parabola, 507
Parallel lines, 360
Percent, 13
Percentage, 13
Perfect square, 441
Perfect square trinomial, 250
Perimeter, 132
Perpendicular lines, 360
Place value, 9
Plotting points, 341
Point-slope form, 366
Polynomial, 188
  addition of, 189
  binomial, 188
  degree of, 188
  descending powers, arrangement in, 188
  division of, 211
  division by monomial, 207
  factoring, 231–32
  monomial, 188
  multiplication of, 195, 202
  numerical coefficient, 87, 187
  subtraction of, 190
  term, 87, 187
  trinomial, 188
Positive number, 43
Power rule for exponents, 165
Powers, descending, 188
Prime factored form, 228
Prime numbers, 1, 227
Product, 65
  cross, 140
Product of the sum and difference of two terms, 203
Product rule
  for exponents, 170
  for radicals, 447
Proper fraction, 1
Properties, 82
  associative, 79
  commutative, 79
  distributive, 81
  identity, 80
  inverse, 80
Proportion, 139
Pythagorean formula, 265, 443

Quadrants, 341
Quadratic equation, 257, 487
  completing the square, 494

  factoring, 257, 260
  square root property, 487
Quadratic formula, 498
Quotient, 3
  of two polynomials, 211–12
Quotient rule
  for exponents, 173
  for radicals, 448

Radical, 441
  like, 455, 465
  product rule for, 447
  quotient rule for, 448
Radical expression, 441
  simplified, 447, 465
Radical sign, 441
Radicand, 441
Ratio, 139
Rational expression, 279
  addition of, 295
  division of, 286
  fundamental property, 280
  in equations, 307
  multiplication of, 285
  subtraction of, 297
Rationalizing the denominator, 459
Rational numbers, 44
Real numbers, 45
  addition of, 51
  division of, 71
  multiplication of, 67
  subtraction of, 59
Reciprocal, 3
Repeating decimal, 13
Root, 441, 444
Rules for exponents, 174, 179

Scientific notation, 183
Second-degree equation. *See* Quadratic equation
Set, 31, 523
  complement, 524
  disjoint, 526
  element of, 523
  empty, 523
  equal, 524
  finite, 523
  infinite, 523
  intersection, 525
  null, 523
  subset, 523
  union, 525
  universal, 523
Set braces, 31, 523
Signed numbers, 43, 74
Simplified form, 447
Simplifying radicals, 447–50, 455–56, 459–61, 465
Slope, 357
Slope-intercept form, 365
Solution of an equation, 31
Solving a linear equation, 31, 115
Solving for a specified variable, 133
Solving a word problem, 123
Square of a binomial, 203

Square root, 441.
   See table on page 536
Square root property, 487
Squaring property of equality, 473
Standard form of a complex number, 530
Subset, 523
Substitution method, 409
Subtraction
   of fractions, 6
   of polynomials, 190
   of radicals, 455–56
   of rational expressions, 297
   of real numbers, 59
Sum, 51
Sum of two cubes, 252
Symbols. *See page 535*
System of linear equations, 389
   addition method for solving, 397
   consistent, 392
   inconsistent, 392
   substitution method for solving, 409

System of linear inequalities, 425

Term, 87, 187
Trinomial, 188
   perfect square, 250

Union, 525
Universal set, 523
Unlike terms, 87

Variable, 29
Variation, 319
   direct, 319
   inverse, 320
Venn diagram, 524
Vertex, 507
Vertical line, 352

Whole numbers, 1, 43
Word problems, steps for solving, 123
Work problems, 319

x-axis, 341
x-intercept, 349

y-axis, 341
y-intercept, 349

Zero exponent, 171
Zero-factor property, 257